Organization of Prokaryotic Cell Membranes

Volume III

Editor

Bijan K. Ghosh, D.Sc.
Professor
Department of Physiology and Biophysics
UMDNJ—Rutgers Medical School
Piscataway, New Jersey

CRC Press
Taylor & Francis Group
Boca Raton London New York

CRC Press is an imprint of the
Taylor & Francis Group, an informa business

First published 1985 by CRC Press
Taylor & Francis Group
6000 Broken Sound Parkway NW, Suite 300
Boca Raton, FL 33487-2742

Reissued 2018 by CRC Press

Library of Congress Cataloging in Publication Data
(Revised for volume 3)
Main entry under title.

Organization of prokaryotic cell membranes

Includes bibliographical references and index
1. Bacterial cell walls 2. Cell membranes.
I. Ghosh, Bijan K., 1935- [DNLM: 1 Bacteria--
Cytology. 2. Cell membrane--Ultrastructure. 3. Cells
QW51 68]
QR77.73 589 9'0875 80-25355
ISBN 0-8493-5653-9 (v. 1)

A Library of Congress record exists under LC control number: 80025355

Publisher's Note
The publisher has gone to great lengths to ensure the quality of this reprint but points out that some imperfections in the original copies may be apparent.

Disclaimer
The publisher has made every effort to trace copyright holders and welcomes correspondence from those they have been unable to contact.

ISBN 13: 978-1-315-89612-0 (hbk)
ISBN 13: 978-1-351-07522-0 (ebk)

Visit the Taylor & Francis Web site at http://www.taylorandfrancis.com and the
CRC Press Web site at http://www.crcpress.com

FOREWORD

The pioneering investigations of Gorter and Grendel (1975) and Danielli and Davson (1935) in which cell membranes were visualized as bimolecular leaflets ushered in what might be termed the modern era of biomembrane research. Electron microscopy played a key role in establishing the universal existence and anatomical features of biological membranes in cells of animal, plant, and microbial origin, and indeed provided the necessary methodology for the isolation and characterization of biomembranes. Moreover, the ultrastructural studies pointed to the essential differences in surface organization and membranous organelles of eukaryotic and prokaryotic cells. The robust bacterial cell walls of Gram-positive organisms and the envelopes of Gram-negative bacteria became amenable to isolation in the 1950s, and soon after, the pioneering work of Weibull (1953) paved the way for the study of prokaryotic cytoplasmic membranes. Three decades of interest in the physiological and biochemical properties of bacterial plasma membranes have witnessed great advances in the state of our knowledge of their structure and functions. Dr. Bijan Ghosh is to be congratulated in bringing together so many distinguished leaders in the field of prokaryotic membrane research in three Volumes devoted to the "Organization of Prokaryotic Cell Membranes." The collection of authoritative articles covering the most active areas of prokaryotic biomembrane investigations into the several volumes has provided a great service not only to those interested in the field but also to microbiologists in general. We are deeply indebted to Dr. Bijan Ghosh for his considerable editorial efforts in assembling truly valuable contributions to our understanding of such basic aspects of bacterial membrane studies as transport functions, energizing membranes, the biochemistry and immunochemistry of membranes, and the structure-function relationships of photosynthetic membranes, gas vacuoles, and the more controversial mesosomes. The extensive reference lists will be invaluable for students and research workers in the various fields of prokaryotic membrane research especially in the "exploding" segments of the molecular and genetic aspects of Gram-negative cell membranes. These monographs will also serve to focus attention on prokaryotic membranes that are so often ignored by eukaryotic "membraneologists" and will provide an excellent reference source for many years to come.

Milton R. J. Salton
Department of Microbiology
New York University
School of Medicine
New York, New York

PREFACE TO VOLUME III

The hallmarks of bacterial physiology are the fast growth rate of these prokaryotes, the minuteness of their size, and a high degree of adaptability to differing growth conditions. Their fast rate of growth and colonization enable the bacterial cells to flourish on limited and transient resources. As a result of their smallness these cells have a high membrane-to-cytoplasm ratio. It is possible that this high ratio is important for the establishment of greater contact between the biochemical machinery of the cells and the biosphere.

It has been frequently suggested in the literature that eukaryotic cells have evolved from prokaryotic ancestors. These forerunners of the eukaryotes might have been basically prokaryotic, but it cannot be denied that contemporary prokaryotic cells have evolved equally as much as the eukaryotic cells. Therefore, a dual direction of cell evolution can be suggested.

Primitive cells ↗ Evolution of membrane → Organized eukaryotic cells
Primitive cells ↘ Evolution of biochemical reactions of intermediary metabolism → Heterotrophic or autotrophic prokaryotic cells

In general, the prokaryotic cell membrane lacks plasticity and ability to differentiate; however, these properties are inherent in the eukaryotic cell membranes. It is possible that the physiological functions related to membrane plasticity and differentiation, e.g., phargocytosis, pinocytosis, organelle formation, etc., did not develop in the prokaryotic cells. Evolution in prokaryotic cells may have progressed towards the diversification of the biochemical reactions for intermediary metabolism. Because of this metabolic diversity the habitat for prokaryotic cells varies widely. The following concepts are consistent with the properties of the prokaryotic membranes: (1) because of the lack of differentiation, the characteristics of the primitive ancestral cell membrane might have been conserved through evolution in prokaryotic cells; (2) the diversity of the intermediary metabolism in different groups of prokaryotic cells is likely to be accompanied by biochemical differences of the membranes. Due to this extreme variability, it is difficult to form a general concept of the structure and function of the prokaryotic cell membrane.

One group of eukaryotic cells (i.e., fungi) does not fit in this simplistic model of cell evolution. Wide diversity of intermediary metabolism is well known in fungal organisms. As regards subcellular morphology, some fungal cells may be richly endowed with a variety of organelles commonly found in plant or animal cells, whereas the others may show such paucity of organelles that hardly any subcellular body could be demonstrated besides the nucleus. This notable variation in subcellular organelle content of fungal cells may be frequently correlated with subtle differences in their growth conditions. One may speculate from these observations that the organelle formation, presumably dependent upon membrane differentiation, is at an intermediate stage of development in fungal cells. This possibility, when taken together with the present view that fungal cells are highly evolved, suggests that membrane differentiation may have failed to progress enough in the ancestral cells of fungal organisms. Hence, both differentiability and primitive property of the membrane has been conserved in fungi. The membrane system in fungal cells may have retained the potential to revert back to its primitive undifferentiated character. Current views on evolution of membrane and intermediary metabolism may be explained by one general model as described above. A critical study of membrane evolution is a rewarding area of research in cell biology.

The above discussion shows the difficulty in organizing a comprehensive properties of diversity of prokaryotic cell membranes. Hence, both general and specialized properties of prokaryotic membranes have been discussed. A general feeling has been expressed by colleagues in the field of membrane research that the work in this area is progressing at a very fast rate. Therefore, the scientific material in a manuscript becomes largely out of date because of the delay inherent in the publication of the book. Hence, the editor's aim is to organize a general text written by experts in the field. The authors have presented a thorough review of the available scientific material within a general conceptual framework and indicated the future direction of research. In addition, an attempt has been made to provide an extensive bibliography.

Frequently, investigators working with a specific organism lose appreciation for the diversity of the prokaryotic membrane. Consequently, discussions on the flow of information from researches on membranes originating from a variety of prokaryotic cells are helpful in formulating a unified approach to study prokaryotic cell membranes.

Bacterial cytoplasm directly interacts with the external environment. Membrane is the interface of this interaction. Therefore, the information exchange between the cytoplasmic material (enzymes, genetic material, and other factors) and the extracellular environment is mediated through the membrane. There is a strong possibility that the prokaryotic cell membrane receives information input from the biosphere and regulates physiological activities accordingly. Vigorous research activity will develop in this area of coupled receptor regulator activity of membranes. A thorough understanding of the regulatory role of membranes will stimulate the development of technology for programming bacterial cells for the production and secretion of industrially important substances into the growth (fermentation) medium.

In fact, uses of bacterial cells are steadily increasing in the field of bioorganic industry. The future holds the possibility of extensive use of bacterial cells in the production of biomedical and agricultural material and biomass utilization. The production of enzymes by microorganisms is already a several hundred million dollar industry.

Recent advances in recombinant DNA technology show the promise that through genetic engineering, strains of bacterial cells, targeted for the synthesis of a wide range of biomedical and agricultural substances, may be constructed. However, uses of these constructed organisms in industrial fermentation are limited because of a variety of problems. One of these is the recovery of material from cell-free fermentation medium. Secretory activity, which is a complex membrane phenomenon, regulates the accumulation of material synthesized inside bacterial cells in the growth medium.

Thus, the study of the prokaryotic cell membrane is rewarding, both for the understanding of basic biological phenomena and for the development of a technology to use prokaryotic cells in industry. I hope the material in these three volumes will stimulate research and help students engaged in the studies on prokaryotic cell physiology.

The publication of the third volume has been delayed for reasons beyond control. The structure of this volume, however, has not been changed. Discussions on the membrane of a highly specialized organism, i.e., hydrocarbon-utilizing bacteria, and general discussions on membrane organization and function, i.e., biosynthesis of outer membrane proteins, functional organization of membrane proteins, and protein secretion have been included in this volume. Through the discussion of twelve topics in these three volumes it has become apparent that the membrane is the hub of physiological and regulatory activity in prokaryotic cells. Within the wide range of microorganisms, there is both unity and diversity in these activities. The growth of information on the bacterial membrane is fast, and new ground is being broken. Hence, a project which started as small requires further expansion. We contemplate, as a consequence, discussions of new topics such as (1) Role of Membrane in Chemotaxis; (2) Sporulation; (3) Organization of Chemoautotrophic Bacterial Membrane; and (4) Membrane Genetics, in a possible future volume.

I must thank all the contributors for their valuable articles. In spite of their extremely busy schedules, they sympathetically considered my proposal and gave their time. With their help, we are making a steady progress in this ambitious project.

THE EDITOR

Bijan K. Ghosh, D.Sc., is Professor in Physiology and Biophysics in the Department of Physiology and Biophysics at the University of Medicine and Dentistry of New Jersey—Rutgers Medical School, Piscataway, and Honorary Professor in Microbiology at the Waksman Institute of Microbiology, Rutgers State University, N.J.

Dr. Ghosh received the B.Sc. and the M.Sc. degrees in physiology from Presidency College, Calcutta University. His doctorate was awarded by Calcutta University while he was working in the Indian Institute of Experimental Medicine in 1963. He engaged in postgraduate study at various institutions including the Woods Hole Oceanographic Institute, Woods Hole, Massachusetts, and the Anatomy Institute of the University of Bern, Switzerland.

Dr. Ghosh was an instructor and subsequently, a junior research fellow, at Presidency College during 1958 and 1959. At the Indian Institute of Experimental Medicine Dr. Ghosh was a Junior Research Fellow from 1959 to 1961, and a Senior Scientific Assistant until 1964. From July 1964 until November 1966 he was a Medical Research Council, Canada, Postdoctoral Fellow at the Department of Bacteriology and Immunology of the University of Western Ontario. He became associated with Rutgers University as Waksman-Merck Postdoctoral Fellow at Waksman Institute of Microbiology in November 1966, and served there as an assistant professor from 1967 to 1973. He was a visiting Professor at the University of Amsterdam, Netherlands in 1973. He moved to the Rutgers Medical School at the end of 1973. He visited China as a visiting Professor on an invitation from the Chinese Academy of Sciences and taught a course on "Cell Ultrastructure and Electron Microscopy" in 1981. He received an International Travel Award in 1984 from the National Science Foundation and visited India on an invitation from the Council of Scientific and Industrial Research of the Government of India, to present a lecture series on cell biology and to initiate a joint U.S.—India research program.

Dr. Ghosh has been a member of the Editorial Board of the *Journal of Bacteriology,* and he was very active in the Morphology and Ultrastructure Division of the American Society for Microbiology. He has organized several symposia in the general area of structure/function interrelationship in microorganisms.

Dr. Ghosh is a member of the Canadian Society of Biochemistry, the American Society for Microbiology, the Electron Microscopic Society of America, and the American Association for the Advancement of Science. He is a fellow of the American Institute of Chemists and a member of the New York Academy of Sciences. He is author or co-author of many original papers including some reviews and chapters on Bacterial and Fungal Ultrastructure in the *CRC Handbook of Microbiology.*

Among Dr. Ghosh's awards are a University Gold Metal from the Calcutta University, the Medical Research Council of Canada Postdoctoral Fellowship, a Waksman-Merck Postdoctoral Fellowship at the Waksman Institute of Microbiology of Rutgers University, and the Research Career Development Award from the National Institute of General Medical Sciences of the National Institutes of Health.

Dr. Ghosh has done extensive research on bacterial fine structure (structure and function of mesosomes), the evolution of subcellular organelles (subcellular organization of minute fungi), and membrane phenomena of enzyme secretion in microorganisms.

CONTRIBUTORS

Joseph M. DiRienzo, Ph.D.
Assistant Professor
Department of Microbiology
School of Dental Medicine
University of Pennsylvania
Philadelphia, Pennsylvania

William R. Finnerty, Ph.D.
Professor and Head
Department of Microbiology
University of Georgia
Athens, Georgia

Arati Ghosh, Ph.D.
Research Specialist
Department of Physiology
and Biophysics
UMDNJ-Rutgers Medical School
Piscataway, New Jersey

Bijan K. Ghosh, D.Sc.
Professor
Department of Physiology
and Biophysics
UMDNJ-Rutgers Medical School
Piscataway, New Jersey

Pamela Green
Department of Biochemistry
State University of New York
at Stony Brook
Stony Brook, New York

Masayori Inouye, Ph.D.
Chairman and Professor
Department of Biochemistry
State University of New York
at Stony Brook
Stony Brook, New York

Mary E. Singer
Research Associate
Department of Microbiology
University of Georgia
Athens, Georgia

Hideo Yamagata, Ph.D.
Assistant Professor
Department of Food Science
and Technology
Nagoya University
Nagoya
Japan

ORGANIZATION OF PROKARYOTIC CELL MEMBRANES

Volume I

The Role of Membranes in the Transport of Small Molecules
The Role of the Membrane in the Bioenergetics of Bacterial Cells
Immunology of the Bacterial Membrane
The Mycoplasma Membrane

Volume II

Bacterial Cell Surface Receptors
The Mesosome
The Gas Vesicle: A Rigid Membrane Enclosing a Hollow Space
Membranes of Phototropic Bacteria

TABLE OF CONTENTS

Volume III

Chapter 1
Membranes of Hydrocarbon-Utilizing Microorganisms 1
W. R. Finnerty and M. E. Singer

Chapter 2
The Biosynthesis of Bacterial Outer Membrane Proteins 45
Pamela Green, Joseph M. DiRienzo, Hideo Yamagata, and Masayori Inouye

Chapter 3
Functional Organization of Bacterial Membrane Proteins.................................. 105
Bijan K. Ghosh and Arati Ghosh

Chapter 4
Protein Secretion in Bacteria 175
Bijan K. Ghosh and Arati Ghosh

Index 255

Chapter 1

MEMBRANES OF HYDROCARBON-UTILIZING MICROORGANISMS

W. R. Finnerty and M. E. Singer

TABLE OF CONTENTS

I. Introduction 2

II. Ultrastructure of Hydrocarbon-Utilizing Microorganisms 3
- A. Hydrocarbon Inclusions 4
- B. Intracytoplasmic Membranes 10

III. Ultrastructure of Alkane-Grown Yeasts 12
- A. Cell Envelope 13
- B. Microbodies 14
- C. Plasma Membrane 14

IV. Membranes of *Acinetobacter* 15
- A. Lipids of Hexadecane-Grown *Acinetobacter* 15
  - 1. Cellular Phospholipids 15
  - 2. Cellular Neutral Lipids 15
  - 3. Extracellular Lipids 16
- B. Isolation of Membranes from *Acinetobacter* sp. H01-N 17
  - 1. Membranes Derived from Nutrient Broth-Grown Cells 18
  - 2. Membranes Derived from Hexadecane-Grown Cells 18
  - 3. Isolation of Hydrocarbon Inclusions 18
  - 4. Isolation of Extracellular Membrane Vesicles 23
- C. Biochemical Composition of *Acinetobacter* Membranes 23
  - 1. Membrane Phospholipids 23
  - 2. Membrane Neutral Lipids 23
  - 3. Membrane Proteins 25
  - 4. Membrane-Localized Cytochromes 25

V. Phospholipid Metabolism in *Acinetobacter* sp. H01-N 26
- A. Phospholipid Biosynthesis 26
  - 1. Stimulation of Phosphatidic Acid Cytidyl Transferase by Exogenous Phospholipid 28
  - 2. Lipid Depletion of UAT Membranes 28
- B. Phospholipid Catabolism 31
  - 1. General 31
  - 2. Outer Membrane Phospholipases of *Acinetobacter* 34

VI. Conclusions 36
- A. Intracytoplasmic Membranes 36
- B. Hydrocarbon Inclusions 36
- C. Extracellular Membrane Vesicles 36
- D. Cytoplasmic Membranes 37
- E. Outer Membrane 39

F. Membrane Biogenesis 40
G. An Overview — Membranes and Hydrocarbon Metabolism 41

References ....

I. INTRODUCTION

Hydrocarbons are organic molecules that contain only the elements of carbon and hydrogen. Such molecules range in size from methane to large polycyclic aromatic molecules of undetermined weight and structure. Most hydrocarbons are, at best, only sparingly soluble in an aqueous environment.

The ability of diverse genera of microorganisms to degrade a wide variety of hydrocarbons is ubiquitous in nature, with microbial degradation of hydrocarbons receiving extensive study over the last three decades.[1] Such studies have been mainly limited to investigations concerning the metabolic interaction between a single, simple hydrocarbon and a single microorganism. Research efforts in microbial hydrocarbon metabolism have largely concentrated on whole cell physiology, primarily describing various substrate-product interrelationships. Several biochemical pathways of hydrocarbon dissimilation and mechanisms of hydrocarbon oxidation have been formulated, although significant gaps in our knowledge remain. Information at the molecular level is notably lacking in hydrocarbon microbiology. The ability of microorganisms to metabolize such hydrophobic molecules poses questions of basic interest regarding the evolution of the genetic information encoding alkane oxidation in some genera but not in others. Further, the organization and regulation of genetic information specifying alkane oxidations is poorly understood, with these genes being plasmid-encoded in *Pseudomonas* species and chromosome-encoded in *Acinetobacter* species.[2] To date, the genetics and regulation of alkane metabolism have been examined only in *Pseudomonas putida,* a bacterium capable of utilizing low-molecular-weight, simple hydrocarbons.

Major questions remain unanswered and often unrecognized concerning the growth of microorganisms at the expense of these water-insoluble substrates:

1. How do hydrocarbons enter the microbial cell?
2. How do microorganisms respond metabolically and biochemically to such physiologically unusual hydrophobic substrates for purposes of energy production and carbon assimilation?
3. What are the mechanisms of genetic regulation of microbial hydrocarbon metabolism?

Accordingly, it is important to recognize that microbial hydrocarbon metabolism requires further research at more sophisticated levels of analysis to resolve these fundamental questions.

Hydrocarbon microbiology has largely focused on studies relating to the cellular physiology of alkane metabolism. A number of basic and fundamental generalizations have emerged which collectively serve to characterize those prokaryotic and eukaryotic microorganisms exhibiting the ability to grow at the expense of alkanes as a sole source of carbon and energy.

First, a large number of bacteria, yeast, and fungi grow at the expense of alkanes varying in carbon number from 1 to 40 carbon atoms. A few representative genera capable of alkane metabolism are *Acinetobacter, Pseudomonas, Brevibacterium, Arthrobacter, Mycobacterium, Nocardia, Corynebacterium, Candida, Cladosporium, Penicillium, Cunninghamella, Fusarium, Mucor, and Aspergillus.* These microorganisms are natural soil inhabitants of ubiquitous distribution. Such alkane-utilizing microorganisms are easily isolated from soil by appropriate selective enrichment procedures.

Second, the mechanism(s) of oxidation employed by these various microorganisms are obligately oxygen-dependent processes involving mono-, di-, and subterminal oxilation of the alkane molecule:

$$RCH_3 \xrightarrow{O_2} RCH_2OH \rightarrow RCHO \rightarrow RCHOOH$$

$$CH_3(CH_2)_nCH_3 \xrightarrow{O_2} CH_3(CH_2)_n\text{–}COOH \xrightarrow{O_2} HOOCH(CH_2)_n\text{–}COOH$$

$$R\text{–}CH_2\text{–}CH_3 \xrightarrow{O_2} R\text{–}CH(OH)\text{–}CH_3 \rightarrow R\text{–}\overset{\overset{\displaystyle O}{\|}}{C}\text{–}CH_3$$

The initial oxidative enzymes are membrane-bound dioxygenases which catalyze the insertion of molecular oxygen into the alkane molecule.[1] This *n*-alkyl hydroperoxide is reduced to fatty alcohol which is oxidized to fatty acid by way of alcohol and aldehyde dehydrogenases.

$$R\text{–}CH_3 \xrightarrow{O_2} RCH_2OOH \rightarrow RCH_2OH \rightarrow RCHO \rightarrow RCOOH$$

Finally, the oxidation of alkanes yield a variety of products, many of which accumulate in the culture broth. Products characterizing alkane oxidations are fatty acids, fatty alcohols, fatty aldehydes, wax esters, ketones, dicarboxylic acids, and hydroxy fatty acids.

This chapter discusses the membrane biochemistry and unique ultrastructural features of alkane-utilizing microorganisms with special attention to *Acinetobacter. Acinetobacter* represents an environmentally ubiquitous and nutritionally diverse genus of Gram-negative bacteria capable of growth on n-alkanes (C10 to C30), fatty alcohols and aldehydes, fatty acids, and dicarboxylic acids, as well as certain amino acids and a few carbohydrates.The physiological attributes exhibited by hexadecane-grown *Acinetobacter* sp. H01-N provide a unique model system for studying the metabolism of hydrophobic, water-insoluble compounds. Physiological changes resulting from the growth of this organism on alkane involve the intracellular sequestering and membranous packaging of alkane, specific synthesis of intracytoplasmic membrane systems, the increase of cellular phospholipid and related membrane-bound phospholipid biosynthetic enzymes, repression of *de novo* fatty acid biosynthesis and quantitative changes in the cellular and extracellular lipid composition during growth on alkane.

II. ULTRASTRUCTURE OF HYDROCARBON-UTILIZING MICROORGANISMS

In *Acinetobacter* sp. H01-N unique ultrastructural modifications result from growth on hydrocarbons: (1) the intracellular pooling of hydrocarbons in membrane-bound cytoplasmic inclusions termed "hydrocarbon inclusions" and (2) the induction of an extensive intracytoplasmic membrane system.[3,4] With the exception of the hydrocarbon

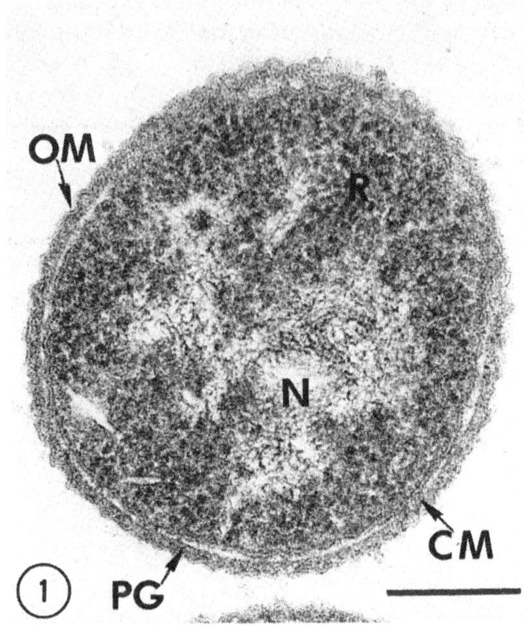

FIGURE 1. Thin section of *Acinetobacter* grown on nutrient broth medium. OM, outer membrane; PG, peptidoglycan; CM, cytoplasmic membrane; R, ribosomes; N, nuclear material. Bar represents 0.5 μm. (From Scott, C. C. L. and Finnerty, W. R., *J. Bacteriol.*, 127, 481, 1976. With permission.)

inclusions and the intracytoplasmic membranes, hexadecane- and nonalkane-grown cells appear morphologically similar, exhibiting the typical cell envelope structure of Gram-negative bacteria in electron micrographs of thin sections or freeze-etched cells (Figures 1 and 5).[5]

A. Hydrocarbon Inclusions

The hydrocarbon inclusions appear in thin sections of hexadecane-grown *Acinetobacter* as electron-transparent, spherical cytoplasmic bodies averaging 0.2 μm in diameter, situated at the periphery of the cell or in contact with the intracytoplasmic membranes (Figure 2).[3,5] These inclusions contain the unmodified growth hydrocarbon, as determined by X-ray diffraction analyses.[3] Hydrocarbon inclusions isolated from hexadecane-grown *Acinetobacter* contain 71.5% hexadecane, 15.8% protein, 9.5% phospholipid, and 3.2% neutral lipid (Table 1).

The limiting membrane of the hydrocarbon inclusion is most easily visualized in thin sections of partially lysed cells containing little cytoplasmic material (Figure 3). The hydrocarbon inclusions are limited by a unique monolayer membrane which does not exhibit the typical bilayer structure of a unit membrane.[5] Freeze-etch studies of isolated hydrocarbon inclusions and of hexadecane-grown cells provide further evidence for the presence of a smooth-surfaced, monolayer inclusion membrane (Figure 4).[5] Freeze-etch studies of nutrient broth-grown cells illustrate the absence of inclusions characteristic of alkane-grown cells (Figure 5).

Since the discovery of these unique cytoplasmic inclusion bodies in *Acinetobacter*, similar structures have been described in other hydrocarbon-grown bacteria and yeasts.[6] Scott and Finnerty[6] documented the presence of hydrocarbon inclusions in hexadecane-grown *Arthrobacter* sp. 80 (Figure 6), *Corynebacterium* sp., *Mycobacterium album* 7E4, *Nocardia rubra, Nocardia* sp. 72 (Figure 7), *Mycobacterium vaccae, Candida lipolytica,* tetradecane-grown *C. tropicalis,* and naphthalene-grown *Pseudo-*

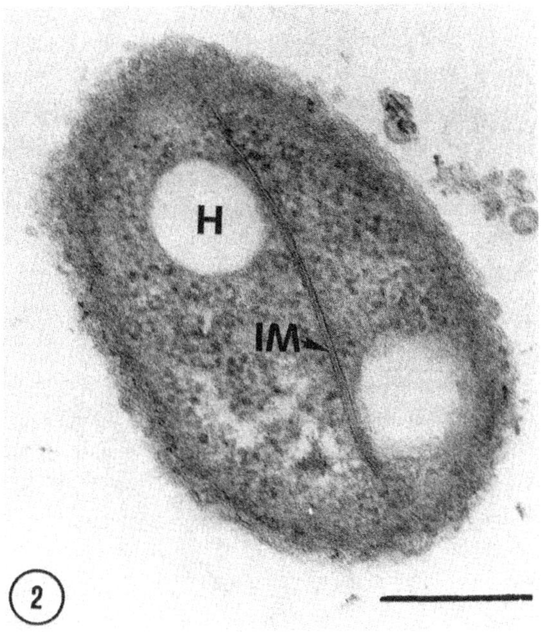

FIGURE 2. Thin section of *Acinetobacter* grown on hexadecane. H, hexadecane inclusion; IM, intracytoplasmic membrane. Bar represents 0.2 μm. (From Scott, C. C. L. and Finnerty, W. R., *J. Bacteriol.*, 127, 481, 1976. With permission.)

Table 1
BIOCHEMICAL COMPOSITION OF HYDROCARBON INCLUSIONS

Component	Total μg	Percentage
Wax ester	140	2.2
Triglyceride	15	0.2
Free fatty acid	8	0.1
Fatty alcohol	22	0.3
Diglyceride	14	0.2
Monoglyceride	2	0.03
Hexadecane	4520	71.5
Phospholipid	600	9.5
Protein	1000	15.8

monas sp. With the exception of *N. rubra*, these microorganisms do not contain electron-transparent inclusions when grown on nonhydrocarbon-containing media. *N. rubra*, however, contains significant amounts of poly-β-hydroxybutyrate when cultured with nutrient broth, but only trace amounts when grown on hexadecane. Both of the *Nocardia* sp. also contain electron-dense bodies as a result of growth on hydrocarbon and nonhydrocarbon substrates. These inclusions most likely represent an accumulation of lipids other than hexadecane. Alkane-grown *Nocardia* characteristically accumulate large amounts of waxes and glycerides.[7] Scott and Finnerty[6] demonstrated that those microorganisms with hydrocarbon inclusions contain intracellularly localized hydrocarbon.

The fine structure of hydrocarbon-grown *C. lipolytica* has been described previ-

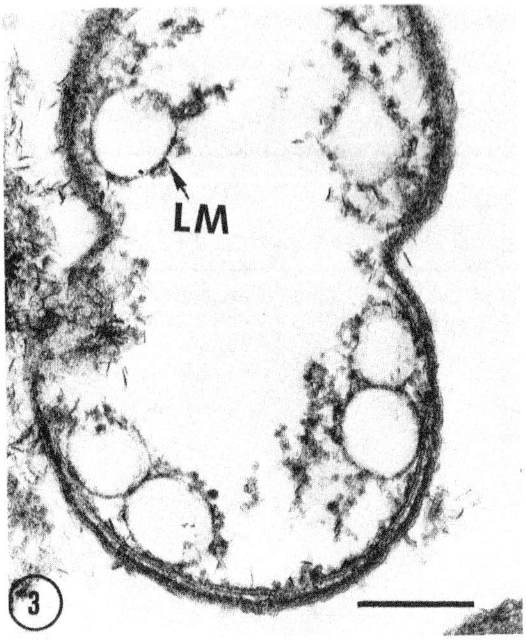

FIGURE 3. Thin section of a partially lysed cell of *Acinetobacter* grown on hexadecane. H, hexadecane inclusion; LM, limiting membrane of hexadecane inclusion. Bar represents 0.2 μm. (From Scott, C. C. L. and Finnerty, W. R., *J. Bacteriol.*, 127, 481, 1976. With permission.)

FIGURE 4. Freeze-etch of *Acinetobacter* grown on hexadecane. H, hexadecane inclusion; LM, limiting membrane of hexadecane inclusion. Bar represents 0.5 μm. (From Scott, C. C. L. and Finnerty, W. R., *J. Bacteriol.*, 127, 481, 1976. With permission.)

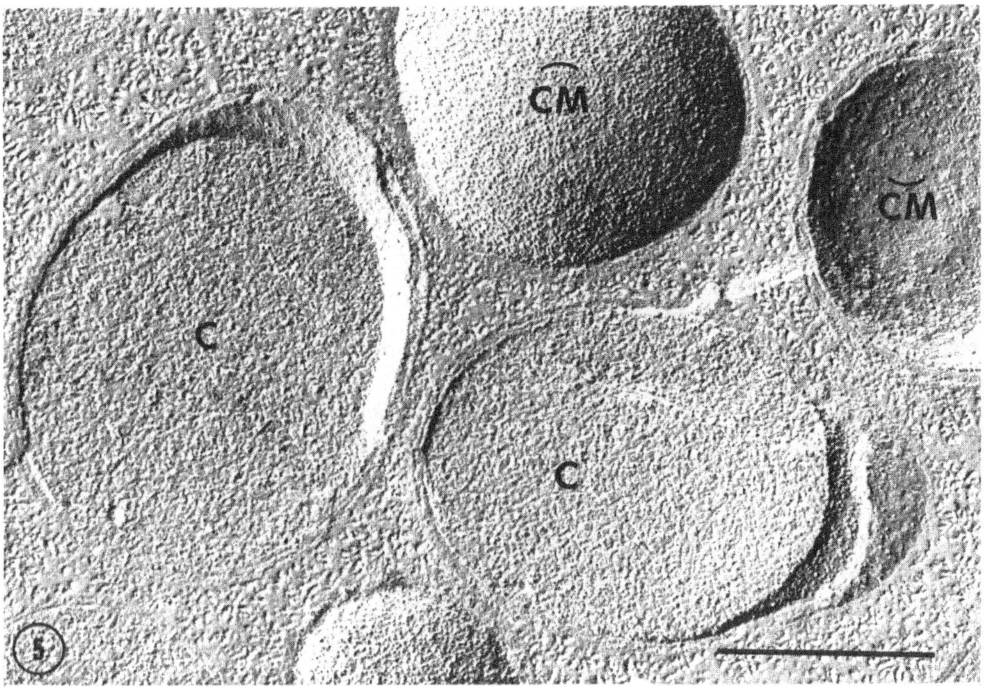

FIGURE 5. Freeze-etch of *Acinetobacter* grown on nutrient broth. $\widehat{CM}$, convex face of cytoplasmic membrane; $\widecheck{CM}$, concave face of cytoplasmic membrane; C, cytoplasm. Bar represents 0.5 μm. (From Scott, C C. L. and Finnerty, W. R., *J. Bacteriol.*, 127, 481, 1976. With permission.)

ously.[8,9] Hexadecane-grown cells were reported to contain a large number of fat vacuoles which were not positively identified. Hexadecane accumulation at the external surface of the cytoplasmic membrane was suggested but not proven. Scott and Finnerty,[6] however, showed no accumulation of hexadecane at the cytoplasmic membrane of *C. lipolytica,* but rather the occurrence of typical cytoplasmic hydrocarbon inclusions (Figure 8). *C. tropicalis* appears to accumulate tetradecane at the exterior surface of the cytoplasmic membrane in small vesicles projecting into the cytoplasm (Figure 9). Occasionally, the vesicles appear within the cytoplasm at the cell periphery, suggesting an endocytotic type of mechanism for hydrocarbon transport. This mechanism was suggested by Ludvik et al.,[8] who observed pinocytotic vesicles at the ends of deep invaginations of the cytoplasmic membrane in hexadecane-grown *C. lipolytica.*

Several other reports document the occurrence of intracytoplasmic structures resembling hydrocarbon inclusions. Bertrand et al.[10] described disc-shaped, electron-transparent cytoplasmic "vesicles" in thin sections and in negatively stained preparations of a hexadecane-grown marine bacterium. These vesicles are not observed in acetate-grown cells. In freeze-etched preparations with fractures through the intracytoplasmic vesicles, the vesicles appear to be membrane limited. The internal structure of the vesicle appears to have a smooth to finely granular texture, indicating the possible hydrocarbon or lipoidal nature of the enclosed material. Although the nature of the material sequestered in these inclusions was not determined, hexadecane-grown cells were shown to contain 20-fold more nonsaponifiable lipid than acetate-grown cells. Results were inconclusive as to whether these cytoplasmic vesicles contain hexadecane, since no distinction was made between externally absorbed and internally accumulated hexadecane. *Thermomicrobium fosteri,* an obligate thermophile capable of growth on hydrocarbons, contains electron-transparent intracytoplasmic inclusions resulting from

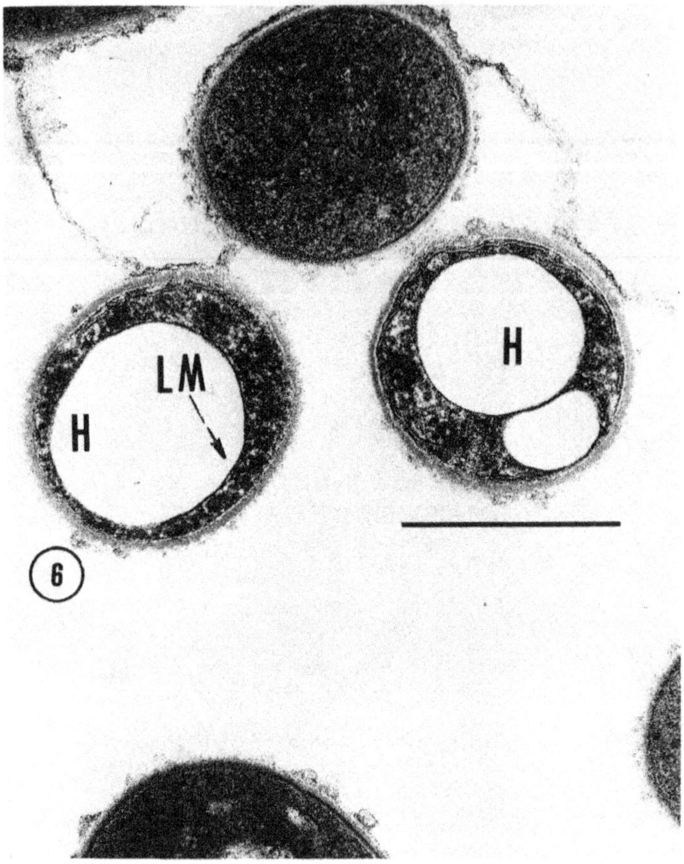

A

FIGURE 6. Thin section of *Arthrobacter* sp. strain 80 grown on (A) hexadecane and (B) nutrient broth medium. H, hydrocarbon inclusions; LM, limiting membrane. Bar represents 0.5 μm. (From Scott, C. C. L. and Finnerty, W. R., *J. Gen. Microbiol.*, 94, 342, 1976. With permission.)

growth on heptadecane.[11] Thus, the sequestering of hydrocarbon appears to be widespread among alkane-grown microorganisms. The cytoplasmic sequestering of alkane poses interesting questions concerning the external vs. internal site of oxidation for such hydrophobic molecules. Further questions arise concerning the transport of alkane across the cytoplasmic membrane and the subsequent intracellular packaging mechanism(s).

Several other types of cytoplasmic inclusion bodies are observed in hydrocarbon-grown bacteria. Atlas and Heintz[12] noted the presence of electron-dense, lipoidal inclusion bodies in two marine bacteria, *Flavobacterium* sp. and *Brevibacterium* sp., grown on crude oil-agar plates. *Flavobacterium* contains similar but smaller inclusions when grown on marine agar plates without oil. The lipoidal nature of the inclusions was verified by positive staining with Sudan Black B.

Spherical, electron-dense inclusion bodies approximately 0.1 μm in diameter are present in thin sections of a cyclohexane-grown *Nocardia* sp.[13] Since these inclusions are absent in succinate-grown cells, the structures are presumed to be hydrocarbon inclusions, although no evidence was presented to indicate whether the inclusions contained cyclohexane or lipoidal material.

Kormendy and Wayman[14] described an unusual ultrastructural modification in *n*-butane- or 1-butanol-grown *Arthrobacter* sp. and *Candida utilis*. These cytoplasmic

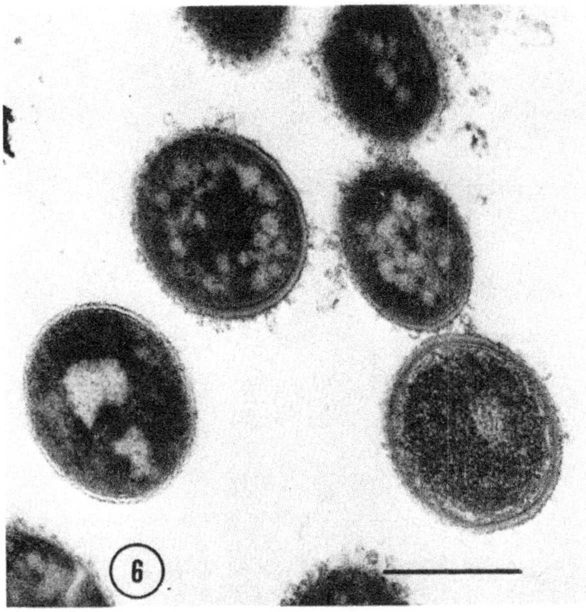

FIGURE 6B

inclusions, termed "oxisomes", consist of a spherical electron-dense body surrounded almost completely by an electron-transparent area. No clearly defined membrane surrounds the oxisome. These structures are not present in glucose-grown cells and are presumed to be the site of butane or butanol oxidation.

B. Intracytoplasmic Membranes

The second ultrastructural feature unique to alkane-grown *Acinetobacter* is the predominance of intracytoplasmic membranes.[3,4] In electron micrographs, the intracytoplasmic membrane appears as multilaminar sheaths of bilayer membranes extending across the length of the cell, apparently continuous with the cytoplasmic membrane, and often appearing to be in direct contact with the hydrocarbon inclusions (Figure 2).[4]

Such extensive intracytoplasmic membrane systems have been noted in only one other hydrocarbon-utilizing bacterium, a cyclohexane-grown *Nocardia* sp.[13] Mesosome-like membranes have been reported for unidentified bacteria grown on ethane, propane, or butane;[15] *n*-butane-grown *Arthrobacter* strain AK19;[14] heptadecane-grown *Thermomicrobium fosteri* sp. nov.;[11] and hexadecane- and nonalkane-grown *Acinetobacter* sp. H01-N.[4] Whether these mesosome-like membranes are related to hydrocarbon metabolism is not known.

In comparison, methylotrophic bacteria possess extensive intracytoplasmic membrane systems when grown on methane or methanol.[15-18] In methylotrophic bacteria, intracytoplasmic membranes are arranged either as stacked or scattered membranous vesicles (type I) or as peripherally located stacked lamellae (type II). The internal membrane systems of methylotrophic bacteria are thought to be involved in specialized energy transduction[16] and in methane hydroxylation.[19]

In *Acinetobacter*, intracytoplasmic membrane synthesis is induced upon exposure of the cells to alkane. No intracytoplasmic membranes are present in nonhydrocarbon-grown *Acinetobacter*. The appearance of intracytoplasmic membranes correlates directly with the ability of cells to oxidize alkane,[4] and with the two to threefold increase in cellular phospholipids described for hexadecane-grown *Acinetobacter*.[20,21] Since the

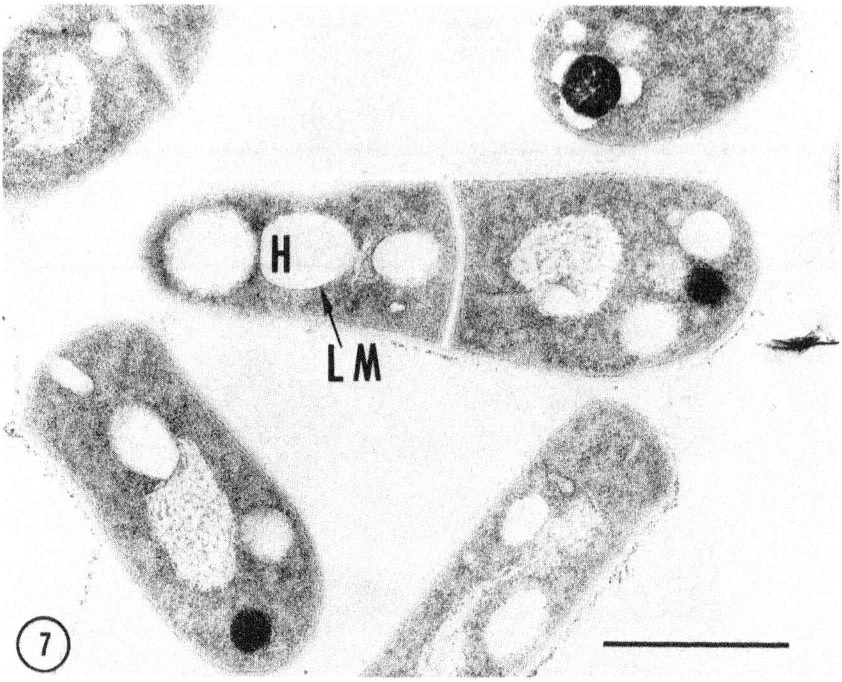

A

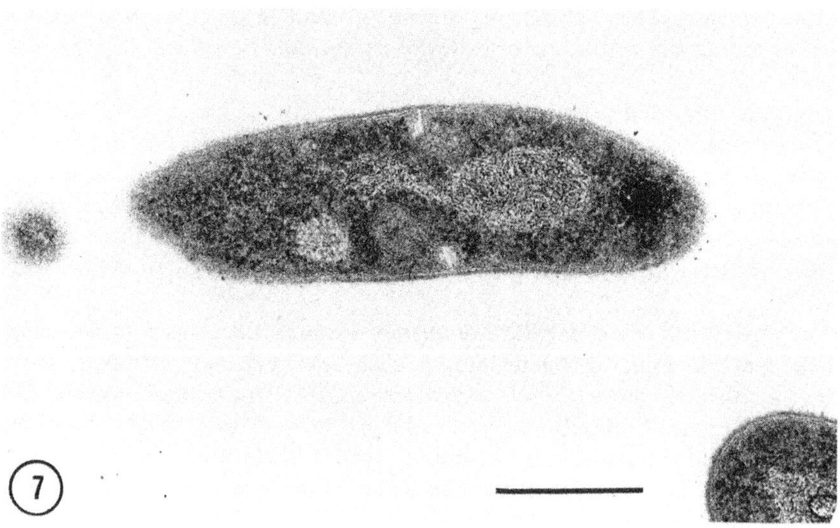

B

FIGURE 7. Thin section of *Nocardia* sp. 72 grown on (A) hexadecane and (B) nutrient broth medium. H, hydrocarbon inclusions; LM, limiting membrane. Bar represents 0.5 μm. (From Scott, C. C. L. and Finnerty, W. R., *J. Gen. Microbiol.*, 94, 342, 1976. With permission.)

intracytoplasmic membranes are formed concomitantly with the enzymatic capacity of the cell for hydrocarbon oxidation, the intracytoplasmic membranes may play one of several roles in hydrocarbon metabolism in *Acinetobacter.*

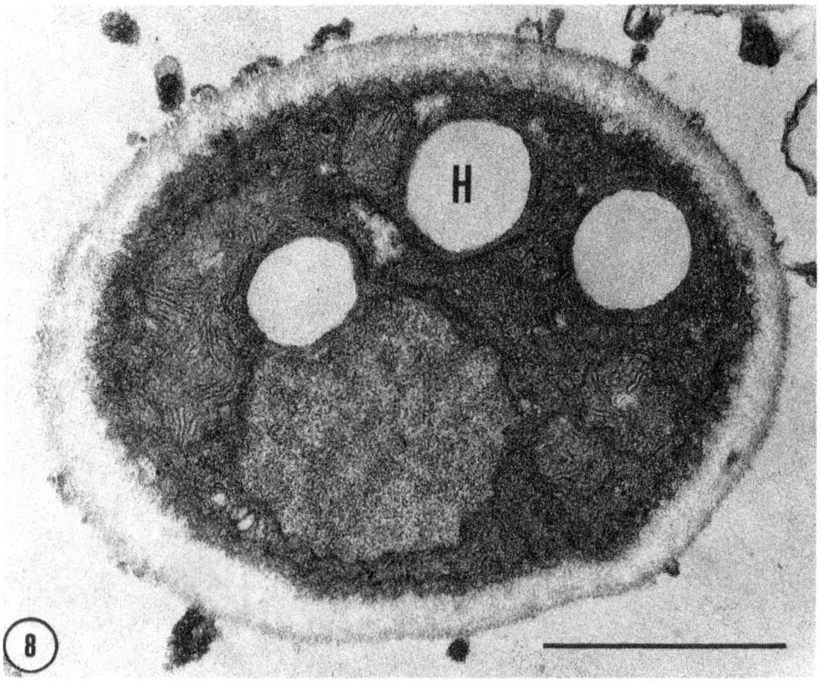

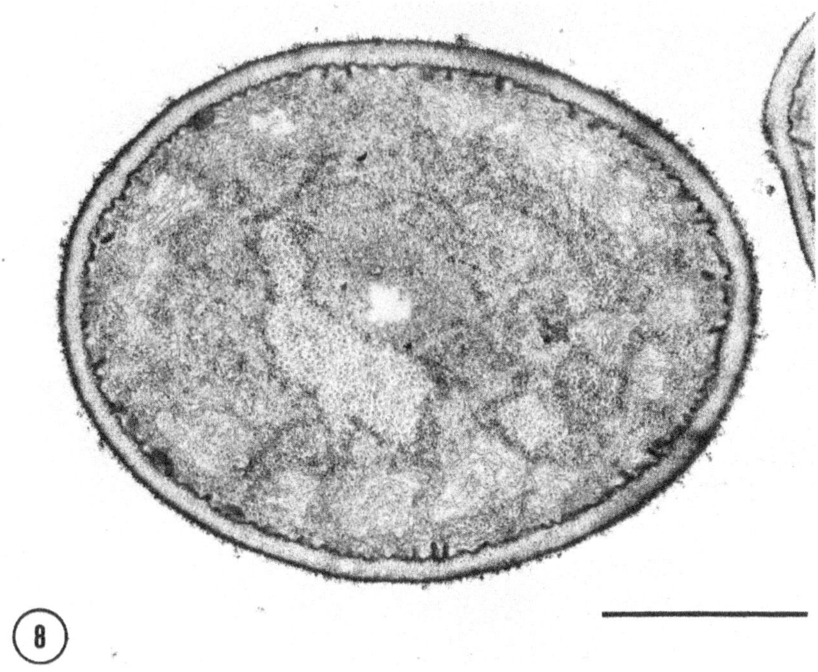

B

FIGURE 8. Thin section of *Candida lipolytica* grown on (A) hexadecane and (B) nutrient broth medium. H, hydrocarbon inclusions. Bar represents 1.0 μm. (From Scott, C. C. L. and Finnerty, W. R., *J. Gen. Microbiol.*, 94, 342, 1976. With permission.)

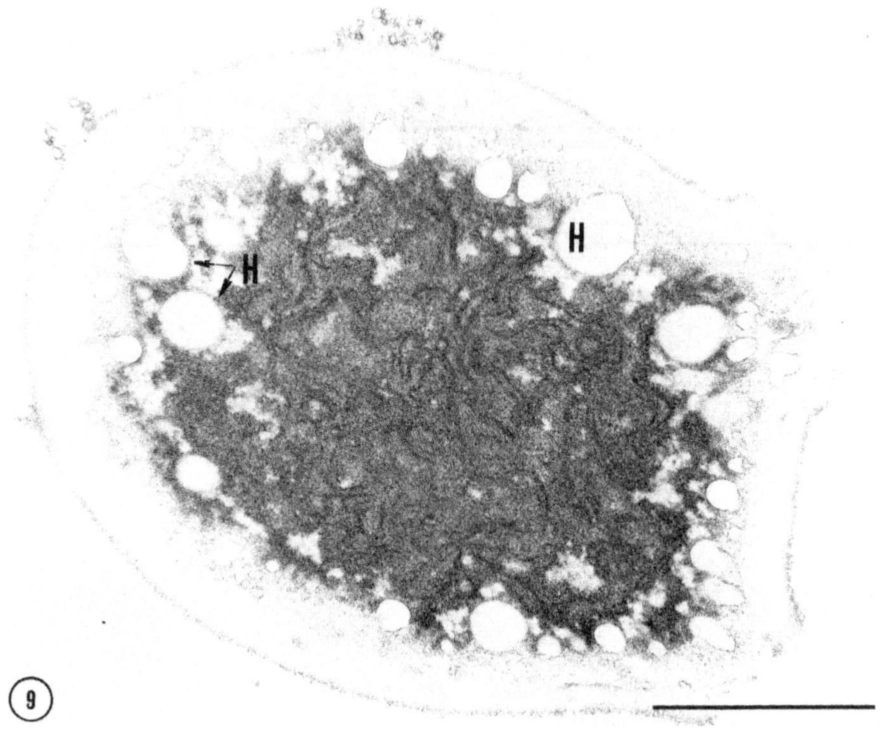

A

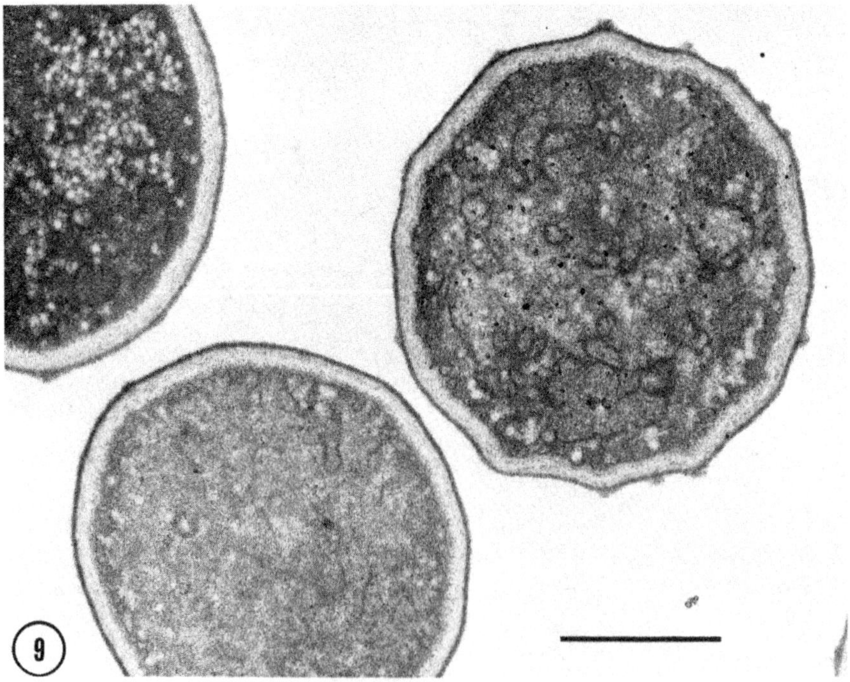

B

FIGURE 9. Thin section of *Candida tropicalis* grown on (A) tetradecane vapors and (B) nutrient broth medium. H, hydrocarbon inclusions. Bar represents 1.0 μm. (From Scott, C. C. L. and Finnerty, W. R., *J. Gen. Microbiol.*, 94, 342, 1976. With permission.)

1. The enzymes responsible for the incorporation of molecular oxygen into the alkane molecule may be associated with or may be a constituent part of, the membrane system.
2. If the enzymes involved in hydrocarbon oxidation are peripheral or integral membrane proteins, the intracytoplasmic membrane may provide a suitable interface for the hydrophobic alkane substrate.
3. One or more of the enzymes involved in the further oxidation of the alkane to the corresponding fatty acid may be localized in the intracytoplasmic membrane.
4. The intracytoplasmic membrane may be the site of increased respiratory activity.
5. The intracytoplasmic membrane may provide a continuous channel with the cytoplasmic membrane for the movement of hydrocarbon to the intracellular site(s) of oxidation.

III. ULTRASTRUCTURE OF ALKANE-GROWN YEASTS

A wide variety of yeasts are capable of growth on *n*-alkanes as the sole source of carbon and energy. The physiology and biochemistry of yeast alkane metabolism are, in many respects, comparable to bacterial alkane metabolism, proceeding by way of a monoterminal oxidation pathway involving a molecular oxygen-dependent oxidation of the alkane molecule to intermediates generally presumed to be the corresponding primary alcohol and aldehyde, which are oxidized to fatty acid via alcohol and aldehyde dehydrogenases. In yeasts, the biochemical steps in the pathway are compartmentalized in various membranous organelles. Cytochrome P-450, which is involved in alkane hydroxylation, has been localized to the yeast microsomal fraction. Alcohol and aldehyde dehydrogenases have been localized to microsomes, microbodies, and to the cytoplasm, while fatty acid metabolism appears to be restricted to microbodies.

Alkane-grown yeasts are characterized by several ultrastructural modifications in comparison to nonhydrocarbon-grown cells: (1) alterations of the cell envelope, (2) appearance of hydrocarbon inclusions, and (3) appearance of numerous microbodies.

A. Cell Envelope

Early investigators observed in thin sections of alkane-grown yeasts the apparent accumulation of alkane at the outer surface of the yeast cytoplasmic membrane.[8,9,22] The cytoplasmic membrane of alkane-grown yeasts often appears highly invaginated and in close contact with small membrane-bound pinocytotic vesicles.[22] Some investigators observed "pores" or "channels" extending across the cell wall of alkane-grown yeasts.[22,23] Thus, a mechanism for alkane uptake by yeasts could involve the adsorption of alkane to the hydrophobic yeast cell envelope, followed by movement of alkane across the rigid cell wall through channels, and pinocytosis at the cytoplasmic membrane.

Physiological studies have shown that alkane uptake in *C. tropicalis* involves two steps: (1) the physical partitioning of alkane to the yeast cell envelope, and (2) the energy-dependent uptake of alkane through the cytoplasmic membrane.[24] Certain yeasts, like bacteria, pool alkane intracellularly. Intracytoplasmic inclusions have been observed in thin sections of alkane-grown *C. tropicalis* and *C. lipolytica.*[6]

The cell surface of alkane-grown *C. tropicalis* ATCC 32113 appears to have a high affinity for alkane. Cells grown on glucose exhibit a lower capacity for adsorption of alkane as compared to hexadecane-grown cells. The interaction between alkane and the cell surface appears to be a process of nonenzymatic adsorption.[25] The cell wall component responsible for binding of alkane is identified as a polysaccharide-fatty acid complex.[26]

The surface-localized polysaccharide of alkane-grown *C. tropicalis* is identified as mannan containing 4% covalently bound fatty acids. Glucose-grown cells lack this mannan-fatty acid complex.[27] In electron micrographs of alkane-grown *C. tropicalis*, mannan is localized by cytochemical techniques on the yeast cell wall.[27] In addition, in freeze-fractured, alkane-grown *C. tropicalis*, radially oriented fibrous projections from the cell surface are visible, which presumably represent the polysaccharide polymer. These projections are not observed in freeze-fractured, smooth-surfaced, glucose-grown cells.[27]

Removal of this cell surface polymer by proteolytic digestion or masking of the mannan by concanavalin A binding significantly reduces the ability of the yeast cell to bind alkane, indicating a possible in vivo role for this polymer in alkane transport by *C. tropicalis*.[26,27]

B. Microbodies

As first reported by Osumi and Fukui,[29] alkane-grown yeasts (*Candida* sp.) contain membrane-bound cytoplasmic organelles, termed "microbodies" or "peroxisomes". In electron micrographs of alkane-grown *C. tropicalis* pk 233, microbodies appear as small, spherical to ovoid bodies with a homogeneous matrix enclosed by a single-unit membrane.[28,29] Microbodies are rarely observed in yeast cells grown on glucose,[28,29] with fewer numbers in cells grown on ethanol or acetate than in alkane-grown cells.[29,30]

In alkane-grown *C. tropicalis* pk 233, the appearance of microbodies is correlated with a significant increase in cellular catalase activity in comparison to cells grown on ethanol or glucose.[31,32,39] Localization of catalase in the microbodies was demonstrated cytochemically by the use of 3,3′-diaminobenzidine.[29-31] Microbodies isolated from alkane-grown *C. tropicalis* pk 233 were shown to possess catalase activity as well as other known peroxisomal marker enzymes.[33,34]

The presence of microbodies has been observed in other alkane-grown yeasts by several investigators: *Candida lipolytica* NRRL Y-6795, *C. albicans* IFO 0583, *C. guilliermondii* IFO 0566; *C. intermedia* NRRL Y-6328-1, *C. tropicalis* IFO 0589,[29] and *C. rugosa*.[35] The "lipid vacuoles" described by Kappeli et al.[36] in alkane-grown *C. tropicalis* KW2 were later referred to as microbodies.[37]

The peroxisomes of alkane-grown yeasts appear to play an essential role in alkane dissimilation. Metabolically, microbodies participate in fatty acid activation, β-oxidation of fatty acids, transfer of C-2 units to mitochondria, and in the synthesis of C-4 units through glyoxylate cycle enzymes. Microbodies are not involved in alkane hydroxylation via the cytochrome P-450 system;[38,39] but may be involved in the conversion of fatty alcohol to fatty acid by way of alcohol and aldehyde dehydrogenases.[40]

C. Plasma Membrane

The plasma membranes from *C. tropicalis* ATCC 32113 grown on glucose or hexadecane were isolated using a method based on differences in the surface charge between the mitochondria and plasma membrane.[41]

Isolated plasma membranes from *C. tropicalis* grown on either glucose or hexadecane are similar with respect to total protein content (60% total dry weight), lipid content (21 to 24% total dry weight), and carbohydrate content (16 to 21% total dry weight).[42] The banding patterns of the proteins derived from membranes of cells grown on these two substrates differed distinctly. The total phospholipid content of plasma membrane isolated from both sources is identical. The neutral lipids of hexadecane- and glucose-grown *C. tropicalis* appear qualitatively similar, composed primarily of sterol and free fatty acid. No significant differences are noted in the carbohydrate composition of plasma membranes isolated from cells grown on either carbon source. Analysis of the cytochrome composition of isolated plasma membrane by reduced-

Table 2
CELLULAR PHOSPHOLIPID COMPOSITION OF *ACINETOBACTER* SP. H01-N

Phospholipid	Percentage of phospholipids in: Hexadecane-grown cells	Acetate-grown cells
Cardiolipin	12.9	14.5
Triacyl-lysocardiolipin	5.7	7.7
Phosphatidylglycerol	19.9	30.6
Phosphatidylserine	Trace	Trace
Phosphatidylethanolamine	61.5	47.2
Acyl-phosphatidylglycerol	Trace	Trace
Phospholipid (μmol/g dry cell weight)	129	60

oxidized difference spectra reveal that the concentration of a b-type cytochrome is threefold higher in membranes isolated from hexadecane-grown cells than in those derived from glucose-grown cells.[42]

IV. MEMBRANES OF *ACINETOBACTER*

A. Lipids of Hexadecane-Grown *Acinetobacter*

1. Cellular Phospholipids

Hexadecane-grown *Acinetobacter* exhibits a two- to threefold increase in cellular phospholipid content over nonhydrocarbon grown cells (Table 2). This alkane-induced synthesis of phospholipid is correlated with the appearance of intracytoplasmic membrane and the ability of the cells to oxidize hexadecane.[4] The cellular phospholipid composition of hexadecane-grown cells is not appreciably different from that of non-hydrocarbon-grown cells (Table 2). In hexadecane-grown *Acinetobacter*, 97% of the phospholipid fatty acids are 16 carbons in chain length, with palmitic acid (C16:0) comprising 29% and palmitoleic acid 68% of the total fatty acid content. Phospholipids of nutrient broth-grown cells contain fatty acids of a wider spectrum of chain lengths, with oleic acid (C18:1) predominating (45%), followed by palmitic acid (41%) and palmitoleic acid (12%).[43,44] These studies affirm that the 16-carbon fatty acid derived from hexadecane oxidation is incorporated directly into cellular lipid and that *de novo* fatty acid biosynthesis is partially repressed in alkane-grown *Acinetobacter*.[45]

The increased phospholipid content in hexadecane-grown *Acinetobacter* sp. H01-N has also been observed in hexadecane-grown *A. lwoffi*,[46] in alkane-grown *Candida tropicalis*,[47,48] and in *Candida* 107.[49]

2. Cellular Neutral Lipids

Hexadecane-grown *Acinetobacter* exhibits an 18-fold increase in the cellular mono- and diglyceride content over that observed in nutrient broth-grown cells (Table 3).[21] In addition, hexadecane-grown cells exhibit a slight increase in cellular triglyceride, free fatty alcohol, and wax ester in comparison to nutrient broth-grown cells (Table 3).[21] The fatty alcohol and wax ester are considered to be end products of hexadecane metabolism in *Acinetobacter*.[50,51] Other investigators have noted increases in cellular triglyceride in propane-grown *Mycobacterium vaccae*,[52] and in the cellular wax ester and glyceride content of hexadecane-grown *Nocardia* sp.[7] An increase in total cellular lipid has been documented in C_{22} to C_{28}-grown *Mycobacterium convolutum*[53] and in alkane-grown *Candida parapsilosis*,[54] while in alkane-grown *Candida* 107, the total lipid content decreases, primarily due to a decrease in the triglyceride fraction.[49]

Table 3
CELLULAR LIPID COMPOSITION OF *ACINETOBACTER* SP. H01-N (IN μMOL/G CELL DRY WEIGHT)

Lipid class	Hexadecane-grown cells	Nutrient broth-grown cells
Triglyceride	2.5	1.8
Mono- and diglyceride	6.8	0.4
Free fatty acid	8.2	7.5
Fatty alcohol	2.6	Trace
Wax ester	18.0	11.5
Hexadecane	360.0	Trace
Phospholipids	129.0	46.0

Table 4
EXTRACELLULAR NEUTRAL LIPID COMPOSITION OF *ACINETOBACTER* SP. H01-N (IN μMOL/*l*)

Neutral lipid class	Hexadecane-grown cells	Nutrient broth-grown cells
Triglyceride	25.6	2.4
Mono- and diglyceride	410.0	0.0
Free fatty acid	60.0	4.0
Fatty alcohol	0.5	0.0
Wax ester	280.0	0.0
Hexadecane	not determined	0.0

[a] Extracellular lipids were obtained from the culture broths of early stationary phase cells.

As with the phospholipid fraction of hexadecane-grown cells, the free and bound fatty acids as well as the free fatty alcohol and the esterified fatty alcohol found in wax ester derived from the neutral lipid of hexadecane-grown cells are predominantly 16 carbons in chain length.[44] In nutrient broth-grown cells, heptanoic acid (C7:0) is the major fatty acid in the neutral lipid fraction. Smaller amounts of other-chain-length fatty acids are detected in the neutral lipid fraction of nutrient broth-grown cells.[43,44]

In *Acinetobacter*, the intracellular pools of hexadecane were shown to comprise 8 to 10% of the cell dry weight of hexadecane-grown cells.[21] In comparison, only trace amounts (less than 0.001% of the cell dry weight) of hexadecane are detectable in nonhydrocarbon-grown cells.[55] The presence of alkanes in nonhydrocarbon-grown cells, however, does indicate that *Acinetobacter* is capable of hydrocarbon synthesis as well as degradation.

3. *Extracellular Lipids*

The extracellular lipids obtained from the culture broths of hexadecane-grown cells are comprised of triglyceride, mono- and diglyceride, and wax ester (Table 4). These lipids are either absent or present in only minor concentrations in the medium of nutrient broth-grown cells. The level of extracellular triglyceride stays relatively constant throughout the growth of *Acinetobacter* on hexadecane. However, the levels of extracellular free fatty acid, mono-, and diglycerides increase rapidly during exponential growth of *Acinetobacter* on hexadecane (Figure 10).[21,44] The extracellular free fatty

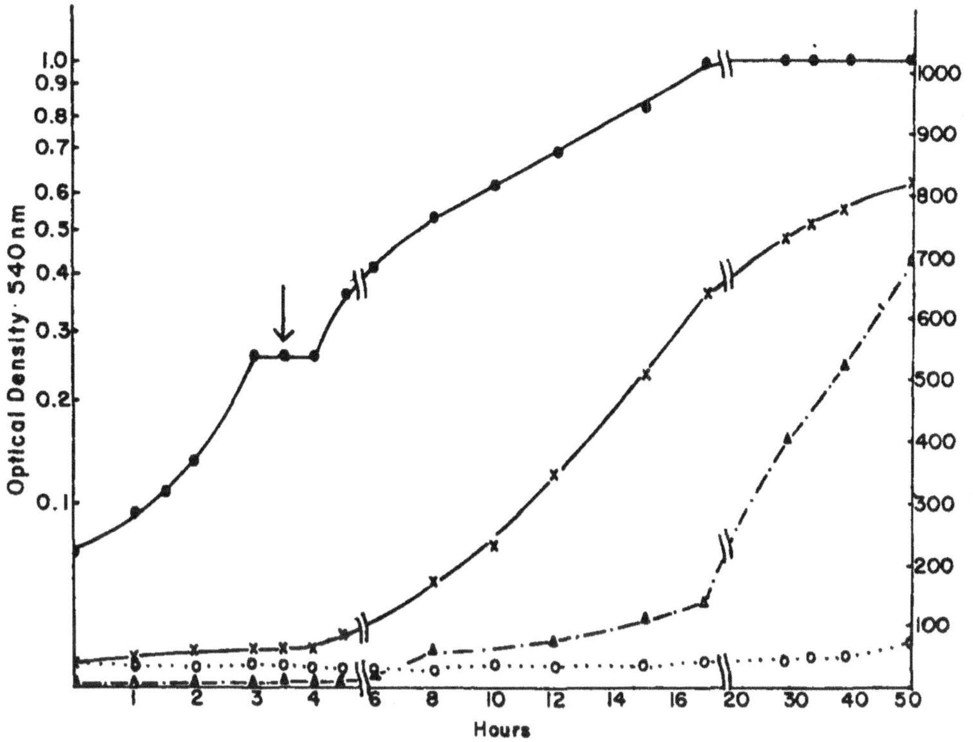

FIGURE 10. Time course for extracellular lipid accumulation by *Acinetobacter* sp. H01-N during a modified diauxic growth curve on a limiting concentration of acetate (10 m*M*) and 0.3% (vol/vol) hexadecane. Arrow indicates addition of hexadecane. Symbols: •, cell growth; X, monoglyceride; Δ, wax ester; O, triglyceride. Results are expressed as total micromoles lipid per liter of culture broth. (From Makula, R. A. et al., *J. Bacteriol.*, 121, 250, 1975. With permission.)

acid is derived directly from hexadecane oxidation, consisting of predominantly palmitic and palmitoleic acids which accumulate at concentrations 12- to 15-fold greater than in the spent growth medium of nonhydrocarbon-grown cells.[44] These extracellular free fatty acids, as well as the mono- and diglycerides, have known surfactant properties and may aid in the solubilization of the hexadecane growth substrate by forming macroemulsions of hexadecane.

The kinetics of extracellular wax ester (hexadecylpalmitate) accumulation in hexadecane-grown *Acinetobacter* differs considerably from the accumulation of fatty acid and glycerides. As shown in Figure 10, wax ester is not detected in the growth medium until early stationary phase and reaches its peak accumulation at late stationary phase.[21] The wax ester hexadecylpalmitate is an end product of hexadecane metabolism formed by the condensation of a fatty alcohol and a fatty acid[50] and is not further metabolized.

B. Isolation of Membranes from *Acinetobacter* sp. H01-N

The cell envelope of nonhydrocarbon-grown *Acinetobacter* resembles that of a typical Gram-negative bacterium (Figure 1). Unlike other Gram-negative organisms, however, the cytoplasmic and outer membranes cannot be separated solely by sucrose gradient centrifugation since the buoyant densities of the two membranes are similar. Hexadecane-grown *Acinetobacter* presents an even more difficult problem for membrane isolation since hexadecane-grown cells contain, in addition to the outer and cytoplasmic membranes, an intracytoplasmic membrane system and a monolayer membrane-bound hydrocarbon inclusion, as well as an extracellular membranous vesicle. We describe below in some detail the procedures developed in our laboratory for the isolation of each membrane component and the subsequent biochemical characterization of each purified membrane.

1. Membranes Derived from Nutrient Broth-Grown Cells

Crude membrane fractions are derived by a gentle lysis procedure utilizing a lysozyme osmotic shock technique developed in our laboratory.[56]

Discontinuous sucrose gradient centrifugation of the crude membranes derived from nutrient broth-grown cells results in the formation of a single band at a density of 1.182 g/cm^3. This band contains enzyme activities normally associated with both the outer and cytoplasmic membranes of Gram-negative bacteria. Differential centrifugation is required to separate the two membranes. Centrifugation of the crude membrane fraction at relatively low speeds (35,000 × g) for a short time period (1 to 3 min) results in sedimentation of the cytoplasmic membrane. The outer membrane, remaining in the supernatant fluid, is then recovered by centrifugation at 65,000 × g for 2 hr. The purity of both membranes was verified by both chemical and enzymological criteria (Table 5). When analyzed on linear sucrose gradients to determine their buoyant densities, little difference is found between the density of the cytoplasmic membrane (1.161 g/cm^3) and the outer membrane (1.147 g/m^3).[56]

2. Membranes Derived from Hexadecane-Grown Cells

Discontinuous sucrose gradient centifugation of the crude membranes derived from hexadecane-grown cells results in the formation of three bands, referred to as Bands 1, 2, and 3. As with nutrient broth-grown cells, the outer and cytoplasmic membranes band at a density of 1.182 g/cm^3 (Band 2). Consequently, the outer and cytoplasmic membranes were isolated from the crude membrane fraction by differential centrifugation. However, centrifugation of the cytoplasmic membrane through a discontinuous sucrose gradient results in Band 1 (density 1.099 g/cm^3) and Band 2 (density 1.182 g/cm^3). Band 1, the low-density membrane, and Band 2 are enzymatically identical with respect to the cytoplasmic membrane markers tested (Table 5). However, several important chemical differences have been noted between the light membrane and the cytoplasmic membrane (Table 5). Band 1 has a twofold higher phospholipid to protein ratio in comparison with the cytoplasmic membrane and contains twofold more membrane-associated hexadecane on a protein basis. Band 1, the low-density, lipid- and hexadecane-rich membrane most likely represents the intracytoplasmic membrane observed in electron micrographs of hexadecane-grown *Acinetobacter.*[56]

The outer membrane of hexadecane-grown cells is recovered by centrifugation of the supernatant fluid obtained by differential centrifugation. However, discontinuous sucrose gradient centrifugation of the outer membrane results in the formation of two bands, Band 2 (density 1.182 g/cm^3) and Band 3 (density 1.258 g/cm^3). Both outer-membrane fractions thus isolated exhibit similar activities for the outer-membrane marker enzymes tested as well as similar concentrations of 2-keto-3-deoxyoctulossonic acid and glucosamine, components of lipopolysaccharide.[56]

Analysis of the membrane fractions by continuous sucrose gradient centrifugation established the buoyant densities as 1.086 g/cm^3 for Band 1, 1.138 g/cm^3 for cytoplasmic membrane (Band 2); 1.141 g/cm^3 for outer membrane (Band 2), and 1.258 g/cm^3 for outer membrane (Band 3).[56] Negative stains of these membrane fractions are shown in Figure 11.

3. Isolation of the Hydrocarbon Inclusions

Centrifugation of cell lysates at 78,000 × g for 8 hr results in the sedimentation of the crude membranes, while the hydrocarbon inclusions rise to the top of the centrifuge

Table 5
PHYSICAL AND BIOCHEMICAL PROPERTIES OF MEMBRANES OF *ACINETOBACTER* SP. H01-N

	Hexadecane-grown cells					Nutrient broth-grown cells	
	Cytoplasmic membrane	Outer membrane	Band 1	Hydrocarbon inclusion	Extracellular membrane vesicle	Cytoplasmic membrane	Outer membrane
Density (g/cm³)	1.138	1.141	1.086	ND[a]	1.158	1.161	1.147
Lipid phosphorus: protein (μmol/mg)	0.30	0.16	0.55	1.0	1.2	0.32	0.23
Hexadecane: protein (μmol/mg)	0.67	0	1.88	20.0	0.92	0.0	0.0
Lipopolysaccharide: protein (pmol/mg)	34.0	203	35.0	95.0	73,400	14	390
Phospholipase (μmol/min/mg)	0.6	10.5	0.21	ND	8.8	0.16	2.71
ATPase (μmol/min/mg)	21.0	0	18.0	ND	ND	22.0	0.0
Phosphatidic acid cytidyl transferase (μmol/mg/hr)	5.9	1.0	7.8	ND	ND	1.7	0.6
NADH oxidase (μmol/min/mg)	53.0	4.0	53.0	ND	ND	60.0	5.0
Number of major membrane polypeptides	20—25	15—20	20—25	14	15—20	20—25	15—20

[a] ND = not determined.

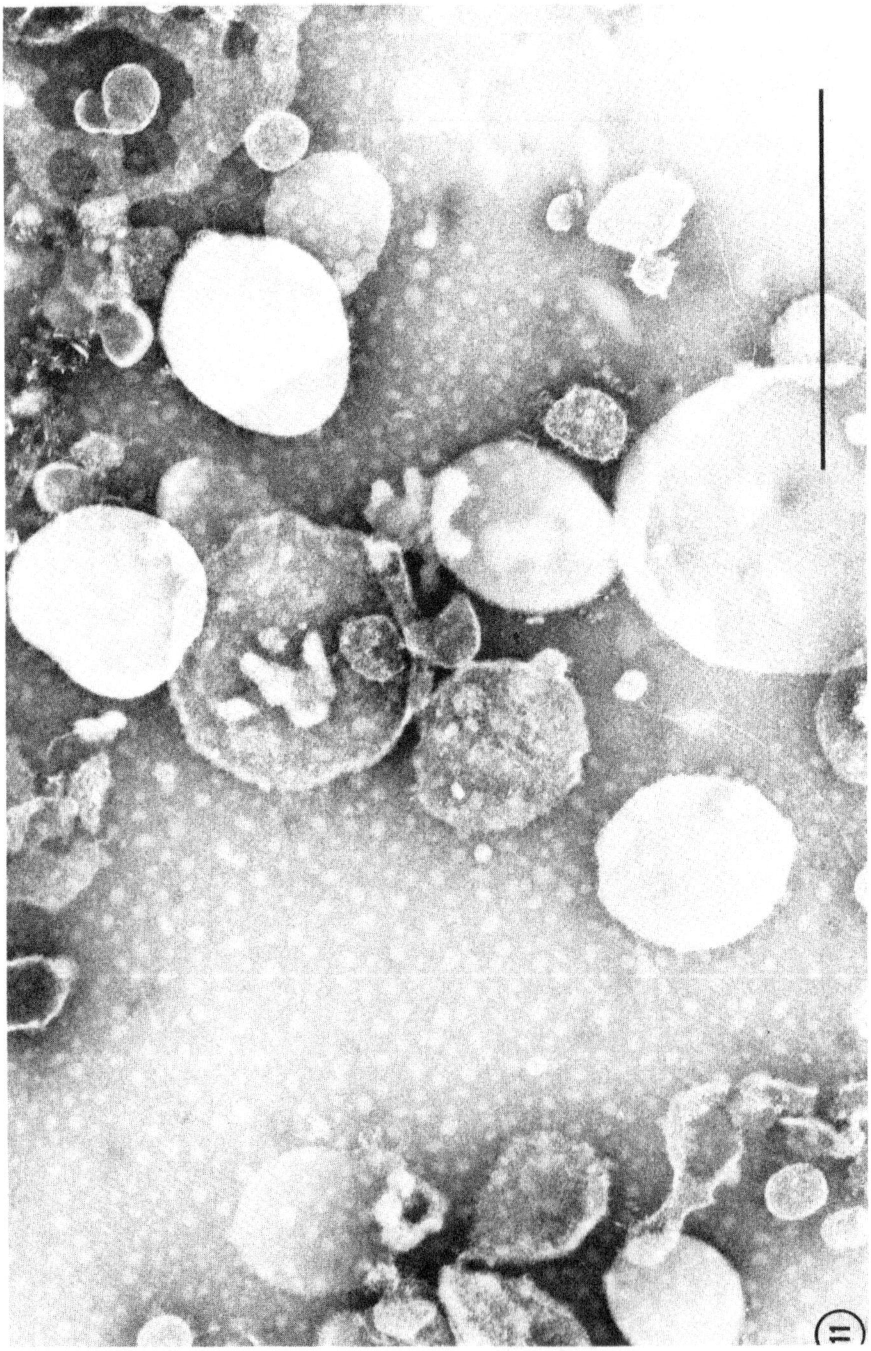

A

FIGURE 11. (A) Negative stain of membrane band 1 derived from hexadecane-grown *Acinetobacter* sp. H01-N; (B) negative stain of cytoplasmic membrane derived from hexadecane-grown *Acinetobacter*; (C) negative stain of outer membrane derived from hexadecane-grown *Acinetobacter*. Bar represents 0.5 μm. (From Scott, C. C. L. et al., *J. Bacteriol.*, 127, 469, 1976. With permission.)

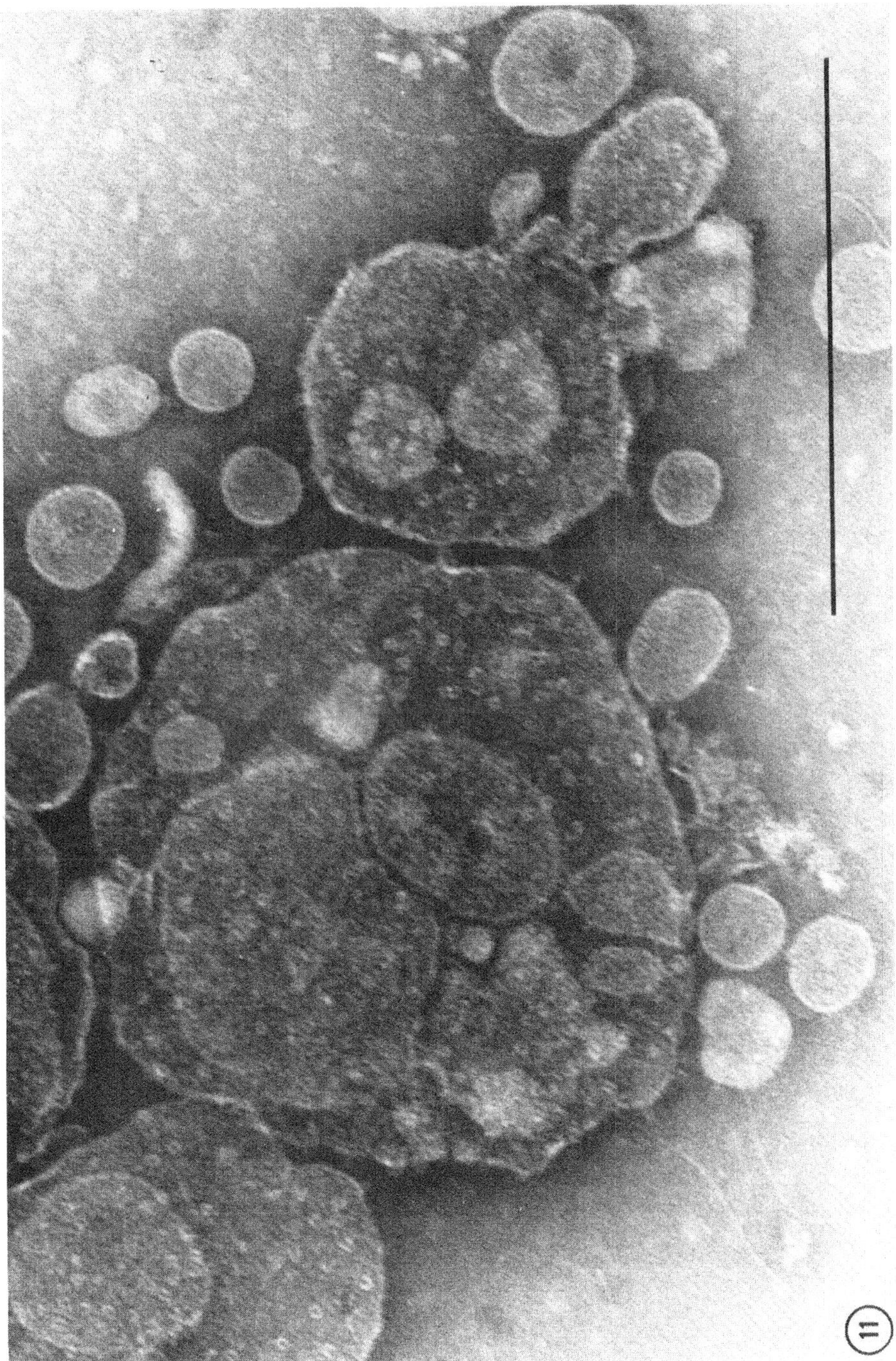

FIGURE 11B

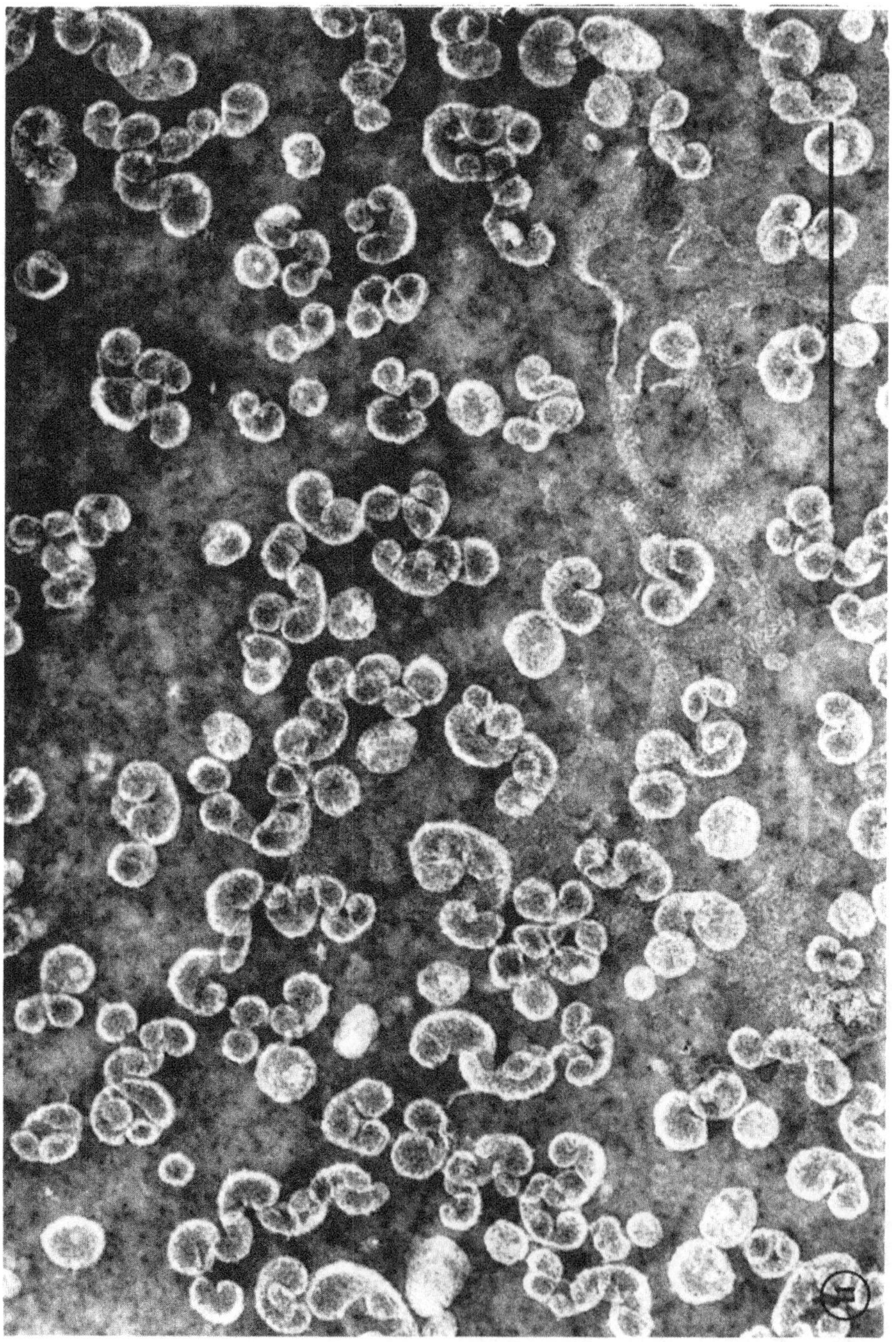

FIGURE 11C

tube due to their high content of hexadecane (Table 5). Soluble protein is separated from hydrocarbon inclusions by chromatography on Sepharose 2B. The inclusions maintain their size and shape after isolation and purification indicating that some form of limiting structure surrounds the hexadecane. Free hexadecane would coalesce, forming large droplets eventually separating from the aqueous phase. These particles retain monodisperse properties characteristic of true hydrophobic colloidal suspensions.[5]

4. *Isolation of the Extracellular Membrane Vesicles*

Hexadecane-grown cells accumulate an extracellular particulate component which appears to be membrane vesicles in electron micrographs of negatively stained preparations. These membrane vesicles are comprised of protein, lipid, and lipopolysaccharide.[57] These vesicles are not detected in the spent culture medium of nonhydrocarbon-grown cells. The isolated vesicles partition exogenously supplied hexadecane in the form of a microemulsion and may be involved in the solubilization of hexadecane in the growth medium in vivo.[57,58]

Isolation of these vesicles involves concentration of the spent growth medium by ultrafiltration. Centrifugation of the concentrated growth medium at 65,000 × g for 5 hr generates a soluble and vesicular fraction. The vesicular fraction partitions sevenfold more hexadecane than the soluble fraction, indicating the solubilization of water-insoluble hexadecane by these membrane vesicles.[57] By analysis of phospholipase activity and polypeptide composition, these membrane vesicles closely resemble the outer membrane (Table 5). These vesicles are highly enriched in KDO, containing 360-fold more KDO than isolated outer membranes from *Acinetobacter* (Table 5). The phospholipid to protein ratio of these membrane vesicles is five- to sevenfold greater than that determined for outer membranes derived from hexadecane-grown *Acinetobacter* (Table 5). Isolated outer membranes contain no bound hexadecane, while extracellular vesicles contain 920 nmol hexadecane, per milligram of protein (Table 5). Analysis of the particle by continuous sucrose gradient centrifugation established a buoyant density of 1.158 g/cm^3.[57]

C. Biochemical Composition of *Acinetobacter* Membranes

1. *Membrane Phospholipids*

The major phospholipid of membranes derived from cells grown on either nutrient broth or hexadecane is phosphatidylethanolamine (Table 6). The outer membrane derived from both sources and Band 1 from hexadecane-grown cells contains higher concentrations of cardiolipin and triacyllysocardiolipin than present in cytoplasmic membranes. Band 1 closely resembles the outer membrane in phospholipid composition. The inclusion membrane closely resembles the cytoplasmic membrane in phospholipid composition with the exception of an increased concentration of phosphatidylglycerol.[5,56] The phospholipid composition of the extracellular membrane vesicle is significantly different from that of Band 1 and the outer membrane. The membrane limiting the hydrocarbon inclusion is similar in phospholipid composition to the extracellular membrane vesicle.[5,57]

The phospholipid to protein ratio of cytoplasmic membranes isolated from either nutrient broth- or hexadecane-grown cells is twice as high as that determined for the outer membrane. Band 1, from hexadecane-grown cells, is lipid-rich, characterized by a phospholipid to protein ratio over twice as high as the cytoplasmic membrane. The hydrocarbon inclusion membrane and the extracellular membrane vesicle are also lipid-rich, having phospholipid protein ratios almost three times higher than that of the cytoplasmic membrane.[5,56,57]

2. *Membrane Neutral Lipids*

The major neutral lipid in cytoplasmic membranes, outer membranes, and Band 1

Table 6
PHOSPHOLIPID COMPOSITION OF MEMBRANES ISOLATED FROM *ACINETOBACTER* SP. H01-N

	Phospholipids in hexadecane-grown cells (%)					Phospholipids in nutrient broth-grown cells(%)	
	Cytoplasmic membrane	Outer membrane	Band 1	Hydrocarbon inclusion membrane	Extracellular membrane vesicle	Cytoplasmic membrane	Outer membrane
Phosphatidylethanolamine	69	40	50	60	67	70	52
Lysophosphatidylethanolamine	3	4	3	5	Not detected	3	3
Cardiolipin	6	20	17	2	4	7	19
Triacyl-lysocardiolipin	3	19	11	6	6	10	15
Phosphatidylglycerol	13	17	8	27	21	10	11
Acyl-phosphatidylglycerol	6	Not detected	11	Not detected	2	Trace	Trace
Phospholipid (μmol/mg membrane protein)	0.30	0.16	0.55	1.0	1.2	0.32	0.23

Table 7
NEUTRAL LIPID COMPOSITION OF MEMBRANES ISOLATED FROM *ACINETOBACTER* SP. H01-N

	Neutral lipids in hexadecane-grown cells (%)			Neutral lipids in nutrient broth-grown cells (%)	
Neutral lipid class	Cytoplasmic membrane	Outer membrane	Band 1	Cytoplasmic membrane	Outer membrane
Wax ester	14	5	11	15	11
Free fatty acid	6	7	3	14	18
Fatty alcohol	64	69	71	48	47
Triglyceride	Not detected	Not detected	Not detected	Not detected	Not detected
Diglyceride	6	9	7	11	16
Monoglyceride	Trace	5	6	11	9
Unidentified lipid	5	4	2	Trace	Trace

isolated from hexadecane-grown cells is hexadecanol (Table 7). Membranes derived from nutrient broth-grown cells contain less free fatty alcohol.[56] Table 1 shows the lipid composition of the hydrocarbon inclusions. The inclusions are comprised predominantly of hexadecane.

Analysis of hexadecane in the various membranes reveal high amounts in Band 1 and the extracellular vesicle (Table 5). No hexadecane is detectable in the outer membrane of hexadecane-grown cells.[56,57]

3. Membrane Proteins

Membranes were analyzed by SDS-polyacrylamide gel electrophoresis for polypeptide composition (Table 5). The cytoplasmic membrane of nutrient broth- and hexadecane-grown cells and Band 1 of hexadecane-grown cells contain between 20 and 25 polypeptide bands. Band 1 and the cytoplasmic membrane of hexadecane-grown cells appear identical except for minor quantitative differences.[56] The outer membrane contains only 15 to 20 polypeptides. No major differences are observed in the respective polypeptide composition of cytoplasmic or outer membranes isolated from nutrient broth- or hexadecane-grown cells.[56]

The limiting membrane of the hydrocarbon inclusion exhibits a polypeptide pattern dissimilar to all other membrane fractions isolated from hexadecane-grown cells.[56] The polypeptide banding pattern of the outer membrane is similar, if not identical, to that of the extracellular membrane vesicle.[57]

4. Membrane-Localized Cytochromes

The cytoplasmic membranes of hexadecane and acetate-grown *Acinetobacter* have been analyzed for carbon-induced differences in cytochrome content and composition.[59] Spectrophotometric analyses reveal that the quantity and type of cytochrome in cytoplasmic membranes change in response to growth under various concentrations of oxygen. Growth on hexadecane as compared to acetate results in only minor differences in cytochrome composition (Table 8). However, under conditions of low oxygen tension, the content of cytochrome o doubles in membranes of hexadecane-grown cells in comparison to membranes derived from acetate-grown cells.[59] Measurements of dissolved oxygen concentration in the growth medium of *Acinetobacter* grown under conditions of high or low aeration on hexadecane show a rapid decrease in dissolved oxygen to undetectable levels during the midexponential growth phase.[59]

Since the total membrane content of hexadecane-grown cells is two to three times greater in comparison to nonhydrocarbon-grown cells, the levels of respiratory pro-

Table 8
CYTOCHROME COMPOSITION OF MEMBRANES IN *ACINETOBACTER* SP. H01-N

Carbon source	Growth conditions	Cytochrome content (nmol/mg protein)			
		b	o	d	a_1
Acetate	High aeration	0.16	0.18	0.03	Trace
	Low aeration	0.56	0.17	0.19	Trace
Hexadecane	High aeration	0.14	0.10	Trace	Not detected
	Low aeration	0.48	0.31	0.17	0.01

teins in *Acinetobacter*, when calculated on the basis of units per whole cells, increases two to threefold during growth on hexadecane. Under conditions of low aeration for both acetate- and hexadecane-grown cells, cytochrome b content increased 3.5-fold, cytochrome o, 3-fold, cytochrome d, 6- to 10-fold, and cytochrome a_1 was synthesized. Cytoplasmic membranes prepared from hexadecane-grown cells exhibit a lower efficiency of energy coupling (P/O = 0.14 to 0.22) as compared to those isolated from acetate-grown cells (P/O = 1.0)[59,60] The reason(s) for the lower efficiency of energy coupling in hexadecane-grown cells remains to be established; however, it has been determined that the decreased coupling efficiency did not result from the differences in respiratory chain composition in hydrocarbon-grown cells. Cells containing only cytochrome o displayed the same P/O ratios as those containing both cytochromes d and o as terminal oxidases.[60]

V. PHOSPHOLIPID METABOLISM IN *ACINETOBACTER* SP. H01-N

The phospholipid composition of microorganisms has received concentrated study in recent years. Collectively, these studies have served to document that phospholipids are dynamic biochemical entities, contributing to a variety of important functional roles in cellular physiology.[61] The realization that numerous biosynthetic and metabolic functions of the living cell are performed by membrane-associated proteins has complemented the fact that cellular lipids are predominantly localized in the membrane. As knowledge of the structural and functional complexity of lipids has developed, it becomes increasingly possible to designate the role of diverse lipids which actively participate in the various aspects of cellular physiology. The following discussion will focus on studies performed with *Acinetobacter* species H01-N as a representative hydrocarbon-oxidizing microorganism. This focus is dictated by the fact that studies with this microorganism in relation to phospholipid metabolism and membranes represent the only extensive analyses of the overall aspects of lipid metabolism with a hydrocarbon-oxidizing microorganism.

A. Phospholipid Biosynthesis

The phospholipid composition of *Acinetobacter* sp. H01-N has been exhaustively analyzed under a variety of cultural conditions.[20,62] The isolation and characterization of the cytoplasmic membrane of *Acinetobacter* has allowed for the study of various phospholipid biosynthetic enzymes and aspects possibly relating to their regulation.[56]

A number of phospholipid biosynthetic enzymes present in *Acinetobacter* and their specific activities are shown in Table 9. McCaman and Finnerty have characterized the enzyme phosphatidic acid cytidyl transferase (PACT) from *Acinetobacter* reporting an unusually high specific activity.[63] A comparative study of PACT activity in seven different genera shows *Acinetobacter* to be especially active, with the genus *Moraxella*

Table 9
SPECIFIC ACTIVITIES OF PHOSPHOLIPID BIOSYNTHETIC ENZYMES FROM *ACINETOBACTER* [a]

Cytoplasmic membrane derived from	Phosphatidic acid cytidyl transferase	α-Glycerol-CMP phosphatidyl transferase	L-Serine-CMP phosphatidyl transferase	Phosphatidyl serine decarboxylase	Phosphatidic acid phosphatase
Hexadecane-grown cells	4.5	6.0	24	3.4	1.5
Nutrient broth-grown cells	1.2	3.2	11	1.6	5.6

[a] In μmol/mg membrane protein per hour.

also exhibiting significantly elevated levels of PACT.[61] PACT is approximately three times greater in hexadecane-grown cells while the other phospholipid biosynthetic enzymes are approximately double in specific activity in comparison to cytoplasmic membrane derived from nutrient broth-grown cells. In contrast, the specific activity of phosphatidic acid phosphatase decreases approximately fivefold in hexadecane-grown cells. The results indicate a significant increase in the specific activity of phospholipid biosynthetic enzymes in the cytoplasmic membrane of hexadecane-grown cells.

The isolation of highly enriched cytoplasmic membrane preparations from *Acinetobacter* exhibiting stable as well as elevated PACT activity prompted a series of purification studies for this enzyme. Numerous procedures used to purify PACT were unsuccessful, including Triton X-100® membrane solubilization and DEAE-cellulose column chromatography in the presence of Triton X-100®. A solubilization of between 10 to 15% of the membrane-bound PACT is achieved with Triton X-100® and extensive sonication. Triton-solubilized PACT retains activity for approximately 36 to 48 hr and removal of Triton X-100® causes the protein to aggregate and inactivate. Possible reasons for the rapid inactivation of detergent-solubilized PACT activity may relate to loss of lipid cofactor(s), enzyme aggregation, conformational changes in the enzyme due to solubilization, dilution inactivation, or other physicochemical parameters unrecognized regarding the stabilization of PACT activity. It is noteworthy, however, that only a small percentage of the total PACT activity is solubilized with the major amount of activity remaining tightly bound to the membrane.

Since purification of enzyme activity was unsuccessful, procedures for the removal of either membrane protein or lipid were developed with the subsequent assessment of membrane-bound PACT activity. The following procedure yields protein- and lipid-deficient membranes with an apparent enrichment of PACT activity. The rationale was to effect as much removal of membrane protein as possible without significant loss of enzyme activity and to deplete membrane-bound phospholipids for assessing a functional role of phospholipids in PACT activity. Since addition of exogenous phospholipid to cytoplasmic membrane neither stimulates nor inhibits PACT activity, such treatments of membranes are necessary for the evaluation of phospholipid effects.

The cytoplasmic and intracytoplasmic membranes of hexadecane-grown *Acinetobacter* are derived as a total crude membrane preparation and suspended in 0.05 *M* Tris-HCl, pH 7.5, containing 0.01 *M* $MgCl_2$, 2*M* urea, and 0.1 *M* $(NH_4)_2SO_4$, stirred for 10 min and centrifuged at 100,000 × g for 60 min. This treatment is repeated and the final membrane pellet is suspended in the same buffer system and dialyzed against 0.05 *M* Tris-HCl, pH 7.5 with 0.01 *M* $MgCl_2$ (THM buffer) and the resulting membrane preparation designated as urea-$(NH_4)_2SO_4$-treated membranes (UAT membranes). The overall total balance of this UAT procedure results in a 68 and 59% loss of protein and a 47 and 37% loss of phospholipid for membranes derived from nutrient broth and hexadecane-grown cells, respectively. An enrichment of membrane-associated PACT is achieved in the UAT-membranes (Table 10).

1. Stimulation of PACT Activity by Exogenous Phospholipid

The effect of exogenous phospholipids on PACT was studied with UAT-membranes derived from hexadecane-grown cells. Figure 12 shows a 100% stimulation of PACT activity at a phospholipid ratio of 0.044. Progressively increasing ratios of exogenous phospholipid causes less stimulation of PACT activity. These results indicate that PACT was maximally stimulated over a narrow range of exogenous to endogenous phospholipid ratios ranging between 0.01 to 0.06. Separate studies demonstrate that radioactive phospholipids at these low concentrations reassociated 100% with the UAT membranes. Ratios above 0.5 did not result in quantitative reassociation of phospholipids to UAT membranes. Stimulation of PACT is specific to phospholipids in that purified neutral lipids derived from *Acinetobacter* are inactive.

2. Lipid Depletion of UAT Membranes

Our results demonstrate that PACT activity associated with intact cytoplasmic membrane is not stimulated by the addition of exogenous phospholipid, whereas UAT membranes exhibit 100% stimulation of PACT activity from the addition of low concentrations of exogenous phospholipids. The treatment of membrane by the urea-ammonium sulfate procedure removed more membrane protein than phospholipid with a corresponding increase in PACT specific activity. A method for further removal of membrane phospholipids was developed to determine the effect of phospholipid depletion on PACT activity. UAT membranes derived from hexadecane-grown cells are chilled in an ice-salt bath and precipitated with 2 volumes of −20°C acetone with stirring over a 5-min time period. The acetone-precipitated membranes are collected by centrifugation, washed twice with THM buffer containing 0.88 *M* sucrose, and stored in the same buffer. Analyses of the biochemical characteristics of membrane fractions discussed are shown in Table 10. An interesting comparison develops with respect to the PACT activity present in cytoplasmic membranes derived from nutrient broth-grown cells. Urea-ammonium sulfate treatment of these membranes results in a labile PACT activity, with a 50% loss of activity in 24 hr as compared to a stable PACT activity in membranes derived from hexadecane-grown cells. Acetone-precipitation of UAT-membranes derived from nutrient broth-grown cells completely inactivates PACT activity. Addition of exogenous phospholipid fails to reactivate the enzyme. The following studies are performed with membranes derived from hexadecane-grown cells since neither urea-ammonium sulfate nor acetone precipitation affects PACT stability.

Stimulation of PACT activity by a mixture of exogenous phospholipid was examined in the acetone-treated membranes (Figure 13). A stimulation of approximately 50% is observed at a ratio of exogenous to endogenous phospholipid of less than 1.0 while a 30% inhibition results at a ratio of 7.5. Analyses of the phospholipids associated with the acetone-treated membrane show a composition of phosphatidylethanolamine (52%), phosphatidylglycerol (26%), and cardiolipin (22%) in comparison to phosphatidylethanolamine (62%), phosphatidylglycerol (20%), and cardiolipin (18%) for UAT membranes. Acetone-treated membrane had 82% less total phospholipid than UAT membranes and a corresponding 65% loss of PACT activity. A precise titration of a mixture of exogenous phospholipid with respect to stimulation of PACT activity demonstrates a 65% stimulation at a ratio of 0.5 (Figure 14).

Individual purified phospholipids were examined for stimulation of PACT activity. Exogenous phospholipid additions were made so that the ratios shown reflect the concentration of the individual exogenous phospholipid to its endogenous counterpart

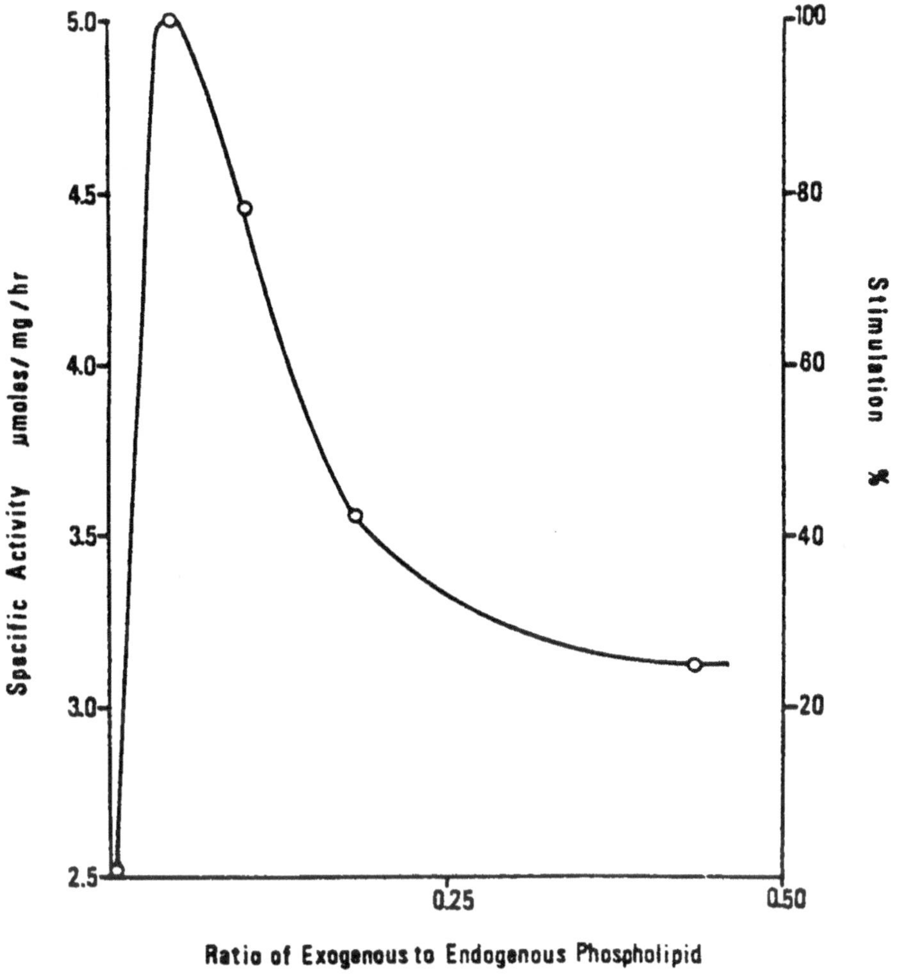

FIGURE 12. Titration of UAT membrane-PACT activity with a mixture of exogenous phospholipid. UAT membrane contains 260 pmol total endogenous phospholipid.

Table 10
BIOCHEMICAL CHARACTERISTICS OF DERIVED MEMBRANE FRACTIONS[a]

Fraction	Phospholipid (µmol/mg protein)	Carbohydrate (nmol/mg)	DNA µg/mg	RNA (µg/mg)	Hexosamine (nmol/mg)	KDO (pmol/ mg)	PACT (µmol/mg/ hr)
Cytoplasmic membrane	0.33	232	4.2	26	101	45	4.99
UAT membranes	0.48	325	0	9	104	77	7.10
Acetone-precipitated membranes	0.09	115	0	0	7	4	2.52

[a] Membranes derived from hexadecane-grown cells.

present in the acetone-treated membrane (Figure 15). Cardiolipin exhibits the least effect on PACT activity, although a 23% stimulation is observed at a ratio of 0.1. Phosphatidylglycerol stimulates PACT activity 35% at a ratio of approximately 0.1, while phosphatidylethanolamine stimulates PACT activity 50 to 60% over the ratio range 0.1 to 0.5.

Exogenous phospholipids were added singly and in combinations for evaluation of

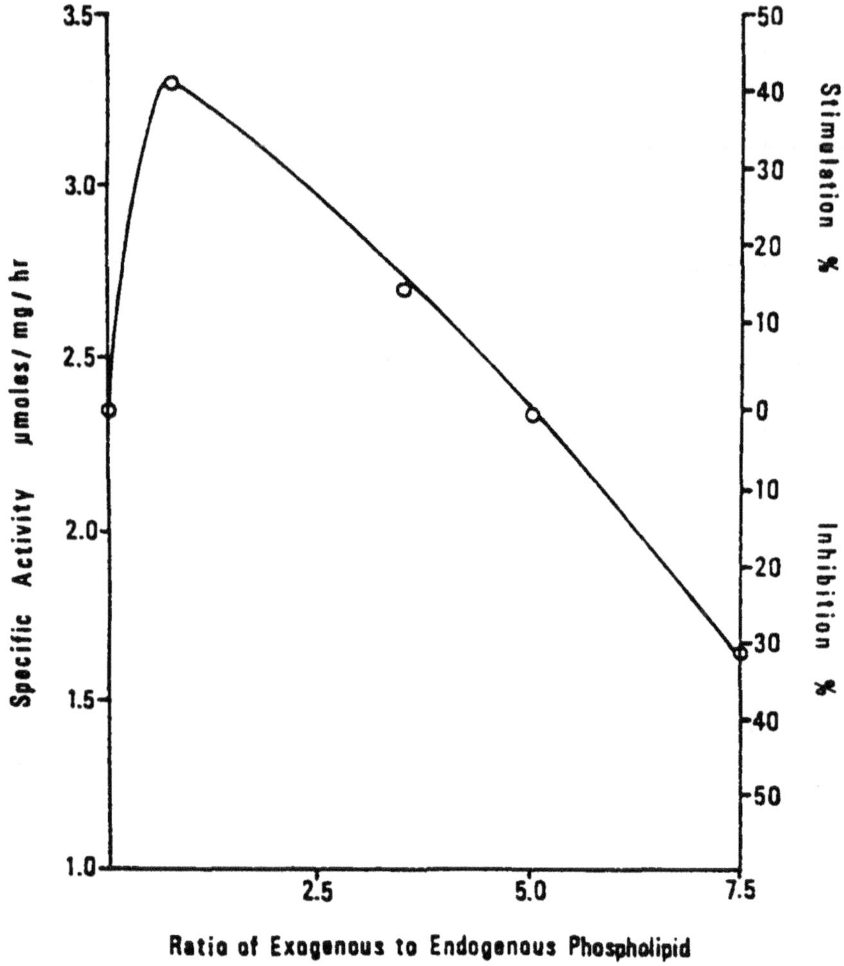

FIGURE 13. Titration of acetone-treated membrane-PACT activity with a mixture of exogenous phospholipid. Acetone-treated membrane contains 17 pmol total endogenous phospholipid.

their specific effect on stimulation of PACT activity (Table 11). All exogenous phospholipid additions were tested at a ratio of 0.1. Phosphatidylserine synthetase activity is included as an additional phospholipid biosynthetic enzyme to evaluate the specificity of the stimulatory effect of phospholipids. Single additions of phosphatidylethanolamine, phosphatidylglycerol, or cardiolipin as exogenous phospholipids demonstrate that phosphatidylethanolamine and phosphatidylglycerol stimulate PACT activity 92 and 64%, respectively, whereas phosphatidylethanolamine and cardiolipin stimulate phosphatidylserine synthetase activity 90 and 110%, respectively. A combination of phosphatidylethanol and phosphatidylglycerol stimulate PACT activity 248%, while phosphatidylethanolamine and cardiolipin stimulate phosphatidylserine synthetase 193%. The stimulation of phosphatidylserine synthetase activity by phosphatidylethanolamine-cardiolipin is additive, while phosphatidylethanolamine-phosphatidylglycerol stimulates PACT activity in excess of the additive values determined for the individual phospholipids.

The ratio of exogenous phospholipid combinations was optimized with respect to maximal stimulation of PACT and phosphatidylserine synthetase (Table 12). The optimum ratio of phosphatidylethanolamine-phosphatidylglycerol is 0.05 for PACT, while the optimum ratio for phosphatidylethanolamine-cardiolipin is 0.25 for phos-

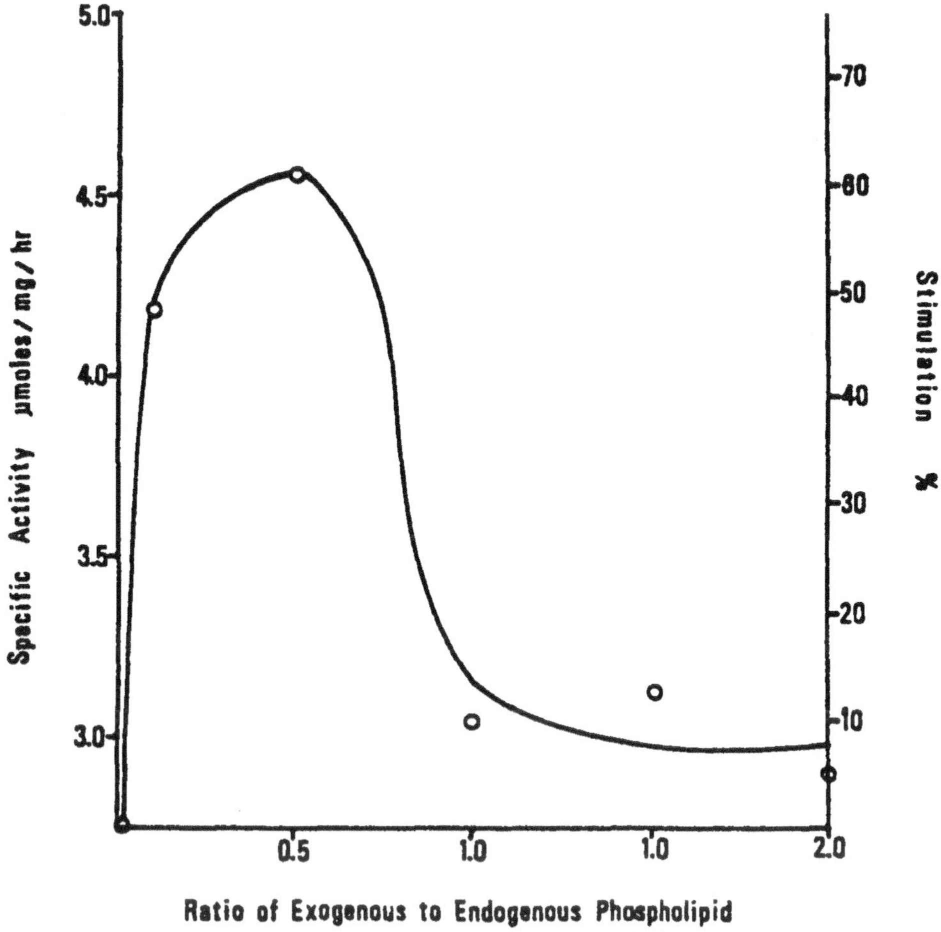

FIGURE 14. Titration of acetone-treated membrane-PACT activity with a mixture of exogenous phospholipid. Acetone-treated membrane contains 69 pmol total endogenous phospholipid.

phatidylserine synthetase. The reassociation of these phospholipid combinations to lipid-depleted membrane is quantitative at the optimum ratios shown, whereas at higher concentrations of exogenous phospholipids reassociation with the membrane is not quantitative.

B. Phospholipid Catabolism

1. General

Phospholipids are metabolically dynamic constituents of membranes which have been studied with respect to turnover of phospholipids, polar end-group conversions, and phospholipid fatty acyltransferase activities. Our studies have attempted to establish a relationship between membrane-localized phospholipids and alkane assimilatory processes in *Acinetobacter*. The turnover kinetics of the major phospholipids of *Acinetobacter* have been studied in the context of defining a functional rather than structural role for individual phospholipids.[21,62] Since total cellular phospholipid is two- to threefold higher in hexadecane-grown cells, it was of interest to assess phospholipid turnover by incorporation of glycerol-1,3-^{14}C, phosphate-^{32}P, and hexadecane-1-^{14}C into the cellular phospholipids of *Acinetobacter*.

Phosphatidylethanolamine represents approximately 62% of the total cellular lipids and exhibits no significant rate of turnover in either growing or nongrowing cultures.[62]

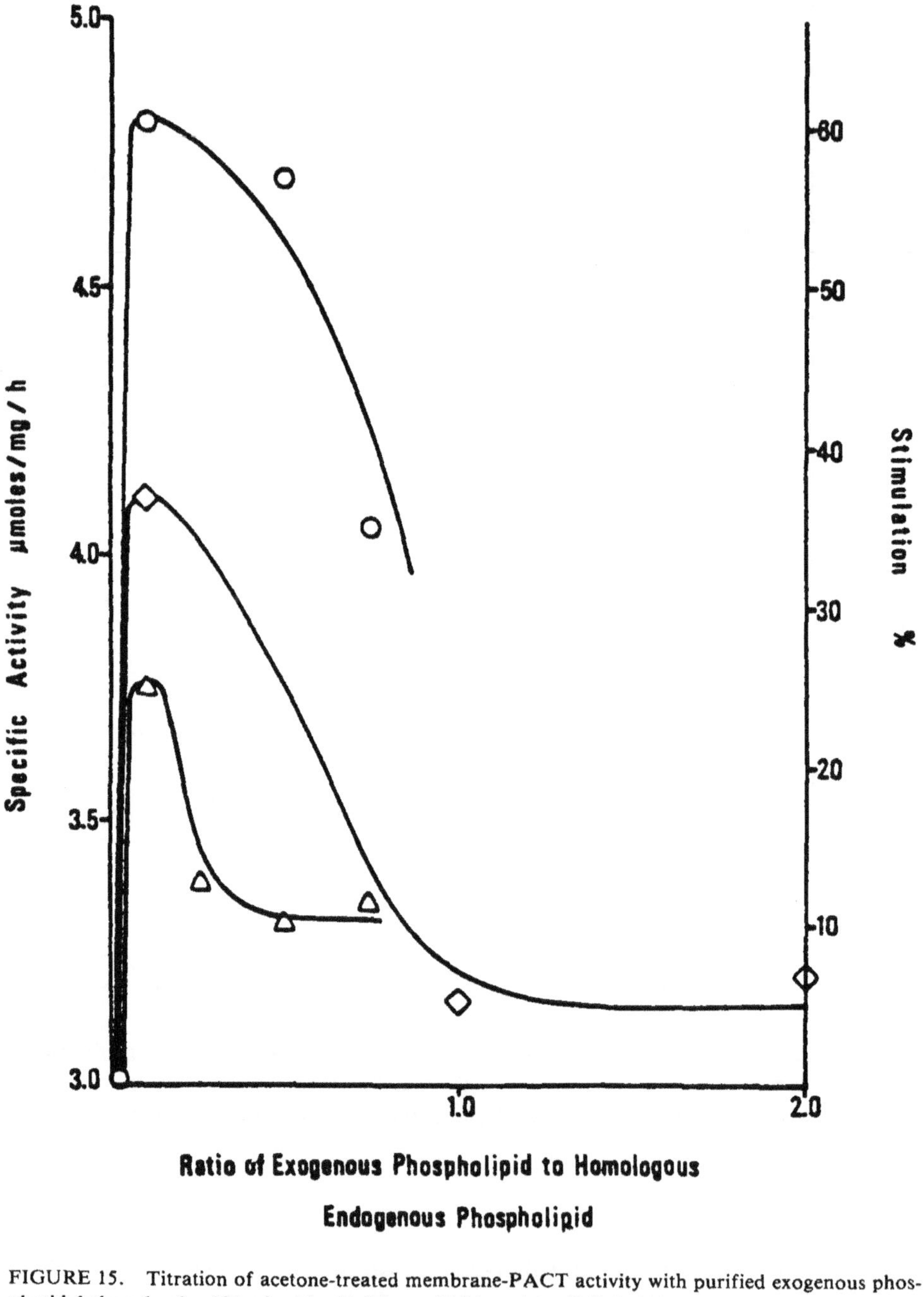

FIGURE 15. Titration of acetone-treated membrane-PACT activity with purified exogenous phosphatidylethanolamine (O), phosphatidylglycerol (◇), and cardiolipin (△). Acetone-treated membrane contains 126 pmol of total endogenous phospholipid consisting of 66 pmol phosphatidylethanolamine, 33 pmol phosphatidylglycerol, and 27 pmol cardiolipin.

The turnover rate of ^{14}C-glycerol-labeled phosphatidylethanolamine is 3.2%/hr and 1.2%/hr for ^{32}P-labeled phosphatidylethanolamine. Phosphatidylglycerol exhibits a turnover rate of 11 and 7.7%/hr for ^{14}C-glycerol-labeled phosphatidylglycerol and ^{32}P-labeled phosphatidylglycerol, respectively, indicating a slow metabolism of phosphatidylglycerol in growing cells.

The turnover kinetics for phosphatidylglycerol in nongrowing cultures demonstrate a rapid reincorporation of ^{14}C-glycerol into phosphatidylglycerol after an initial decrease.[62] This labeling pattern could result from the action of phospholipase D on

Table 11
STIMULATION OF PHOSPHOLIPID BIOSYNTHETIC ENZYMES BY PHOSPHOLIPIDS

Phospholipid addition	Exogenous/endogenous[a] phospholipid ratio	PACT (μmol/mg/hr)	PS synthetase (nmol/mg/hr)
—	0	2.5	10
PE	0.1	4.8 (92%)	19.1 (90%)
PG	0.1	4.1 (64%)	11.4 (10%)
CL	0.1	3.7 (48%)	21.4 (110%)
PE + PG	0.1	8.7 (248%)	16.5 (65%)
PE + CL	0.1	4.6 (84%)	29.3 (193%)
PG + CL	0.1	3.9 (56%)	20.5 (105%)

[a] 126 pmol of endogenous phospholipid: PE, 66 pmol; PG, 33 pmol; CL, 27 pmol.

Table 12
PHOSPHOLIPID STIMULATION OF PHOSPHATIDIC ACID CYTIDYL TRANSFERASE AND PHOSPHATIDYLSERINE SYNTHETASE

Phospholipid addition	Exogenous/endogenous phospholipid ratio	PACT (μmol/mg/hr)	PS synthetase (nmol/mg/hr)
PE + PG	0.01	4.4	—
	0.05	9.1	—
	0.10	8.9	—
	0.25	3.0	—
	0.50	2.4	—
	1.00	2.1	—
PE + CL	0.01	—	13.4
	0.05	—	19.1
	0.10	—	28.7
	0.25	—	33.1
	0.50	—	21.1
	1.00	—	12.2

phosphatidylglycerol producing phosphatidic acid and glycerol, with glycerol being reincorporated into the head group of phosphatidylglycerol without dilution through a pool. Our kinetic labeling studies indicate a preferential 75% incorporation of glycerol-1,3-^{14}C into the nonacylated portion of the phosphatidylglycerol molecule over incorporation into the acylated portion.

The turnover of cardiolipin is quite slow in growing cells; whereas, nongrowing cells exhibit a 30 and 36% increase per hour for ^{14}C-glycerol-labeled cardiolipin and ^{32}P-labeled cardiolipin, respectively, for the initial 3 hr.[62] Separate studies demonstrate that glycerol is incorporated only into the water-soluble backbone portion of cellular phospholipids without conversion to fatty acid carbon.[62]

Studies with hexadecane-1-^{14}C demonstrate that 97% of the phospholipid fatty acids derived from hexadecane-grown cells are 16-carbon fatty acids. Further, these 16-carbon fatty acids are derived from the oxidation of hexadecane and direct assimilation into the cellular phospholipids.[44] Phospholipid fatty acid radioactivity increases 50-fold over the growth cycle through the direct oxidation of hexadecane to 16-carbon fatty acids and their subsequent incorporation into phospholipids. Phosphatidylethanolamine fatty acyl residues are extensively labeled within 2 hr, while phosphatidylglycerol and cardiolipin fatty acid labeling patterns remain low initially and increase with cellular growth. The data indicate a rapid and extensive turnover of phospholipid fatty

Table 13
PHOSPHOLIPASE A_1 ACTIVITIES OF *ACINETOBACTER* OUTER MEMBRANE

Enzyme	Culture conditions	Specific activity (nmol/min/mg)	pH optimum	Triton X-100® conc (%)	K_m (m*M*)	V_{max}
Phosphatidylethanol-amine	Nutrient broth-grown	80	9.5	0.5	1.8	380
phospholipase A_1	Hexadecane-grown	214	9.5	0.5	2.2	655
Phosphatidylglycerol	Nutrient broth-grown	77	7.0	0.5	4.6	250
phospholipase A_1	Hexadecane-grown	149	7.5	0.5	1.5	500
Cardiolipin	Nutrient broth-grown	40	6.0	0.5	1.9	170
phospholipase A_1	Hexadecane-grown	90	7.0	0.5	2.2	390

acyl groups in *Acinetobacter* growing at the expense of hexadecane.[62] Whether the phospholipid turnover data supports net membrane synthesis or breakdown and reassembly of existing membrane subunits is difficult to assess since the experiments were conducted with hexadecane-grown cells where new membrane synthesis had been completed. However, separate experiments relating the induction of new membrane synthesis to alkane utilization demonstrate that membrane synthesis precedes cellular growth at the expense of hexadecane.[4]

2. Outer Membrane Phospholipases of Acinetobacter

The isolation and purification of the outer membrane of *Acinetobacter* sp. H01-N has enabled the study of three phospholipases.[64-66] Table 13 shows a summary of the physical constants of phosphatidylethanolamine phospholipase A_1, phosphatidylglycerol phospholipase A_1, and cardiolipin phospholipase A_1. The specific activities of these phospholipases are two to three times higher in the outer membrane derived from hexadecane-grown cells. A similar observation is noted for phospholipid biosynthetic enzymes derived from hexadecane-grown cells. The optimal pH and K_m values are similar for membranes derived from either nutrient broth- or hexadecane-grown cells with the exception of the K_m determined for phosphatidylglycerol phospholipase A_1. Reasons for this higher K_m in nutrient-broth grown cells remains unclear.

The products resulting from the action of phosphatidylethanolamine phospholipase, phosphatidylglycerol phospholipase, and cardiolipin phospholipase are lysophosphatidylethanolamine, lysophosphatidylglycerol, and triacyllysocardiolipin. With the exception of pH optima and substrate saturation curves, the enzyme activities have similar characteristics. In addition, these phospholipase A activities are similar in outer membranes prepared from either hexadecane- or nutrient broth-grown cells, with the notable exception that specific activities are two to three times higher when cells are grown on hexadecane as the sole source of carbon and energy. The highest specific activities observed for phospholipase A from *Acinetobacter* outer membranes are 20- to 2000-fold higher than that reported for particulate preparations from *Escherichia coli*,[67-69] *Mycobacterium phlei*,[70] or *Neisseria gonorrhaeae*.[71] These three activities are similar with respect to Triton X-100® requirements, heat lability, lack of metal ion requirement, and sensitivity to ionic detergents, suggesting a single polypeptide with broad substrate specificity. Preliminary purification studies indicate the presence of a single polypeptide with activity towards phosphatidylethanolamine, phosphatidylglycerol, and cardiolipin. However, the physiological presence of triacyllysocardiolipin, but not lysophosphatidylethanolamine or lysophosphatidylglycerol, suggests possible regulation of enzyme activity or the nonaccessibility of the substrate to the membrane-bound enzyme. Thus, cardiolipin is hydrolyzed in vivo; phosphatidylethanolamine and

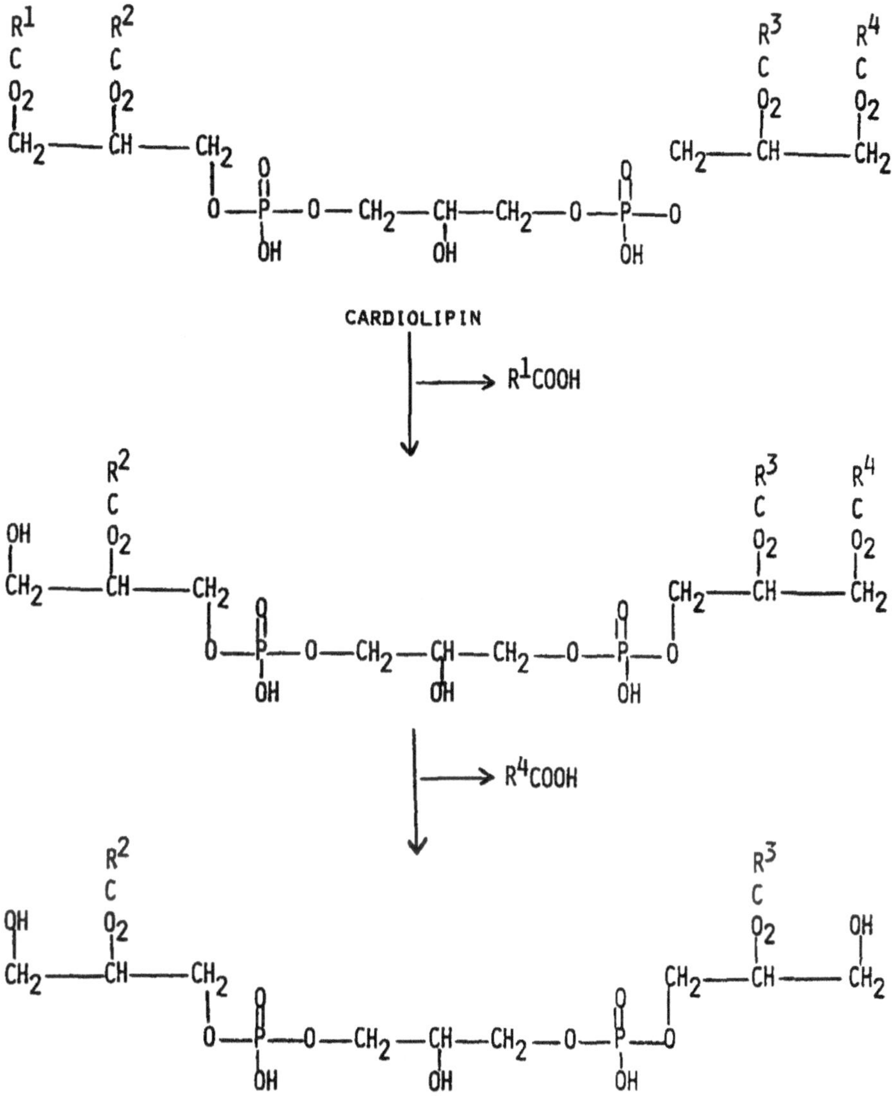

FIGURE 16. Hydrolysis of cardiolipin by cardiolipin phospholipase.

phosphatidylglycerol are hydrolyzed only under the conditions of in vitro assay.

The cardiolipin phospholipase derived from *Acinetobacter* has been extensively analyzed with respect to product formation.[64] Triacyllysocardiolipin and diacyllysocardiolipin are routinely detected in whole cell lipid extracts of the bacterium, representing 2 to 12% of the total cellular phospholipids, depending on the age of the culture. The structures determined for triacyl- and diacyllysocardiolipin isolated from *Acinetobacter* are shown in Figure 16. These structures were established through extensive chemical, physical, and enzymatic procedures.[64]

The specificity of cardiolipin phospholipase A_1 indicates preferential hydrolyses of the 1-sn-acyl ester group of cardiolipin (Figure 16), further corroborating the non-equivalent character of the two ends of cardiolipin.[65] The formation of diacyllysocardiolipin from triacylcardiolipin involves the specific hydrolysis of 1″ fatty acyl residue. The pathway shown in Figure 16 is based on products formed in in vitro studies of cardiolipin hydrolysis by cardiolipin phospholipase, as well as the structures determined for triacyl- and diacyllysocardiolipin isolated from cellular lipid extracts.

VI. CONCLUSIONS

This discussion of membrane systems in hydrocarbon-oxidizing microorganisms has focused on studies with *Acinetobacter*, primarily for reasons relating to the extensive data base that has been developed concerning membranes and associated lipids in *Acinetobacter*.

Aside from the outer and cytoplasmic membranes typical of Gram-negative bacteria, *Acinetobacter* produces three additional types of membranous structures as a consequence of growth on alkane: (1) the intracytoplasmic membrane system, (2) the hydrocarbon inclusion membrane, and (3) the extracellular membrane vesicles.

A. Intracytoplasmic Membranes

The intracytoplasmic membrane system is a unique subcellular unit membrane structure implicated in hydrocarbon oxidation. This membrane system represents, to our knowledge, the only example of carbon-induced membrane synthesis in heterotrophic microorganisms. These internal membranes are synthesized only when cells are grown on alkane as the sole source of carbon and energy. One might consider why a bacterium would synthesize internal membrane in response to a specific carbon source. Obvious possibilities meriting consideration are, firstly, that the enzyme(s) catalyzing the initial oxidation of alkanes are extrinsic and/or intrinsic proteins localized in the induced membrane system. Thus, the internal membranes are formed concomitantly with the enzymatic ability of the cells to oxidize an alkane. Secondly, the intracytoplasmic membrane provides a site for intracellular hydrocarbon inclusion biogenesis. Our studies demonstrate a direct physical relationship between the hydrocarbon inclusion and the intracytoplasmic membrane system forming a hydrocarbon inclusion-membrane complex. Whether the hydrocarbon inclusion associates with the intracytoplasmic membrane through lipid-lipid, protein-protein, or protein-lipid interactions is unknown. Studies on the origin and mechanism of hydrocarbon inclusion formation will aid in understanding the structure-function relationships that explain this membrane complex. The induction of an internal membrane system coupled to intracellular alkane accumulation appears to be related to the induction of alkane-oxidizing enzymes, and represents a prerequisite for growth of the bacterium on alkanes. Net membrane synthesis represents an obligatory biochemical response that occurs prior to exponential cell growth on alkane as the sole source of carbon and energy.

B. Hydrocarbon Inclusions

Acinetobacter sequesters the unmodified hydrocarbon growth substrate in a unique cytoplasmic inclusion termed the "hydrocarbon inclusion," which is limited by a unique monolayer membrane consisting of phospholipid, protein, and trace concentrations of neutral lipids. The major constituent of the hydrocarbon inclusion is alkane, amounting to approximately 71.5% of the total weight of the inclusion. The remaining components are membrane protein (15.8%), phospholipid (9.5%), and neutral lipid (3.2%). The major neutral lipid is wax ester (2.2%), representing only a minor component of the hydrocarbon inclusions. Whether the neutral lipids represent true structural components of the hydrocarbon inclusion or merely trace contaminants resulting from isolation remains unknown.

C. Extracellular Membrane Vesicles

The extracellular membrane vesicles are phospholipid-rich, bilayer membrane particles which originate from the outer membrane and accumulate in the growth medium of hydrocarbon-grown *Acinetobacter*. The membrane vesicles partition alkane into optically clear microemulsions by a strictly physical, nonenzymatic process. Published results with whole cells demonstrate the major concentrations of hexadecane are associated with the hydrocarbon inclusion and intracytoplasmic membrane, with significantly less hexadecane associated with the cytoplasmic membrane and no hydrocarbon associated with the outer membrane. Such studies indicate a structure-function relationship between hexadecane and the intracytoplasmic membrane-inclusion complex and extracellular vesicle membranes. A protein-protein interaction between the vesicle and the cell surface exists which is destroyed by protease treatment of either the cell surface or the vesicle. When extracellular membrane vesicles labeled with ^{32}P-phospholipids and ^{3}H-amino acids and partitioned with ^{14}C-hexadecane are added exogenously to whole cells, 100% of the vesicle-associated ^{14}C-hexadecane becomes packaged intracellularly in the hydrocarbon inclusion, while 10% of the vesicle ^{32}P-phospholipids is incorporated directly into the inclusion membrane.[78] Vesicle-derived ^{3}H-protein is not translocated to the hydrocarbon inclusion membrane. These results suggest that the extracellular membrane vesicles function as "transport" mediators, serving to solubilize alkane for cellular uptake and metabolism. These observations imply that the bacterium responds metabolically only to "soluble" alkane, and that the extraphasic, bulk alkane remains refractory to cellular recognition and metabolic processing.

Separate studies have demonstrated that the uptake and intracellular packaging of hexadecane in hydrocarbon inclusions is an energy-dependent process.[79] Procedures have been devised to measure the uptake of hexadecane by induced whole cells. The kinetic parameters and conditions for uptake of the water-insoluble hexadecane are similar if not identical to those parameters established for the active transport of water-soluble solutes. Whether the translocation of hexadecane from the extracellular vesicle to the cytoplasmic membrane is an energy-dependent process or strictly physical partitioning remains unresolved. The packaging of alkane is energy dependent in that hydrocarbon inclusions are not formed in the presence of metabolic inhibitors which block energy generation.

D. Cytoplasmic Membranes

The cytoplasmic membranes of hexadecane- and nutrient broth-grown *Acinetobacter* are similar, if not identical, with respect to ATPase, NADH oxidase, major membrane polypeptides, and the amount of phospholipid per unit of membrane protein. The significant dissimilarities that exist between these two membrane entities are density, phosphatidic acid cytidyl transferase activity, and the amount of bound hexadecane in cytoplasmic membranes derived from hexadecane-grown cells. The latter possibility merits further consideration concerning the translocation of hexadecane from the external milieu to its ultimate site in the inclusion complex. Select phospholipid biosynthetic enzymes exhibit at least twofold greater specific activities in cytoplasmic membranes derived from hexadecane-grown cells. Whether such membranes contain quantitatively increased amounts of these enzymes or, alternatively, the enzymes are functionally more active due to different regulatory parameters is equivocal at this time. Interestingly, urea-ammonium sulfate treatment of cytoplasmic membranes derived from hexadecane- and nutrient broth-grown cells results in only 10% greater PACT activity in the hexadecane-derived membranes, indicating that PACT remains tightly bound in both membrane preparations. A major difference that appears between the two membrane preparations occurs following acetone treatment. The cytoplasmic membranes derived from nutrient broth-grown cells inactivated rapidly, whereas the hexadecane-derived membranes remain active following acetone treatment, although lower in PACT specific activity. The observation that exogenously supplied phospholipids, but not neutral lipids, stimulates PACT activity in only hexadecane-derived membranes suggest a possibly stabilizing role for phospholipids or perhaps a regulatory function.

The role of specific lipids as obligate structural components of proteins has assumed greater significance and importance in protein chemistry. Many enzymes require specific lipids acting as co-factors. Interestingly, enzymes exhibiting specific lipid requirements are not all membrane associated or, alternatively, involved with lipid metabolism.[61] Therefore, a heretofore unrecognized interaction between lipids and proteins to yield an inactive or active enzyme depending on the absence or presence of a lipid has posed many new questions as well as problems.

Our studies demonstrate a significant stimulation of UAT membrane-localized PACT by the addition of exogenous phospholipid. The development of this observation was extended to membranes depleted of phospholipid by acetone precipitation to enable a demonstration of possible phospholipid(s) requirements for PACT. The removal of 82% membrane-bound phospholipids with a commensurate 65% decrease in PACT activity proved significant when the readdition of exogenous phospholipid caused a corresponding 65% restoration of PACT activity. The stimulation of PACT appears to be neither phospholipid specific nor additive, although phosphatidylethanolamine exhibits the greatest enhancement of PACT activity as a single addition. Paired combinations of specific phospholipids (phosphatidylethanolamine and phosphatidylglycerol) stimulate PACT activity approximately 250% for reasons not clearly understood. The stimulation appears phospholipid specific in that phosphatidylserine synthetase is stimulated 110% by cardiolipin and 193% by the phosphatidylethanolamine-cardiolipin combination. Definitive reasons for such stimulations are unclear. Obvious possibilities for consideration include enzyme regulation with phospholipids acting as specific positive effectors of enzyme activity or, alternatively, the surface active properties of the phospholipids serve to stimulate enzyme activity. Interestingly, CDP-diglyceride, the product of PACT activity, is the obligate precursor for all succeeding phospholipid biosynthetic enzymes. PACT as a branch point enzyme would be most sensitive to regulatory effectors. The low concentration of exogenous phospholipid relative to endogenously-bound membrane phospholipid appears critical to obtaining maximal stimulation of enzyme activity. The addition of exogenous phospholipids at greater than optimal ratios are progressively inhibitory, arguing for a strictly physical surface active phenomenon. However, the observation that specific but different phospholipids are maximally stimulatory to different enzymes draws criticism to strictly physical effects. Control of phospholipid biosynthetic enzymes may reside in the varying of specific ratios of phospholipids rather than the result of absolute phospholipid concentrations. Since membrane phospholipid compositions are variable as well as subject to individual disproportionate turnover characteristics, regulation of specific membrane-bound enzymes may well constitute an example of concerted regulation as described for amino and biosynthetic pathways.[72]

Physical, biochemical, and enzymological differences support the conclusion that alkane-grown cells possess membranes which differ from nonhydrocarbon-grown cells. Whether membrane proteins, lipids, or a combination of both confer distinctive characteristics is undetermined. Alkane-grown cells are characterized by membranes containing phospholipid fatty acid compositional patterns that are highly type specific with respect to fatty acid chain length and unsaturation. In contrast, nonhydrocarbon-grown cells possess typical phospholipid fatty acid compositions characteristic of Gram-negative bacteria. The intrinsic value of these fatty acids in alkane-grown cells with respect to membrane structure-function relationships could be important when one considers that 98 to 99% of all cellular fatty acids are 16-carbon fatty acids in hexadecane-grown cells and that these fatty acids are derived exclusively from the oxidation of hexadecane and their assimilation into cellular lipids.

Experimental approaches addressing the relationships that exist between lipid physical properties and membrane-associated physiological processes have attempted to answer the question of how a microorganism regulates and controls membrane composition as related to specific structure-function characteristics. A considerable body of information is available where alterations in membrane lipid composition appear to influence physiological functionality of certain membrane-associated phenomena. The regulation of the functional aspects of membranes has been addressed through studies involving altered phase transitions, assessment of the specificity of acyltransferases and fatty acid synthetases, regulation of fatty acid chain length, regulation of the saturated to unsaturated fatty acid ratio, and the effects of temperature in the control of phospholipid fatty acid composition.[73,74] Considerable attention has been focused on the fatty acyl groups although changes in the polar head groups of phospholipids could, as well, greatly influence the physical properties of lipids. Experimental alteration of phospholipid head groups with reference to functional membrane changes has yet to be studied. Whether the stimulatory effects of phospholipids on certain membrane-bound phospholipid biosynthetic enzymes are the result of specific regulation at the enzyme level or more reflective of strictly physical alterations of the enzymes in their hydrophobic environment remain undetermined. Regardless, the membrane systems of alkane-grown *Acinetobacter* represent advantageous model systems for studying specific structure-function relationships associated with membrane physiology.

E. Outer Membrane

The outer membranes derived from either hexadecane- or nutrient broth-grown cells are essentially identical in context of defined biochemical parameters with the exception of phospholipase activity. Hexadecane-grown cells exhibit two- to fivefold greater phospholipase A_1 activity towards phosphatidylethanolamine, phosphatidylglycerol, and cardiolipin. With the exception of pH, these activities exhibit similar physical characteristics. The different pH optima may reflect a "preference" by the enzyme for a particular ionic state of the substrate or substrate-detergent complex. *Acinetobacter* cardiolipin phospholipase A_1 hydrolyzed cardiolipin and phosphatidylglycerol at pHs near neutrality in the presence of nonionic detergent. Under these conditions, the substrate-detergent micelle would be negatively charged. The pH optimum for phosphatidylethanolamine hydrolysis is 9.5. At this pH, the amino nitrogen of phosphatidylethanolamine would predominantly exist in the unprotonated form, resulting in phosphatidylethanolamine bearing a net negative charge. Alternatively, the different pH optima may reflect the presence of different phospholipases, each specific for one of three phospholipids. The similarities in Triton requirement, heat lability, and response to Ca^{2+}, EDTA, and ionic detergents would argue against this possibility. Apparent K_m and V_{max} values are given for comparative purposes only. Lipolytic reactions can be formally treated mathematically to follow Michaelis-Menten kinetics. It cannot be expected, however, that the calculated K_m will have similar importance and meaning as for homogeneous aqueous reactions. The principal difficulty encountered in kinetic analyses of lipolytic enzymes is in defining the substrate concentration. In lipid-water emulsions or lipid-detergent micelles, the amount of substrate available to the enzyme is a function of interfacial surface area. Lipolytic reactions will, however, conform to the Michaelis-Menton equation of the substrate concentration and Michaelis constant is measured in units of interfacial area per unit volume.[75]

The in vivo role of phospholipases remains unclear. Phospholipid degradation in bacteria appears to be exceedingly complex with respect to the number of enzymes involved, their regulation, and their localization within the cell. The fatty acyl chains of *Acinetobacter* phospholipids appear to exhibit rapid turnover in hexadecane growing cells. The phospholipase A activities could thus be involved in phospholipid fatty acyl turnover by way of a deacylation-reacylation cycle. A monoacyl-diacyl cycle has been demonstrated to effect the turnover of certain phospholipid fatty acids in mam-

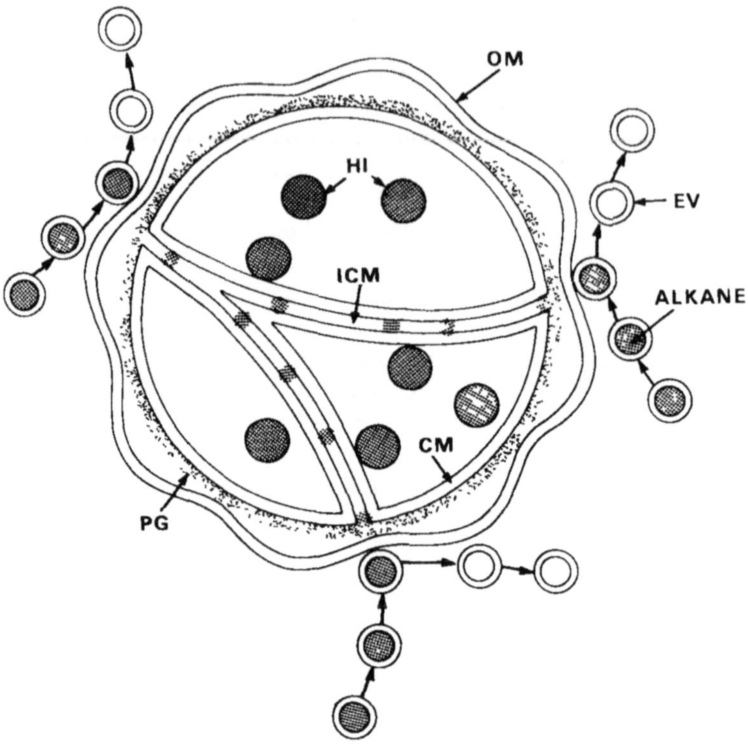

FIGURE 17. Diagrammatic model for alkane uptake and packaging in *Acinetobacter* sp. H01-N. HI, hydrocarbon inclusions; ICM, intracytoplasmic membrane; CM, cytoplasmic membrane; OM, outer membrane; PG, peptidoglycan; EV, extracellular vesicle.

malian tissue.[76] Our studies have indicated a lysophospholipid acyltransferase in *Acinetobacter*. Thus, *Acinetobacter* possesses the requisite enzymatic capacity for the operation of such a cycle. In addition to phospholipid turnover, other possible in vivo functions for phospholipases involve nonhydrolytic reactions. In the hydrophobic core of the membrane, water may not be readily accessible to serve as an acceptor. Accordingly, the phospholipases may also function as a transacylase under certain conditions.[77] Our studies demonstrate that triacyllysocardiolipin is present in vivo. A structural and/or functional role for this lysophospholipid has not been established, although its presence implies a metabolically functional phospholipase. It is noteworthy that there is an apparent translocation of triacyllysocardiolipin from its origin in the outer membrane to the cytoplasmic and intracytoplasmic membranes.

F. Membrane Biogenesis

The origin and biosynthesis of the intracytoplasmic membrane complex appears to represent a carbon-induced synthesis and extension of the cytoplasmic membrane. Isolated intracytoplasmic membranes and cytoplasmic membranes are similar enzymologically and biochemically with the exception of the high phospholipid to protein ratios and hexadecane content noted for intracytoplasmic membranes. Whether the synthesis of intracytoplasmic membrane from the cytoplasmic membrane is a process of single- or multiple-site formation of new membrane remains unresolved. A hypothetical model can be conceptualized that serves to explain the origin of the internal membrane system (Figure 17). This model requires an invagination and extension of the cytoplasmic membrane bilayer where the inner leaflet of the cytoplasmic membrane becomes the outer leaflet of the intracytoplasmic membrane. Such site-specific invaginations of the cytoplasmic membrane represent a mechanism whereby a hydrophobic membrane continuum with exposure to the external layers of the cell is functionally related to the partition and diffusion of extracellular alkane to the intracellular site of oxidation. Proof for this model resides, in part, in the determination of the asymmetric distribution of phospholipids and proteins of the cytoplasmic and intracytoplasmic membranes as well as the demonstration of the enzyme systems responsible for alkane oxidation on the intracytoplasmic membrane.

Hydrocarbon inclusions always appear in close proximity to intracytoplasmic membranes. The polypeptide composition of the monolayer inclusion membrane is dissimilar to any other membrane species isolated from hexadecane-grown *Acinetobacter.* The phospholipid composition, however, resembles that of the cytoplasmic membrane. In comparison to the other membrane species, the inclusion membrane is phospholipid rich. Hypothetically, the monolayer inclusion membrane could originate from one leaflet of the intracytoplasmic membrane bilayer.

The extracellular membrane vesicles most likely originate from the outer membrane by an active biosynthetic process and not by a simple "blebbing" phenomenon. Alternatively, the possibility exists where self assembly of extracellular vesicles occurs with the assembly of lipopolysaccharide, phospholipid, and protein into a membranous structure. These extracellular vesicles are highly enriched in lipopolysaccharide and phospholipid, most closely resembling the hydrocarbon inclusion membrane in phospholipid composition and the outer membrane in polypeptide composition. It is unlikely, however, that these vesicles represent extruded hydrocarbon inclusions, or that hydrocarbon inclusions originate from these vesicles during vesicle-cell interaction, since only minor amounts of vesicle-associated ^{32}P-phospholipid and no vesicle-associated ^{3}H-protein is found in inclusion membranes subsequent to cell-vesicle interaction.

G. An Overview — Membranes and Hydrocarbon Metabolism

The above-designated membrane species are unique structures resulting from the growth of *Acinetobacter* on alkane. The relationship of these membrane structures to the process of hydrocarbon metabolism by *Acinetobacter* poses interesting and novel physiological consequences for *Acinetobacter.* Our interpretation of physiological mechanisms underlying the phenomenon of hydrocarbon metabolism by *Acinetobacter* indicates a complexity heretofore unrecognized, emphasizing the necessity for analyses at the molecular level. A diagrammatic representation of alkane transport and uptake by *Acinetobacter* in relationship to defined membrane systems is shown in Figure 17.

The extracellular vesicle is envisaged as a particle which physically partitions alkane from the bulk phase forming a microemulsion. These alkane-rich particles associate with the cell surface through a protein-protein interaction with the subsequent translocation of alkane to the cytoplasmic membrane. The alkane diffuses intracellularly from the cytoplasmic membrane by way of the intracytoplasmic membrane system which forms a continuum with the cytoplasmic membrane, allowing translocation of alkane to the intracellular site of alkane oxidation localized in the intracytoplasmic membrane. There the alkane-oxidizing enzymes catalyze the oxidation of alkane to fatty acid by a pathway not fully understood. Alkane is packaged intracellularly by an unknown mechanism. Our working hypothesis is that as the alkane exceeds its solubility in the intracytoplasmic membrane, an energy-dependent step occurs where one leaflet of the phospholipid bilayer forms a monolayer membrane around the alkane, forming the hydrocarbon inclusion structure, which remains in direct physical contact with the intracytoplasmic membrane or separates to form an intracellular hydrocarbon inclusion. Evidence supporting this mechanism resides primarily in the fact that intracellular alkane inclusions do not form either in the absence of oxygen or when energy production is blocked by metabolic inhibitors such as azide or cyanide. Such a mechanism insures a continuous intracellular supply of the sole carbon and energy source which, in context of physical diffusional restrictions, would become rate limiting with respect to cell growth and division.

REFERENCES

1. Singer, M. E. and Finnerty, W. R., Microbial metabolism of straight chain and branched alkanes, *Petroleum Microbiology*, Atlas, R., Ed., Macmillan, New York, 1984, 1.
2. Singer, J. T. and Finnerty, W. R., The genetics of hydrocarbon-utilizing microorganisms, *Petroleum Microbiology*, Atlas, R., Ed., Macmillan, New York, 1984, 299.
3. Kennedy, R. S., Finnerty, W. R., Sudarsanan, K., and Young, R. A., Microbial assimilation of hydrocarbons. I. The fine structure of a hydrocarbon oxidizing *Acinetobacter* sp., *Arch. Mircobiol.*, 102, 75, 1975.
4. Kennedy, R. S. and Finnerty, W. R., Microbial assimilation of hydrocarbons. II. Intracytoplasmic membrane induction in *Acinetobacter* sp., *Arch. Microbiol.*, 102, 83, 1975.
5. Scott, C. C. L. and Finnerty, W. R., Characteristics of intracytoplasmic hydrocarbon inclusions from the hydrocarbon-oxidizing *Acinetobacter* sp. H01-N, *J. Bacteriol.*, 127, 481, 1976.
6. Scott, C. C. L. and Finnerty, W. R., A comparative analysis of the ultrastructure of hydrocarbon-oxidizing microorganisms, *J. Gen. Microbiol.*, 94, 342, 1976.
7. Raymond, R. L. and Davis, J. B., n-Alkane utilization and lipid formation by a *Nocardia*, *Appl. Microbiol.*, 8, 329, 1960.
8. Ludvik, J., Munk, V., and Dostalek, M., Ultrastructural changes in the yeast *Candida lipolytica* caused by penetration of hydrocarbons into the cell, *Experientia*, 24, 1066, 1968.
9. Munk, V., Dostalek, M., and Volfova, O., Cultivation of yeast on gas oil, *Biotechnol. Bioeng.*, 11, 383, 1969.
10. Bertrand, J. C., Mutaftschiev, S., Henkel, H. G., Bazin, H., and Azoulay, E., Isolement et'etude d'une bacterie marine se developpant sur hydrocarbures. I. Caracteres physiologiques, *Ann. Microbiol. (Inst. Pasteur)*, 127B, 373, 1976.
11. Phillips, W. E., and Perry, J. J., *Thermomicrobium fosteri* sp. nov., a hydrocarbon-utilizing obligate thermophile, *Int. J. Syst. Bacteriol.*, 26, 220, 1976.
12. Atlas, R. M. and Heintz, C. E., Ultrastructure of two species of oil-degrading marine bacteria, *Can. J. Microbiol.*, 19, 43, 1972.
13. Stirling, L. A., Watkinson, R. J., and Higgins, I. J., Microbial metabolism of alicyclic hydrocarbons: isloation and properties of a cyclohexane-degrading bacterium, *J. Gen. Microbiol.*, 97, 119, 1977.
14. Kormendy, A. C. and Wayman, M., Characteristic cytoplasmic structures in microorganisms utilizing n-butane and 1-butanol, *Can. J. Microbiol.*, 20, 225, 1974.
15. Davies, S. L. and Whittenbury, R., The structure of methane-utilizing bacteria, *J. Gen. Microbiol.*, 61, 227, 1970.
16. Proctor, H. M., Norris, J. R., and Ribbons, D. W., Fine structure of methane-utilizing bacteria, *J. Appl. Bacteriol.*, 32, 118, 1969.
17. Smith, V., Ribbons, D. W., and Smith, O. S., The fine structure of *Methylococcus capsulatus*, *Tissue Cell*, 2, 513, 1970.
18. DeBoer, W. E. and Hazeu, W., Observations on the fine structure of a methane-utilizing bacterium, *Antonie van Leuwenhoek, J. Microbiol Serol.*, 38, 33, 1972.
19. Ribbons, D. W. and Michalover, J. W., Methane oxidation by cell free extracts of *Methylococcus capsulatus*, *FEBS Lett.*, 11, 41, 1970.
20. Makula, R. A. and Finnerty, W. R., Microbial assimilation of hydrocarbons: identification of phospholipids, *J. Bacteriol.*, 103, 348, 1970.
21. Makula, R. A., Lockwood, P. J., and Finnerty, W. R., Comparative analysis of the lipids of *Acinetobacter* sp. grown on hexadecane, *J. Bacteriol.*, 121, 250, 1975.
22. Meissel, M. N., Medvedeva, G. A., Koslova, T. M., Pomoshnikova, N. A., Zaikina, A. I., and Fedoseeva, G. E., Regularities and penetration into yeast cells of higher fatty acids and hydrocarbons, their intracellular migration and concentrations, in *Proc. Third Int. Spec. Symp. Yeasts*, Suomalainen, H. and Waller, C., Eds., Otaniemi, Helsinki, 1973, 144.

23. Tyurin, V. S., Il'Chenbo, A. P., and Dmitriev, V. V., Comparative study of the ultrastructural characteristics of the yeast *Torulopsis candida* cultured on glucose and hexadecane, *Microbiology*, 44, 425, 1975.
24. Kappeli, O., and Fiechter, A., Properties of hexadecane uptake by *Candida tropicalis*, *Curr. Microbiol.*, 6, 21, 1981.
25. Kappeli, O. and Fiechter, A., The mode of interaction between the substrate and cell surface of the hydrocarbon-utilizing yeast, *Candida tropicalis*, *Biotechnol. Bioeng.*, 28, 967, 1976.
26. Kappeli, O. and Fiechter, A., Component from the cell surface of the hydrocarbon-utilizing yeast *Candida tropicalis* with possible relation to hydrocarbon transport, *J. Bacteriol.*, 131, 917, 1977.
27. Kappeli, O., Muller, M., and Fiechter, A., Chemical and structural alterations at the cell surface of *Candida tropicalis*, induced by hydrocarbon substrate, *J. Bacteriol.*, 133, 952, 1978.
28. Osumi, M. and Fukui, S., Relation between the function and the ultrastructure of profusely appearing microbodies in hydrocarbon-utilizable yeasts, in Proc. 30th Annu. Mtg. Electron Microscopical Society of America, Los Angeles, 1972, p. 322.
29. Osumi, M., Miwa, N., Teranishi, Y., Tanaka, A., and Fukui, A., Ultrastructure of *Candida* yeasts grown on n-alkanes, *Arch. Microbiol.*, 99, 181, 1974.
30. Teranishi, Y., Tanaka, A., Osumi, M., and Fukui, S., Catalase activities of hydrocarbon-utilizing *Candida* yeasts, *Agric. Biol. Chem.*, 38, 1213, 1974.
31. Teranishi, Y., Kawamoto, S., Tanaka, A., Osumi, M., and Fukui, S., Induction of catalase activity by hydrocarbons in *Candida tropicalis* pk 233, *Agric. Biol. Chem.*, 38, 1221, 1974.
32. Osumi, M., Fukuzumi, F., Teranishi, Y., Tanaka, A., and Fukui, S., Development of microbodies in *Candida tropicalis* during incubation in a *n*-alkane medium, *Arch. Microbiol.*, 103, 1975.
33. Osumi, M., Imaizumi, F., Imai, M., Sato, H., and Yamaguchi, H., Isolation and characterization of microbodies from *Candida tropicalis* pk 233 cells grown on normal alkanes, *J. Gen. Appl. Microbiol.*, 21, 375, 1975.
34. Kawamoto, S., Tanaka, A., Yamamura, M., Teranishi, Y., and Fukui, S., Microbody of *n*-alkane grown yeast, *Arch. Microbiol.*, 112, 1, 1977.
35. Tanaka, K. and Iida, M., Microbodies of the hydrocarbon-assimilating yeast, *Candida rugosa*, *J. Gen. Appl. Microbiol.*, 23, 201, 1977.
36. Kappeli, O., Aeschbach, H., Schneider, H., and Fiechter, A., A comparative study of carbon energy reserve metals of *Candida tropicalis* growing on glucose and on hydrocarbons, *Eur. J. Appl. Microbiol.*, 1, 199, 1975.
37. Schneider, H., Gmunder, F., and Fiechter, A., n-Alkane oxidation by *Candida tropicalis*, *Eur. J. Appl. Microbiol.*, 2, 221, 1976.
38. Fukui, S. and Tanaka, A., Peroxisomes of alkane- and methanol-grown yeasts. Metabolic functions and practical applications, *J. Appl. Biochem.*, 1, 171, 1979.
39. Fukui, S. and Tanaka, A., Yeast peroxisomes, *Trends Biochem. Sci.*, 4, 246, 1979.
40. Yamada, T., Nawa, H., Kawamoto, S., Tanaka, A., and Fukui, S., Subcellular localization of alcohol dehydrogenase and aldehyde dehydrogenase in n-alkane-grown *Candida tropicalis*, *Arch. Microbiol.*, 128, 145, 1979.
41. Schneider, H., Fiechter, A., and Fuhrmann, G. F., Plasma membranes from *Candida tropicalis* grown on glucose or hexadecane. I. Isolation, identification and purification, *Biochem. Biophys. Acta*, 512, 495, 1978.
42. Schneider, H., Fuhrmann, G. F., and Fiechter, A., Plasma membrane from *Candida tropicalis* grown on glucose or hexadecane. II. Biochemical properties and substrate-induced alterations, *Biochem. Biophys. Acta*, 554, 309, 1979.
43. Makula, R. and Finnerty, W. R., Microbial assimilation of hydrocarbons. I. Fatty acids derived from normal alkanes, *J. Bacteriol.*, 95, 2102, 1968.
44. Makula, R. A. and Finnerty, W. R., Microbial assimilation of hydrocarbons: cellular distribution of fatty acids, *J. Bacteriol.*, 112, 398, 1972.
45. Sampson, K. L. and Finnerty, W. R., Regulation of fatty acid biosynthesis in the hydrocarbon oxidizing microorganism, *Acinetobacter* sp., *Arch. Microbiol.*, 99, 203, 1974.
46. Vachon, V., McGarrity, J. T., Breuil, C., Armstrong, J. B., and Kushner, D. J., Cellular and extracellular lipids of *Acinetobacter lwoffi* during growth on hexadecane, *Can. J. Microbiol.*, 28, 660, 1982.
47. Hug, H., Blanch, H., and Fiechter, A., The functional role of lipids in hydrocarbon assimilation, *Biotechnol. Bioeng.*, 16, 965, 1974.
48. Mishina, M., Isurgi, M., Tanaka, A., and Fukui, S., Lipids of *Candida tropicalis* and *Candida lipolytica* grown on *n*-alkanes and glucose, *Agric. Biol. Chem.*, 41, 517, 1977.
49. Thorpe, R. F. and Ratledge, C., Fatty acid distribution in triglyceride of yeasts grown on glucose and n-alkane, *J. Gen. Microbiol.*, 72, 151, 1972.

50. Finnerty, W. R. and Kallio, R. E., Origin of palmitic acid carbon in palmitates formed from hexadecane-1-C^{14} and tetracecane-1-C^{14} by *Micrococcus cerificans, J. Bacteriol.*, 87, 1261, 1964.
51. Finnerty, W. R., The biochemistry of microbial alkane oxidation: new insights and perspectives, *Trends Biochem. Sci.*, 2, 73, 1977.
52. Vestal, J. R., and Perry, J. J., Effect of substrate on the lipids of the hydrocarbon-utilizing *Mycobacterium vaccae, Can. J. Microbiol.*, 17, 445, 1971.
53. Hallas, L. E. and Vestal, J. R., The growth of *Mycobacterium convolutum* on solid n-alkane substrates: effect on cellular lipid composition, *Can. J. Microbiol.*, 24, 1197, 1978.
54. Omar, S. H. and Rehm, H. J., Physiology and metabolism of two alkane oxidizing and citric acid producing strains of *Candida parapsilosis, Eur. J. Appl. Microbiol. Biotechnol.*, 11, 42, 1980.
55. Cohn, C. J. and Finnerty, W. R., unpublished data, 1981.
56. Scott, C. C. L., Makula, R. A., and Finnerty, W. R., Isolation and characterization of membranes from a hydrocarbon-oxidizing *Acinetobacter* sp., *J. Bacteriol.*, 127, 469, 1976.
57. Käppeli, O. and Finnerty, W. R., Partition of alkane by an extracellular vesicle derived from hexadecane-grown *Acinetobacter.*, *J. Bacteriol.*, 140, 707, 1979.
58. Käppeli, O. and Finnerty, W. R., Characteristics of hexadecane partition by the growth medium of *Acinetobacter* sp., *Biotechnol. Bioeng.*, 22, 495, 1980.
59. Ensley, B. D. and Finnerty, W. R., Influence of growth substrates and oxygen on the electron transport system in *Acinetobacter* sp. H01-N, *J. Bacteriol.*, 142, 859, 1980.
60. Ensley, B. D., Irwin, R. M., Carreira, L. A., Hoffman, P. S., Morgan, T. V., and Finnerty, W. R., The effects of growth substrate and respiratory chain composition on bioenergetics in *Acinetobacter* H01-N, *J. Bacteriol.*, 148, 508, 1981.
61. Finnerty, W. R., Physiology and biochemistry of bacterial phospholipid metabolism, in *Adv. Bacterial Physiol.*, 18, 177, 1980.
62. Makula, R. A. and Finnerty, W. R., Microbial assimilation of hydrocarbons: phospholipid metabolism, *J. Bacteriol.*, 107, 806, 1971.
63. McCaman, R. E. and Finnerty, W. R., Biosynthesis of cytidine diphosphate-diglyceride by a particulate fraction from *Micrococcus cerificans, J. Biol. Chem.*, 243, 5074, 1968.
64. Torregrossa, R. E., Makula, R. A., and Finnerty, W. R., Characterization of lysocardiolipin from *Acinetobacter* sp. H01-N, *J. Bacteriol.*, 131, 486, 1977.
65. Torregrossa, R. E., Makula, R. E., and Finnertv, W. R. Outer membrane phospholipase A from *Acinetobacter* sp. H01-N., *J. Bacteriol.*, 131, 493, 1977.
66. Torregrossa, R. E., Makula, R. E., and Finnerty, W. R., Hydrolysis of phosphatidylethanolamine and phosphatidylglycerol by outer membrane phospholipase A from *Acinetobacter* sp. H01-N, *J. Bacteriol.*, 136, 803, 1978.
67. Albright, F. R., White, D. A., and Lennarz, W. J., Studies on enzymes involved in the catabolism of phospholipids in *Escherichia coli, J. Biol. Chem.*, 248, 3968, 1973.
68. Nishijima, M., Nakaike, S., Tamari, Y., and Nojima, S., Detergent-resistant phospholipase A of *Escherichia coli* K-12. Purification and properties, *Eur. J. Biochem.*, 73, 115, 1977.
69. Scandella, C. J. and Kornberg, A., A membrane bound phospholipase A_1 purified from *Escherichia coli, Biochemistry*, 10, 4447, 1971.
70. Nishijima, M., Akamatsu, Y., and Nojima, S., Purification and properties of a membrane-bound phospholipase A_1 from *Mycobacterium phlei, J. Biol. Chem.*, 249, 5658, 1974.
71. Senff, L. M., Wegener, W. S., Brooks, G. F., Finnerty, W. R., and Makula, R. A., Phospholipid composition and phospholipase A activity of *Neisseria gonorrhea, J. Bacteriol.*, 127, 874, 1976.
72. Datta, P., Regulation of branched biosynthetic pathways in bacteria, *Science*, 165, 557, 1969.
73. Cronan, J. E. and Gelmann, E. P., Physical properties of membrane lipids: biological relevance and regulation, *Bacteriol. Rev.*, 39, 232, 1975.
74. Cronan, J. E., Molecular biology of membrane lipids, *Annu. Rev. Biochem.*, 47, 163, 1978.
75. Benzonana, G. and Desnuelle, P., Etude cinetique de l'action de la lipase pancreatique sur des triglycerides on emulsion. Essai d ûre enzymologie an milieu heterogene, *Biochem. Biophys. Acta*, 105, 121, 1965.
76. Van den Bosch, H., Phosphoglyceride metabolism, *Ann. Rev. Biochem.*, 43, 243, 1974.
77. Doi, O., Ohki, M., and Nojima, S., Two kinds of phospholipase A and lysophospholipase in *Escherichia coli, Biochim. Biophys. Acta*, 260, 244, 1972.
78. Käppeli, O., and Finnerty, W. R., unpublished observations, 1978.
79. Käppeli, O., Verntullo, R., and Finnerty, W. R., unpublished observations, 1978.

Chapter 2

THE BIOSYNTHESIS OF BACTERIAL OUTER MEMBRANE PROTEINS

Pamela Green, Joseph M. DiRienzo, Hideo Yamagata, and Masayori Inouye

TABLE OF CONTENTS

I. Introduction 46

II. Genes Involved in the Synthesis of Outer-Membrane Proteins 46
  A. Major Protein Structural and Regulatory Genes 46
  B. Minor Protein Genes 50
  C. Genes Presumed to Be Involved in Protein Secretion 51

III. Gene Expression of the Major Outer-Membrane Proteins 52
  A. Lipoprotein 52
  B. OmpA Protein 57
  C. Matrix Proteins — Osmoregulation 64
    1. Gene Structures 64
    2. Osmoregulation 66
    3. The *ompB* Locus 66
    4. Other Regulatory Factors 68
  D. Other Proteins 70
    1. λ Receptor Protein 70
    2. PhoE Protein 71
    3. OmpT Protein 72

IV. Post-Transcriptional Events and Final Assembly 72
  A. Preliminary Considerations Concerning the Early Stages of Outer-Membrane Protein Biosynthesis 73
  B. Translocation across the Cytoplasmic Membrane 74
    1. Translocation Models 74
    2. Secretory Machinery 80
    3. Energy for Translocation 80
  C. Cleavage of the Signal Peptide 81
    1. Temporal Mode 81
    2. Signal Peptidases 82
    3. Secretory Protein Mutations Affecting Cleavage 84
  D. Post-Translational Modification 84
  E. Outer-Membrane Protein Assembly 87

V. Conclusions 90

Acknowledgments 91

References 91

I. INTRODUCTION

The outer membrane of Gram-negative bacteria is a distinct membrane structure which differs from the cytoplasmic (also called the plasma or inner) membrane in terms of location, function, and structure. As the primary barrier between the external environment and the cell, the outer membrane must shield the cell from potentially toxic chemicals while allowing the uptake of nutrient compounds. The outer membrane also serves to confine hydrolytic enzymes and specialized binding proteins localized within the periplasmic space between the cytoplasmic membrane and the outer membrane. Other processes involving the outer membrane include the interaction with other cells, phages, colicins, and macromolecules.

Although under the electron microscope the outer membrane appears to be similar to the cytoplasmic membrane,[1,2] the molecular architecture of the two membranes is quite different. In contrast to the cytoplasmic membrane, the outer leaflet of the outer membrane bilayer is composed of lipopolysaccharide (LPS) while phospholipids make up the inner leaflet.[3,4] The protein profile of the outer membrane reveals an array of "minor" proteins and a small number of characteristic "major" proteins. These proteins are intimately involved in the specialized functions of the outer membrane. The lipoprotein, the OmpA protein, and the matrix proteins, OmpC protein, and OmpF protein, are traditionally referred to as the major outer membrane proteins of *Escherichia coli.* The structure, function, and genetics of the major outer membrane proteins have been studied in great detail (see References 5 through 13 for a detailed discussion of these proteins). *E. coli* contains 1 to 7 × 10^5 molecules of each major outer membrane-protein per cell.[14-16] Other outer-membrane proteins such as λ receptor protein, PhoE protein, and OmpT protein can be classified as major proteins under special growth conditions.

In this chapter, we will discuss the biosynthesis of outer-membrane proteins from transcription on through to the final stages of assembly. In some specific instances, data will be presented on the biosynthesis of other classes of proteins, such as the periplasmic proteins, in order to provide insight into those stages or biosynthetic events about which information is lacking for outer-membrane protein synthesis. Emphasis will be placed on the genetic aspects of the biosynthesis of the major outer-membrane proteins primarily because recent technical advances have facilitated the molecular cloning, deoxyribonucleic acid (DNA) sequencing, and the physical manipulation of the genes involved in their synthesis. Such studies are leading to a better understanding of how membrane protein synthesis is regulated at the genetic level and how the bacterial secretory machinery functions.

II. GENES INVOLVED IN THE SYNTHESIS OF OUTER-MEMBRANE PROTEINS

The first step toward understanding the mechanism(s) of biosynthesis of outer-membrane proteins is the identification and characterization of the genes involved. Figure 1 shows the chromosomal locations which have been determined so far for genes involved in the biosynthesis of *E. coli* outer-membrane proteins. In this section, we will briefly summarize the characteristics of these genes and some of their counterparts in *Salmonella typhimurium* to facilitate subsequent discussion of the synthesis and assembly of bacterial outer-membrane proteins. For more detailed information, see several recent reviews (References 7, 9, and 17).

A. Major Protein Structural and Regulatory Genes

The loci *ompB, ompF,* and *ompC* of *E. coli* K-12 are responsible for the synthesis

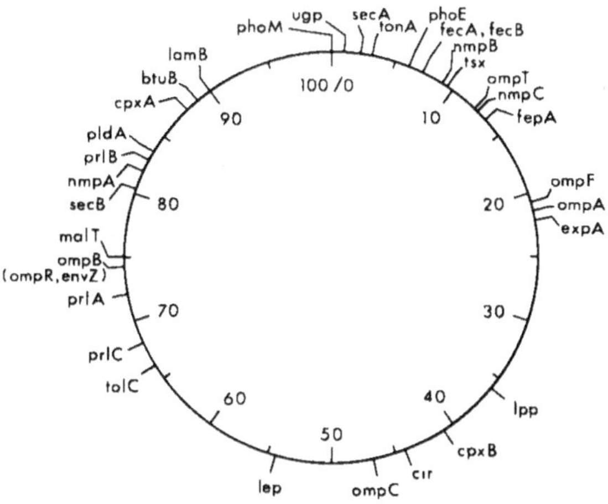

FIGURE 1. Genes involved in the synthesis of outer-membrane proteins of *E. coli*. Genetic symbols are based on the genetic map of Bachmann and Low[17] and are as described in the text. Gene symbols that may be synonymous with those shown are as follows: *fluA, (tonA); cbt, cbr, feuB (feA); cmlB, tolF, coa, pompF); tolG, con, tut(ompA); lpo(lpp); feuA(cir); par, meoA(ompC); kmt, cry(ompB); perA, tpo, (envZ); secY(prlA); phoS, phoT(nmpA); bte, cer(btuB).*

of matrix proteins OmpF protein and OmpC protein, formerly called Ia and Ib.[18] These proteins are characterized by their tight but noncovalent association with the peptidoglycan. OmpF protein of mol. wt. 37,100[19] was first isolated by an elegant method described by Rosenbusch.[16] Nakae[20] and later Nikaido and Rosenberg[21] showed that the matrix proteins function as passive diffusion pores, and because of this property they are also called "porins." The OmpC and OmpF proteins of *E. coli* K-12 share various characteristics and can be separated in sodium dodecyl sulfate (SDS) gel electrophoresis systems only under special conditions such as those used in the disc gel system containing 8 *M* urea[22] or in slab gel systems[18,23,24] based on the method of Laemmli.[25] Schmitges and Henning[24] compared the cyanogen bromide fragments of these two proteins. Among six fragments, only one was found to differ in electrophoretic mobility. The contention that the matrix proteins represent basically the same polypeptide is further supported by analyses of the cyanogen bromide fragments[18,26] and amino-terminal amino acid sequence analyses.[26] *E. coli* mutants lacking OmpF protein have been shown to contain mutations at a single locus designated *ompF (tolF)* which is located at 21 min on the *E. coli* chromosomal map (Figure 1).[17,27] These *ompF* mutants are tolerant to colicins K, L, A, S4, N, E2, and E3.[27-30] The *ompF* locus is probably identical to the *cmlB* locus, which determines increased resistance to the antibiotics chloramphenicol and tetracycline,[27] and to the *coa* locus.[31] Mutations resulting in the lack of only OmpC protein have been mapped at the *ompC (par, meoA)* locus located at 47 min on the *E. coli* chromosomal map (Figure 1).[17,18] These *ompC* mutants are resistant to bacteriophages PA2,[32,33] Tulb,[26,34] Mel,[35] 434,[36] or Db10.[37]

In contrast to the above mutations, an *ompB* mutation can affect both matrix proteins. Some *ompB* mutants lack both OmpF protein and OmpC protein,[18,38] while in others only one matrix protein is affected.[39] The *ompB* mutations map at 74 min on the *E. coli* chromosome (Figure 1),[17,38] and mutants which have previously been designated *tol* IV, *tol* XIV, *tol* XV,[38] *Kmt*,[40] or *cry*[41] affect this locus. Typical *ompB* mutants show resistance to colicins or bacteriophages similar to *ompC* and *ompF* mutants. Thus the three loci *ompB, ompC,* and *ompF* are initimately involved in the synthesis of the matrix proteins in *E. coli* K12. Since OmpC and OmpF proteins are chemically quite similar, *ompB* was considered to be the structural gene for both proteins while the other two genes were thought to be responsible for some modification of the polypeptide which is encoded by *ompB*. However, Ichihara and Mizushima[42] showed that the amino acid sequences of the amino-terminal fragments of OmpF and OmpC proteins are different indicating that the two proteins are coded for by different structural genes. Sato and Yura[43] found that a pair of outer membrane proteins of *S. typhimurium* was produced instead of OmpF protein when the 21-min region of the *S. typhimurium* chromosome was transduced into *E. coli* by bacteriophage P1. They also discovered that a strain of *S. typhimurium*, harboring an F′ plasmid which carries the *ompF* region of the *E. coli* chromosome, produces a protein which is indistinguishable from OmpF protein. In another study, two mutations at the *ompC* locus in *E. coli* were found to lead to the alteration of OmpC protein either in electrophoretic mobility or in the ability of the purified protein to neutralize bacteriophage Me1.[44] Hall and Silhavy[45] constructed gene fusions of the *lac* operon and the *ompC* promoter and found that the expression of β-galactosidase was reduced 14- to 31-fold when an *ompB* mutation was introduced into these fusion strains. These observations strongly suggest that the *ompC* and *ompF* genes are the structural genes for OmpC protein and OmpF protein, respectively, and that the *ompB* gene is a positive regulatory gene controlling *ompC* and *ompF*. As will be discussed in detail in Section III.C, the recent characterization of these genes at the nucleotide level leaves no doubt that this is the case. In addition, such analyses have shown that the *ompB* regulatory locus is an operon containing two genes, *ompR* and *envZ*, which control *ompC* and *ompF* expression. Mutations in *envZ* map identically with *tpo* mutations,[46,47] which affect the transcription of *lamB* (the structural gene for λ receptor protein, a 55,000-mol wt outer-membrane protein involved in the transport of maltose and maltodextrans),[48-50] and *malE* (the structural gene for the periplasmic maltose binding protein). Mutations in *perA* also map identically with *envZ* and affect the production of alkaline phosphatase[46,51] and outer-membrane iron-inducible proteins.[52] In addition mutations in *perA* result in a decrease in the amount of OmpT protein,[52,53] which is described below. It has also been suggested that a mutation called *cpr* may map in the *envZ* gen[54] but as will be discussed in Section III.C.4, *cpr* and *envZ* are most likely separate genes with similar mutant phenotypes.

Another gene, *tolC*, maps at 66 min on the *E. coli* chromosome[17,55] and appears to affect the expression of the porin proteins OmpF, NmpC, and protein 2.[56] Mutants of the *tolC* locus are tolerant to colicin E1, have an altered bacteriophage-sensitivity pattern, and are hypersensitive to dyes, detergents, and certain antibiotics.[29] Unlike the *ompB* operon which appears to encode a positive control element affecting the transcription of both *ompC* and *ompF*,[45,57] the *tolC* gene product may affect the biosynthesis of the OmpF protein at a post-transcriptional level.[56] Morona et al.,[56,58,59] using the cloned *tolC* gene, identified the gene product as a 52,000-mol-wt outer-membrane protein.

Mutations in the genes *nmpA* (83 min),[60,61] *nmpB* (9 min),[61] and *nmpC* (12 min)[61] give rise to new outer-membrane proteins A, B, and C, respectively, in those strains lacking the matrix proteins, such as those with mutations in *ompB* or *ompC* and *ompF*. Since strains that are missing the matrix proteins grow very slowly in many media, *nmp* mutants can be easily obtained by selecting for revertants which grow more rapidly. Similar to the matrix proteins, the new membrane proteins act as porins and make cells sensitive to colicins E2 and E3.[61] Biochemical analyses indicated that *nmpA* and *nmpB* mutations could facilitate the expression of the same new outer-membrane protein called NmpAB.[62] Some genetic data prompted the suspicious suggestion that NmpAB protein had two structural genes, *nmpA* and *nmpB*.[63] However, subsequent studies showed that a single structural gene for this protein exists at 6 min on the *E. coli* chromosome.[64,65] NmpAB protein is now known as PhoE protein.

Tommassen and Lugtenberg[66] showed that the production of PhoE protein co-regulated with products of the *pho* regulon, such as alkaline phosphatase, which are only expressed in wild-type cells during phosphate limitation. Mutations in *nmpA* and *nmpB* appear to be identical to known mutations effecting the expression of the *pho* regulon.[66] The constitutive synthesis of PhoE protein, alkaline phosphatase, and a discrete set of phosphate-regulated periplasmic proteins is a characteristic phenotype of both *nmpB* and *phoR* mutants. The *nmpB* and *phoR* genes, which have similar map positions, are assumed to be identical. Classes of *nmpA* mutants are indistinguishable from *phoS* and *phoT* mutants. The role of *phoS* and *phoT* in regulation may be an indirect result of their role in phosphate transport. One mutation, selected for its ability to confer resistance to a phage specific for PhoE protein was found to affect *phoB*.[66] The product of the *phoB* gene (which is tightly linked to but probably transcribed separately from *phoR*) is likely to be an activator of *phoE* expression as is the product of the *phoM* gene.[67,68] A mutation in another gene, *ugp*, which maps at 1 min[69] causes constitutive synthesis of PhoE protein and other *pho* regulon products.[70] A current model for how many of these genes are involved in *pho* regulation is discussed in Section III.D.2.

Another conditionally produced porin is called protein 2. When *E. coli* is lysogenized by bacteriophage PA-2, OmpC protein and OmpF protein are supplanted by protein 2.[33] This protein is not identical to the other *E. coli*[32,62] porins although it can confer sensitivity to colicins E2 and E3. The structural gene for protein 2 is apparently contained within the phage genome.

S. typhimurium has three major outer-membrane proteins, designated OmpD, OmpF, and OmpC, which are associated with peptidoglycan and act as porins.[71] Mutations which lead to the disappearance of OmpF and OmpC proteins have been mapped at 21 (*ompF*)[43] and 46 min (*ompC*),[72] respectively, on the *S. typhimurium* chromosomal map similar to the *E. coli* situation. The *ompD* gene maps at 28 min.[72] An *ompB* locus has also been characterized in *S. typhimurium* and found to contain two genes which have been designated *ompR* and *envZ* after their *E. coli* counterparts.[73,74]

The *ompA* gene, which maps at 21 min on the *E. coli* chromosome,[17,30,75] is the structural gene for OmpA protein. This protein has a mol wt of 35,200[76] and is the only protein among the major outer-membrane proteins which can be partially digested with trypsin or pronase when the whole cell envelope or outer membrane is enzymatically treated.[77-79] This indicates that part of OmpA protein is exposed on the outer surface of the membrane. OmpA protein exhibits anomalous heat modifiability on SDS-polyacrylamide gels with its apparent molecular weight increasing when the membrane is solubilized in SDS above 50°C.[22,77,80,81] Purified OmpA protein inhibits conjugation,[82,83] inactivates the phages K3[83] and TuII*,[84] and *ompA* mutants are found among those cells which are tolerant to colicin L,[85] resistant to phages K3[86] or TuII*,[34] or are defective recipients in conjugation.[87,88] Although most of these mutants lack OmpA protein, several of them produce an OmpA protein which is altered in electrophoretic mobility or phage-receptor activity. A structural alteration has also been shown based on the analysis of cyanogen bromide fragments.[89] The transfer of an F′ plasmid carrying the *ompA* region of the *E. coli* K12 chromosome to *S. typhimurium* or *Proteus mirabilis* gives rise to a heat-modifiable protein in these species with the same electrophoretic mobility as the *E. coli* OmpA protein[90] These results indicate that the *ompA* gene is the structural gene for OmpA protein. Recently, the gene for this major outer-membrane protein has been isolated and analyzed in great detail (see Section III.B). The *S. typhimurium* OmpA protein, which has a size similar to that of the *E. coli* OmpA protein, is also trypsin sensitive and heat modifiable.[91] Mutations which lead to the lack of OmpA protein have been mapped at 21 min on the *S. typhimurium* chromosome. Therefore, the *S. typhimurium ompA* gene is situated in the same position as the *E. coli ompA* gene.[72]

The *lpp* gene is the structural gene for lipoprotein, the most abundant protein in the *E. coli* cell in terms of the number of molecules. Braun and Rehn[14] found that lipoprotein (mol wt approximately 7000) is covalently linked to the peptidoglycan. The complete chemical structure of this protein has been determined by Braun and his co-workers.[92-94] Inouye and his co-workers[15,95] independently found that lipoprotein also exists in the *E. coli* membrane without a covalent attachment to the peptidoglycan. A mutant carrying an altered structural gene for lipoprotein was found among a collection of temperature-sensitive mutants[96] and it was later demonstrated that the arginine residue at position 57 in this protein had been replaced by a cysteine residue.[97] The mutation *(lpp-1)* was found to map at 36 min on the *E. coli* chromosomal map and is not related to the temperature-sensitive phenotype.[96] Another lipoprotein mutant, designated *lpo (lpp-2)*, in which the lipoprotein is completely missing was found by Hirota et al.[98] Chromosomal DNA from this mutant does not contain the restriction fragment which normally hybridizes with purified radioactive mRNA for the lipoprotein.[99] This mutation, which is most likely a deletion of the *lpp* gene, also maps at 36 min on the *E. coli* chromosome.[98] Although the *lpp-2* mutant appears to grow and divide normally, it is sensitive to ethylenediaminetetraacetic acid (EDTA) and leaks a considerable amount of periplasmic enzymes into the medium. Two groups have isolated lipoprotein mutants by a suicide selection method which is based on the specific incorporation of radioactive amino acids into the lipoprotein during starvation for the specific amino acids that are lacking in the lipoprotein molecule.[100] In one such mutant, isolated by Wu and Lin,[101] the lipoprotein lacks the covalently linked diglyceride moiety, and the amount of the bound form of the lipoprotein is greatly reduced. Amino acid sequence analysis of the lipoprotein isloated from this mutant showed that a glycine residue at position -7 within the signal sequence had been replaced by aspartic acid.[102] In a separate study, Torti and Park[103] isolated a mutant, temperature-sensitive in cell division, which forms filamentous cells at 42°C. This mutant produces very low amounts of lipoprotein at 42°C but normal amounts at permissive temperatures. Revertants of this mutant grow and make lipoprotein normally at 42°C. Apparently, the mutation is not in the lipoprotein structural gene since it maps at 74 min.[104] More recent lipoprotein mutant isolation techniques (reviewed in Reference 105) involve site-specific mutagenesis or the use of globomycin selection. The mode of action of the antibiotic globomycin and the mutants isolated by these methods all discussed in Section IV. The *lpp* gene of *E. coli* K12 has been cloned and the nucleotide sequence has been determined.[106] Several interesting features concerning the mechanism of expression of the *lpp* gene are presented in Section III.A. It should be noted that several putative lipoprotein genes have been identified which hybridize to the *lpp* gene.[107] Some or all of these genes may code for the new, minor lipoproteins described in Section IV.D.

Recently the structural gene for OmpT protein (also referred to as protein a or 3b) has been identified and designated *ompT*.[53] This gene is located at 12.5 min on the *E. coli* chromosome. OmpT protein is produced as a 42,000-mol-wt precursor protein which is processed to a 40,000-mol-wt outer-membrane protein.[108] The appearance of the *ompT* gene product in the outer membrane is temperature dependent.[109] The possibility that the *envZ (perA)* gene product may regulate the synthesis of OmpT protein has been discussed above. Earlier studies suggested that the *fepA* gene might also regulate *ompT* expression since this protein was absent from *fepA* mutants carrying deletions in the 12.5-min region of the chromosome.[110] However, hybridization data indicate that the *ompT* gene has been deleted in these mutants. Therefore, the *fepA* locus does not appear to be involved in the regulation of OmpT protein.[53]

The *lamB* gene is the structural gene for λ receptor protein and is located within the *malB* region at 90 min on the *E. coli* chromosomal map.[17,111] The λ receptor protein is involved in the transport of maltose and maltodextrans.[112,113] As described in Section III.D.1, *lamB* is positively regulated by the product of the *malT* gene which maps at 75 min.[114] Fusions of the *lamB* gene to the *lacZ* gene have been constructed, and provide a means for studying outer-membrane protein localization[115] (see Section IV).

B. Minor Protein Genes

In addition to the identification of the structural and regulatory genes for the major outer-membrane proteins, many genes involved in the synthesis of minor outer-membrane proteins have been characterized. For instance, the *tsx* gene maps at 9 min and determines resistance to the T6 bacteriophage. In *tsx* mutants, an outer-membrane protein with a molecular weight of 25,000 is missing.[116,117] These mutants are also resistant to colicin K and show a reduced uptake of nucleosides.[116] The *btuB (bfe, btuA, cer)* gene maps at 88 min and determines resistance to phage BF23 and the E group of colicins.[118,119] The *btu* mutants lack a 60,000-mol-wt outer-membrane protein.[120] This protein acts as a receptor for vitamin B12 as well as for the phage and colicins mentioned above. The genes *tonA (fhuA), fepA(feuB),* cir*(feuA), fecA,* and *fecB* determine the production of outer-membrane proteins involved in iron uptake. The *tonA* gene maps at 3 min and its alteration can establish resistance to the phages T1, T5, and 80, as well as to colicin M.[121] Mutations in *tonA* also confer resistance to albomycin and a ferrichrome analogue.[122] TonA protein, with a molecular weight of 78,000, has been characterized.[123,124] The *fepA (cbt, cbr, feuB)* gene which maps at 13 min affects ferric enterochelin uptake and resistance to colicins B and D.[125] An outer-membrane protein of mol wt 81,000, missing in *fepA* mutants, is considered to be a receptor for ferrienterochelin as well as colicins B and D.[126] The *cir (feuA)* gene, which maps at 44 min, determines the production of a receptor for colicin I. The *cir* mutants lack an outer-membrane protein that has a molecular weight of 74,000.[127] Production of Cir protein is repressed by the presence of iron in the growth medium.[128] The *fecA* gene maps at 7 min on the bacterial chromosome, and mutations in this locus prevent growth on ferric citrate as the sole source of iron.[129] An outer-membrane protein (FecA, formerly Cit) having a molecular weight of 81,000 which is produced in the presence of citrate appears to be the gene product.[130,131] This protein functions as a receptor for ferric citrate. Other ferric citrate transport mutants which are unaltered in the FecA protein also map at approximately 7 min on the chromosome and comprise a second class of mutants, designated *fecB*. Finally, the *pldA* gene is involved in the synthesis of phospholipase A1 and maps at 84 min on the *E. coli* chromosome.[132]

C. Genes Presumed to Be Involved in Protein Secretion

Several genes appear to be associated with the translocation or assembly of outer-membrane proteins. On such example is the *secA* gene identified in a conditional lethal mutant of *E. coli* which is pleiotropically defective in the translocation of a number of outer-membrane proteins.[133] At the nonpermissive temperature this mutant accumulates outer-membrane precursor proteins in the cytoplasm. The *secA* locus maps at approximately 2.5 min on the bacterial chromosome and codes for a protein product of 92,000 mol wt.[134] Mutations in another gene designated *secB*, which map at 80.5 min,[135,136] cause less severe export defects. In *secB* mutants, the localization of maltose binding protein and sometimes OmpF protein is defective. This effect becomes much more pronounced in *secA secB* double mutants.[136] Three other genes, *prlA, prlB,* and *prlC,* have been identified by mutations which suppress the defective export of λ receptor protein caused by certain lesions in *lamB*.[137] The *prlA* gene appears to be identical to *secY*,[138] which maps in the promoter distal region of the *spc* operon at 72 min.

Mutations in *prlB* and *prlC*, which map as indicated in Figure 1, suppress fewer *lamB* mutations that do *prlA* mutations. Michaelis and Beckwith[6] have previously reviewed many of the mutations which affect secretory protein localization. The more recent observations concerning the most thoroughly studied *secA* and *prlA (secY)* genes are presented in Section IV.B.2 of this chapter.

McEwen and Silverman[139] have shown that mutations in two loci together, *cpxA* and *cpxB*, alter the protein composition of the cytoplasmic and outer membranes of *E. coli*. Their preliminary analysis indicated that the OmpF protein and the lipoprotein are deficient or absent in *cpxA cpxB* mutants, while the OmpC and OmpA proteins are unaffected. The sensitivity of the mutants to detergents or hydrophobic antibiotics is altered. Subsequent experiments[140] showed that the overall synthesis of lipoprotein observed during a short (5-min) pulse is unaffected in the mutant. However, using longer labeling periods, much less lipoprotein is found in the *cpxA cpxB* mutants than in the wild type, perhaps due to an increased turnover rate. McEwen et al.[140] suggest that the lipoprotein which is synthesized is not localized in the outer membrane because they fail to detect the presence of bound-form lipoprotein. They have also shown, by monitoring the β-galactosidase levels in *ompF-lacZ* fusions, that *cpxA cpxB* mutations decrease *ompF* expression by three- to sixfold. Yet the relatively modest decrease in gene expression does not explain the virtual absence of OmpF protein in the outer membrane as detected using phage sensitivity. These data may indicate that *cpx* mutations negatively influence translocation across the outer membrane. The *cpxA* and *cpxB* genes map at 87.8 and 41 min, respectively, on the *E. coli* chromosome and are known to influence DNA donor activity and surface exclusion.[141]

Another mutation which reportedly affects the amounts of outer-membrane proteins is called *expA*.[142] This mutation, which maps at 22 min, does not appear to effect inner-membrane proteins, but the amounts of three outer-membrane proteins and ten periplasmic proteins in the *expA* mutant are decreased. Whether *expA* affects the expression of certain secretory protein genes or the export of their products is unknown.

III. GENE EXPRESSION OF THE MAJOR OUTER-MEMBRANE PROTEINS

As discussed earlier, the genes for outer-membrane proteins appear to be controlled by their own specific regulatory mechanisms. Compared to other cellular proteins, so-called major outer-membrane proteins in *E. coli* such as lipoprotein, OmpA protein, and the matrix proteins are present in extremely large quantities. The major lipoprotein exists at a level of approximately 7×10^5 molecules per cell, while OmpA protein and the matrix proteins exist at a level of 2×10^5 molecules per cell. How can the cell manage to produce outer-membrane proteins in such a large quantity? How is their gene expression regulated? In an effort to answer these questions, many of the genes involved in the synthesis of *E. coli* major outer-membrane proteins have been cloned and sequenced. The isolation of most of these genes was a difficult task due to the lack of suitable selection methods. Cloning strategies often had to be designed to avoid the lethal effects of increased gene dosage. Overcoming these obstacles has facilitated not only the isolation and structural analysis of the major outer-membrane protein genes, but also the development of novel cloning methods[106,143] and controlled gene expression vehicles.[144-146]

A. Lipoprotein

The unusual stability,[147] abundance, and size of the lipoprotein mRNA made possible its purification[148] and subsequent sequencing.[99,149,150] Using ^{32}P-lipoprotein mRNA, the *lpp* gene was cloned into a phage vector. The DNA sequence determined from this clone[106] is shown in Figure 2A. The structure revealed many interesting features. The

```
1   ATCTGGCACACAAAGGTGACGTTGTAGTTATCGTTTCTGGTGCACTGGTACCGAGCGGCACTACTAACACCGCATCTGTTCACGTCCTGTAATATTGCTT 100

101 TTGTCAATTAATTTGTATATCGGCGCTTTTTTTATTTAATCGATAACCAGAAGCCAATAAAAAATCAAATCGGATTTCACTATATAATCTCACTTTATCTA 200

201 AGATGAATCCGATGGAAGCATCCTGTTTTCTCTCAATTTTTTTATCTAAAACCCAGCGTTCGATGCTTCTTTGAGCGAACGATCAAAAATAAGTGCCTTC 300

301 CCATCAAAAAAATATTCTCAACATAAAAAACTTTGTCTAATACTTGTAACGCTACATGGAGATTAACTCAATCTAGAGGGTATTAATAATGAAAGCTACT 400
                     RNA Polymerase          Pribnow                                                 MetLysAlaThr
                     Recognition Site        Box                                                     -20
401 AAACTGGTACTGGGCGCGGTAATCCTGGGTTCTACTCTGCTGGCAGGTTGCTCCAGCAACGCTAAAATCGATCAGCTGTCTTCTGACGTTCAGACTCTGA 500
    LysLeuValLeuGlyAlaValIleLeuGlySerThrLeuLeuAlaGlyCysSerSerAsnAlaLysIleAspGlnLeuSerSerAspValGlnThrLeuA
                                                  -1 +1
501 ACGCTAAAGTTGACCAGCTGAGCAACGACGTGAACGCAATGCGTTCCGACGTTCAGGCTGCTAAAGATGACGCAGCTCGTGCTAACCAGCGTCTGGACAA 600
    snAlaLysValAspGlnLeuSerAsnAspValAsnAlaMetArgSerAspValGlnAlaAlaLysAspAspAlaAlaArgAlaAsnGlnArgLeuAspAs
         20                                                        40
601 CATGGCTACTAAATACCGCAAGTAATAGTACCTGTGAAGTGAAAAATGGCGCACATTGTGCGACATTTTTTTTGTCTGCCGTTTACCGCTACTGCGTCAC 700
    nMetAlaThrLysTyrArgLys

701 GCGTAACATATTCCCTTGCTCTGGTTCACCATTCTGCGCTGACTCTACTGAAGGCGCATTGCTGGCTGCGGGAGTTGCTCCACTGCTCACCGAAACCGG  800
```

A

FIGURE 2. DNA sequences of the *lpp*,[106] *ompA*,[76] *ompF*,[19] *ompC*,[204] *ompR*,[212] *envZ*,[214] *lamB*,[224] and *phoE*[205] genes from *E. coli*: (A) *lpp*; (B) *ompA*; (C) *ompF*; (D) *ompC*; (E) *ompR*; (F) *envZ*; (G) *lamB*; (H) *phoE*. Solid lines above the sequences point out significant ATG codons. Termination codons are underlined. The amino acid sequences deduced from the DNA sequences are also shown. Negative numbers have been assigned to amino acid residues contained within the signal peptide. Dots (.) above the sequences indicate homology with the 3′ end of the 16S rRNA (Shine-Dalgarno sequences). For known promoters, the RNA polymerase recognition sequence and Pribnow box are underlined and labeled. Large arrows (▲) indicate transcription initiation sites and small arrows (▼) indicate signal peptide cleavage sites. Horizontal arrows indicate the inverted (→←) or direct (→→) repeats referred to in the text. In (B) the components of the *ompA* P_1 promoter are indicated with broken lines (- - →) above the sequence, while the *ompA* P_2 promoter is indicated as described above. The *ompA* initiation codons ATG-I to ATG-III, described in the text, are labeled. In (C) and (D) the conserved sequence CATCt/aATAG, found upstream of the translation initiation sites in *ompF* and *ompC*, is boxed. In (E) the last line of the sequence includes the amino-terminal coding region of the *envZ* gene. This line appears again at the beginning of (F) to illustrate the colinearity of the *ompR* and *envZ* genes.

45-base-pair region preceding the transcription initiation site of the *lpp* gene is very AT rich. The *lpp* promoter resides within this region. The sequence from positions 336 to 341, TgTAAT, is almost identical to the Pribnow box or −10 homology generalized as TATAAT.[151,152] (The nucleotides in lower-case letters indicate differences from the consensus sequence.) In the region about 35 base pairs upstream of the transcription initiation site, the sequence TTctCA from 315 to 320 shows good homology with the consensus sequence for the RNA polymerase recognition site or −35 region, TTGACA. The positions of mutations which adversely affect promoter function attest to the importance of specific base sequences in the −10 and the −35 regions.[151-153]

The uneven distribution of AT base pairs in the *lpp* gene is most striking. The promoter region from 306 to 350 has the highest AT content (80%). The two stretches of 7As and 6As flanking the RNA polymerase recognition site are especially noteworthy. The AT content of the mRNA coding region and the segment following the transcription termination site is approximately the same as the average AT content of *E. coli* (49%).[154] In contrast, the 5′ and 3′ untranslated regions of the mRNA have AT contents of 66 and 64%, respectively. The AT richness of the promoter is likely to destabilize the helix structure and facilitate the strand unwinding by RNA polymerase which is necessary for transcription initiation.[155] Transient melting of base pairs in AT-rich

```
   1 TCTCTTCTGTAAATTGTCGCTGACAAAAAAGATTAAACATACCTTATACAAGACTTTTTTTTCATATGCCTGACGGAGTTCACACTTGTAAGTTTTCAAC  100
                                                                                   RNA Polymerase
                                                                                 Recognition Sites

 101 TACGTTGTAGACTTTACATCGCCAGGGGTCCTCGGCATAAGCCGAAGATATCGCTAGAGTTAATATTGAGCAGATCCCCCGGTGAAGGATTTAACCGTGT  200
        Pribnow       ▲
         Boxes
          ....     *III       II        ....      I
 201 TATCTCGTTGGAGATATTCATGGCGTATTTTCGATGATAACGAGGCGCAAAAAATGAAAAAGACAGCTATCGCGATTGCAGTGGCACTGGCTGGTTTCGC  300
                                                          MetLysLysThrAlaIleAlaIleAlaValAlaLeuAlaGlyPheAl
                                                             -20

 301 TACCGTAGCGCAGGCCGCTCCCAAAGATAACACCTGGTACACTGGTGCTAAACTGGGCTGGTCCCAGTACCATGACACTGGTTTCATCAACAACAATGGC  400
     aThrValAlaGlnAla▼AlaProLysAspAsnThrTrpTyrThrGlyAlaLysLeuGlyTrpSerGlnTyrHisAspThrGlyPheIleAsnAsnAsnGly
                -1 +1                                                         20

 401 CCGACCCATGAAAACCAACTGGGCGCTGGTGCTTTTGGTGGTTACCAGGTTAACCCGTATGTTGGCTTTGAAATGGGTTACGACTGGTTAGGTCGTATGC  500
     ProThrHisGluAsnGlnLeuGlyAlaGlyAlaPheGlyGlyTyrGlnValAsnProTyrValGlyPheGluMetGlyTyrAspTrpLeuGlyArgMetP
                                  40                                                           60

 501 CGTACAAAGGCAGCGTTGAAAACGGTGCATACAAAGCTCAGGGCGTTCAACTGACCGCTAAACTGGGTTACCCAATCACTGACGACCTGGACATCTACAC  600
     roTyrLysGlySerValGluAsnGlyAlaTyrLysAlaGlnGlyValGlnLeuThrAlaLysLeuGlyTyrProIleThrAspAspLeuAspIleTyrTh
                                                          80

 601 TCGTCTGGGTGGCATGGTATGCCGTGCAGACACTAAATCCAACGTTTATGGTAAAAACCACGACACCGGCGTTTCTCCGGTCTTCGCTGGCGGTGTTGAG  700
     rArgLeuGlyGlyMetValTrpArgAlaAspThrLysSerAsnValTyrGlyLysAsnHisAspThrGlyValSerProValPheAlaGlyGlyValGlu
              100                                                    120

 701 TACGCGATCACTCCTGAAATCGCTACCCGTCTGGAATACCAGTGGACCAACAACATCGGTGACGCACACACCATCGGCACTCGTCCGGACAACGGCATGC  800
     TyrAlaIleThrProGluIleAlaThrArgLeuGluTyrGlnTrpThrAsnAsnIleGlyAspAlaHisThrIleGlyThrArgProAspAsnGlyMetL
                                 140                                                         160

 801 TGAGCCTGGGTGTTTCCTACCGTTTCGGTCAGGGCGAAGCAGCTCCAGTAGTTGCTCCGGCTCCAGCTCCGGCACCGGAAGTACAGACCAAGCACTTCAC  900
     euSerLeuGlyValSerTyrArgPheGlyGlnGlyGluAlaAlaProValValAlaProAlaProAlaProAlaProGluValGlnThrLysHisPheTh
                                                              180

 901 TCTGAAGTCTGACGTTCTGTTCAACTTCAACAAAGCAACCCTGAAACCGGAAGGTCAGGCTGCTCTGGATCAGCTGTACAGCCAGCTGAGCAACCTGGAT 1000
     rLeuLysSerAspValLeuPheAsnPheAsnLysAlaThrLeuLysProGluGlyGlnAlaAlaLeuAspGlnLeuTyrSerGlnLeuSerAsnLeuAsp
              200                                                    220

1001 CCGAAAGACGGTTCCGTAGTTGTTCTGGGTTACACCGACCGCATCGGTTCTGACGCTTACAACCAGGGTCTGTCCGAGCGCCGTGCTCAGTCTGTTGTTG 1100
     ProLysAspGlySerValValValLeuGlyTyrThrAspArgIleGlySerAspAlaTyrAsnGlnGlyLeuSerGluArgArgAlaGlnSerValValA
                                 240                                                         260

1101 ATTACCTGATCTCCAAAGGTATCCCGGCAGACAAGATCTCCGCACGTGGTATGGGCGAATCCAACCCGGTTACTGGCAACACCTGTGACAACGTGAAACA 1200
     spTyrLeuIleSerLysGlyIleProAlaAspLysIleSerAlaArgGlyMetGlyGluSerAsnProValThrGlyAsnThrCysAspAsnValLysGl
                                                         280

1201 GCGTGCTGCACTGATCGACTGCCTGGCTCCGGATCGTCGCGTAGAGATCGAAGTTAAAGGTATCAAAGACGTTGTAACTCAGCCGCAGGCTTAAGTTCTC 1300
     nArgAlaAlaLeuIleAspCysLeuAlaProAspArgArgValGluIleGluValLysGlyIleLysAspValValThrGlnProGlnAla
              300                                                    320

1301 GTCTGGTAGAAAAACCCCGCTGCTGCGGGGTTTTTTTTGCCTTTAGTAAATTGAACTGACTTTCGTCAGTTATTCCTTACCCAGCAATGCCTGCAGATCC 1400

1401 TGCTTCAGAGAAGACATTTTATTCGCGTATTTCTCTTTGTTTTCCGCATCTTCAATC
```

FIGURE 2B

regions could also aid RNA polymerase in recognizing the *lpp* promoter and forming a stable interaction with it.[156-157] The AT richness of the *lpp* promoter may therefore be essential for the efficient transcription of the *lpp* gene.

The 322-base-pair sequence of the *lpp* mRNA has several other notable features. Translation initiates at the AUG on the mRNA corresponding to the ATG located from positions 389 to 391 in Figure 2A. A second ATG codon from positions 356 to 358 has been shown to be inactive.[158] The Shine-Dalgarno sequence,[159] GAGG located at positions 376 to 379, is likely to be the ribosome binding site. Mutations affecting this region decrease expression, presumably by making the translation initiation process less efficient.[158] The eukaryotic ribosome-binding site found in Brome mosaic virus RNA4 is identical with the *lpp* mRNA sequence GUAUUAAUA*AUG*.[160] Both mRNAs use the underlined initiation codon. Considering this homology it is not surprising that the lpp mRNA could be translated in a wheat-germ-cell-free system.[161] Mutagenesis studies indicate that the structure of the 5′-untranslated region preceding the Shine-Dalgarno sequence is also important for efficient expression of the *lpp* gene. A muta-

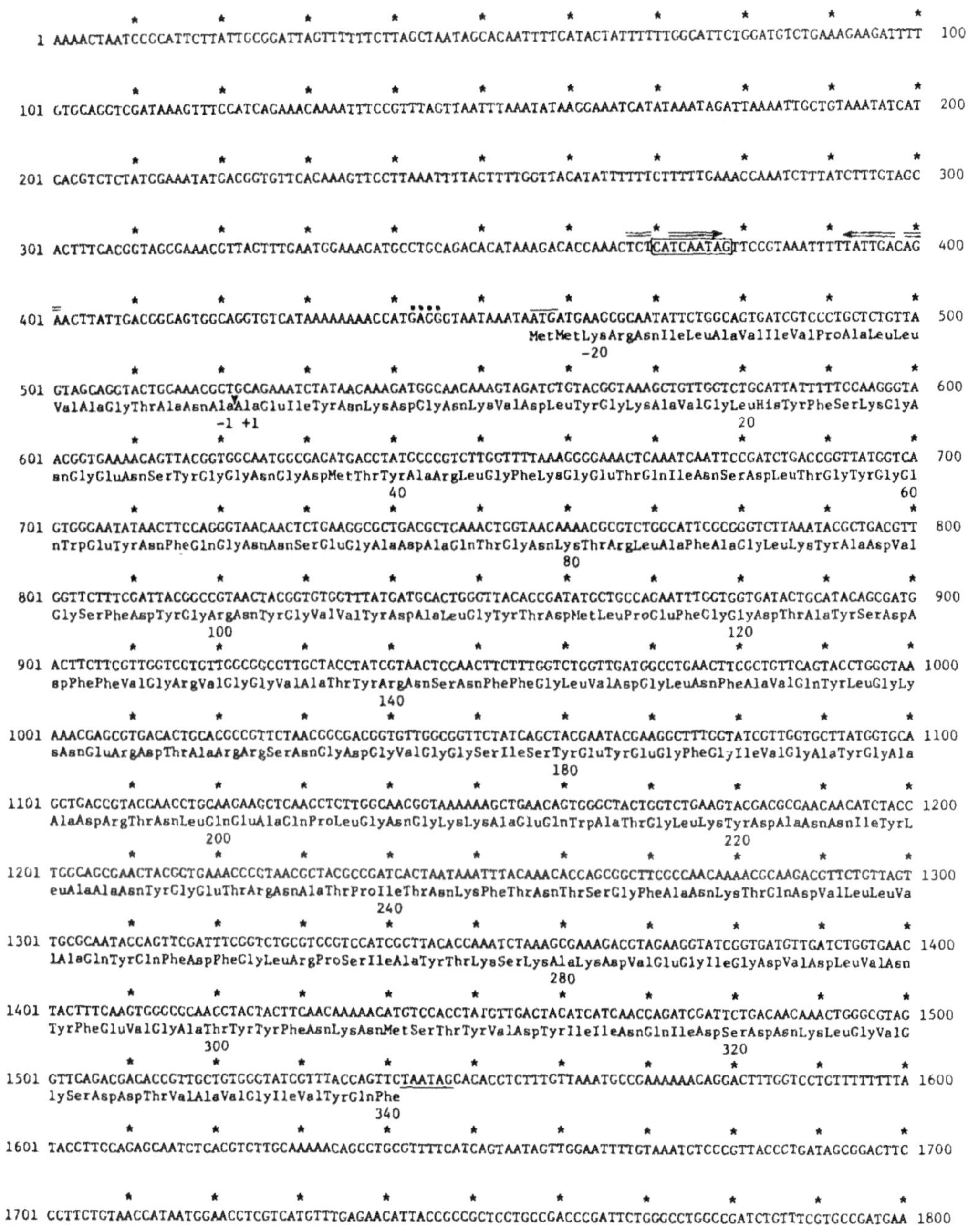

FIGURE 2C

tion allowing the formation of a stable secondary structure ($\Delta G = -7.35$ kcal)[162] in the area of the ribosome-binding site results in a threefold reduction in *lpp* expression even though the Shine-Dalgarno sequence remains intact.[158] The parental sequence can only give rise to the formation of a very weak structure in this region ($\Delta G = -3.05$ kcal). Maintaining an initiation region which is free of stable secondary structure may contribute favorably to ribosome binding.[163]

The lack of stable secondary structure in the *lpp* mRNA extends only to nucleotide 65 (position 415 in Figure 2A). Beyond this point, 85% of the remaining nucleotides are involved in potential hairpin structures (see Figure 5 in Reference 99). The highly

```
1    CGCTTGAAATACGGGTAAACAGACATTCAGAAATCAATGACGGTAATAAATAAAGTTAATGATGATAGCGGGAGTTATTCTACTTGCGAGTGAAGGTTTT  100

101  GTTTTGACATTCAGTGCTGTCAAATACTTAAGAATAAGTTATTGATTTTAACCTTGAATTATTATTGCTTGATGTTAGGTGCTTATTTCGCCATTCCGCA  200

201  ATAATCTTAAAAAGTTCCCTTGCATTTACATTTTGAAACATCTATAGCGATAAATGAAACATCTTAAAAGTTTTAGTATCATATTCGTGTTGGATTATTC  300

301  TGCATTTTTGGGGAGAATGGACTTGCCGACTGATTAATGAGGGTTAATCAGTATGCAGTGGCATAAAAAAGCAAATAAAGGCATATAACAGAGGGTTAAT  400

401  AACATGAAAGCTAAAGTACTGTCCCTCCTGGTCCCAGCTCTGCTGGTAGCAGGCGCAGCAAACGCTGCTGAAGTTTACAACAAAGACGGCAACAAATTAG  500
        MetLysValLysValLeuSerLeuLeuValProAlaLeuLeuValAlaGlyAlaAlaAsnAlaAlaGluValTyrAsnLysAspGlyAsnLysLeuA
        -20                                                        -1 +1
501  ATCTGTACGGTAAAGTAGACGGCCTGCACTATTTCTCTGACAACAAAGATGTAGATGGCGACCAGACCTACATGCGTCTTGGCTTCAAAGGTGAAACTCA  600
     spLeuTyrGlyLysValAspGlyLeuHisTyrPheSerAspAsnLysAspValAspGlyAspGlnThrTyrMetArgLeuGlyPheLysGlyGluThrGl
                    20                                                          40
601  GCTTACTGACCAGCTGACCGGTTACGGCCACTGGAATATCAGATCCAGGGCAACAGCGCTGAAAACGAAAACAACTCCTGGACCCGTGTGGCATTCGCA  700
     nValThrAspGlnLeuThrGlyTyrGlyGlnTrpGluTyrGlnIleGlnGlyAsnSerAlaGluAsnGluAsnAsnSerTrpThrArgValAlaPheAla
                                    60
701  GGTCTGAAATTCCAGGATGTGGGTTCTTTCGACTACGGTCGTAACTACGGCGTTGTTTATGACGTAACTTCCTGGACCGACGTACTGCCAGAATTCGGTG  800
     GlyLeuLysPheGlnAspValGlySerPheAspTyrGlyArgAsnTyrGlyValValTyrAspValThrSerTrpThrAspValLeuProGluPheGlyG
       80                                                100
801  GTGACACCTACGGTTCTGACAACTTCATGCAGCAGCGTGGTAACGGCTTCGCGACCTACCGTAACACTGACTTCTTCGGTCTGGTTGACGGCCTGAACTT  900
     lyAspThrTyrGlySerAspAsnPheMetGlnGlnArgGlyAsnGlyPheAlaThrTyrArgAsnThrAspPhePheGlyLeuValAspGlyLeuAsnPh
                    120                                                         140
901  TGCTGTTCAGTACCAGGGTAAAAACGGCAACCCATCTGGTGAAGGCTTTACTAGTGGCGTAACTAACAACGGTCGTGACGCACTGCGTCAAAACGGCGAC  1000
     eAlaValGlnTyrGlnGlyLysAsnGlyAsnProSerGlyGluGlyPheThrSerGlyValThrAsnAsnGlyArgAspAlaLeuArgGlnAsnGlyAsp
                                    160
1001 GGCGTCGGCGGTTCTATCACTTATGATTACGAAGGTTTCGGTATCGGTGGTGCGATCTCCAGCTCCAAACGTACTGATGCTCAGAACACCGCTGCTTACA  1100
     GlyValGlyGlySerIleThrTyrAspTyrGluGlyPheGlyIleGlyGlyAlaIleSerSerSerLysArgThrAspAlaGlnAsnThrAlaAlaTyrI
       180                                               200
1101 TCGGTAACGGCGACCGTGCTGAAACCTACACTGGTGGTCTGAAATACGACGCTAACAACATCTACCTGGCTGCTCAGTACACCCAGACCTACAACGCAAC  1200
     leGlyAsnGlyAspArgAlaGluThrTyrThrGlyGlyLeuLysTyrAspAlaAsnAsnIleTyrLeuAlaAlaGlnTyrThrGlnThrTyrAsnAlaTh
                    220                                                         240
1201 TCGCGTAGGTTCCCTGGGTTGGGCGAACAAAGCACAGAACTTCGAAGCTGTTGCTCAGTACCAGTTCGACTTCGGTCTGCGTCCGTCCCTGGCTTACCTG  1300
     rArgValGlySerLeuGlyTrpAlaAsnLysAlaGlnAsnPheGluAlaValAlaGlnTyrGlnPheAspPheGlyLeuArgProSerLeuAlaTyrLeu
                                    260
1301 CAGTCTAAAGGTAAAAACCTGGGTCGTGGCTACGACGACGAAGATATCCTGAAATATGTTGATGTTGGTGCTACCTACTACTTCAACAAAAACATGTCCA  1400
     GlnSerLysGlyLysAsnLeuGlyArgGlyTyrAspAspGluAspIleLeuLysTyrValAspValGlyAlaThrTyrTyrPheAsnLysAsnMetSerT
       280                                               300
1401 CCTACGTTGACTACAAAATCAACCTGCTGGACGACAACCAGTTCACTCGTGACGCTGGCATCAACACTGATAACATCGTAGCTCTGGGTCTCGTTTACCA  1500
     hrTyrValAspTyrLysIleAsnLeuLeuAspAspAsnGlnPheThrArgAspAlaGlyIleAsnThrAspAsnIleValAlaLeuGlyLeuValTyrGl
                    320                                                         340
1501 GTTCTAATCTCGATTGATATCGAACAAGGGCCTGCGGGCCCTTTTTTCATTGTTTTCAGCGTACAAACTCAGTTTTTTGGTGTACTCTTGCGACCGTTCG  1600
     nPhe

1601 CATGAGGATAATCACGTACGGAAATA
```

FIGURE 2D

structured nature of the *lpp* mRNA is probably related to its long half-life. The fact that most of the third bases of the lipoprotein codons are such that they can form stable structures may be of evolutionary significance. The most 3′ hairpin in the mRNA has the features characteristically found in prokaryotic transcription termination sequences.[151,164]

As shown in Table 1, *lpp* mRNA exhibits unusual codon usage. From the amino acid sequence, it can be calculated that the synthesis of lipoprotein could involve up to 50 different codons, but only 25 are used. Most of the codons chosen correspond to the major species of isoaccepting tRNAs.[165] Such unique codon usage is not coincidental. Similar patterns are characteristic of highly expressed genes. When Gouy and Dautier[166] analyzed the codon usage for 83 *E. coli* genes, they calculated that *lpp* was

```
1    XCGGGTAACCAGGCGCCTTTTCATCTCGTTGATTCCCTTTGTCTGTTTGATAATGCGCACATTGGGTATAACGTGATCATATCAACAGAATCAATAATGT  100

101  TTCGCCGAATAAATTGTATACTTAAGCTGCTGTTTAATATGCTTTGTAACAATTTAGGCTGAAATTCATACCAGATTTAGCTGGTGACGAACGTGAGCTT  200
         RNA Polymerase        Pribnow
         Recognition Site      Box
201  TTTTAAGAATACACGCTTACAAATTGTTGCGAACCTTTGGGAGTACAAACAATGCAAGAGAACTACAAGAATCTGGTGGTCGATGACGACATGCGCCTGC  300
                                                      MetGlnGluAsnTyrLysAsnLeuValValAspAspAspMetArgLeuA
                                                      1
301  GTGCGCTGCTGGAACGTTATCTCACCGAACAAGGCTTCCAGGTTCGAAGCGTCGCTAATGCAGAACAGATGGATCGCCTGCTGACTCGTGAATCTTTCCA  400
     rgAlaLeuLeuGluArgTyrLeuThrGluGlnGlyPheGlnValArgSerValAlaAsnAlaGluGlnMetAspArgLeuLeuThrArgGluSerPheHi
         20                                                      40
401  TCTTATGCTACTGGATTTAATGTTACCTGGTGAAGATGGCTTGTCGATTTGCCGACGTCTTCGTAGTCAGAGCAACCCGATGCCGATCATTATGGTGACG  500
     sLeuMetValLeuAspLeuMetLeuProGlyGluAspGlyLeuSerIleCysArgArgLeuArgSerGlnSerAsnProMetProIleIleMetValThr
                     60                                                                      80
501  GCGAAACGGGAAGAAGTGGACCGTATCGTAGGCCTGGAGATTGGCGCTGACGACTACATTCCAAAACCGTTTAACCCGCGTGAACTCCTGGCCCGTATCC  600
     AlaLysGlyGluGluValAspArgIleValGlyLeuGluIleGlyAlaAspAspTyrIleProLysProPheAsnProArgGluLeuLeuAlaArgIleA
                                                 100
601  GTCCGGTGCTGCGTCGTCAGCCGAACGAACTGCCAGGCGCACCGTCACAGGAAGAGGCGGTAATTGCTTTCGGTAAGTTCAAACTTAACCTCGCTACCCG  700
     rgAlaValLeuArgArgGlnAlaAsnGluLeuProGlyAlaProSerGlnGluGluAlaValIleAlaPheGlyLysPheLysLeuAsnLeuGlyThrAr
         120                                                     140
701  CGAAATGTTCCGCGAAGACGAGCCCATGCCGCTCACCAGCGGTGAGTTTGCGGTACTGAAGGCACTGGTCAGCCATCCGCGTGAGCCGCTCTCCCGCGAT  800
     gGluMetPheArgGluAspGluProMetProLeuThrSerGlyGluPheAlaValLeuLysAlaLeuValSerHisProArgGluProLeuSerArgAsp
                     160                                                                     180
801  AAGCTGATGAACCTTGCCCGTGGTCGTGAATATTCCGCAATGGAACGCTCCATCGACGTGCAGATTTCGCGTCTGCGCCGCTGGTGGAAGAAGATCCAGC  900
     LysLeuMetAsnLeuAlaArgGlyArgGluTyrSerAlaMetGluArgSerIleAspValGlnIleSerArgLeuArgArgTrpTrpLysLysIleGlnA
                                                 200
901  GCATCCGCGTTACATTCAGACCGTCTGGGGTCTGGCTACGTCTTTGTACCGGACGGCTCTAAAGCATGAGGCGATTGCGCTTCTCGCCACGAAGTTCAT  1000
     rgIleArgValThrPheArgProSerGlyValTrpAlaThrSerLeuTyrArgThrAlaLeuLysHisGluAlaIleAlaLeuLeuAlaThrLysPheIl
         220                                                     240
1001 TTGCCCGTACGTTATTGCTCATCGTCACCTTGCTGTTCGCCAGCCTCGTGACGACTTATCTCGGTGGTGCTGAACTTCGCGATTTTGCCGAGCCTCCAGCA  1100
     eCysProTyrValIleAlaHisArgHisLeuAlaValArgGlnProGlyAspAspLeuSerGlyGlyAlaGluLeuArgAspPheAlaGluProProAla
                     260                                                                     280
1101 GTTTAATAAAGTCCTCGCGTACGAAGTGCGTATGTTGATGACCGACAAACTGCAACTGGAGGACGGCACGCAGTTGGTTGTGCCTCCCGCTTTCCGTCGG  1200
     Val                         MetLeuMetThrAspLysLeuGlnLeuGluAspGlyThrGlnLeuValValProProAlaPheArgArg
     284                         1                                                                   20
```

FIGURE 2E

among the genes whose translation required the fewest tRNA discriminations per elongation cycle. In an independent study, Konigsberg and Godson[167] chose 29 genes to analyze for rare codon usage. The *lpp* gene was one of the two genes in their sample found to contain no rare codons (the other gene was *rpsU*). These observations suggest that the translation of *lpp* mRNA is much faster than average. Gouy and Gautier also calculated that *lpp* had the highest P2 index. This parameter is related to the fraction of codons which have made the "right choice" between a U or a C in the third position to give a codon-anticodon binding energy nearest the mean. Large P2 indexes were shown to be related to high gene expressivity.

The DNA sequence of the *lpp* gene from *Serratia marcescens*,[168] *Erwinia amylovora*,[169] and *Morganella morganii*[170] have also been determined and features similar to those in the *E. coli lpp* gene have been found. The cumulative effects of the features discussed above are most likely responsible for the very efficient expression of the *lpp* gene.

It should be noted that the expression of the *lpp* gene is constitutive and a gene dosage effect has been clearly demonstrated.[171] Therefore, cloning of the intact *lpp* gene in pBR322 was unsuccessful presumably because of the lethal effects of lipoprotein overproduction. It appears that *E. coli* can tolerate about fourfold the normal amount of lipoprotein.[145,172]

When excessive lipoprotein production is engineered using specialized inducible expression vectors,[146] cell lysis occurs.[145] Short-lived lipoprotein precursors are also detected under these conditions. No accumulation of other outer-membrane protein

```
1101 GTTTAATAAAGTCCTCGCGTACGAAGTGCGTATGTTGATGACCGACAAACTGCAACTGCAGGACGGCACCGCAGTTGCTTGTGCCTCCCGCTTTCCGTCGG 1200
     Val                             MetLeuMetThrAspLysLeuGlnLeuGluAspGlyThrGlnLeuValValProProAlaPheArgArg
                                      1                                                         20
1201 GAGATCTACCGTGAGCTGGGATCTCTCTCTACTCCAACGAGGCTGCCCAAGAGGCAGGTCTCCGTTGGGCGCAACACTATGAATTCTTAAGCCATCAGA 1300
     GluIleTyrArgGluLeuGlyIleSerLeuTyrSerAsnGluAlaAlaGluGluAlaGlyLeuArgTrpAlaGlnHisTyrGluPheLeuSerHisGlnM
                                                    40
1301 TGGCGCAGCAACTGGCCGGCCCGACGGAAGTGCGCGTTGAGGTCAACAAAAGTTCGCCTGTCGTCTGGCTGAAAACCTGGCTGTCGCCCAATATCTGGGT 1400
     etAlaGlnGlnLeuGlyGlyProThrGluValArgValGluValAsnLysSerSerProValValTrpLeuLysThrTrpLeuSerProAsnIleTrpVa
          60                                                      80
1401 ACGCGTGCCGCTGACCGAAATTCATCAGGGCGATTTCTCTCCGCTGTTCCGCTATACGCTGGCGATTATGCTATTGGCGATAGGCGGGGCGTGGCTGTTT 1500
     lArgValProLeuThrGluIleHisGlnGlyAspPheSerProLeuPheArgTyrThrLeuAlaIleMetLeuLeuAlaIleGlyGlyAlaTrpLeuPhe
                                  100                                                   120
1501 ATTCGTATCCAGAACCGACCGTTGGTCGATCTCGAACACGCAGCCTTGCAGGTTGGTAAAGGATTATTCCGCCGCCGCTGCGTGAGTATGGCGCTTCGG 1600
     IleArgIleGlnAsnArgProLeuValAspLeuGluHisAlaAlaLeuGlnValGlyLysGlyIleIleProProProLeuArgGluTyrGlyAlaSerG
                                                   140
1601 AGGTGCGTTCCGTTACCCGTGCCTTTAACCATATGGCGGCTGGTGTTAAGCAACTGGCGGATGACCGCACGCTGCTGATGGCGGGGGTAAGTCACGACTT 1700
     luValArgSerValThrArgAlaPheAsnHisMetAlaAlaGlyValLysGlnLeuAlaAspAspArgThrLeuLeuMetAlaGlyValSerHisAspLe
           160                                                     180
1701 GCGCACGCCGCTGACGCGTATTCGCCTGGCGACTGAGATGATGAGCGACCAGGATGCTATCTGGCAGAATCGATCAATAAAGATATCGAAGAGTGCAAC 1800
     uArgThrProLeuThrArgIleArgLeuAlaThrGluMetMetSerGluGlnAspGlyTyrLeuAlaGluSerIleAsnLysAspIleGluGluCysAsn
                                  200                                                   220
1801 GCCATCATTGAGCAGTTTATCGACTACCTGCGCACCGGCCAGGAGATGCCGATGGAAATGGCGGATCTTAATGCAGTACTCGGTGAGGTGATTGCTGCCG 1900
     AlaIleIleGluGlnPheIleAspTyrLeuArgThrGlyGlnGluMetProMetGluMetAlaAspLeuAsnAlaValLeuGlyGluValIleAlaAlaG
                                                   240
1901 AAAGTGGCTATGAGCGGCAAATTGAAACCGCGCTTTACCCCGGCAGCATTGAAGTGAAAATGCACCCGCTGTCGATCAAACGCGCGGTGGCGAATATGGT 2000
     luSerGlyTyrGluArgGluIleGluThrAlaLeuTyrProGlySerIleGluValLysMetHisProLeuSerIleLysArgAlaValAlaAsnMetVa
           260                                                     280
2001 GGTCAACGCCGCCCGTTATGGCAATGGCTGGATCAAAGTCAGCAGCGGAACGGAGCCGAATCGCGCCTGGTTCCAGGTGGAAGATGACGGTCCGGGAATT 2100
     lValAsnAlaAlaArgTyrGlyAsnGlyTrpIleLysValSerSerGlyThrGluProAsnArgAlaTrpPheGlnValGluAspAspGlyProGlyIle
                                  300                                                   320
2101 GCGCCGGAACAACGTAAGCACCTGTTCCAGCCGTTTGTCCGCGGCGACAGTGCGCGCACCATTAGCGGCACGGGATTAGGCTGGCAATTGTGCAGCGTA 2200
     AlaProGluGlnArgLysHisLeuPheGlnProPheValArgGlyAspSerAlaArgThrIleSerGlyThrGlyLeuGlyLeuAlaIleValGlnArgI
                                                   340
2201 TCGTGGATAACCATAACGGGATGCTGGAGCTTGGCACCAGCGAGCGGGGCGGCTTTCCATTCGCCCCTGGCTGCCAGTCCCGCTAACGCGGCCGCACGG 2300
     leValAspAsnHisAsnGlyMetLeuGluLeuGlyThrSerGluArgGlyGlyLeuSerIleArgAlaTrpLeuProValProValThrArgAlaGlnGl
           360                                                     380
2301 CACGACAAAAGAAGGCTAAATAAACGGGAGGCGAAGGTGCCTCCCGTTTTGCTTTCTATAAGATACTGGATAGATATTCTCCAGCTTCAAATCATTACAG 2400
     yThrThrLysGluGly

2401 TTTCGGACCAGCCGCTACCACCGCGGCACCCGCAGGCGTGTCGGTGTATTTATCGAAGTTGTCGATAAACAGTTTCGCCAGGGTTTCGGCTTTTTCCTGC 2500

2501 CACTGTTCCGCAGAAGCGTACCTGTTACGCGGATCGAGAATCTTCGTCTCTACGCCCGGCAGTTCGGTTGGATCGCCAGGTTAAACATCGGCAGAGTGA 2600

2601 AGCTTTCTGCATTATCCACCAACCGTTGAGGATGGCGTCGGATAATGGCGCGGCTATCTTTAATCGACATACGTTTGCCTGTCCCGTTCCAGCCAGTGTT 2700
```

FIGURE 2F

precursors are found to accumulate even when lipoprotein accounts for more than 40% of total membrane protein synthesis. However, the synthesis of several membrane proteins decreases. Further analysis of these effects will provide useful information as to the nature of the molecular events involved in the translocation and assembly of membrane proteins.

B. OmpA Protein

Like lipoprotein, it was assumed that the overproduction of OmpA protein was lethal, prohibiting the cloning of *ompA* in multicopy plasmids.[173] Therefore, the cloning of the wild-type *E. coli ompA* gene was accomplished using low-copy-number plasmids[174,175] or phage vectors.[76] Recently, the lethality of OmpA protein overproduction was confirmed when the *E. coli ompA* structural gene was cloned into a multicopy expression vector where *ompA* expression is controlled by the *lac* promoter and

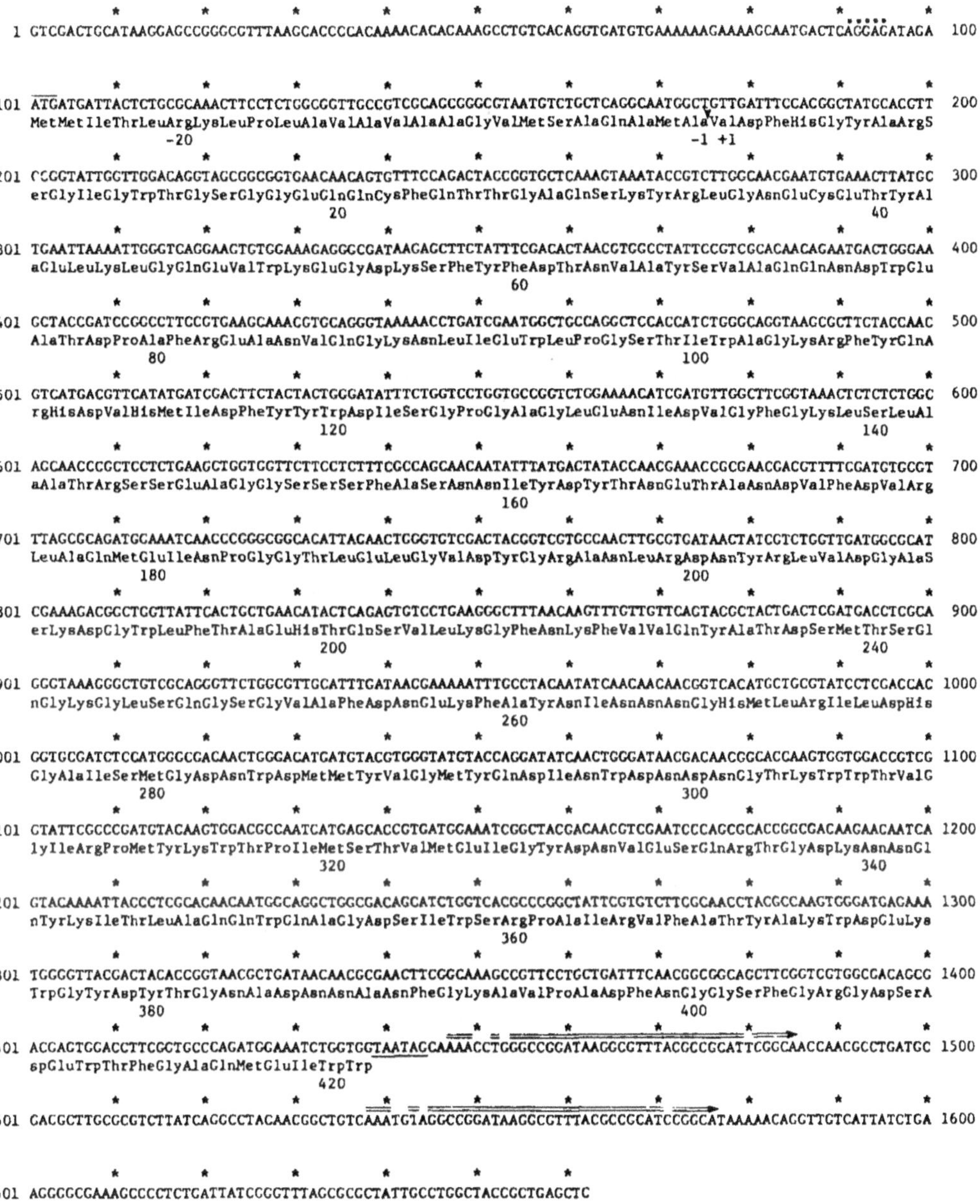

FIGURE 2G

operator. When a *lac* inducer is added to the media of cells harboring this plasmid, the increased synthesis of OmpA protein causes disruption of normal outer-membrane protein biogenesis and cell lysis.[176] *E. coli* has evolved some way to deal with small increases in *ompA* gene dosage without effect.[90] Yet, it appears that these mechanisms break down when the *ompA* copy number is greatly increased.

Despite these gene dosage problems, the entire sequence of the *ompA* gene from *E. coli* was determined by both Movva et al.[76] and by Beck and Bremer.[177] Aside from a few inconsistencies,[178] the sequences correspond well to each other and to the amino acid sequence previously determined by Chen et al.[179] The *ompA* nucleotide sequence is shown in Figure 2B. This sequence has several unique features which may be involved in the efficient expression of *ompA*.

```
   1 CTCTGCCATTCAATTTTAGGAAAAATGATATCAAACGAACGTTTTAGCAGGACTGTCGTCGGTTGCCAACCATCTGCGAGCAAAGCATGGCGTTTTGTTG  100

 101 CGCGGGATCAGCAAGCCTAGCGGCAGTTGTTTACGCTTTTATTACAGATTTAATAAATTACCACATTTTAAGAATATTATTAATCTGTAATATATCTTTA  200

 201 ACAATCTCAGGTTAAAAACTTTCCTGTTTTCAACGGGACTCTCCCGCTGAATATTCGCCCGTTAATTAAAATCAGGAATGAAAATGAAAAAGAGCACTCT  300
                                                                                         MetLysLysSerThrLe
                                                                                             -20
 301 GGCATTACTGGTGATGGGCATTGTGGCATCTGCATCTGTACAGGCTGCAGAAATATATAATAAAGACGGTAATAAACTGGATGTCTATGGCAAAGTTAAA  400
     uAlaLeuValValMetGlyIleValAlaSerAlaSerValGlnAlaAlaGluIleTyrAsnLysAspGlyAsnLysLeuAspValTyrGlyLysValLys
                                                 -1 +1
 401 GCCATGCATTATATGAGTGATAACGCCAGTAAAGATGGCGACCAGAGTTATATCCGTTTTGGTTTCAAAGGCGAAACACAAATTAACGACCAACTGACTG  500
     AlaMetHisTyrMetSerAspAsnAlaSerLysAspGlyAspGlnSerTyrIleArgPheGlyPheLysGlyGluThrGlnIleAsnAspGlnLeuThrG
       20                                                       40
 501 GTTATGGTCGTTGGCAAGCAGAGTTTGCCGGTAATAAAGCAGAGAGTGATACTGCACAGCAAAAAACGCGTCTCGCTTTTGCCGGGTTGAAATATAAAGA  600
     lyTyrGlyArgTrpGluAlaGluPheAlaGlyAsnLysAlaGluSerAspThrAlaGlnGlnLysThrArgLeuAlaPheAlaGlyLeuLysTyrLysAs
                        60                                                     80
 601 TTTGGGTTCTTTCGATTATGGTCGTAACCTGGGGGCGTTGTATGACGTGGAAGCCTGGACCGATATGTTCCCGGAATTTGGTGGCGATTCCTCGGCGCAC  700
     pLeuGlySerPheAspTyrGlyArgAsnLeuGlyAlaLeuTyrAspValGluAlaTrpThrAspMetPheProGluPheGlyGlyAspSerSerAlaGln
                                               100
 701 ACCGACAACTTTATGACCAAACGCGCCAGCGGTCTGGCGACGTATCGGAACACCGACTTCTTCGGCGTTATCGATGGCCTGAACTTAACCCTGCAATATC  800
     ThrAspAsnPheMetThrLysArgAlaSerGlyLeuAlaThrTyrArgAsnThrAspPhePheGlyValIleAspGlyLeuAsnLeuThrLeuGlnTyrG
      120                                                      140
 801 AAGGGAAAAACGAAAACCGCGACGTTAAAAAGCAAAACGGCGATGGCTTCGGCACGTCATTGACATATGACTTTGGCGGCAGCGATTTCGCCATTAGTGG  900
     lnGlyLysAsnGluAsnArgAspValLysLysGlnAsnGlyAspGlyPheGlyThrSerLeuThrTyrAspPheGlyGlySerAspPheAlaIleSerGl
                       160                                                     180
 901 GGCCTATACCAACTCAGATCGCACCAACGAGCAGAACCTGCAAAGCCGTGGCACAGGCAAGCGTGCAGAAGCATGGGCAACAGGTCTGAAATACGATGCC 1000
     yAlaTyrThrAsnSerAspArgThrAsnGluGlnAsnLeuGlnSerArgGlyThrGlyLysArgAlaGluAlaTrpAlaThrGlyLeuLysTyrAspAla
                                               200
1001 AATAATATTTATCTGGCAACTTTCTATTCTGAAACACGCAAAATGACGCCAATAACTGGCGGCTTTGCCAATAAGACACAGAACTTTGAAGCGGTCGCTC 1100
     AsnAsnIleTyrLeuAlaThrPheTyrSerGluThrArgLysMetThrProIleThrGlyGlyPheAlaAsnLysThrGlnAsnPheGluAlaValAlaG
      220                                                      240
1101 AATACCAGTTTGACTTTGGTCTGCGTCCATCGCTGGGTTATGTCTTATCGAAAGGGAAAGATATTGAAGGTATCGGTGATGAAGATCTGGTCAATTATAT 1200
     lnTyrGlnPheAspPheGlyLeuArgProSerLeuGlyTyrValLeuSerLysGlyLysAspIleGluGlyIleGlyAspGluAspLeuValAsnTyrIl
                       260                                                     280
1201 CGACGTCGGTGCTACGTATTATTTCAACAAAAATATGTCACCGTTTGTTGATTATAAAATCAACCAACTGGATAGCGATAACAAATTGAATATTAATAAT 1300
     eAspValGlyAlaThrTyrTyrPheAsnLysAsnMetSerAlaPheValAspTyrLysIleAsnGlnLeuAspSerAspAsnLysLeuAsnIleAsnAsn
                                               300
1301 GATGATATTGTCGCGGTTGGCATGACGTATCAGTTTTAATGAATATTGCCGGATGTGATGCATCCCGCAGATTTCACTCACCCCGTTAACTTCACCGGCT 1400
     AspAspIleValAlaValGlyMetThrTyrGlnPhe
      320
1401 TCGTCACGAAAATCATCCGCCGGTTCCAGC
```

FIGURE 2H

In the region about 160 base pairs upstream of the translation initiation codon, two sequences resembling prokaryotic promoters can be found. The promoter farthest from the site of translation initiation is called P1. P1 overlaps the slightly downstream promoter P2. Preceding the RNA polymerase recognition site is a long AT-rich region followed by a short GC-rich region. The *ompA* promoter region also contains the sequence TCACACTT (from positions 80 to 87 in Figure 2B) found in the *gal* promoter where it is known to interact with cAMP receptor protein (CRP).[180] In vitro, this sequence can facilitate the binding of cAMP-CRP to the *ompA* promoter region and under certain conditions, negatively influence *ompA* transcription.[178] The applicability of these findings to the in vivo situation is not clear. The influence of cAMP on the biosynthesis of *E. coli* outer-membrane proteins has been investigated but no significant effects of cAMP on the production of OmpA protein[181,182] or on β-galactosidase produced from an *ompA-lacZ* fusion[183] have been observed.

There is a striking homology between the *ompA* P1 promoter and both the *gal* and *lac* promoters.[184] Curiously, this promoter does not appear to function in vivo[185,186]

Table 1

CODON USAGE IN THE *lpp, ompA, ompF, ompC, ompR* AND *envZ* GENES OF *E. COLI* K-12

SECOND BASE

FIRST BASE	U	a	b	c	d	e	f	g	C	a	b	c	d	e	f	g	A	a	b	c	d	e	f	g	G	a	b	c	d	e	f	g	THIRD BASE
U	Phe	0	2	7	2	3	4		Ser	3	4	6	6	4	2		Tyr	0	2	9	5	2	6		Cys	0	1	0	0	0	0		U
		0	7	12	17	7	6	*		2	8	5	8	3	3			1	15	20	24	4	4	*		1	1	0	0	2	1		C
	Leu	0	1	2	1	3	2			0	0	1	0	1	0		Term	1	1	1	1	1	1		Term	0	0	0	0	0	0		A
		0	0	0	0	2	6			0	0	0	0	2	5			0	0	0	0	0	0		Trp	0	5	2	4	3	8		G
C	Leu	0	0	3	1	7	4		Pro	0	1	2	0	3	2		His	0	2	1	0	4	4		Arg	3	10	10	12	16	11	*	U
		0	0	0	1	5	3			0	0	0	0	0	3			0	3	0	1	1	5			1	3	2	1	12	12	*	C
		0	0	0	0	1	1			0	3	1	3	3	1	*	Gln	0	2	6	1	2	5			0	0	0	0	1	1		A
		9	22	18	24	16	24	*		0	15	2	1	11	15	*		5	15	7	20	8	13	*		0	0	0	0	1	4		G
A	Ile	0	1	1	0	9	14		Thr	4	10	7	12	1	1	*	Asn	0	1	4	0	2	6		Ser	0	0	1	1	1	4		U
		2	15	13	10	6	11	*		0	12	11	12	2	8	*		6	18	28	32	6	8	*		2	4	3	2	4	7		C
		0	0	0	0	0	1			0	1	1	0	1	1		Lys	6	15	16	17	3	9	*	Arg	0	0	0	0	1	0		A
	Met	3	6	5	4	11	14			0	0	3	0	5	10			1	4	3	0	8	2			0	0	0	0	0	0		G
G	Val	3	17	16	12	5	5	*	Ala	8	22	16	18	8	5	*	Asp	2	5	13	9	6	8		Gly	2	24	33	29	8	5	*	U
		0	1	1	2	4	7			0	1	2	0	4	9			6	17	14	23	9	7			1	14	15	19	5	16	*	C
		2	7	4	9	4	4	*		3	11	11	8	5	5	*	Glu	0	10	13	11	15	15	*		0	0	0	0	0	3		A
		1	2	5	2	5	12	*		1	3	5	3	8	16	*		0	3	1	0	8	18			0	0	1	0	2	9		G

FIRST BASE

THIRD BASE

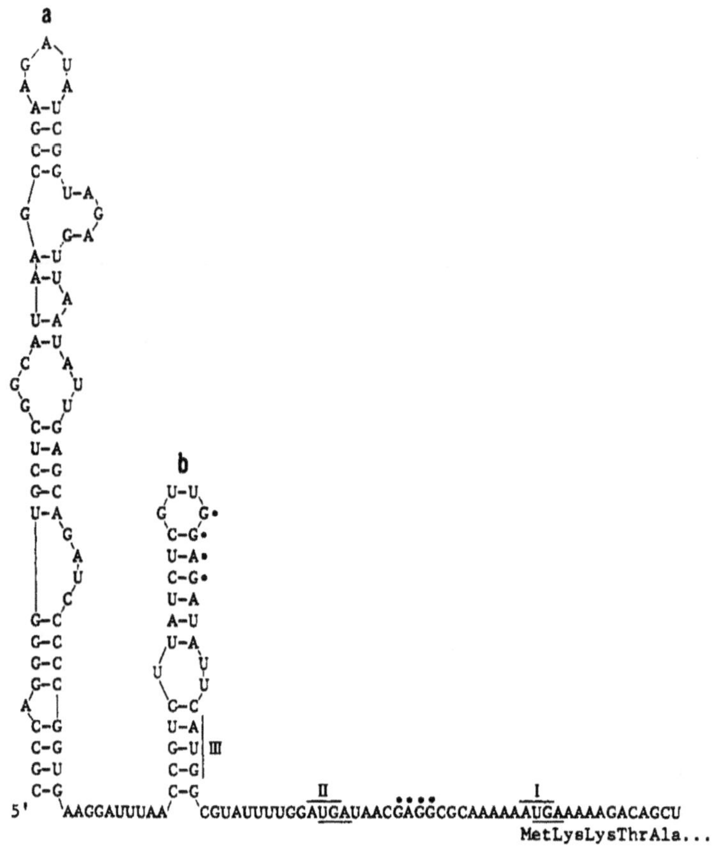

FIGURE 3. Possible secondary structure in the 5′ leader region of the *ompA* mRNA. Translation of pro-OmpA protein begins as shown from AUG-I. Lines under the sequence indicate termination codons while lines over or to the right of the sequence indicate initiation codons. Dots (•) indicate the Shine-Dalgarno sequences. Calculated ΔG values[162] are (a) −28.0 and (b) −10.8 kcal.

although in vitro *ompA* can be expressed from P1, albeit at a lower efficiency than from P2.[185] Cole et al.[185] showed that in vivo, transcription begins with the C at position 120 just following P2, giving rise to an *ompA* mRNA with a 134-nucleotide-long so-called "5′-untranslated leader region" preceding the AUG initiation codon. The significance of this unusually long and highly structured (see Figure 3) leader region has been investigated through the construction of a series of *ompA-lacZ* fusion plasmids which contain the *ompA* promoter and varying lengths of the DNA corresponding to the 5′ leader region of the *ompA* mRNA.[186] β-Galactosidase expression from the fusions was found to differ greatly. The most β-galactosidase is produced by the clone containing the longest leader region while another fusion, which contains just a few bases of the *ompA* leader, is only able to synthesize 4% as much β-galactosidase. Fusions with leaders intermediate in length produce 6 to 20% of the activity of the most efficient fusion. Upon analysis of the *lacZ* mRNA levels, it was found that the amount of mRNA in the fusion strains closely resembles their individual enzyme activities. This strongly suggests that the 5′ leader region of the mRNA is very important for efficient gene expression and that it exerts all or most of its influence at the RNA level. Further study is required to determine whether the *ompA* leader RNA or DNA interacts with a positive effector to increase transcription or if it enhances gene expression by increasing the stability of the mRNA.

Another interesting part of the *ompA* gene is the region just preceding the translation initation site. As indicated in Figure 2B., the synthesis of the secretory precursor of OmpA protein (pro-OmpA protein) initiates at a position on the mRNA corresponding to ATG-I from positions 254 to 256. The Shine-Dalgarno sequence GAGG is located upstream of ATG-I from positions 342 to 345. There are two other initiation codons (II and III) upstream from the initiation codon (I) for the pro-OmpA protein. These correspond to ATG-II from positions 234 to 236, and ATG-III from positions 220 to 222. Initiation codon III is also accompanied by the Shine-Dalgarno sequence GGAG from positions 210 to 213. Therefore, the AUG in the mRNA corresponding ATG-III could be a functional initiation codon. If this is true, a pentapeptide, Met-Ala-Ile-Phe-Gly, would be synthesized. Since the termination codon (TGA) for this peptide overlaps ATG-II, it is likely that the ribosome which completes the synthesis of the pentapeptide immediately restarts from the ATG-II initiation codon without being released from the mRNA. This would start the production of a heptapeptide Met-Ile-Thr-Arg-Arg-Lys-Lys. The termination codon (TGA) of this heptapeptide overlaps the initiation codon (ATG-I) in exactly the same way as described for ATG-II. Therefore ribosomal reinitiation without release could again occur at the ATG-I initiation codon. Such reinitiation events from an initiation codon overlapping or located close to a termination codon have been well documented.[187-189] These tandemly repeated initiation codons may play an important role in the efficient translation of the *ompA* mRNA by increasing the initiation rate. It is equally possible that small peptides are produced which may have important cellular functions. It should be pointed out that both AUG-III and its Shine-Dalgarno sequence may be sequestered in a fairly stable secondary structure in the mRNA (see Structure b in Figure 3), and therefore be unavailable for ribosome binding when this structure forms. Some preliminary data obtained in our laboratory indicate that this may occur.[183] When the *lacZ* gene was fused to ATG-II (in the proper reading frame) no β-galactosidase activity was produced. One Lac^+ derivative of this fusion was sequenced and found to contain a mutation destabilizing this structure. This may mean that OmpA protein synthesis can be increased by the reinitiation mechanism described earlier, only if the formation of structure II is antagonized. However, the final assessment of this situation awaits the characterization of more Lac^+ ATG-II fusion mutants.

At the 5′ end of the 134-base leader sequence of the *ompA* mRNA, two very stable secondary structures can be constructed as shown in Figure 3. Structure a has a ΔG value of −28.0 kcal.[162] Structure b, referred to above, has a ΔG value of −10.8 kcal. The coding region of the mRNA is also highly structured.[76] These secondary structures may account for the very long half-life of the *ompA* mRNA[147,190] as well as be involved in the regulation of *ompA*. The secondary structures found in the 5′ leader regions of many bacterial mRNAs are known play an important role in regulating gene expression.[191-193] The base sequence after structure b, where the pro-OmpA initiation codon and ribosome binding site are located, appears to be devoid of secondary structure. This may facilitate ribosome-binding to the *ompA* mRNA. From the DNA sequence of the *ompA* gene from *Shigella dysenteriae* it seems that the secondary structure of the *ompA* mRNA has been conserved during evolution.[194] Sequence analysis of the *ompA* genes from the other bacteria which have recently been cloned should reveal whether this is true.[195]

The synthesis of OmpA protein does not exhibit a gene dosage effect[90] as does the synthesis of lipoprotein.[171] This suggests that the regulation of *ompA* expression may involve a feedback mechanism.[90] Since *ompA* copy number does not effect the transcription of *ompA-lacZ* fusions,[196] it appears that the *ompA* promoter is not autogenously regulated; at least not without the involvement of downstream sequences. Recently, it was shown that *crp* autoregulation involves an interaction between CRP-cAMP and a DNA site more than 20 base pairs downstream of the transcription initiation site.[197] In light of this finding, the regulatory importance of *ompA* promoter distal sequences should be investigated.

It has been reported that LPS may play a role in the regulation of *ompA*.[196] When a *ompA-lacZ* operon fusion was introduced into several LPS mutants, β-galactosidase levels were elevated when compared to wild-type stains containing the fusion. However, elevated levels of β-galactosidase are also made by an operon fusion between a foreign *ompA* promoter and *lacZ* in a strain where OmpA protein is not expressed due to the existence of some kind of LPS/*ompA* incompatibility. The incompatibility of the foreign (clinically isolated) *ompA* allele with K-12 LPS[198] does not result in the accumulation of pro-OmpA protein. These observations indicate that transcription effects alone cannot explain how LPS is involved in *ompA* regulation. It has been suggested that LPS influences the pool size of stable *ompA* mRNA.[196] It is interesting to note that a *S. typhimurium* mutant with a temperature-sensitive defect in lipid A synthesis experiences a transient increase in OmpA protein synthesis at the restrictive temperature which may be related to *ompA* mRNA stability.[199] Perhaps secondary structure in the 5′ leader region of the *ompA* mRNA is involved in these phenomena.

C. Matrix Proteins — Osmoregulation

1. Gene Structures

The matrix protein genes, which encode the major outer-membrane porins OmpF protein and OmpC protein, were cloned using different strategies. The *ompF* gene was isolated by Mutoh et al.[200] from a λ transducing phage carrying the closely linked *asnS* gene. Mizuno et al.[201] used a λ phage containing an *ompC-lacZ* fusion[45] as a hybridization probe to isolate the *ompC* gene.

Unlike *lpp* and *ompA*, the matrix protein genes could be cloned into multicopy plasmids without alteration. This was possible due to the existence of *ompB* mutants which allow very little expression of *ompF* and *ompC*. Efforts to introduce these plasmids into wild-type hosts resulted in unhealthy transformants which formed very small colonies.[202] This provides further evidence that the overproduction of the major outer-membrane proteins is deleterious to *E. coli* cells.

Structural[18,24,26] and functional[203] analyses of OmpC and OmpF protein have indicated that the major outer-membrane porins are quite similar. This was confirmed when their DNA sequences,[19,204] shown in Figures 2C and 2D, were compared.[204] Within the protein coding regions, the similarity between *ompF* and *ompC* is particularly striking. At the amino acid level, 210 (61%) are conserved between OmpC protein and OmpF protein. Furthermore, 42 of the residues that differ between the two porins are functionally identical. The DNA sequences of *ompF* and *ompC* are 69% homologous. This extensive homology indicates that *ompC* and *ompF* evolved from a common ancestral gene.

In several regions of the structural gene, the amino acid sequences of the matrix proteins diverge considerably. The corresponding DNA sequences suggest that such differences arose via deletions or insertions. Perhaps these events were associated with the direct or inverted repeats which in most cases flank the involved regions (see Figure 4b for an example).

Another recently sequenced[205] gene in *E. coli* known as *phoE* is quite homologous to *ompF* and *ompC*. The *phoE* gene product has been identified as a porin specific for anion uptake. Some salient features of the *phoE* DNA sequence (shown in Figure 3h) are discussed in Section III.D. Although *ompC* and *ompF* are more homologous to each other than either is to *phoE*,[204] it seems obvious that all three genes evolved from a common ancestor. Examples of homology and divergence among the three porin genes are shown in Figures 4b and c. Outside the coding regions, *ompF*, *ompC*, and

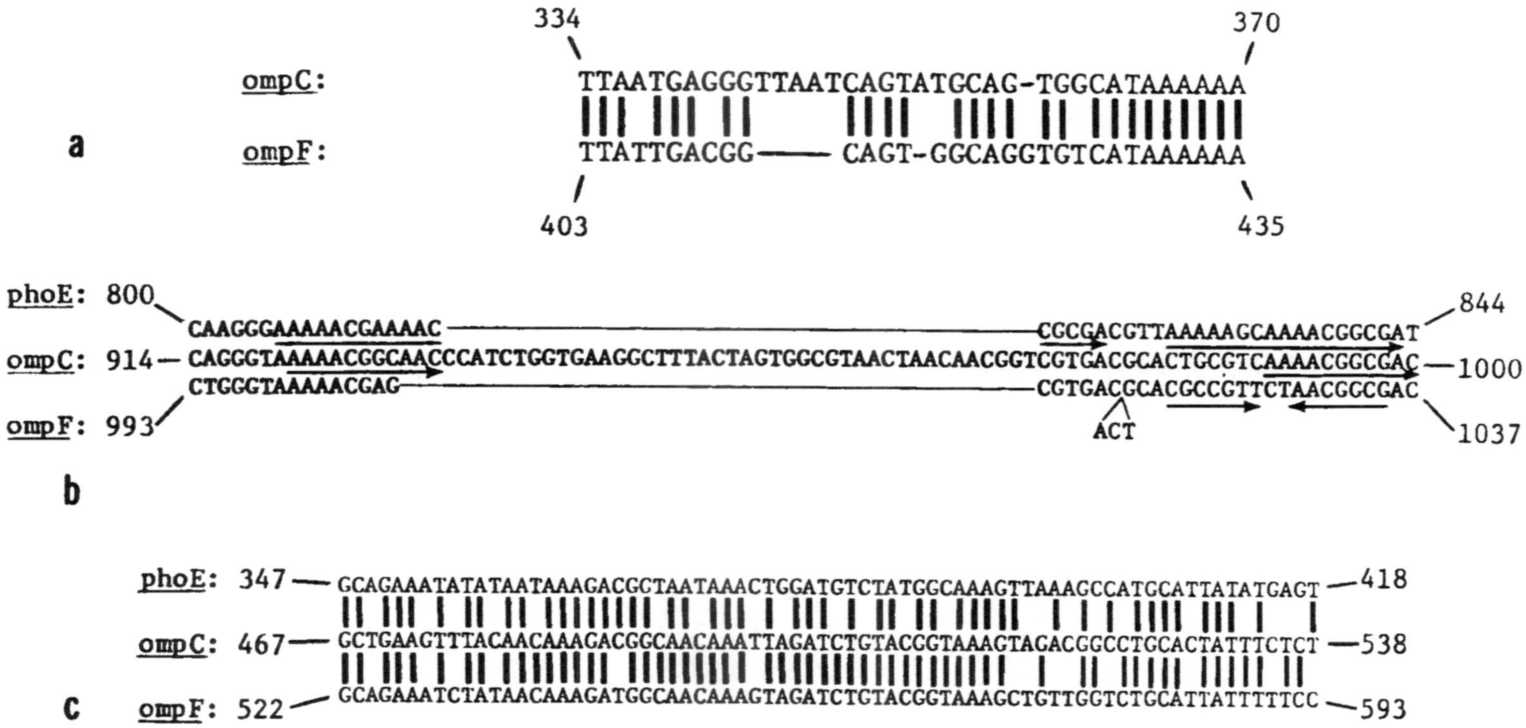

FIGURE 4. DNA sequence comparison[204] among the *ompF*,[19] *ompC*,[204] and *phoE*[205] genes. The nucleotide positions are numbered as in Figure 2. (a) The homologous region in the 5′ end noncoding region of *ompC* (upper line) and *ompF* (lower line) are shown. Identical nucleotides are connected with bars. (b) The DNA sequences of the *phoE* gene (upper line), the *ompC* gene (middle line) and the *ompF* gene (lower line) corresponding to a segment of the amino acid coding region are shown. Direct (→→) and inverted (→←) repeat sequences are indicated by arrows. (c) The DNA sequences encoding the first 24 amino acid residues at the amino termini of *PhoE*, *OmpC*, and *OmpF* proteins are shown on the top, middle, and bottom lines, respectively. Identical nucleotides are connected with bars.

phoE bear little resemblance to each other at the DNA sequence level except for the promoter regions of *ompC* and *ompF*. Both *ompF* and *ompC* contain a nine-base-pair sequence which is identical except for one mismatch in the center. As shown boxed in Figures 2C and 2D, the sequence CATCt/aATAG is located at slightly different positions upstream of the translation initiation site in each case and is part of an inverted repeat. Another homologous region is shown in Figure 4a. These sequences are found 34 and 21 bases upstream of the *ompC* and *ompF* initiation codons, respectively.

As will be discussed later, both *ompC* and *ompF* are sensitive to the positive regulation exerted by the *ompR* gene product. Either of the two conserved regions may serve as the binding site for the *ompR* gene product. The slight differences in structure and location of the homologous sequences may explain the differential expression of *ompF* and *ompC*. This situation is similar to that for the maltose regulon, where a seven-base-pair sequence is found in the control region of each operon under the positive control of the *malT* gene product (see Section III.D.3).[206]

In addition to the sequences described above, *ompC* and *ompF* have the identical sequence "GAGGGT" in their ribosome-binding sites. Curiously, in the *ompC* gene, the same sequence and the four bases that follow are repeated upstream (from positions 339 to 348 in Figure 2D.). The configuration of the surrounding sequences is such that in the *ompC* mRNA, the homology to the 3′ end of the 16S RNA, GAGG, could be exposed in the loop of a hairpin with a $\Delta G = -8.1$ kcal.[162] Adjacent to this putative structure lies another AUG codon corresponding to the ATG from positions 353 to 355 in Figure 2D. Translation initiating at this position within the leader region of the *ompC* mRNA would produce a peptide containing 11 amino acids and terminate just upstream of the putative Shine-Dalgarno sequence used for pro-OmpC protein translation. The 5′ leader region of the *ompA* mRNA has similar characteristics.

Several features appear to be shared by all the major outer-membrane protein genes, and *ompC* and *ompF* are no exception. These include the nonrandom distribution of AT base pairs around the promoter region, the unusual codon usage (Table 1), and the presence of similar termination signals. These traits undoubtedly contribute to the very efficient expression of the genes encoding the major outer-membrane proteins of *E. coli.*

2. Osmoregulation

The regulation of the *ompF* and *ompC* genes is most unusual. The expression of both genes has been shown to be controlled by the osmolarity of the culture medium. This osmoregulation is mediated by an unlinked locus, *ompB*, which maps at 74 min (see Figure 1). Cells grown in nutrient broth or yeast broth synthesize decreased amounts of OmpC protein and increased amounts of OmpF protein.[18,109] Complete medium supplemented with a fermentable carbon source such as glucose or lactate brings about the opposite response,[18] i.e., an increase in the synthesis of OmpC protein and a concomitant decrease in OmpF protein synthesis. Media of low osmotic strength causes an effect similar to yeast or nutrient broth. When the osmolarity of the culture media is increased by the addition of sodium chloride, potassium chloride, or sucrose, OmpF protein levels decrease and OmpC protein levels increase.[207,208]

3. The ompB Locus

Hall and Silhavy[45,47,57] monitored the expression of *ompF-lacZ* and *ompC-lacZ* fusions to elucidate the nature of the media-dependent fluctuation of the matix proteins. They also studied the effects of the three types of *ompB* mutations, $OmpC^-OmpF^-$, $OmpC^-OmpF^+$, and $OmpC^+OmpF^-$, on the expression of *ompC* and *ompF* using fusions to β-galactosidase. Finally, they analyzed the nature of *ompB* mutations with fine-structure mapping and complementation studies. The results of these experiments led to the conclusions listed below.

1. In addition to being required to simply facilitate the transcription of *ompC* and *ompF*, *ompB* adjusts the relative transcription levels of the major outer-membrane porin genes to suit the media conditions.
2. The *ompB* locus contains two genes referred to as *ompR* and *envZ*. The *ompR* gene product has two genetically separate functions which are defined by mutations conferring either an $OmpF^-OmpC^-$ phenotype *(ompR1)* or an $OmpF^+OmpC^-$ phenotype *(ompR2)*. Mutations in *envZ* give rise to an $OmpF^-OmpC^+$ phenotype.
3. *ompB* mutants which cannot express one of the two matrix protein genes, express the other porin in a constitutive manner.
4. *envZ* alleles are codominant while *ompR2* mutations are dominant to *ompR1* mutations and to the wild type.

Hall and Silhavy have proposed a model which is consistent with these findings.[47] In this model, the *ompR* gene product can exist in a monomeric or multimeric form. The former is necessary to activate *ompF* transcription while the latter is required to stimulate *ompC* transcription. Mutations of the *ompR1* moiety block the *ompR* gene product from binding to the regulatory regions of both *ompC* and *ompF*. Multimerization is blocked by *ompR2* mutations. The *envZ* gene product is an essential envelope protein which senses the external environment and conveys a signal to the cell by shifting the equilibrium between the monomeric and multimeric forms of OmpR protein to induce the synthesis of the appropriate porin.

This model can account for much of the data pertaining to *ompB*-mediated regulation, but many other observations must still be explained. These include the stimulation of *ompF* expression by cAMP,[209] the decreased production of OmpF protein and the constitutive synthesis of OmpC protein in adenylate cyclase mutants,[202] the pleiotropic nature of *envZ* mutations,[46,51,52] and the curious finding that *ompF* mutants produce OmpC protein constitutively, while in *ompC* mutants the production of OmpF protein is sometimes sensitive to osmolarity.[56,210] Taylor et al.[211] recently isolated two new classes of *envZ* mutants which must also be explained. One type exhibits an $OmpF^+OmpC^-$ phenotype. The other expresses both matrix proteins with little regard for the growth conditions. Furthermore, the model offers no information as to the exact function of EnvZ protein.

Recently, Wurtzel et al.[143,212] and Mizuno et al.[213,214] cloned and sequenced the entire *ompB* locus. They found that *ompB* contains two genes, which is consistent with the genetic data. The protein products of *ompR* and *envZ* were identified in minicells as 32,500- and 44,000-mol-wt polypeptides, respectively. The primary structure of OmpR protein deduced from the nucleotide sequence[212] revealed that the protein has a very basic carboxy-terminal region (see Figure 2E, residues 207 to 263 where 13 residues are positively charged [i.e. arginine or lysine] and only one is negatively charged). Another region of the protein was found to show sequence homology with two known DNA-binding proteins,[215] cro protein, and the P22 repressor. Perhaps these regions are related to the two functional domains postulated by Hall and Silhavy.[47] When the amino acid sequence of the *envZ* gene product was analyzed,[214] it was found to contain a very acidic region (see Figure 2F, residues 202 to 274 where 18 residues are negatively charged and only three are positively charged). It should be interesting to see whether the aforementioned oppositely charged domains are required for the interaction between the *ompR* and *envZ* gene products.

Sequence analysis, in vitro transcription, and subcloning experiments have confirmed that *ompB* is an operon containing the *ompR* and *envZ* genes.[212,214] Figure 2E shows that sequences of the two genes are separated by only 28 base pairs. Transcription initiates at a promoter upstream of the *ompR* gene.[212] A potential stem and loop structure characteristic of a prokaryotic transcription termination site is located after the *envZ* gene from positions 2318 to 2355 in Figure 2F. Cloning of the *ompB* promoter 5′ to the β-galactosidase structural gene indicated that transcription of the *ompB* operon is under the positive control of cAMP.[212] A sequence homologous to the *lac* CRP binding site[151] can be found upstream of the RNA polymerase recognition site. The presence of this sequence, from positions 93 to 100 in Figure 2E, suggests that the positive effect of cAMP on *ompB* expression is the result of increased transcriptional activity mediated by the CRP protein. It is not clear at this point how the increased expression of *ompB* stimulated by cAMP can affect porin production when the increased *ompB* expression due to gene dosage does not.[214] However, the recent isolation of both the *ompF* and the *ompC* genes and the purification of the *ompR* gene product should provide us with the means to answer this and many of the other remaining questions concerning *ompB*-mediated regulation.

Based on the results of in vitro transcription studies,[212] a putative *ompB* mRNA start site has been positioned at the G at 146 in Figure 2E. The *ompB* mRNA appears to have a long (106 base pairs) 5′-untranslated leader region of unknown function. Like the *ompA* mRNA leader region, the *ompB* leader has the potential to form secondary structures. The AT content in the leader region (62%) and the promoter region (72%) is very high when compared to the coding regions. In this respect *ompB* resembles the genes coding for major outer-membrane proteins. However, the codon usage in *ompR* and *envZ* is more like that found for other regulatory genes. The data in Table 1 indicate this is especially true for *envZ*. It can be calculated that the frequency of rare codon usage in *ompR* and *envZ* is 4.9 and 8.1%, respectively. In contrast, the average among the major outer-membrane protein genes described so far is only 1.5%. In a recent analysis, Konigsberg and Godson[167] presented data indicating that rare codon usage was higher in regulatory than in nonregulatory genes. They suggested that rare codon usage may play a role in modulating the differential expression of multicistronic mRNAs.

Although the *ompR* and *envZ* cistrons appear to be transcribed as one mRNA, the *envZ* gene product is synthesized in much smaller amounts that the *ompR* gene product.[213] The *envZ* gene had to be subcloned and expressed from a pair of strong promoters in order to identify the EnvZ protein.[214] One reason for the disproportionate expression of *ompR* and *envZ* could be the differences in rare codon usage noted above. Yet, the subcloning experiments and sequence analysis support the idea that a translational barrier exists between *ompR* and *envZ*. A candidate for such a barrier is the secondary structure shown in Figure 5 which can form in the *envZ* translation initiation region. The unfavorable effects of similarly located secondary structures on translation initiation have been described.[163] It is not known why an expression barrier exists between the two genes of the *ompB* operon. Perhaps further study will reveal that the *envZ* gene product is required in much smaller amounts than the *ompR* gene product to achieve optimum expression of *ompB* sensitive genes.

It is interesting to note that the *ompB* region from *S. typhimurium* has also been cloned.[73] The *E. coli* and *Salmonella ompR* gene products have the same apparent molecular weight and limited proteolysis patterns.[73] The *envZ* clone was shown to complement *envZ* mutations in *E. coli*.[74] However, Lijestrom et al.[74] reported that *envZ* and *ompR* are transcribed in opposite directions in *S. typhimurium* indicating that the expression of *ompB* in *Salmonella* is markedly different from that in *E. coli*. These authors further suggest that the synthesis of the *ompR* gene product is regulated by *envZ* rather than the multimerization.

4. *Other Regulatory Factors*

Although media-dependent fluctuation and cAMP effects on matrix protein synthesis are mediated by *ompB*, for several other factors which influence porin production,

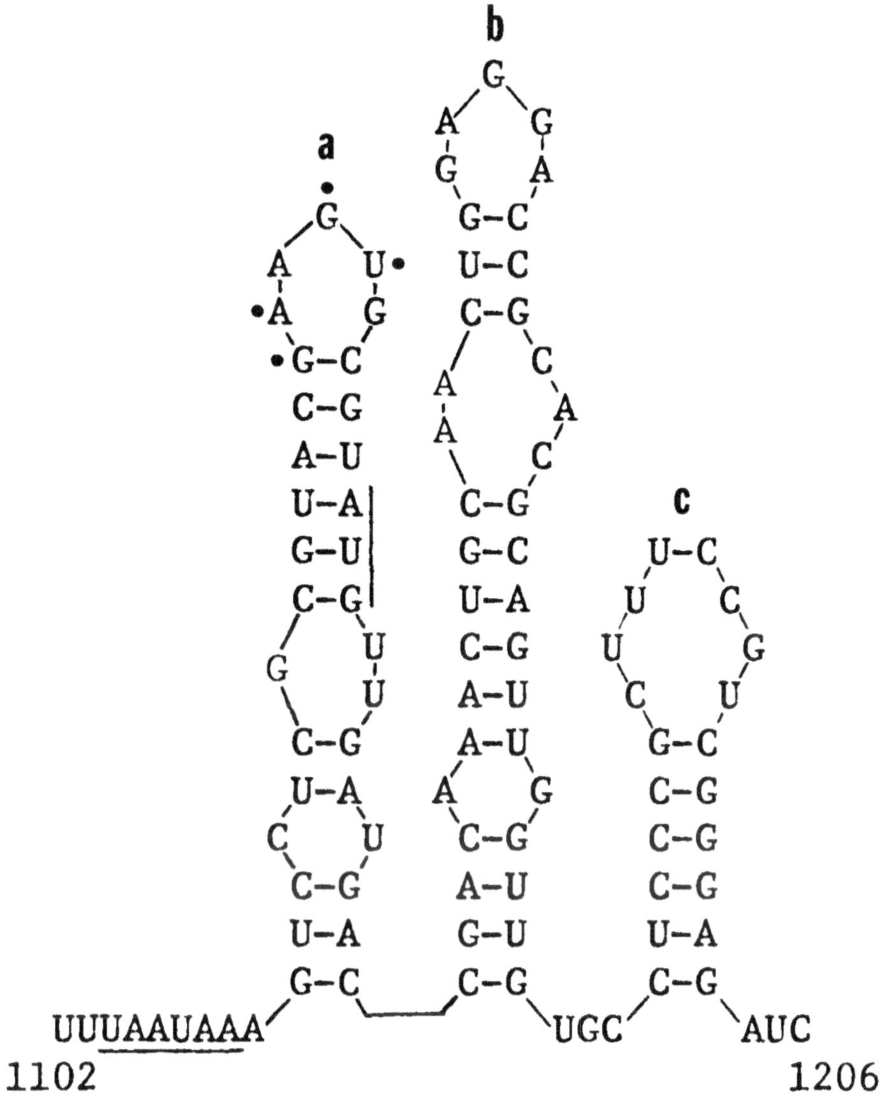

FIGURE 5. Possible secondary structure in the translation initiation region of the *envZ* mRNA. The translation termination codons for *ompF* are underlined. The line to the right of structure a highlights the *envZ* translation initiation codon. Calculated ΔG values are (a) −8.9, (b) −11.1, and (c) −13.4 kcal.

the *ompB* association is absent or unclear. These include the growth temperature,[109,216] the state and structure of membrane lipids[217] and lipopolysaccharide,[109,196] lysogeny by phage PA-2,[33,218,219] and the presence of local anesthetics.[220,221] Lundrigan and Earhardt[216] recently reported that cells grown at 27°C contain more *ompF* protein (and receptor protein) and less OmpC protein than cells grown at 37°C. These observations are consistent with an earlier report by Lugtenberg et al.[109] concerning the effect of temperature on OmpF protein synthesis. Using a mutant which did not exhibit this temperature effect, a functionally complementing region of the chromosome was cloned.[216] In minicells, this clone produced a 28,000-dalton polypeptide at 27°C, but not at 37°C. These data suggest the presence of a gene involved in the temperature-dependent regulation of the matrix proteins. Apparently this gene, located adjacent to *ompT*, maps well away from *ompB* and may exert an independent effect.

The biosynthesis of the matrix proteins can be negatively influenced by membrane fluidity[217] or by alterations in the structure of LPS.[109] Curiously, mutants with defective LPS synthesize large amounts of β-galactosidase from *ompF-lacZ* or *ompC-lacZ* fusions.[196] The relationship of these findings to *ompB*-mediated regulation has not been investigated. Another outer-membrane component that influences matrix protein production is the OmpA protein. When one of these proteins is absent, sometimes there is compensatory production of another.[34,109,222] This effect may be related to the greatly reduced amounts of the matrix proteins found in the membranes of plasmid-bearing cells which overproduce *OmpA* protein.[90,175] Perhaps the biosynthesis and stable assembly of these proteins in the outer membrane involves a common limiting step.

Matrix protein levels are also influenced by lysogenic conversion. When *E. coli* is infected by bacteriophage PA-2, OmpC protein and OmpF protein are supplanted by a new porin called protein 2.[33] Gene-fusion experiments have shown that lysogeny by PA-2 does not affect transcription of *ompC* or transcription or translation of *ompF*.[47,57] Yet some mutants, isolated for their defects in protein 2 production, are unable to produce either OmpC protein or OmpF protein.[219] This report indicates that further investigation of lysogenic conversion could provide important information about matrix protein regulation.

Local anesthetics like procaine[221] or phenethyl alcohol[220] can exert still another regulatory effect on porin protein synthesis. Pages and Lazdunski[221] found that low concentrations of these reagents affect the transcription of *ompF* and *ompC*. The fact that the effects of procaine can be used to isolate *envZ* mutants[211] may indicate that the drug response is *ompB*-mediated. A protease exposed to external surface of the outer membrane has been identified by Cavard et al.[223] which cleaves colicin A into two fragments. The finding that this colicin protease (Cpr) is inhibited by procaine or high salt concentrations prompted the isolation of a *cpr* mutant.[54] Cavard et al.[54] have proposed that the *cpr* mutation, which resembles an *envZ* mutation, is itself a mutation in *envZ* or affects *envZ* expression. However, it is unlikely that the *cpr* mutation is in the *envZ* gene because it cannot be complemented with a plasmid containing the entire *ompB* operon.

D. Other Proteins

Certain cultural conditions increase the number of major proteins found in the *E. coli* outer membrane. PhoE protein and λ receptor (LamB) protein become major outer membrane proteins in low-phosphate media or when maltose serves as the carbon source, respectively. During growth at high temperature, OmpT protein is found in the cell envelope in amounts comparable to other major outer-membrane proteins. The isolation and characterization of the genes encoding each of these outer-membrane proteins have made important contributions to our understanding of the mechanisms by which they are expressed.

1. λ Receptor Protein

The DNA sequence of the entire structural gene for λ receptor protein *lamB* has been determined[224] and is shown in Figure 2G. Although λ receptor is a functional porin, the *lamB* gene structure does not appear to be related to the *ompC*, *ompF*, and *phoE* genes. The preferential use of certain codons can be observed in *lamB*, but not to the extent of most of the major outer-membrane proteins shown in Table 1. Another notable feature of the *lamB* gene is the absence of sequences having the potential to form a rho-independent transcription termination structure following the protein-coding region. However, other secondary structures may form in the *lamB* mRNA beyond the translation termination codon partially due to the presence of the repeated sequences located from positions 1446 to 1484 and from 1537 to 1575. At present, it is not known whether transcription continues beyond this region through the adjacent *molA* gene, which may encode a secretory protein. The coding sequences of *lamB* and *molA* are separated by 170 base pairs so if this region is transcribed, its conformation could foreseeably influence the expression of either or both genes.[224]

The *lamB* gene is expressed as part of the *malK-lamB* operon from a promoter located more than 1 kilobase 5′ of the sequence shown in Figure 2G.[206] Like the other two operons in the maltose regulon, induction of *malK-lamB* is mediated by the *malT* gene product.[114,225] The MalT protein is thought to activate transcription when maltose is present. The expression of *lamB* is also positively regulated by CRP which, in the presence of cAMP, enhances the transcription of both the *malT* gene and the *malK-lamB* operon.[206,226] Recently, Bedouelle and Hofnung[227] determined the DNA sequence of the promoter regions of the *malK-lamB* operon and the adjacent transcriptionally divergent operon *malEFG*. Each of the operons was found to contain a putative MalT protein binding site "GGAGGAT" upstream of an RNA polymerase recognition site. These sequences appear to be joined by four possible CRP binding sites and, therefore, the operons may be regulated coordinately.[206]

2. PhoE Protein

As mentioned previously, the determination of the DNA sequence of the *phoE* structural gene[205] (shown in Figure 2H) revealed that the primary structure of PhoE protein is quite homologous to the matrix proteins, OmpC protein and OmpF protein. When PhoE protein and OmpF protein are compared, 212 amino acid residues (64%) are found to be identical. Including those residues which are functionally identical brings the total to 268 or 81%. Similarly, 210 amino acid residues (63%) are conserved between PhoE and OmpC protein (79% or 262, including the functionally identical residues). A close comparison of the primary structures of PhoE, OmpF, and OmpC proteins has enabled Mizuno et al.[204] to classify portions of the polypeptide chains into categories based on whether they are similar in all three proteins, similar in only two of the proteins or different in each protein. Perhaps the most interesting finding of this analysis was that the regions which are composed of different kinds of amino acids in each of the three proteins are basic in nature in PhoE protein. This property may contribute to the anion specificity of the PhoE pore.

The control region of the *phoE* gene is much more homologous to *phoA* than to *ompC* and *ompF*. This is expected since *phoE* and *phoA* are regulated by the *phoB* gene product while *ompC* and *ompF* are controlled by the products of the *ompB* operon. A comparison of *phoA* and *phoE* has focused attention towards the three homologous regions as possible interaction sites. The first is from positions 180 to 186 in Figure 2H (which is preceded by a fairly homologous region from positions 148 to 157), the second is from position 205 to 236, and the third is from 136 to 159. Overbeeke et al.[205] have speculated that the latter region may be recognized by the *phoB* gene product because it is followed by a sequence homologous to the Pribnow box from positions 175 to 180, and transcription of *phoA* is proposed to initiate in a homologous position.[228] A *phoA* mutation resulting in *phoB* independent expression also occurs in this region.[229] Beyond the *phoE* coding sequence, a possible transcription termination site can be found. The mRNA corresponding to this region (from 1340 to 1376) can form a very stable stem and loop structure but it is not followed by the poly-U characteristic of rho-independent terminators.

The regulation of *phoE* and *phoA* is fairly complex because the expression of the regulatory gene *phoB* appears to be regulated by the products of the two other genes, *phoM* and *phoR*. Recently, Tommassen et al.[67] proposed a model, based on the available data, to explain how *phoE* and other genes of the *pho* regulon are induced by phosphate limitation. According to their model, the *phoB* gene product is the positive activator necessary for *phoE* expression. The *phoB* gene is controlled by the bifunctional *phoR* gene product which represses *phoB* in the presence of inorganic phosphate and induces it during phosphate limitation. Another regulatory protein, the *phoM* gene product, can also activate *phoB* expression. Under high-phosphate conditions, this protein must compete with the PhoR repressor for the same binding site resulting in a limitation of *phoB* expression. The model explains that induction of PhoE protein and the other proteins of the *pho* regulon occurs when phosphate levels are low because the positive activator, PhoB protein, is induced by the gene products of *phoR* (activator form) and *phoM.* The closely linked *phoB* and *phoR* genes have recently been cloned and their products have been identified[67] so this model can soon be tested.

3. OmpT Protein

Less is known about the expression of the OmpT protein, which is a major outer-membrane protein in cells grown at elevated temperatures. Although the *ompT* gene has been cloned,[53] the DNA sequence has not yet been reported. Gayda et al.[230] have presented evidence that temperature influences OmpT protein synthesis at a post-transcriptional level. Their data indicate that at low temperature the OmpT precursor protein is processed to a 20,500-mol-wt form instead of the 40,000-mol-wt mature form of OmpT protein which is the product of a trypsin-like cleavage at 37°C.

Several observations suggest that the synthesis of capsular polysaccharide and OmpT protein are related.[108] In *capR* strains which produce large quantities of capsular polysaccharides, processing of the OmpT protein precursor is decreased compared to the wild type. Another interesting finding is that plasmids containing the *ompT* gene can supress the mucoid phenotype of *capR* strains. The expression of the plasmid-encoded *ompT* gene appears to be necessary to inhibit capsular polysaccharide synthesis in such mutants. If the synthesis of mature OmpT protein (40,000-mol-wt form) is important to repress capsular polysaccharide synthesis, insufficient OmpT precursor processing at low temperatures may explain why wild type *E. coli* are mucoid under such conditions.[108]

The exact role of *ompT* gene product is somewhat obscure. OmpT protein appears to process the ferric enterobacitin receptor to a lower-molecular-weight form in vitro[231] but the relationship of this finding to the in vivo role of the protein is unknown. The isolation and characterization of *ompT* point mutants may lead to a better understanding of the physiological significance of this activity. This coupled with the determination of the *ompT* DNA sequence should provide insight into the unique expression and processing of OmpT protein.

IV. POST-TRANSCRIPTIONAL EVENTS AND FINAL ASSEMBLY

A current understanding of how the biosynthesis of outer membrane proteins is regulated at the level of the gene and how nucleic acid structure may control or contribute to gene expression was presented in the preceding sections of this chapter. This section will discuss the post-transcriptional events involved in outer-membrane protein biosynthesis and assembly. These events can be generally categorized as (1) translation on membrane-bound (associated) polyribosomes, (2) translocation across the cytoplasmic membrane, (3) proteolytic processing of the precursor to the mature form of the protein, (4) post-translational modification (in some cases), or (5) insertion into the outer-membrane structure and possible post-assembly reorientation. Since these processes can proceed at an extremely rapid rate, it is often difficult to experimentally determine whether some of them occur simultaneously, as coupled reactions, rather than in a strict chronological order where the completion of one step is required for the next stage of synthesis to begin.

A. Preliminary Considerations Concerning the Early Stages of Outer-Membrane Protein Biosynthesis

An interesting approach in the study of outer-membrane protein biosynthesis was to examine the effects of ribosomal-directed antibiotics on the biosynthesis of outer-membrane and cytoplasmic proteins.[147] Distinct differences were found between the two classes of proteins suggesting that polyribosomes may be compartmentalized in the bacterial cell on the basis of function. It was also noted that the synthesis of the outer-membrane proteins was generally more resistant to the mRNA-directed antibiotic rifampicin, indicating that the mRNAs that code for these proteins have relatively long half-lives when compared to those mRNAs coding for cytoplasmic proteins. The stability of the outer-membrane protein mRNAs could then account for the abundance of several of these proteins, such as the matrix proteins, OmpA protein, and lipoprotein. In Section III, the roles of the secondary structures of these mRNAs were discussed in relation to mRNA stability and expression. However, the increased mRNA stability represents only one unique aspect of outer-membrane protein biosynthesis. The former observation concerning the differential effects of ribosomal-mediated antibiotics suggests that inherent differences also exist between polyribosome populations. Based on these data and earlier ribosomal work in eukaryotic[232,233] as well as prokaryotic[234,235] systems, Randall and Hardy[236] were able to separate bacterial polyribosomes physically into two classes, those that were associated with the cytoplasmic membrane and those that were found free in the cytoplasm. In an in vitro system the membrane-bound polyribosome-enriched fraction directed the synthesis of predominantly secretory proteins while the membrane-free fraction was found to produce primarily cytoplasmic proteins.

The detailed investigation of outer-membrane protein biosynthesis led to a possible explanation of polyribosomal differentiation in bacterial cells. When examining toluene-treated cells, which require an exogenous supply of adenosine triphosphate (ATP) for protein synthesis, it was found that membrane proteins were exclusively synthesized[237] and immunoprecipitation of the proteins accumulated in these cells with anti-lipoprotein serum revealed the presence of a new, higher-molecular-weight form of the lipoprotein.[238] Peptide mapping and double-labeling experiments, using radioactive amino acids, confirmed that this new protein was compositionally and structurally similar to the mature lipoprotein except for the presence of a peptide extension at the amino-terminal end of the protein. When this new protein was compared to a similar protein which was synthesized in a cell-free, mRNA-directed system, the two proteins were found to be identical.[239,240] Sequence analysis of the amino-terminal portion of the protein synthesized in the cell-free system and comparison to the known amino acid sequence of the mature lipoprotein[93] established that this new protein represented a precursor lipoprotein, designated prolipoprotein, having an extra peptide sequence of 20 amino acid residues at the amino-terminus.[240] The structure of the prolipoprotein will be discussed in more detail later in this section.

The results obtained by specific immunoprecipitation of toluene-treated cells with an anti-matrix protein serum and utilization of the fact that the OmpA protein is heat modifiable[22,77] indicated that these major outer-membrane proteins also accumulated as higher-molecular-weight precursors.[241] Subsequent to the above study, by using cells treated with phenethylalcohol (PEA),[220] it was verified that the matrix and OmpA proteins are synthesized as precursor molecules. Since both the PEA and membrane fluidity changes represent milder treatments of the cells as compared to the toluene system, it was possible to show a specific in vivo precursor-product relationship between the pro-matrix and pro-OmpA proteins and their corresponding mature products by employing radioactive pulse-chase experiments.[217,220] These studies confirmed that each of the primary translation products of the major outer-membrane protein genes of *E. coli* is a higher-molecular-weight precursor form.

The inherent structure of outer-membrane precursor proteins is considered to be one of the determining factors in ribosomal differentiation. However, the primary role of the extra peptide sequence still remains to be determined. Is this portion of the protein responsible for (1) the recognition of the secretory machinery, (2) the direct binding of the ribosome to the membrane, (3) directing translocation of the protein across the membrane, or (4) a combination of several or all of these properties? Since it is apparent that the structure of the precursor protein plays such an important role, it would be worthwhile to review what is known about the structure and properties of these molecules. It has now been demonstrated by protein characterization and/or DNA sequence analysis that virtually all exported bacterial proteins are synthesized as higher-molecular-weight precursors containing amino-terminal peptide extensions. Table 2 shows the amino acid sequences of the extra peptides (signal peptides) determined to date. Several structural homologies are evident among the signal sequences listed. The amino-terminal section is basic in nature and positively charged due to the presence of one to three basic amino acids. Following this basic region, the content of hydrophobic amino acids increases significantly. This hydrophobic core, which contains 10 to 15 residues, is followed by a serine and/or threonine residue. The amino acid just preceding the cleavage site is, with one exception, either glycine or alanine. Finally, a glycine or proline residue is located within the hydrophobic domain in most signal peptides. The functional significance of these common features will be discussed later.

B. Translocation across the Cytoplasmic Membrane

1. Translocation Models

One of the most difficult problems to approach in the study of membrane protein synthesis, yet one of the most intriguing, is the mechanism by which proteins are secreted across the membrane. Although some elegant studies have begun to provide information on some of the detailed events which are involved, the basic overall mechanism is still speculative.

The "signal hypothesis" was the first model describing the events in the translocation of proteins across membranes in eukaryotic systems. However, it has features which are directly applicable to the prokaryotic system.[232,242,243] This model suggests that in prokaryotes, the hydrophobic leader sequence of the secretory precursor protein directs the polyribosome to bind to a receptor at the inner surface of the cytoplasmic membrane which initiates or stimulates the aggregation of a peptide-transport channel. The polyribosome itself then binds directly to a membrane-receptor protein which is part of the newly formed transport channel and the nascent peptide is pushed through the channel by the force generated by the elongation process.

An alternative model describing the function of the signal peptide and the translocation event has been termed the "membrane trigger hypothesis".[244,245] Briefly, this model is based on the inherent ability of the cytoplasmic membrane phospholipids to trigger or stimulate the elongating peptide to adopt the proper conformation for insertion into the unit bilayer structure. The signal sequence is proposed to provide the stimulus for the proper folding of the protein. Thus, the presence of the signal sequence allows the elongating peptide to initally conform to a secondary structure which is adapted to a hydrophilic environment such as the cytoplasm. Membrane recognition in this model is based on this accquired conformation which is influenced or directed by the signal sequence. The membrane receptor can be protein or lipid in nature or can be a property of the physical state of the membrane itself; that is the hydrophobic environment.

Table 2
SIGNAL SEQUENCES OF PROKARYOTIC SECRETORY PROTEINS

Gram-negative bacteria[a]

		Ref.
Outer-membrane proteins		
Lipoprotein	MetLysAlaThrLysLeuValLeuGlyAlaValIleLeuGlySerThrLeuLeuAlaGly▼CysSer −20 … +1	106
Lipoprotein (*S. marcescens*)	MetAsnArgThrLysLeuValLeuGlyAlaValIleLeuGlySerHisSerAlaGly▼CysSer −19 … +1	168
Lipoprotein (*E. amylovora*)	MetAsnArgThrLysLeuValLeuGlyAlaValIleLeuGlySerThrLeuLeuAlaGly▼CysSer −20 … +1	169
Lipoprotein (*M. morganii*)	MetGlyArgSerLysIleValLeuGlyAlaValValLeuAlaSerAlaLeuLeuAlaGly▼CysSer −20 … +1	170
OmpA protein	MetLysLysThrAlaIleAlaIleAlaValAlaLeuAlaGlyPheAlaThrValAlaGlnAla▼AlaPro −21 … +1	174
OmpA protein (*S. dysenteriae*)	MetLysLysThrAlaIleAlaIleThrValAlaLeuAlaGlyPheAlaThrValAlaGlnAla▼AlaPro −21 … +1	194
λ Receptor protein	MetMetIleThrLeuArgLysLeuProLeuAlaValAlaValAlaAlaGlyValMetSerAlaGlnAlaMetAla▼ValAsp −25 … +1	349
PhoE protein	MetLysLysSerThrLeuAlaLeuValValMetGlyIleValAlaSerAlaSerValGlnAla▼AlaGlu −21 … +1	205
OmpF protein	MetMetLysArgAsnIleLeuAlaValIleValProAlaLeuLeuValAlaGlyThrAlaAsnAla▼AlaGlu −22 … +1	350
OmpC protein	MetLysValLysValLeuSerLeuLeuValProAlaLeuLeuValAlaGlyAlaAlaAsnAla▼AlaGlu −21 … +1	201
Periplasmic proteins		
Alkaline phosphatase	MetLysGlnSerThrIleAlaLeuAlaLeuLeuProLeuLeuPheThrProValThrLysAla▼ArgThr −21 … +1	229
AmpC β-lactamase	MetPheLysThrThrLeuCysAlaLeuLeuIleThrAlaSerCysSerThrPheAla▼AlaPro −19 … +1	351
TEM β-lactamase	MetSerIleGlnHisPheArgValAlaLeuIleProPhePheAlaAlaPheCysLeuProValPheAla▼HisPro −23 … +1	352
Arabinose-binding protein	MetLys ? ThrLysLeuValLeuGlyAlaValIleLeuThrAlaGlyLeuSer ? GlyAla ? Ala▼GluAsn −23 … +1	353
Lysine-arginine-ornithine binding protein	MetLysLysThrValLeuAlaLeuSerLeuLeuIleGlyLeuGlyAlaThrAlaAlaSerTyrAla▼AlaLeu −22 … +1	354

Table 2 (continued)
SIGNAL SEQUENCES OF PROKARYOTIC SECRETORY PROTEINS

Protein	Signal sequence	Ref.
Gram-negative bacteria[a]		
Histidine-binding protein	MetLysLysLeuAlaLeuSerLeuSerLeuValLeuAlaPheSerSerAlaThrAlaAlaPheAla▼AlaIle −22 … +1	354
Leucine-binding protein	MetLysAlaAsnAlaLysThrIleIleAlaGlyMetIleAlaLeuAlaIleSerHisThrAlaMetAla▼AspAsp −23 … +1	355
Isoleucine-valine-binding protein	MetAsnIleLysGlyLysAlaLeuLeuAlaGlyCysIleAlaLeuAlaPheSerAsnMetAlaLeuAla▼ −23	356
Maltose-binding protein	MetLysIleLysThrGlyAlaArgIleLeuAlaLeuSerAlaLeuThrThrMetMetPheSerAlaSerAlaLeuAla▼LysIle −26 … +1	357
Viral proteins		
Major coat protein (phage M13)	MetLysLysSerLeuValLeuLysAlaSerValAlaValAlaThrLeuValProMetLeuSerPheAla▼AlaGlu −23 … +1	358
Minor coat protein (phage M13)	MetLysLysLeuLeuPheAlaIleProLeuValValProPheTyrSerHisSer▼AlaGlu −18 … +1	359
Enterotoxins		
Heat labile toxin, subunit B	MetAsnLysValLysCysTyrValLeuPheThrAlaLeuLeuSerSerLeuTyrAlaHisGly▼AlaPro −21 … +1	360
Heat labile toxin, subunit A	MetLysAsnIleThrPheIlePhePheIleLeuLeuAlaSerProLeuTyrAla▼AsnGly −18 … +1	361
Gram-positive bacteria		
Penicillinase *(B. Licheniformis)*	MetLysLeuTrpPheSerThrLeuLysLeuLysLysAlaAlaAlaValLeuLeuPheSerCysValAlaLeuAlaGly▼CysAla −26 … +1	362
β-Lactamase (*S. aureus*)	MetLysLysLeuIlePheLeuIleValIleAlaLeuValLeuSerAlaCysAsnSerAsnSerSerHisAla▼LysGlu −24 … +1	363
α-Amylase (*B. amyloliquefaciens*) (*B. amyloliquefaciens*)	MetIleGlnLysArgLysArgThrValSerPheArgLeuValLeuMetCysThrLeuLeuPheValSerLeuProIleThrLysThrSerAla▼ValAsn −31 … +1	364

[a] Proteins are from *E. coli* unless otherwise noted

In both of the hypotheses, the functions of the signal peptide are not well defined and very obscure. Inouye and co-workers[7,13,246] have proposed a possible function of the signal peptide for protein secretion, now called the "loop model", which is in contrast to the "linear translocation model" originally proposed in the signal hypothesis. Briefly, in this model as the nascent peptide is elongating, the positively charged amino terminus of the molecule recognizes the inner surface of the cytoplasmic membrane which is negatively charged due to the presence of phosphatidyl glycerol and interacts ionically with it. The next segment of the signal peptide is extremely hydrophobic; consequently it can be readily inserted into the membrane bilayer. The signal peptide can bend or loop at the glycine or proline residue while the amino-terminal segment remains attached as an anchor point. The elongating peptide is then secreted through the cytoplasmic membrane as a loop rather than in a linear fashion.

A series of studies has been performed in order to examine the nature of ribosomal attachment and secretion of precursor proteins across the cytoplasmic membrane. The results obtained from these analyses tend to support some of the proposed aspects of the models discussed above. In the first of these studies Smith et al.[247] were able to demonstrate that in spheroplasts of *E. coli*, nascent alkaline phosphatase (a periplasmic enzyme) peptides are secreted across the cytoplasmic membrane as growing peptide chains and that the polyribosomes which direct the synthesis of this protein are functionally attached to the membrane through peptidyl tRNA. This result was extended by the same group with a later finding which indicated that the nascent chain is the only source of attachment of most or all polyribosomes to the membrane.[248] When isolated membrane-polyribosome complexes were treated with puromycin, 75% of the polyribosomes were released. The freed polyribosomes were found to still contain bound mRNA so it seems unlikely that membrane attachment is directly mediated by the mRNA. Curiously, a small population of membrane-bound polyribosomes was resistant to very high levels of puromycin. It is not known whether this is an artifact or if it indicates the presence of a class of polyribosomes involved in the synthesis of secreted proteins. The *E. coli* puromycin-resistant polyribosomes do not appear to be related to the class of puromycin-resistant membrane-bound ribosomes found in eukaryotic cells[249] since the former are not released by salt.[248] The relationship between the fraction of *E. coli* polyribosomes not released from the membrane with puromycin, and the puromycin-resistant synthesis of outer-membrane proteins[147,250] has not been investigated. In another study using pronase-treated *Bacillus subtilis* protoplasts, Smith et al.[251] found that the length of the protected nascent peptides corresponded to the thickness of the cytoplasmic membrane. In addition, the removal of the external section of the elongating peptides during the pronase digestion did not result in the release of the polyribosome-nascent peptide complex. From these findings the authors suggested that (1) the membrane-bound polyribosome in bacteria is attached to the cytoplasmic membrane only through the elongating precursor peptide, (2) the processes of translation and secretion coincide, and (3) the membrane plays an active role in anchoring the elongating peptide during translation. However, in a recent report, Smith[252] presented data which may indicate that proteins located on the cytoplasmic surface of the inner membrane may be required for polyribosome attachment.

Over the past few years, several periplasmic[247,253] and two outer-membrane proteins[253,254] have now been shown to be synthesized on membrane-bound polyribosomes isolated from *E. coli*. Co-translational secretion in other bacterial systems has also been demonstrated.[255] However, the assumption that the mere presence of a signal sequence is sufficient to dictate synthesis on membrane-bound polyribosomes and/or co-translational secretion should be viewed with caution. Wickner and co-workers[256] have shown that the signal bacteriophage M13 coat protein is preferentially synthesized on free polyribosomes and can be post-translationally incorporated into the cytoplasmic membrane. It has also been suggested that both co- and post-translational translocation of outer-membrane proteins can occur.[257] Furthermore, the studies presented above do not conclusively discount the existence of ribosomal binding proteins, analogous to the eukaryotic "ribophorins",[258] in bacteria. Some of the genetic studies reviewed later in this chapter suggest that certain outer-membrane proteins cannot be co-translationally translocated in the absence of specific protein factors.[133] In eukaryotic cells, it has been established that a specific ribonucleoprotein complex, containing a 7S RNA species, functions to recognize and guide the emerging nascent signal sequence along with its associated polyribosome to a specific receptor at the site of translocation on the endoplasmic reticulum.[259,260] An analogous system in prokaryotes has not been described as yet, but its existence cannot be discounted.

Another approach to study the role of the signal peptide is to isolate mutants which are defective in secretion of outer-membrane proteins. A mutation in the signal peptide region of prolipoprotein changes the amino acid at position −7 (see Table 2) from glycine to aspartic acid.[102,261] Only 60% of the mutant prolipoprotein was found to be located in the outer membrane while 20% was found in each of the cytoplasmic membrane and soluble protein fractions. An interpretation of these results suggests that the mutation inhibited proteolytic processing of the precursor but only slightly affected translocation. Apparently the single amino acid alteration from a neutral to a charged residue is not enough to affect the translocation mechanism. In separate studies, a series of mutants has been isolated which affect the translocation and assembly of the λ receptor protein. The characterization of these mutants and the methodology employed in their isolation has been reviewed elsewhere[262] and will only be briefly discussed here. DNA sequencing revealed that in all cases alterations in the signal peptide sequence had occurred. It is especially interesting to note that the chosen selection technique (where mutants which block the export of a secretion-jamming fusion protein were selected for) led to the isolation of many identical point mutations. These mutations result in the substitution of a charged amino acid for one of four hydrophobic or weakly hydrophilic amino acids as follows: at position −12 valine is replaced by aspartic acid, at position −11 alanine is replaced by glutaminc acid, at position −10 alanine is replaced by glutamic acid, or at position −7 methionine is replaced by arginine (Table 2). Deletion mutations were also isolated which could effect the overall hydrophobicity of the signal peptide or its secondary structure. All of the described mutations result in the accumulation of the pro-λ receptor in the cytoplasm. These mutations and similar ones isolated in the genes encoding periplasmic proteins[6] lend support to the contention that the hydrophobic region of the signal peptide is important in the process of translocation. Since only a small number of residues are affected, the conformation of the hydrophobic core may be more important than the overall hydrophobicity.[262] Recently, von Heijne[263] pointed out that signal sequences are not uniformly hydrophobic. Instead, the more hydrophobic amino acids appear to be concentrated in the central region. It was suggested that this property may give rise to a U-shaped signal sequence configuration consistent with the loop model. If this is true, then three of the four λ receptor protein signal sequence point mutations may block export by disrupting loop formation.

The importance of the positively charged amino-terminal portion of the signal peptide has also been studied by mutant isolation and characterization. An extensive analysis of the amino terminus was recently reported by Inouye et al.[264] and Vlasuk et al.[265] These investigators used mutants isolated by oligonucleotide-guided site-specific mutagenesis to study the influence of amino-terminal charge on the early stages of propoprotein secretion. From their study they were able to make three conclusions. First, the synthesis of prolipoprotein is facilitated by the presence of a positively charged amino terminus. Second, the existence of basic amino acids in the amino terminus is not absolutely required for the translocation of prolipoprotein across the cytoplasmic membrane. Finally, a net negative charge at the amino terminus causes an initial cytoplasmic accumulation of prolipoprotein which is translocated across the cytoplasmic membrane post-translationally at a rate which is dependent on the number of positive charges in this region.

The results of another study provide further evidence that the charged amino terminus plays an important role in the initial interaction occurring between the signal sequence and the membrane. Hall et al.[266] characterized a mutation in the λ receptor protein signal sequence which reduces the charge of the amino terminus from +2 to +1. They found that this alteration decreases the overall synthesis of the λ receptor protein by 75%. Since this mutation does not appear to affect transcription or translation initiation, it was suggested that the arginine residue at position −20 (which is changed to serine in the mutant) is involved in the coupling of the synthesis of receptor protein to export. Thus they propose that a "stop translation" sequence is synthesized early in λ receptor protein gene product which blocks translation until the signal melts into the membrane. They feel that protein synthesis is impeded in the aforementioned mutant because the altered signal sequence can only facilitate this interaction with 25% efficiency. A similar model could also explain the reduction in lipoprotein synthesis resulting from mutations in the basic region of the signal peptide.[264,265] Perhaps these effects are mediated in a manner similar to that described for the translation arrest of eukaryotic secretory protein synthesis by the signal recognition particle (SRP).[267]

The importance of a functional signal sequence is stressed most in translocation models while the roles of the mature protein sequence are not clear. In the membrane trigger hypothesis, where translation is completed before translocation begins, the structure of the entire precursor protein is thought to be important to achieve the necessary tertiary conformation to facilitate export. For proteins secreted according to the linear or loop models, progressive folding of the newly transported portion of mature protein outside of the cytoplasmic membrane may function to pull the remaining carboxy-terminal region through the bilayer. For this task it would appear that the amino terminal part of the processed (mature) protein may play an important role.

During the cloning of the *ompA* gene, Movva et al.[76] showed that OmpA protein could still function as a phage receptor in the outer membrane when 96 carboxy-terminal amino acids were removed and replaced by four residues coded for by vector sequences. Bremer et al.[268] have also shown the effect of carboxy-terminal deletions on OmpA protein localization. They found that *OmpA* protein fragments missing up to 122 carboxy-terminal amino acids can be efficiently incorporated into the outer membrane. This implies that the entire *OmpA* protein is not required for translocation. Besides OmpA protein, many other secretory proteins such as the lipoprotein[269] do not require their carboxy-terminal regions for translocation. Koshland and Botstein[270] originally reported that the carboxy-terminal region of β-lactamase is required for the protein to cross the inner membrane. Later this finding was revised when it was shown that translocation can take place without the carboxy terminus,[271,272] but the release of β-lactamase from the outer surface of the cytoplasmic membrane does not occur.[271]

Thus far, all identified *E. coli* signal peptides are present as amino-terminal extensions. However, if proteins are secreted according to the loop model, there is no obligatory requirement of the signal sequence to be at the amino terminus. Therefore, one way to obtain supporting evidence for the loop model is to examine whether a protein can be secreted across the membrane even if the signal sequence is internalized within the protein. Recently, Coleman et al.[273] constructed a β-galactosidase-prolipoprotein fusion protein in order to determine the functionality of an internalized signal sequence. Mature lipoprotein was found in the membrane of strains producing the fusion protein even though the signal sequence is 145 amino acids from the amino terminus.

Similar results were obtained with prolipoprotein-prolipoprotein and pro-β-lactamase-prolipoprotein fusions.[273] Further evidence that an internalized signal sequence can function comes from the behavior of a β-galactosidase-preproinsulin fusion protein in *E. coli.* Talmadge et al.[274] have shown that pro-insulin secretion through the cytoplasmic membrane can be mediated by a signal peptide beginning 18 amino acids from the amino terminus of the fusion protein. It is difficult to imagine the translocation of these proteins proceeding in a linear fashion, especially in the first case, where up to 145 amino acids derived from a soluble protein would have to pass through the membrane before the signal peptide can emerge. These results can most easily be explained by the loop model because with the amino-terminal end of the internalized signal peptide remaining on the inside of the cytoplasmic membrane along with any soluble peptide sequences, the secretory protein could be translocated and processed normally.

2. Secretory Machinery

As described in Section II, mutational analysis has been used to identify several genetic loci involved in secretion events. The two which have been most thoroughly characterized are *prlA* and *secA*. A temperature-sensitive *secA* mutant has been isolated which accumulates at least two periplasmic proteins, alkaline phosphatase and maltose binding protein, as well as some outer-membrane proteins.[133] The *secA* gene product has been identified as a 92,000-mol-wt polypeptide[134] which appears to be peripherally associated with the cytoplasmic membrane.[275] Curiously, the synthesis of SecA protein increases in response to blocked secretion. Oliver and Beckwith[275] have suggested that components of the secretory apparatus like SecA protein may be regulated by the secretory needs of the cell. They further suggest that the *secA* gene product may be involved in the coupling of translation and secretion because, for maltose-binding protein biosynthesis, SecA protein appears to be required for both processes.

Emr et al.[276] identified the *prlA* gene by isolating suppressor mutations which allow the transport of a λ receptor protein with a defective signal sequence. These mutations can suppress the export defects caused by signal sequence mutations in both *lamB* and *malE.*[137] The *prlA* mutations exhibit allelic specificity indicating that a direct interaction between the *prlA* gene product and the signal peptides may occur. The *prlA* gene has recently been mapped with the *spc* ribosomal operon.[277] It lies promoter distal of the last known ribosomal gene and is not thought to code for a ribosomal protein. Although the synthesis and export of normal secretory proteins is not defective in *prlA* strains, *secA prlA* double mutants grow poorly and appear mucoid.[137] The presence of both mutations may overpower the export apparatus even at the permissive temperature.

Independently, Ito et al.[138] have identified a gene called *secY* which they feel is identical to *prlA*. Both genes map in similar locations in the *spc* operon. In contrast to the *prlA* mutations, the *secY* lesion is a temperature-sensitive conditional lethal mutation. This suggests that the *secY* gene product is essential. The *secY* mutation appears to more severely alter the export machinery than do the *prlA* mutations. At the nonpermissive temperature, the processing of OmpA protein, OmpF protein, and maltose binding protein is hampered in *secY* mutants. Like SecA protein, the *secY* gene product is probably directly involved in the export process. The identification and characterization of *secA* and *prlA (secY)* is just the first step in defining the components required for the translocation of proteins across the inner membrane. It will be interesting to see if there are functional correlations between these gene products and the components of the eukaryotic secretory apparatus such as the SRP and the docking protein.[259]

3. Energy for Translocation

Another aspect of the translocation mechanism currently under investigation is the energy source. Most available evidence points to a role for the membrane potential in the export of outer-membrane proteins.[257,278,279] Pro-OmpA protein has been found to accumulate in the cytoplasm of cells treated with the uncoupler carbonyl cyanide *M*-chlorophenylhydrazone (CCCP).[257] CCCP-induced accumulation of the secretory precursors of OmpF protein and λ receptor protein has also been reported, but in these cases it was not determined if translocation, cleavage, or both is inhibited.[279] Lipoprotein translocation, however, is not sensitive to CCCP[279] unless the presence of a mutation forces it to proceed post-translationally by way of a soluble intermediate[265] in a manner similar to that described for M13 coat protein.[280] CCCP treatment prevents the processing of several periplasmic protein in a manner similar to that observed for the outer-membrane proteins.[278,279]

Interpreting uncoupler-mediated inhibition as conclusive evidence that the proton motive force is the required energy source for protein export should be done with caution. CCCP is known to dissipate the membrane potential; yet it can also perturb membrane structure and therefore this compound may exert multiple indirect effects. Studies using an *E. coli* mutant *(uncA)* unable to maintain the proton motive force have demonstrated that dissipating the membrane potential without the use of CCCP can inhibit the processing of the precursor of maltose-binding protein.[279] This effect does not appear to be related to cellular ATP levels.

Since the membrane potential can be thought to orient the signal peptide and/or the remainder of the protein within the membrane, a role for the proton motive force can be accommodated by the loop model, the membrane trigger hypothesis, or the linear model. Alternatively, in the membrane trigger hypothesis, energy may be derived from protein-protein and protein-lipid interactions within the membrane through the process of self-assembly.[244] Although this would seem to be a reasonable mechanism for the insertion of cytoplasmic membrane proteins, it is difficult to envision how an amphipathic protein destined for assembly in the outer membrane would pass through the cytoplasmic membrane. Another argument against the process of self-assembly supplying the necessary energy for translocation can be obtained from experiments involving membrane perturbation. In cells treated with PEA, the pro-matrix protein is synthesized and accumulated external to the cytoplasmic membrane indicating that the translocation of this percursor protein was unaffected.[220] In another study, when the fluidity of the membrane was greatly reduced, the pro-OmpA protein was synthesized and also localized external to the cytoplasmic membrane.[217] According to the conditions used in each of these experiments, the physical state of the membrane would have been vastly different than the normal or equilibrium state. Integral membrane proteins and phospholipids would tend to segregate during reduced fluidity and the entire membrane would become more rigid in terms of fatty-acid conformation.[281-284] Any of these changes in the physical state of the membrane would surely affect the self-assembly or conformation of the translocating protein, yet these outer membrane proteins are still secreted through the membrane.

C. Cleavage of the Signal Peptide

1. Temporal Mode

After the initiation of translocation, proteins destined to be localized in the outer membrane are processed to their mature forms by specific signal peptidases. Processing does not appear to be necessary for translocation across the cytoplasmic membrane. As presented above, when outer-membrane protein biosynthesis is examined after PEA treatment[220] or following a reduction in membrane fluidity,[217] the pro-matrix and pro-OmpA proteins, respectively, are found to accumulate exterior to the cytoplasmic membrane, indicating that translocation can be completed when processing is blocked. In addition, several mutant major outer-membrane proteins can be secreted even though their processing is inhibited.[102,261,285,286] However, in wild-type cells under physiological conditions, cleavage of the signal peptide and translocation may occur in concerted fashion.

Perturbing the membrane with PEA or decreasing the membrane fluidity also separates the translation process from the proteolytic removal of the signal peptide. Post-translational processing of the secretory precursors which accumulate under these special conditions can be observed when PEA is removed or when the membrane is restored to a more fluid state.[217,220] The OmpA protein precursor which accumulates in the presence of CCCP can also be processed after the uncoupler is inactivated with 2-mercaptoethanol.[257] Crowlesmith and Gamon[287] have demonstrated the post-translational processing of OmpF protein and OmpA protein precursors at low temperature. The studies of mutant lipoproteins provide still other examples of precursor processing after the completion of translation.[285,286]

Yet many of the aforementioned systems or approaches were designed to dissect the steps involved in outer-membrane protein biogenesis[288] and therefore may separate events that occur simultaneously in the cell. To determine when the signal sequences of several proteins are removed in vivo, Josefsson and Randall[289] designed a system to monitor precursor processing during translation.

For each protein analyzed, these authors isolated a pulse-labeled population of precursor, mature, and nascent polypeptide species. The amino termini of each species were then characterized by partial proteolysis and two-dimensional gel electrophoresis. From the results obtained it was possible to determine what percentage (if any) of the protein was processed co-translationally. They were also able to deduce the critical length a nascent chain would have to achieve before processing could be initiated. Using this method, Josefsson and Randall analyzed the processing modes of two outer-membrane proteins, OmpA protein and λ receptor protein, and five periplasmic proteins. One periplasmic protein, AmpC β-lactamase was found to be processed entirely co-translationally while TEM β-lactamase was found to be processed post-translationally. Both modes of processing were exhibited by each of the other proteins studied. In most cases, polypeptide elongation to 80% of full length was required to facilitate signal peptide cleavage. Approximately 60% of the OmpA protein was processed co-translationally. Only nascent polypeptides 30,000 daltons or larger were found to be processed, indicating that OmpA polypeptides chains must elongate to 86% of full length before cleavage can occur. Furthermore, most of the λ receptor protein was shown to be processed post-translationally and removal of the signal peptide was a relatively late event.

Since different modes of processing can be observed for proteins destined to be localized in the periplasm and because most proteins appear to be processed both co-translationally and post-translationally, the authors suggest that the mode of processing is not involved in the final localization into cell envelope compartments. They offer three explanations for the "critical molecular weight" required for processing:

1. A critical size may be required for recognition by the signal peptidase.
2. Perhaps a sequence in the polypeptide or mRNA at the critical point must be present or form a structure for protease activation.
3. The critical length may be necessary to provide access to the signal peptidase (i.e., to facilitate translocation).

2. Signal Peptidases

To further understand the complexities involved in the removal of the signal peptide, an effort has been made to characterize the signal peptidases. To date only two signal peptidases have been identified in *E. coli*. The different temporal modes of cleavage and the PEA[220] and membrane fluidity[217] studies described previously provide indirect evidence that differences exist in either the specificity of the processing enzymes or in the mechanism of processing itself; that is, the availability or conformation of the proprotein substrate. For instance, when cells were treated with 0.3% PEA, pro-matrix protein was specifically accumulated while the inhibition of processing of pro-OmpA protein did not occur until the PEA concentration reached 0.4%.[220] Higher concentrations of PEA (up to 1.0%) failed to cause the inhibition of the processing of prolipoprotein. A similar differential inhibition of precursor processing was also observed by DiRienzo and Inouye.[217] During conditions of reduced membrane fluidity pro-OmpA protein was accumulated. However, at no time could the appearance of pro-matrix protein or prolipoprotein be detected.

Wickner's group has shown that the purified leader peptidase (M13 pro-coat protein signal peptidase), known to convert M13 procoat to coat protein, can process the precursor of OmpA protein in vitro.[257] It may also cleave other outer-membrane protein precursors to their mature forms (unpublished data cited in References 290, 291, and 292). Furthermore, some periplasmic proteins can be cleaved by leader peptidase.[245,293] It is interesting to note that the leader peptidase activity fractionates with both the inner and outer membranes in about equal amounts.[292] Whether this enzyme exists in both membranes or in the zones of adhesion[294] between them is unclear.[292] Each of the proteins known to be processed by leader peptidase has alanine-X at the cleavage site[291] (with X representing an amino acid with a small side chain). At present, it is not known what determines the substrate specificity of this enzyme.

The leader peptidase gene *(lep)* has recently been cloned into pBR322 by Date and Wickner.[295] Genetic evidence described elsewhere [296] indicates that the leader peptidase is an essential enzyme. Two regions of DNA, approximately 1.3 and 0.35 kb in length, are required for leader peptidase expression.[295,296] These are called the L and the P regions, respectively. The L region apparently corresponds to the structural gene[296] which can be expressed from the P_R promoter in the absence of the P region.[290] The P region cannot function in the opposite orientation and is therefore assumed to act in *cis* towards the L region. Date[296] has suggested that it serves as the promoter of the *lep* gene. The function of the 1.5-kb stretch of DNA which separates the L and the P regions is unknown, but leader peptidase synthesis still occurs in its absence. The *lep* gene maps between 54 and 55 min on the *E. coli* chromosome (see Figure 1).[297] Further characterization of the *lep* gene and the isolation of *lep* mutants should broaden our understanding of the role of leader peptidase in outer-membrane protein maturation.

Using an in vitro assay, Tokunaga and co-workers have shown the *E. coli* lipoprotein precursor is not cleaved by leader peptidase[291] even though it contains the sequence alanine-glycine just before the known cleavage site (see Table 1). Prolipoprotein is processed by another signal peptidase referred to as prolipoprotein signal peptidase (PLP signal peptidase).[291,298] PLP signal peptidase and leader peptidase can be distinguished by many properties. Unlike leader peptidase, PLP signal peptidase is active at pH 5 and sensitive to the cyclic antibiotic globomycin.[291,299] The two enzymes also appear to be compartmentalized differently. PLP signal peptidase fractionates with the inner membrane[291] and as mentioned above, leader peptidase fractionates with both the inner and the outer membranes. PLP signal peptidase is specific for the glycine-cysteine cleavage site in prolipoprotein and may cleave some of the recently identified new minor lipoproteins[300] and the peptidoglycan associated lipoprotein (PAL)[301-303] in a similar manner. For cleavage to occur in vitro, prolipoprotein must first be modified at the cysteine residue with glycerol and two fatty acids.[304] This glyceride modification will be described in more detail in Section IV.D. When *E. coli* cells are treated with globomycin, the glyceride modified precursor forms of the major lipoprotein PAL and the new minor lipoproteins accumulate in the cell envelope.[300] This could be explained by the inhibition of a single globomycin-sensitive signal peptidase species. Further evidence that the PLP signal peptidase processes this diverse set of lipoproteins comes from the analysis of a temperature-sensitive PLP signal peptidase mutant.[305] This mutant is not only defective in processing the major lipoprotein but many minor ones as well. The isolation of such mutants is an important approach to gain a better understanding of the relationship among the PLP signal peptidase, the leader peptidase, and any other peptidases involved in cleavage of the signal peptides from outer membrane protein precursors.

3. Secretory Protein Mutations Affecting Cleavage

In order to determine what residues are required for processing by PLP signal peptidase, Inouye et al.[285,286] created several mutants using site-specific mutagenesis (for a review of this technique see Reference 105). Changing the cysteine (at the +1 position as indicated in Table 2) at the cleavage site to glycine results in the accumulation of unmodified prolipoprotein in the outer membrane,[286] while substituting alanine for the glycine (at the −1 position) has no effect.[285] Yet, creating the same alanine-cysteine cleavage site by deleting rather than changing the glycine (at the −1 position) completely eliminates the modification (see Section IV.D) and subsequent cleavage of the mutant lipoprotein precursor.[285] This suggests that in addition to the cysteine at +1, the region of the signal peptide from −1 to −6 is important for proper modification of prolipoprotein because the deletion of the glycine at −7 or −12 caused no defect[285] in this process (although replacement of the glycine at −7 with aspartic acid is inhibitory).[102]

Mutational alterations in other secretory proteins support the contention that processing enzymes recognize more than just a cleavage site. A mutant λ receptor protein missing 12 amino acids from the central portion of the signal peptide is not properly processed (when the export block is removed by a *prlA* mutation) although an intact cleavage site is present.[137] Mutations located upstream of the β-lactamase cleavage site have also been characterized which slow or block maturation.[306] Last, the cleavage defects resulting from the combined effect of three amino acid substitutions (glutamic acid to leucine at +2, proline to serine at +6, and asparagine to aspartic acid at +12) in the mature portion of M13 coat protein suggests that sequences beyond the cleavage site may also be important in vivo.[307] This mutant is particularly interesting because the altered procoat is cleaved efficiently in vitro by purified leader peptidase.[308]

D. Post-Translational Modification

The post-transcriptional events involved in the biogenesis of most outer-membrane proteins can be adequately described in terms of the mechanisms involved in translation, translocation across the cytoplasmic membrane, signal peptide cleavage (processing), and assembly into the outer membrane. However, a meaningful discussion of lipoprotein maturation cannot be limited to these events because this protein must undergo extensive post-translational modification. The amino-terminal cysteine of the mature lipoprotein is thioether-linked to diglyceride and amide-linked to a fatty acid. One third of these molecules are covalently attached to the peptidoglycan by the amino group of the carboxy-terminal lysine.[15] The chemical events involved in lipoprotein modification can be broken up into the steps outlined in Figure 6. In this pathway, steps 1 and 2 precede the proteolytic removal of the signal peptide (step 3) discussed in the last section while steps 4 and 5 follow this event. The processing and modification of lipoprotein have been recently reviewed by Wu.[309]

The primary translation product of the lipoprotein mRNA has been designated prolipoprotein I. This unmodified prolipoprotein has been synthesized in vitro and extensively characterized.[240] Several mutations which block the first modification

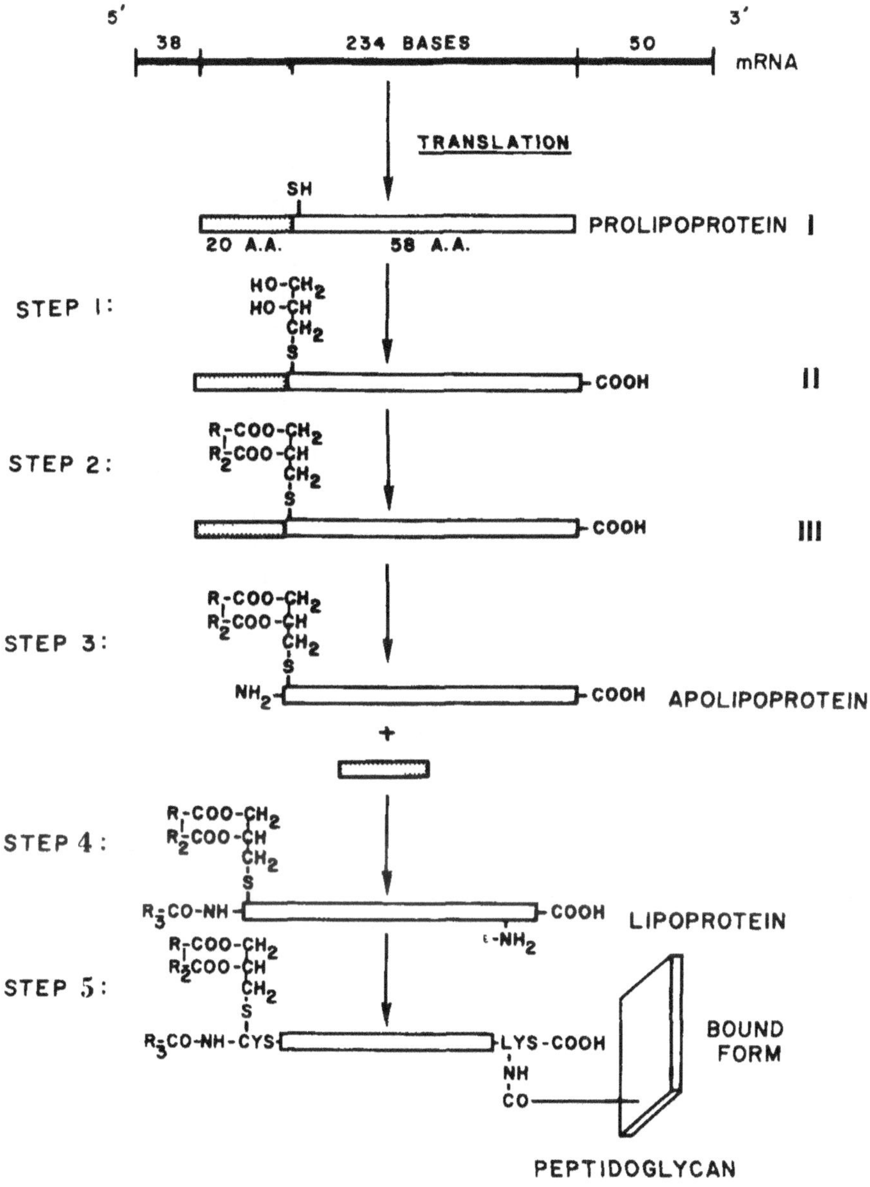

FIGURE 6. A proposed mechanism for the biogenesis of lipoprotein. All of the lipoprotein precursors shown have been detected with the exception of prolipoprotein II.

step[285,286,310] cause the accumulation of prolipoprotein I as does the presence of a particular *malE-lacZ* fusion.[304] The first modification step appears to involve the transfer of glycerol from phosphatidylglycerol[311,312] to prolipoprotein I forming prolipoprotein II. Making use of the different pH profiles of the glyceryltransferase and signal peptidase activities in an in vitro system, Tokunaga et al.[304] demonstrated that the cleavage of the signal peptide required the prior synthesis of the glyceride-containing prolipoprotein III. Since a mutant has been isolated with a free[313] or partially free[96] glycerylcysteine, it seems likely that prolipoprotein II is the first modified intermediate synthesized. Prolipoprotein II has so far escaped detection in vitro and in vivo apparently because it is short lived. Phospholipids serve as donors for the subsequent rapid acylation of this species, converting it to the glyceride modified form, prolipoprotein III,[314,315] as shown in step 2.

When the in vivo processing of prolipoprotein is inhibited by the antibiotic globomycin, the intermediate which accumulates is prolipoprotein III.[316] Similarly, globomycin in vitro does not affect modification steps 1 and 2. It is likely that globomycin competitively inhibits the PLP signal peptidase which catalyzes step 3 because of its hydrophobic nature and structural similarity to the prolipoprotein signal peptide. All known globomycin-resistant mutants do not synthesize lipoprotein[317] or contain lipid deficient prolipoprotein.[310] No mutants have been isolated which produce an unmodified, cleaved lipoprotein species. These observations further support the correctness of the scheme depicted in Figure 6 where the glyceride modification must precede cleavage. Accordingly, the *lpp* mutations described in Section IV.C.3 which allow only unmodified prolipoprotein synthesis may block the cleavage step by altering a prolipoprotein determinant required for recognition by glyceryltransferase.

The apolipoprotein product of the step 3 cleavage reaction has been identified and characterized by Inukai et al.[318] and Hussain et al.[298] These groups showed that apolipoprotein could be observed when lipoprotein was overproduced[318] or when membranes containing prolipoprotein III were incubated at high temperature.[298,318] Presumably under these conditions the acyltransferase activity is insufficient to catalyze the step 4 reaction. As expected, the apolipoprotein migrates slightly faster than the mature fully modified lipoprotein on SDS polyacrylamide gels. The results of sequential Edman degradation of the apolipoprotein species were consistent with the direct cleavage of prolipoprotein III by PLP signal peptidase between the glycine (at the −1 position) and the cysteine (at the +1 position) as shown in Table 2, generating a free amino terminus.[318] The apolipoprotein was shown to be a true intermediate in lipoprotein biosynthesis rather than a breakdown product because it could be chased into mature lipoprotein and its production could be blocked by globomycin.[318] At high temperatures, the accumulation of apolipoprotein was accompanied by the appearance of a smaller-molecular-weight polypeptide assumed to be the intact signal peptide.[298]

In the last step of lipoprotein biosynthesis, the free amino group of the glyceride-cysteine at the amino terminal of apolipoprotein is acylated by a fatty acid. An additional post-translational modification of the lipoprotein is the covalent attachment of one third of the protein (bound-form) to the peptidogylcan[15] as shown in step 5. By employing radiolabeling pulse and pulse-chase techniques it was determined that the free form of the lipoprotein is synthesized first and then converted to the bound form.[15] Interruption of protein synthesis by amino acid starvation or treatment with chloramphenicol and the inhibition of energy production did not affect the conversion process. On this basis it was proposed that the reaction proceeds by transpeptidation where (1) the terminal D-alanine on the peptidoglycan cross bridge is cleaved by carboxypeptidase I, and (2) the remaining D-alanine is then cleaved by carboxypeptidase II followed by the attachment of the lipoprotein to the diaminopimelic acid residue.[8] Kinetic experiments performed by Wensink and Witholt[319] showed that the conversion of pulse-labeled free-form lipoprotein to bound form was 50% complete within one sixth of a generation time and 90% complete within half a generation time. Bicyclomycin, a penicillin derivative, appears to selectively inhibit the biosynthesis of lipoprotein, especially the bound form; however, it is not known which stage this antibiotic specifically affects.[320] Braun and co-workers have indicated that the lipoprotein is bound very slowly to newly inserted peptidoglycan subunits and the growth of the peptidoglycan layer proceeds at sites where there is no bound lipoprotein.[321,322] Somewhat different results were obtained by Wensink et al.[323] who were unable to detect differences between peptidoglycan at the lipoprotein attachment sites and the rest of the peptidoglycan. Their data imply that growth of the peptidoglycan and lipoprotein insertion are not necessarily related.

Some *S. typhimurium* mutants have been isolated that contain extremely low levels of the bound form of the lipoprotein.[324] At the time of septum formation in these cells, the outer membrane cannot properly invaginate, although the normal ingrowth of the cytoplasmic membrane and peptidoglycan layer occurs. The absence of the bound form of lipoprotein may also give rise to increased detachment of excess outer-membrane material in the form of vesicles. This idea is consistent with the finding that outer-membrane vesicles released from *E. coli* during normal growth have less free-form lipoprotein than expected and very little bound form.[325] The role of the bound form of the lipoprotein may be to anchor the outer membrane to the peptidoglycan.[326]

In terms of the post-translational modification of other outer-membrane proteins there has been some previous indication of the presence of glycosyl residues or unusual amino acid derivatives. It has been reported that the matrix protein(s) contain(s) 1 mol of 2-amino-2-deoxyglucose per mole of peptide[327] and may also contain α-amino adipic acid and δ-semialdehyde (allysine).[328] However, both of these findings were refuted when the amino acid sequence of a matrix protein (OmpF protein from *E. coli* B/r) was determined.[329] This does not discount contradictions which can arise due to strain differences or the fact that only a small proportion of the molecules are modified. It has also been reported that the OmpA protein contains 2 mol of reducing sugar per mole of protein.[330] It is likely that this originated from the LPS known to complex with OmpA protein.[331] Amino acid sequencing of the OmpA protein revealed no evidence of glycosylation.[179]

E. Outer-Membrane Protein Assembly

Following the process of translocation across the cytoplasmic membrane, some determining factor is necessary to direct the protein to the outer membrane. It is not possible that the signal peptide sequence can guide the protein to the outer membrane, since there are no apparent major structural differences among the extra peptide sequences of cytoplasmic membrane, periplasmic, and outer-membrane proteins as well as among eukaryotic and prokaryotic membrane precursor proteins.[13]

The LPS,[87] and outer-membrane-specific component, may play an essential role in the assembly of the outer-membrane proteins. However, the following results indicate that at least concomitant synthesis of outer-membrane proteins and LPS is not required for outer-membrane protein assembly. Under conditions of reduced membrane fluidity, normal LPS synthesis is inhibited and a newly synthesized lipid A precursor accumulates.[332] The lipid A precursor as well as the pro-OmpA protein in this system can each be chased into their mature forms following the restoration of a normal fluid state. However, the coincident addition of metabolic uncouplers such as potassium cyanide or sodium arsenate inhibit the conversion of the LPS precursor to its mature form while processing the pro-OmpA protein and assembly of newly formed OmpA protein is unaffected, suggesting that this protein is translocated and assembled independently of any LPS. In addition, the rate of the chase of pro-OmpA to mature OmpA protein is much more rapid than the chase of LPS precursor to mature LPS. These kinetics of conversion would argue against molecular co-assembly. It is not possible to determine in this system whether this situation also exists for the other outer-membrane proteins.

Turning towards the role of the secretory protein, it is possible that structural intermediates exist which recognize the outer membrane and must then be altered for proper assembly. Evidence for outer membrane protein intermediates can be obtained by examining the results from cells treated with PEA.[220] When cells are treated with concentrations of PEA that were lower than those that resulted in the inhibition of processing, specific processed outer-membrane proteins accumulate in the periplasm rather than in the outer membrane. Specifically, at 0.3% PEA the OmpA protein and the lipoprotein are localized within the periplasmic space and upon the removal of PEA the OmpA protein can be chased into the outer-membrane fraction. The trypsin sensitivity of the soluble lipoprotein that accumulates in the periplasm suggests that this protein was not fully modified with fatty acid.[96,220,333] The fact that these soluble intermediates are as yet unassembled indicates that the signal peptide sequence does not necessarily play a role in outer-membrane recognition or assembly. It should be noted that PEA may also affect the structure or nature of the outer membrane so that assembly of the proteins cannot occur until the PEA is removed, thus returning the membrane to its normal state.

Silhavy and co-workers have attempted to examine the assembly process through genetic means.[334] They fused portions of a gene encoding an outer-membrane protein, λ receptor protein, to portions of the *lacZ* gene encoding the cytoplasmic enzyme β-galactosidase. Hybrid proteins were formed which retained enzymatic activity. The highlights of this work and its possible implications are presented below. For a more detailed discussion, the reader is referred to previous reviews.[9,115]

Four classes of fusions were obtained. The class I fusions produce hybrid proteins which remain entirely in the cytoplasm. One of the most extensively studied fusion proteins in this group has been sequenced and found to contain only two amino acids coded for by *lamB*,[335] which explains its cytoplasmic location. Another type of fusion has been isolated which is similar to the class I fusions in that it remains in the cytoplasm. In this hybrid protein, designated a class II fusion, the signal sequence and 15 amino acids from the λ receptor protein are insufficient to export β-galactosidase from the cytoplasmic compartment. Class III fusions contain a larger portion of the λ receptor protein. Although they contain enough information to be exported to some degree, class III fusions are lethal to the cell when induced by maltose (maltose-sensitive). Class IV fusions, which contain more than half the *lamB* gene, are not as maltose sensitive as Class III fusions, presumably because they are more efficiently transported to the outer membrane.

Based on the behavior of the four different fusion classes, Silhavy, and co-workers[115] have proposed a model for the export of λ receptor protein. According to their model the λ receptor protein is translocated and processed co-translationally according to the loop model until the ribosome translates a "dissociation-sequence". This sequence causes the release of the ribosome from the inner membrane which stops vectoral transfer of the growing polypeptide. The absence of such a sequence can be used to explain the maltose-sensitive phenotype of class III fusions because the continued translocation of β-galactosidase sequence appears to jam the cellular export machinery.[115] The class IV fusions are thought to contain the "dissociation sequence" which the export machinery of the cell recognizes as a signal for outer-membrane localization. If only outer-membrane proteins have dissociation sequences, as Silhavy's group proposes, then fusions between periplasmic proteins and β-galactosidase should behave like class III fusions. This has been found for *malE-lacZ*[336] fusions and *phoA-lacZ*[337] fusion strains which are sensitive to maltose and low phosphate levels, respectively. One of the *malE-lacZ* fusions includes nearly all of the *malE* structural gene so it is unlikely that *malE* contains a dissociation sequence. Blobel[338] has proposed a role for similar "stop transfer" sequences in eukaryotic secretion.

The "dissociation sequence" renders the polypeptide imbedded in the cytoplasmic membrane with its amino terminus in the periplasm. Silhavy and co-workers[9,115] envision that the transfer of λ receptor protein to the outer membrane occurs by way of a vesicle intermediate such that in class IV fusions, β-galactosidase would never need to pass through the membrane. Without reorientation, proteins exported by this model would have their carboxy-terminal ends facing the extracellular environment. It should be pointed out that this arrangement is inconsistent with the known orientation of the fully assembled OmpA protein and the lipoprotein, suggesting that some aspects of the assembly model proposed for λ receptor are not easily applied to other outer-membrane proteins.

Proteins directed to the outer membrane by other mechanisms must be inserted into its hydrophobic matrix. Several studies demonstrate how this may occur. Electron microscopic observations of plasmolyzed Gram-negative bacteria has revealed the presence of localized areas of adhesion between the cytoplasmic and outer membrane. Approximately 200 to 400 such sites are usually found per cell.[294,339] These sites, referred to as Bayer's junctions, have been implicated in the translocation process of LPS,[340] capsular polysaccharide,[341] and most notably the porin proteins of *S. typhimurium*.[342] By utilizing the technique of labeling specific molecules with ferritin coupled to antibody and combining this labeling with pulse-chase experiments, it was demonstrated that newly inserted porin proteins are assembled as discrete patches which are uniformly distributed over the entire cell surface. These newly inserted protein molecules could be specifically labeled at zones of adhesion. On the basis of these results it was proposed that the porin proteins are inserted at the sites representing Bayer's junctions. Whether other-membrane proteins are also inserted into the outer membrane through Bayer's junctions remains to be seen. As inferred earlier, the λ receptor protein may be transported to the outer membrane by a different mechanism. It appears to be inserted into the cell envelope only during the final quarter of each generation and predominantly at the regions of septum formation.[343] Consequently, the synthesis of some proteins may be synthesized in a cell-cycle-dependent fashion and assembled by zonal growth as opposed to the diffuse intercalation mechanism proposed for the porin proteins.[342,343]

Presumably cleavage of the signal peptide facilitates the release of outer-membrane proteins from the cytoplasmic membrane. Yet in some cases, such as that of prolipoprotein I, the affinity for the outer membrane appears to be sufficient to dissociate the signal sequence from the cytoplasmic membrane. As noted earlier, in *lpp* cleavage site mutants, unmodified prolipoprotein I can be assembled in the outer membrane.[285,286] This process cannot be completed when the glyceride-modified prolipoprotein III accumulates in the presence of globomycin. Prolipoprotein III appears to have been directed towards the outer membrane but it still maintains an attachment to the inner membrane.

Inukai and Inouye[344] used a variety of methods to determine the orientation of this intermediate in the cell envelope. They found that prolipoprotein III was associated with the cytoplasmic membrane when sucrose gradient fractionation was used. However, when the inner membrane was solubilized with detergent, prolipoprotein III was associated with the insoluble outer-membrane fraction. In addition, labeling experiments using ^{3}[H]-dinitrophenylfluorobenzene indicated that the two amino-terminal lysine residues were left in the cytoplasm. This indicates that prolipoprotein III which accumulates in the presence of globomycin has an affinity for the outer membrane and associates with it, but its signal sequence remains imbedded in the cytoplasmic membrane with the positively charged amino-terminal region in the cytoplasm as proposed in the loop model. In contrast to prolipoprotein I, prolipoprotein III is unable to pull its signal peptide out of the cytoplasmic membrane, most likely due to the presence of the glyceride moiety which may imbedded in the cytoplasmic membrane.[344]

Once a protein is inserted into the outer-membrane bilayer either by diffuse intercalation or by zonal growth it must then be correctly organized both spatially and functionally. For example, OmpA protein must be assembled in a transmembrane fashion with its amino-terminal half partly in the membrane and partly outside[345] and its carboxy-terminal portion in the periplasm.[268] Although fractionation techniques can indicate whether a mutant protein has been properly localized in the outer membrane, determining the exact topology of a protein in this compartment is more difficult. Protease sensitivity has been used as a tool to make this type of evaluation.[345,346] Inukai et al.[346] constructed a lipoprotein mutant which is properly modified, processed, and directed to the outer membrane. However, trypsin sensitivity differences and the deleterious effect of the mutant protein on cell growth lead to the conclusion that the mutant lipoprotein does not have the same arrangement in the outer membrane as the wild-type lipoprotein. These types of studies can indicate which domains of the protein are important for proper orientation.

As discussed above, lipoprotein maturation involves a series of modifications and a processing step. Several of the mutations engineered by Inouye et al.[285,286] to block all these steps (described in Section IV.C.3) still allow transport of unmodified prolipoprotein I to the outer membrane. However, the cell growth is inhibited when the synthesis of such defective proteins is induced. In the case of lipoprotein mutants, this lethality may be indicative of an abnormal spatial organization within the outer membrane. In the absence of mutations, many of the outer-membrane proteins have a strong association with other envelope macromolecules and this association could be the factor which determines their proper assembled orientation. On this basis, the outer-membrane proteins can be classified according to their interactions with either (1) peptidoglycan or (2) LPS.

The matrix proteins of *E. coli* are well known for their ability to tightly bind noncovalently to the peptidoglycan in vivo.[16,207] This binding has been shown to be due to ionic interactions between the matrix proteins and the peptidoglycan. It is interesting to note that the bound form of the lipoprotein plays a role in this interaction.[347] Both the bound and free forms of the lipoprotein have also shown an ionic association with the peptidoglycan and the matrix proteins.[347] These ionic interactions are thought to participate in the stabilization of the matrix protein pores.[7,8] Whether these interactions directly contribute to the organization and assembly of the matrix proteins and the lipoprotein is not known. However, the covalent attachment of lipoprotein to the peptidoglycan would appear to dictate a specific order to the structure of the outer membrane.

There are also apparently specific and important interactions between outer-membrane proteins and LPS. Yu and Mizushima[348] have shown that LPS stimulates the binding of matrix proteins to the peptidoglycan as well as the inactivation of bacteriophages by the matrix proteins.[84] The OmpA protein also demonstrates a functional association with LPS since it has been shown that LPS protects this protein from proteolytic attack,[331] and the presence of LPS is required for this protein to function as a bacteriophage receptor[83,84,331] and in F-pilus mediated conjugation.[82] These studies indicate that molecular interactions between the outer-membrane components are essential for the final structural orientation and assembly of these proteins as well as their biological function.

V. CONCLUSIONS

Monumental advancements have been made in elucidating the mechanisms of bacterial outer-membrane protein biosynthesis, regulation, and assembly since these proteins were first identified at the beginning of the last decade. As a comprehensive understanding of these general mechanisms emerges, it is becoming clear that basic similarities exist between the biosynthesis of prokaryotic outer-membrane proteins and eukaryotic secretory proteins. There are several aspects of outer-membrane protein biosynthesis that remain to be clarified and which may reveal fundamental differences or more similarities between the prokaryotic and eukaryotic situations. For instance, the roles of the genetically identified components of the bacterial export machinery have yet to be determined. Another controversial aspect of membrane protein biosynthesis is the source of energy which is required to drive the translocation machinery. It should also be determined how the outer-membrane proteins are directed to their final location. Increasing our knowledge in these areas should, in addition, provide the details of the translocation mechanism itself.

Although the study of bacterial membrane protein biosynthesis is no longer in its infancy, a great deal of work remains to be accomplished. As is expressed as the prevading theme of this chapter, the advent of molecular genetics and its application to membrane protein research has facilitated and will continue to facilitate the studies of membrane protein biosynthesis, regulation, and perhaps the most intriguing aspect of all, membrane differentiation.

ACKNOWLEDGMENTS

We would like to thank Donna Ryan and Laura Antonacci for expertly typing the manuscript. We are also grateful to George Vlasuk for critically reading this review.

REFERENCES

1. DePetris, S., Ultra structure of the cell wall of *Escherichia coli* and chemical nature of its constituent layers, *J. Ultrastruct. Res.*, 19, 45, 1967.
2. Murray, R. G. E., Steed, P., and Elson, H. H., The location of mucopeptide in sections of the cell wall of *Escherichia coli* and other gram-negative bacteria, *Can. J. Microbiol.*, 11, 547, 1965.
3. Funahara, Y. and Nikaido, H., Asymmetric localization of lipopolysaccharides on the outer membrane of *Salmonella typhimurium*, *J. Bacteriol.*, 141, 1463, 1980.
4. Smit, J., Kamio, Y., and Nikaido, H., Outer membrane of *Salmonella typhimurium*: chemical analysis and freeze-fracture studies with lipopolysaccharide mutants, *J. Bacteriol.*, 124, 942, 1975.
5. Reeves, P., The genetics of outer membrane proteins, in *Bacterial Outer Membranes. Biogenesis and Functions*, Inouye, M., Ed., John Wiley & Sons, New York, 1980, 255.
6. Michaelis, S. and Beckwith, J., Mechanism of incorporation of cell envelope proteins in *Escherichia coli*, *Annu. Rev. Microbiol.*, 36, 435, 1982.
7. DiRienzo, J. M., Nakamura, K., and Inouye, M., The outer membrane proteins of gram-negative bacteria: biosynthesis, assembly and functions, *Annu. Rev. Biochem.*, 47, 481, 1978.
8. Inouye, M., Lipoprotein of the outer membrane of *Escherichia coli*, in *Biomembranes*, Vol. 10, Mansen, L. A., Ed., Plenum Press, New York, 1979, 141.
9. Hall, M. N. and Silhavy, T. J., Genetic analysis of the major outer membrane proteins of *Escherichia coli*, *Annu. Rev. Genet.*, 15, 91, 1981.
10. Osborn, M. J. and Wu, H. C. P., Proteins of the outer membrane of gram-negative bacteria, *Annu. Rev. Microbiol.*, 34, 369, 1980.
11. Inouye, M., Lipoprotein from the bacterial outer membranes. Their gene structures and assembly mechanism, in *Membrane and Transport*, Vol. 1, Martonosi, A. N., Ed., Plenum Press, New York, 1982, 299.
12. Vlasuk, G. P., Ghrayeb, J., and Inouye, M., The major outer membrane lipoprotein of *Escherichia coli*: secretion, modification and processing, in *The Enzymes of Biological Membranes*, Vol. 2, 2nd ed., Martonosi, A., Ed., Plenum Press, New York, 1985, 309.
13. Inouye, M. and Halegoua, S., Secretion and membrane localization of proteins in *Escherichia coli*, *CRC Crit. Rev. Biochem.*, 7, 339, 1980.
14. Braun, V. and Rehn, K., Chemical characterization, special distribution and function of a lipoprotein (murein-lipoprotein) of the *E. coli* cell wall. The specific effect of trypsin on the membrane structure, *Eur. J. Biochem.*, 10, 426, 1969.
15. Inouye, M., Shaw, J., and Shen, C., The assembly of a structural lipoprotein in the envelope of *Escherichia coli*, *J. Biol. Chem.*, 247, 8154, 1972.
16. Rosenbusch, J. P., Characterization of the major envelope protein from *Escherichia coli*. Regular arrangement in the peptidoglycan and unusual dodecyl sulfate binding, *J. Biol. Chem.*, 249, 8019, 1974.

17. Bachmann, B. J. and Low, K. B., Linkage map of *Escherichia coli* K-12, Edition 6, *Microbiol. Rev.*, 44, 1, 1980.
18. Bassford, P. J., Jr., Diedrich, D. L., Schnaitman, C. A., and Reeves, P., Outer membrane proteins of *Escherichia coli*. VI. Protein alteration in bacteriophage-resistant mutants, *J. Bacteriol.*, 131, 608, 1977.
19. Inokuchi, K., Mutoh, N., Matsuyama, S., and Mizushima, S., Primary structure of the *ompF* gene that codes for a major outer membrane protein of *Escherichia coli* K-12, *Nucleic Acid Res.*, 10, 6957, 1982.
20. Nakae, T., Identification of the outer membrane protein of *E. coli* that produces transmembrane channels in reconstituted vesicle membranes, *Biochem. Biophys. Res. Commun.*, 71, 877, 1976.
21. Nikaido, H. and Rosenberg, E., Porin channels in *Escherichia coli:* studies with liposomes reconstituted from purified proteins, *J. Bacteriol.*, 153, 241, 1983.
22. Uemura, J. and Mizushima, S., Isolation of outer membrane proteins of *Escherichia coli* and their characterization on polyacrylamide gels, *Biochem. Biophys. Acta*, 413, 163, 1975.
23. Lugtenberg, B., Meyers, J., Peters, R., van der Hoek, P., and van Alphen, L., Electrophoretic resolution of the major outer membrane protein of *Escherichia coli* K12 into 4 bands, *FEBS Lett.* 58, 254, 1975.
24. Schmitges, C. J. and Henning, U., The major proteins of the *Escherichia coli* outer cell-envelope membrane. Heterogeneity of protein. I, *Eur. J. Biochem.* 63, 47, 1976.
25. Laemmli, U. K., Cleavage of structural proteins during the assembly of the head of bacteriophage T4, *Nature (London)*, 222, 680, 1970.
26. Henning, U., Schmidmayr, W., and Hindennach, I., Major proteins of the outer cell envelope membrane of *Escherichia coli* K12: multiple species of protein. I, *Mol. Gen. Genet.*, 154, 293, 1977.
27. Foulds, J., *tolF* locus in *Escherichia coli:* chromosomal location and relationship to loci *cmlB* and *tolD*, *J. Bacteriol.*, 128, 604, 1976.
28. Chai, T-J. and Foulds, J., *Escherichia coli* K12 *tolF* mutants: alterations in protein composition of the outer membrane, *J. Bacteriol.*, 130, 781, 1977.
29. Davies, J. K. and Reeves, P., Genetics of resistance to colicins in *E. Coli* K12: cross-resistance among colicins of group A, *J. Bacteriol.*, 123, 102, 1975.
30. Foulds, J. and Barrett, C., Characterization of *Escherichia coli* mutants tolerant to bacteriocin JF246: two new classes of tolerant mutants, *J. Bacteriol.*, 116, 885, 1973.
31. Lavori, M. and Mathieu, L. G., Isolation and partial characterization of an *Escherichia coli* mutant resistant to colicin A[12], *Can J. Microbiol.*, 21, 1595, 1975.
32. Diedrich, D. L., Summers, A. O., and Schnaitman, C. A., Outer membrane proteins of *E. coli*. V. Evidence that protein 1 and bacteriophage-directed protein 2 are different polypeptides, *J. Bacteriol.*, 131, 598, 1977.
33. Schnaitman, C. A., Smith, D., and Forn de Salsas, M., Temperate bacteriophage which causes the production of a new major outer membrane protein by *Escherichia coli*, *J. Virol.*, 15, 1121, 1975.
34. Henning, U. and Haller, I., Mutants of *Escherichia coli* K12 lacking all "major" proteins of the outer cell envelope membrane, *FEBS Lett.*, 55, 161, 1975.
35. Verhoef, C., de Graaf, P. J., and Lugtenberg, E. J. J., Mapping of a gene for a major outer membrane protein of *Escherichia coli* K12 with the aid of a newly isolated bacteriophage, *Mol. Gen. Genet.*, 150, 103, 1977.
36. Hantke, K., Major outer membrane proteins of *E. coli* K12 serve as receptors for the phage T2 (Protein la) and 434 (Protein lb), *Mol. Gen. Genet.* 164, 131, 1978.
37. Siden, I. and Boman, H. G., *Escherichia coli* mutants with an altered sensitivity to cecropin D, *J. Bacteriol.*, 154, 170, 1983.
38. Sarma, V. and Reeves, P., Genetic locus *(ompB)* affecting a major outer membrane protein in *Escherichia coli* K12, *J. Bacteriol.*, 132, 23, 1977.
39. Verhoef, C., Lugtenberg, B., van Boxtel, R., de Graaff, P., and Verheij, H., Genetics and biochemistry of the peptidoglycan-associated proteins b and c of *Escherichia coli* K12, *Mol. Gen. Genet.* 169, 137, 1979.
40. von Meyerburg, K., Transport-limited growth rates in a mutant of *Escherichia coli*, *J. Bacteriol.*, 107, 878, 1971.
41. Beacham, I. R., Haas, D., and Yagil, E., Mutants of *Escherichia coli* "cryptic" for certain periplasmic enzymes: evidence for an alteration of the outer membrane, *J. Bacteriol.*, 129, 1034, 1977.
42. Ichihara, S. and Mizushima, S., Characterization of major outer membrane proteins 0-8 and 0-9 *Escherichia coli* K12: evidence that structural gene for the two proteins are different, *J. Biochem. Tokyo*, 83, 1095, 1978.
43. Sato, T. and Yura, T., Chromosomal location and expression of the structural gene for major outer membrane protein Ia of *Escherichia coli* K12 and of the homologous gene of *Salmonella typhimurium*, *J. Bacteriol.*, 139, 468, 1979.

44. van Alphen, W., Lugtenberg, B., von Boxtel, R., Hack, A., Verhoef, C., and Havekas, L., *meoA* is the structural gene for the outer membrane protein c of *Escherichia coli* K12, *Mol. Gen. Genet.*, 169, 147, 1979.
45. Hall, M. N. and Silhavy, T. J., Transcriptional regulation of *Escherichia coli* K12 major outer membrane protein Ib, *J. Bacteriol.*, 140, 342, 1979.
46. Wandersman, C.,. Moreno, F., and Schwartz, M., Pleiotropic mutations rendering *Escherichia coli* K12 resistant to bacteriophage TP1, *J. Bacteriol.*, 143, 1374, 1981.
47. Hall, M. N. and Silhavy, T. J., Genetic analysis of the *ompB* locus in *Escherichia coli* K12, *J. Mol. Biol.*, 151, 1, 1981.
48. Ferenci, T. and Boos, W., The role of the *Escherichia coli* λ receptor in the transport of maltose and maltodextrins, *J. Supramol. Struct.*, 13, 101, 1980.
49. Szmelcman, S. and Hofnung, M., Maltose transport in *Escherichia coli* K12: involvement of the bacteriophage lambda receptor, *J. Bacteriol.*, 124, 112, 1975.
50. Nakae, T. and Ishii, J., Permeability properties of *Escherichia coli* outer membrane containing, pore-forming proteins: comparison between lambda receptor protein and porin for saccharide permeation, *J. Bacteriol.*, 142, 735, 1980.
51. Wanner, B., Sarthy, A., and Beckwith, J., *Escherichia coli* pleiotropic mutant that reduces amounts of several periplasmic and outer membrane proteins, *J. Bacteriol.*, 140, 229, 1979.
52. Lundrigan, M. and Earhart, C. F., Reduction in three iron-regulated outer membrane proteins and protein a by the *Escherichia coli* K12 *perA* mutation, *J. Bacteriol.*, 146, 804, 1981.
53. Rupprecht, K. R., Gordon, G., Lundrigan, M., Gayda, R. C., Markovitz, A., and Earhart, C., *ompT: Escherichia coli* K12 structural gene for protein a(3b), *J. Bacteriol.*, 153, 1104, 1983.
54. Cavard, D., Pages, J. M., and Lazdunski, C. J., A protease as a possible sensor of environmental conditions in the *E. coli* outer membrane, *Mol. Gen. Genet.*, 188, 508, 1982.
55. Whitney, E. M., The *tolC* locus in *Escherichia coli* K12, *Genetics*, 67, 39, 1971.
56. Morona, R. and Reeves, P., The *tolC* locus of *Escherichia coli* affects the expression of three major outer membrane proteins, *J. Bacteriol.*, 150, 1016, 1982.
57. Hall, M. and Silhavy, T., The *ompB* locus and the regulation of the major outer membrane proteins of *Escherichia coli* K12, *J. Mol. Biol.*, 146, 23, 1981.
58. Morona, R. and Reeves, P., Molecular cloning of the *tolC* locus of *Escherichia coli* K12 with the use of the transposon Tn10, *Mol. Gen. Genet.*, 184, 430, 1981.
59. Morona, R., Manning, P. A. and Reeves, P., Identification and characterization of the TolC protein, an outer membrane protein from *Escherichia coli*, *J. Bacteriol.*, 153, 693, 1983.
60. Foulds, J. and Chai, T.-J., New major outer membrane protein found in an *Escherichia coli tolF* mutant resistant to bacteriophage Tulb, *J. Bacteriol.*, 133, 1478, 1978.
61. Pugsley, A. P. and Schnaitman, C. A., Identification of three genes controlling production of new outer membrane pore proteins in *Escherichia coli* K12, *J. Bacteriol.*, 135, 1118, 1978.
62. Lee, D. R., Schnaitman, C. A., and Pugsley, A., Chemical heterogeneity of major outer membrane pore proteins of *Escherichia coli*, *J. Bacteriol.*, 138, 861, 1979.
63. Pugsley, A. P., Lee, D. R., and Schnaitman, C. A., Genes affecting the major outer-membrane proteins of *Escherichia coli* K12: mutations at *nmpA* and *nmpB*, *Mol. Gen. Genet.*, 177, 681, 1980.
64. Tommassen, J. and Lugtenberg, B., Localization of *phoE*, the structural gene for major outer membrane protein e of *Escherichia coli* K12, *J. Bacteriol.*, 147, 118, 1981.
65. Tommassen, J., Overduin, P., Lugtenberg, B., and Bergmans, H., Cloning of *phoE*, the structural gene for the *Escherichia coli* phosphate limitation-inducible outer membrane pore protein, *J. Bacteriol.*, 149, 668, 1982.
66. Tommassen, J. and Lugtenberg, B., Outer membrane protein e of *Escherichia coli* K12 is co-regulated with alkaline phosphatase, *J. Bacteriol.*, 143, 151, 1980.
67. Tommassen, J., deGeus, P., Lugtenberg, B., Hackett, J., and Reeves, P., Regulation of the *pho* regulon of *Escherichia coli* K12. Cloning of the regulatory genes *phoB* and *phoR* and identification of their gene products, *J. Mol. Biol.*, 157, 265, 1982.
68. Wanner, B. L. and Latterelle, P., Mutants affected in alkaline phosphatase expression: evidence for multiple positive regulators of the phosphate regulon in *Escherichia coli*, *Genetics*, 96, 353, 1980.
69. Argast, M., Ludtke, D., Silhavy, T. J., and Boos, W., A second transport system for sn-glycerol-3-phosphate in *Escherichia coli*, *J. Bacteriol.*, 136, 1070, 1978.
70. Argast, M. and Boos, W., Co-regulation in *Escherichia coli* of a novel transport system for *sn*-Glycerol-3-phosphate and outer membrane proteln Ic (e,E) with alkaline phosphatase and phosphate binding protein, *J. Bacteriol.*, 143, 142, 1980.
71. Nakae, T. and Ishii, J., Transmembrane permeability channels in vesicles reconstituted from single species of porins from *Salmonella typhimurium*, *J. Bacteriol.*, 133, 1412, 1978.
72. Sanderson, K. E. and Hartman, P. E., Linkage map of *Salmonella typhimurium*, Edition V, *Microbiol. Rev.*, 42, 471, 1978.

73. Lijestrom, P., Maattanen, P. L., and Palva, E. T., Cloning of the regulatory locus *ompB* of *Salmonella typhimurium* LT-2. I. Isolation of the *ompR* gene and identification of its gene product, *Mol. Gen. Genet.*, 188, 184, 1982.
74. Lijestrom, P., Maattanen, P. L., and Palva, E. T., Cloning of the regulatory locus *ompB* of *Salmonella typhimurium* LT-2. II. Identification of the *envZ* gene product, a protein involved in the expression of the porin proteins, *Mol. Gen. Genet.*, 188, 190, 1982.
75. Foulds, J., Chromosomal location of the *tolG* locus for tolerance to bacteriocin JF246 in *Escherichia coli* K12, *J. Bacteriol.*, 117, 1354, 1974.
76. Movva, R., Nakamura, K., and Inouye, M., Gene structure of the OmpA protein, a major surface protein of *E. coli* required for cell-cell interaction, *J. Mol. Biol.*, 143, 317, 1980.
77. Bragg, P. D. and Hou, C., Organization of proteins in the native and reformed outer membrane of *Escherichia coli, Biochim. Biophys. Acta*, 274, 478, 1972.
78. Garten, W. and Henning, U., Cell envelope and shape of *Escherichia coli* K12 isolation and preliminary characterization of the major ghost-membrane proteins, *Eur. J. Biochem.*, 47, 343, 1974.
79. Inouye, M. and Yee, M., Specific removal of proteins from the envelope of *Escherichia coli* by protease treatment, *J. Bacteriol.*, 112, 585, 1972.
80. Inouye, M. and Yee, M., Homogeneity of envelope proteins of *Escherichia coli* separated by gel electrophoresis in sodium dodecyl sulfate, *J. Bacteriol.*, 113, 304, 1973.
81. Schnaitman, C. A., Outer membrane proteins of *Escherichia coli.* III. Evidence that the major protein of *Escherichia coli* 0111 outer membrane consists of four distinct polypeptide species, *J. Bacteriol.*, 118, 442, 1974.
82. Schweizer, M. and Henning, U., Action of a major outer cell envelope membrane protein in conjugation of *Escherichia coli* K12, *J. Bacteriol.*, 129, 1651, 1977.
83. van Alphen, L., Havekes, L., and Lugtenberg, B., Major outer membrane protein D of *Escherichia coli* K12. Purification and *in vitro* activity of bacteriophage K3 and F pilus mediated conjugation, *FEBS Lett.*, 75, 285, 1977.
84. Datta, D. B., Arden, B., and Henning, U., Major proteins of the *Escherichia coli* outer cell envelope membrane as bacteriophage receptors, *J. Bacteriol.*, 131, 821, 1977.
85. Chai, T.-J. and Foulds, J., Demonstration of a missing outer membrane protein in *tolG* mutants of *E. coli, J. Mol. Biol.*, 85, 465, 1974.
86. Skurray, R. A., Hancock, R. E. W., and Reeves, P., Con⁻ mutants: class of mutants in *Escherichia coli* K12 lacking a major cell wall protein and defective in conjugation and adsorption of a bacteriophage, *J. Bacteriol.*, 119, 726, 1974.
87. Havekes, L. M. and Hoekstra, W. P. M., Characterization of an *Escherichia coli* F con⁻ mutant, *J. Bacteriol.*, 126, 593, 1976.
88. Manoil, C. and Rosenbusch, J., Conjunction-deficient mutants of *Escherichia coli* distinguish classes of functions of the outer membrane OmpA protein, *Mol. Gen. Genet.* 187, 148, 1982.
89. Henning, U., Hindennach, I., and Haller, I., The major proteins of the *Escherichia coli* outer cell envelope membrane: evidence for the structural gene protein II*, *FEBS Lett.*, 61, 46, 1976.
90. Datta, D. B., Kramer, C., and Henning, U., Diploidy for a structural gene specifying a major protein of the outer cell envelope membrane from *Escherichia coli* K12, *J. Bacteriol.*, 128, 834, 1976.
91. Stocker, B. A. D., Nurminen, M., and Makela, P. H., Mutants defective in the 33K outer membrane protein of *Salmonella typhimurium, J. Bacteriol.*, 139, 376, 1979.
92. Braun, V. and Bosch, V., Sequence of the murein-lipoprotein and the attachment site of the lipid, *Eur. J. Biochem.*, 28, 51, 1972.
93. Braun, V. and Bosch, V., Repetitive sequences in the murein-lipoprotein of the cell wall of *Escherichia coli, Proc. Natl. Acad. Sci. U.S.A.*, 69, 970, 1972.
94. Hantke, K. and Braun, V., Covalent binding of lipid to protein. Diglyceride and amide-linked fatty acid at the N-terminal end of the murein-lipoprotein of the *Escherichia coli* outer membrane, *Eur. J. Biochem.*, 34, 284, 1973.
95. Hirashima, A., Wu, H. C., Venkateswaran, P. C., and Inouye, M., Two forms of a structural lipoprotein in the envelope of *E. coli.* Further characterization of free form, *J. Biol. Chem.*, 248, 5654, 1973.
96. Inouye, S., Lee, N., Inouye, M., Wu, H. C., Suzuki, H., Nishimura, Y., Iketani, H., and Hirota, Y., Amino acid replacement in a mutant lipoprotein of the *Escherichia coli* outer membrane, *J. Bacteriol.*, 132, 308, 1977.
97. Suzuki, H., Nishimura, Y., Iketani, H., Campisi, J., Hirashima, A., Inouye, M., and Hirota, Y., Novel mutation that causes a structural change in a lipoprotein in the outer membrane of *Escherichia coli, J. Bacteriol.*, 127, 1494, 1976.
98. Hirota, Y., Suzuki, H., Nishimura, Y., and Yasuda, S., On the process of cellular division in *Escherichia coli:* a mutant of *E. coli* lacking a murein-lipoprotein, *Proc. Natl. Acad. Sci. U.S.A.*, 74, 1417, 1977.

99. Nakamura, K., Pirtle, R. M., Pirtle, I. L., Takeishi, K., and Inouye, M., Messenger ribonucleic acid of the lipoprotein of the *Escherichia coli* outer membrane. II. The complete nucleotide sequence, *J. Biol. Chem.*, 255, 210, 1980.
100. Hirashima, A. and Inouye, M., Specific biosynthesis of an envelope protein of *Escherichia coli*, *Nature (London)*, 242, 405, 1973.
101. Wu, H. C. and Lin, J. J.-C., *Escherichia coli* mutants altered in murein lipoprotein, *J. Bacteriol.*, 126, 147, 1976.
102. Lin, J. J.-C., Kanazawa, H., Ozols, J., and Wu, H. C., An *Escherichia coli* mutant with an amino acid alteration within the signal sequence of outer membrane prolipoprotein, *Proc. Natl. Acad. Sci. U.S.A.* 75, 4891, 1978.
103. Torti, S. V. and Park, J. T., Lipoprotein of gram-negative bacteria is essential for growth and division, *Nature*, 263, 323, 1976.
104. Torti, S. and Park, J., personal communication, 1979.
105. Coleman, J., Inouye, S., and Inouye, M., Isolation of mutants of the major outer membrane lipoprotein of *Escherichia coli* for the study of its assembly, *Meth. Enzymol.*, 97, 124, 1983.
106. Nakamura, K. and Inouye, M., DNA sequence of the gene for the outer membrane lipoprotein of *Escherichia coli:* an extremely, A-T rich protomer, *Cell*, 18, 1109, 1979.
107. Huang, Y. X., Ching, H., Yamagata, and Inouye, M., Existence of several homologous sequences in the *Escherichia coli* chromosome to the gene for the major outer membrane lipoprotein, *FEBS Lett.*, 137, 168, 1982.
108. Gayda, R. C., Avni, H., Berg, P. E., and Markovitz, A., Outer membrane protein a and other polypeptides regulate capsular polysaccharide synthesis *in E. coli* K12, *Mol. Gen. Genet.*, 175, 325, 1979.
109. Lugtenberg, B., Peters, R., Bernheimer, H., and Berendsen, W., Influence of cultural conditions and mutations on the composition of the outer membrane proteins of *Escherichia coli*, *Mol. Gen. Genet.*, 147, 251, 1976.
110. Earhart, C. F., Lundrigan, M., Pickert, C. L., and Pierce, J. R., *Escherichia coli* K12 mutants that lack major outer membrane protein a, *FEBS Lett.*, 6, 277, 1979.
111. Hofnung, M., Jezierka, A., and Braun-Breton, C., *lamB* mutations in *Escherichia coli* K12: growth of λ host range mutants and effect of nonsense suppressors, *Mol. Gen. Genet.*, 145, 207, 1976.
112. Hazelbauer, G. L., Role of the receptor for bacteriophage lambda in the functioning of the maltose chemoreceptor of *Escherichia coli*, *J. Bacteriol.*, 124, 119, 1975.
113. Szmelcman, S. and Hofnung, M., Maltose transport in *Escherichia coli* K12: involvement of the bacteriophage lambda receptor, *J. Bacteriol.*, 124, 112, 1975.
114. Debarboulle, M., Shuman, H. A., Silhavy, T. J., and Schwartz, M., Dominant constitutive mutations in *malT*, the positive regulator of the maltose regulon in *Escherichia coli*, *J. Mol. Biol.*, 124, 359, 1978.
115. Emr, S. D., Hall, M., and Silhavy, T., A mechanism of protein localization: the signal hypothesis and bacteria, *J. Cell Biol.*, 86, 701, 1980.
116. Hantke, K., Phage T6 — colicin K receptor and nucleoside transport in *Escherichia coli*, *FEBS Lett.*, 70, 109, 1976.
117. Manning, P. A. and Reeves, P., Outer membrane of *Escherichia coli* K12: tsx mutants (resistant to bacteriophage T6 and colicin K) lack an outer membrane protein, *Biochem. Biophys. Res. Commun.*, 71, 466, 1976.
118. Bradbur, C., Woodrow, M. L., and Khalifah, L. I., Transport of vitamin B12 in *Escherichia coli:* common receptor system for vitamin B12 and bacteriophage BF23 on the outer membrane of the cell envelope, *J. Bacteriol.*, 125, 1032, 1976.
119. DiMasi, R. D., White, J. C., Schnaitman, C. A., and Bradbeer, C., Transport of vitamin B12 in *Escherichia coli:* common receptor sites for vitamin B12 and the E colicins on the outer membrane of the cell envelope, *J. Bacteriol.*, 115, 506, 1973.
120. Sabet, S. F. and Schnaitman, C. A., Purification and properties of the colicin E3 receptor of *Escherichia coli*, *J. Biol. Chem.*, 248, 1797, 1973.
121. Fredericq, P. and Smarda, J., Complexite du facteur colicinogene B, *Ann. Inst. Pasteur (Paris)*, 118, 767, 1970.
122. Hantke, K. and Braun, V., Membrane receptor dependent iron transport in *Escherichia coli*, *FEBS Lett.*, 49, 301, 1975.
123. Guterman, S. K., Colicin B: mode of action and inhibition by enterochelin, *J. Bacteriol.*, 114, 1217, 1973.
124. Wayne, R. and Neilands, J. B., Evidence for common binding sites for ferrichrome compounds and bacteriophage 80 in the cell envelope of *Escherichia coli*, *J. Bacteriol.*, 121, 497, 1975.
125. Pugsley, A. P. and Reeves, P., Characterization of group B colicin-resistant mutants of *Escherichia coli* K12: colicin resistance and the role of enterochelin, *J. Bacteriol.*, 127, 218, 1976.

126. McIntosh, M. A., Chenault, S. S., and Earhart, C. F., Genetic and physiological studies on the relationship between colicin B resistant and ferrienterochelin uptake in *Escherichia coli* K12, *J. Bacteriol.*, 137, 1979.
127. Pugsley, A. P. and Reeves, P., The role of colicin receptors in the uptake of ferrienterochelin by *Escherichia coli* K12, *Biochem. Biophys. Res. Commun.*, 74, 903, 1977.
128. McIntosh, M. A. and Earhart, C. F., Coordinate regulation by iron of the synthesis of phenolate compounds and three outer membrane proteins in *Escherichia coli*, *J. Bacteriol.*, 131, 331, 1977.
129. Woodrow, G. C., Langman, L., Young, I, G., and Gibson, F., Mutations affecting the citrate-dependent iron uptake system in *Escherichia coli*, *J. Bacteriol.*, 133, 1524, 1978.
130. Hancock, R. E. W., Hantke, K., and Braun, V., Iron transport in *Escherichia coli* K12, involvement of the colicin B receptor and of a citrate-inducible protein, *J. Bacteriol.*, 127, 1370, 1976.
131. Wagegg, W. and Braun, V., Ferric citrate transport in *Escherichia coli* requires outer membrane receptor protein FecA, *J. Bacteriol.*, 145, 156, 1981.
132. Abe, M., Okamoto, N., Doi, O., and Nojima, S., Genetic mapping of the locus for detergent-resistant phospholipase A *(pldA)* in *Escherichia coli* K12, *J. Bacteriol.*, 119, 543, 1974.
133. Oliver, D. B. and Beckwith, J., *E. coli* mutant pleiotropically defective in the export of secreted protein, *Cell*, 25, 765, 1981.
134. Oliver, D. B. and Beckwith, J., Identification of a new gene *(secA)* and gene product involved in the secretion of envelope proteins in *Escherichia coli*, *J. Bacteriol.*, 150, 686, 1982.
135. Oliver, D., Kumamoto, C., Quinlan, M., and Beckwith, J., Pleiotropic mutants affecting the secretory apparatus of *Escherichia coli*, *Ann. Microbiol. (Inst. Pasteur)*, 133A, 105, 1982.
136. Kumamoto, C. A. and Beckwith, J., Mutations in a new gene, *secB*, cause defective protein localization in *Escherichia coli*, *J. Bacteriol.*, 154, 253, 1983.
137. Emr, S. D. and Bassford, P. J. Jr., Localization and processing of outer membrane and periplasmic proteins in *Escherichia coli* strains harboring export-specific suppressor mutations, *J. Biol. Chem.*, 257, 5852, 1982.
138. Ito, K., Wittekind, M., Nomura, M., Miura, A., and Nashimoto, H., A temperature-sensitive mutant of *E. coli* exhibiting slow processing of exported proteins, *Cell*, 32, 789, 1983.
139. McEwen, J. and Silverman, P. M., Mutations in genes *cpxA* and *cpxB* alter the protein composition of *Escherichia coli* inner and outer membranes, *J. Bacteriol.*, 151, 1553, 1982.
140. McEwen, J., Sambucetti, L., and Silverman, P., Synthesis of outer membrane proteins in *cpxA cpxB* mutants of *Escherichia coli* K12, *J. Bacteriol.*, 154, 375, 1983.
141. McEwen, J. and Silverman, P. M., Genetic analysis of *Escherichia coli* K12 chromosomal mutants defective in expression of F-plasmid functions: identification of genes *cpxA* and *cpxB*, *J. Bacteriol.*, 144, 60, 1980.
142. Dassa, E. and Boquet, P. L., *expA:* a conditional mutation affecting the expression of a group of exported proteins of *Escherichia coli* K12, *Mol. Gen. Genet.*, 181, 192, 1981.
143. Wurtzel, E., Movva, N. R., Ross, F. L., and Inouye, M., Two-step cloning of the *Escherichia coli* regulatory gene *ompB*, employing phage Mu, *J. Mol. Appl. Genet.*, 1, 61, 1981.
144. Masui, Y., Coleman, J., and Inouye, M., Multipurpose expression cloning vehicles in *E. coli*, in *Experimental Manipulation of Gene Expression*, Inouye, M., Ed., Academic Press, New York, 1983, 15.
145. Nakamura, K., Masui, Y., and Inouye, M., Use of a *lac* promoter-operator fragment as a transcriptional control switch for expression of the constitutive *lpp* gene in *Escherichia coli*, *J. Mol. Appl. Genet.*, 1, 289, 1982.
146. Nakamura, K. and Inouye, M., Construction of versatile expression cloning vehicles using the lipoprotein gene of *Escherichia coli*, *EMBO J.*, 1, 771, 1982.
147. Hirashima, A., Childs, G., and Inouye, M., Differential inhibitory effects of antibiotics on the biosynthesis of envelope proteins of *Escherichia coli*, *J. Mol. Biol.*, 79, 373, 1973.
148. Wang, S., Pirtle, R., Pirtle, I., Small, M., and Inouye, M., Purification of the messenger ribonucleic acid for the lipoprotein of the *Escherichia coli* outer membrane, *Biochemistry*, 18, 4270, 1979.
149. Pirtle, R. M., Pirtle, I. L., and Inouye, M., Homologous nucleotide sequences between prokaryotic and eukaryotic mRNAs: the 5′-end sequence of the mRNA of the lipoprotein of the *Escherichia coli* outer membrane, *Proc. Natl. Acad. Sci. U.S.A.*, 75, 2190, 1978.
150. Pirtle, R. M., Pirtle, I. L., and Inouye, M., Messenger ribonucleic acid of the lipoprotein of the *Escherichia coli* outer membrane. I. Nucleotide sequence at the 3′ terminus and sequences of oligonucleotides derived from complete digests of the mRNA, *J. Biol. Chem.*, 225, 199, 1980.
151. Rosenberg, M. and Court, D., Regulators sequences involved in the promotion and termination of RNA transcription, *Annu. Rev. Genet.*, 13, 319, 1979.
152. Siebenlist, U., Simpson, R. B., and Gilbert, W., *E. coli* RNA polymerase interacts homologously with two different promoters, *Cell*, 20, 269, 1980.

153. Reznikoff, W. S. and Abelson, J. H., The *lac* promoter, in *The Operon*, Miller, J. H. and Reznikoff, W. S., Eds., Cold Spring Harbor Laboratory, Cold Spring Harbor, N.Y., 1978, 221.
154. Buchanan, R. E. and Gibbons, N. E., *Bergey's Manual of Determinative Bacteriology*, 8th ed., Williams Wilkins, Baltimore, 1974.
155. Chamberlin, M. J., Interaction of RNA polymerase with the DNA template, in *RNA Polymerase*, Losick, R. and Chamberlin, M., Eds., Cold Spring Harbor Laboratory, Cold Spring Harbor, N.Y., 159, 1976.
156. Vollenweider, H. J., Fiandt, M., and Szybalski, W., A relationship between DNA helix stability and recognition sites for RNA polymerase, *Science*, 205, 509, 1979.
157. Vollenweider, H. J. and Szybalski, W., Electron microscopic mapping of RNA polymerase binding to coliphage lambda DNA, *J. Mol. Biol.*, 123, 485, 1978.
158. Coleman, J., Nakamura, K., and Inouye, M., Mutations upstream of the ribosome-binding site affect translational efficiency, *J. Mol. Biol.*, in press.
159. Shine, J. and Dalgarno, L., Determinant of cistron specificity in bacterial ribosomes, *Nature (London)*, 254, 34, 1975.
160. Dasgupta, R., Shih, D., Saris, C., and Kaesberg, P., Nucleotide sequences of a viral RNA fragment that binds to eukaryotic ribosomes, *Nature (London)*, 256, 624, 1975.
161. Wang, S., Marcu, K. B., and Inouye, M., Translation of a specific mRNA from *Escherichia coli* in a eukaryotic cell-free system, *Biochem. Biophys. Res. Commun.*, 68, 1194, 1976.
162. Salser, W., Globin mRNA sequences: analysis of base pairing and evolutionary implications, *Cold Spring Harbor Symp. Quant. Biol.*, 42, 985, 1977.
163. Gold, L., Pribnow, D., Schneider, T., Shinedling, S., Singer, B. S., and Stormo, G., Translational initiation in prokaryotes, *Annu. Rev. Microbiol.*, 35, 365, 1981.
164. Wu, A. and Platt, T., Transcription termination: nucleotide sequence at 3′ end of tryptophan operon in *Escherichia coli*, *Proc. Natl. Acad. Sci. U.S.A.*, 75, 5442, 1978.
165. Ikemura, T., Correlation between the abundance of *Escherichia coli* transfer RNAs and the occurence of the respective codons in its protein genes: a proposal for a synonymous codon choice that is optimal for the *E. coli* translational system, *J. Mol. Biol.*, 151, 389, 1981.
166. Gouy, M. and Gautier, C., Codon usage in bacteria: correlation with gene expressivity, *Nucleic Acid Res.*, 10, 7055, 1982.
167. Konigsberg, W. and Godson, G. N., Evidence for the use of rare codons in the dnaG gene and other regulatory genes of *Escherichia coli*, *Proc. Natl. Acad. Sci. U.S.A.*, 80, 687, 1983.
168. Nakamura, K. and Inouye, M., DNA sequence of the *Serratia marcescens* lipoprotein gene, *Proc. Natl. Acad. Sci. U.S.A.*, 77, 1369, 1980.
169. Yamagata, H., Nakamura, K., and Inouye, M., Comparison of the lipoprotein gene among the *Enterobacteriaceae:* DNA sequence of *Erwinia amylovora* lipoprotein gene, *J. Biol. Chem.*, 256, 2194, 1981.
170. Huang, Y.-X., Ching, G., and Inouye, M., Comparison of the lipoprotein gene among *Enterobacteriaceae:* DNA sequence of *Morganella morganii* lipoprotein gene and its expression in *Escherichia coli*, *J. Biol. Chem.*, 258, 8139, 1983.
171. Movva, R., Katz, E., Asdourian, R. L., Hirota, Y., and Inouye, M., Gene dosage effects of the structural gene for a lipoprotein of the *Escherichia coli* outer membrane, *J. Bacteriol.*, 133, 81, 1978.
172. Lee, N., Nakamura, K., and Inouye, M., Expression of the *Serratia marcescens* lipoprotein gene in *Escherichia coli*, *J. Bacteriol.*, 146, 861, 1981.
173. Bremer, E., Beck, E., Hindennach, I., Sonntag, I., and Henning, U., Cloned structural gene *(ompA)* for an integral outer membrane protein of *Escherichia coli* K12, *Mol. Gen. Genet.*, 179, 13, 1980.
174. Movva, R., Nakamura, K., and Inouye, M., Amino acid sequence of the signal peptide of ompA protein, a major outer membrane protein of *E. coli*, *J. Biol. Chem.*, 255, 27, 1980.
175. Henning, U., Royer, H.-D., Teather, R. M., Hindennach, I., and Hallenberg, C. P., Cloning of the structural *(ompA)* for an integral outer membrane protein of *Escherichia coli* K12, *Proc. Natl. Acad. Sci. U.S.A.*, 76, 4360, 1979.
176. Green, P., Sperrazza, C., and Inouye, M., manuscript in preparation.
177. Beck, E. and Bremer, E., Nucleotide sequence of the gene *ompA* coding the outer membrane protein II* of *Escherichia coli* K12, *Nucleic Acids Res.*, 8, 3011, 1980.
178. Movva, R., Green, P., Nakamura, K., and Inouye, M., Interaction of the cAMP receptor protein with the *ompA* gene, a gene for a major outer membrane protein of *Escherichia coli*, *FEBS Lett.*, 128, 186, 1981.
179. Chen, R., Schmidmayr, W., Kramer, C., Chen-Schmeisser, W., and Henning, U., Primary structure of major outer membrane protein II* (ompA protein) of *Escherichia coli* K=12, *Proc. Natl. Acad. Sci. U.S.A.*, 77, 1592, 1980.
180. Taniguchi, T., O'Neill, M., and de Crombrugghe, B., Interaction site of *Escherichia coli* cyclic AMP receptor protein on DNA of galactose operon promoters, *Proc. Natl. Acad. Sci. U.S.A.*, 76, 5090, 1979.

181. Aono, R., Yamasaki, M., and Tomura, G., Changes in composition of envelope proteins in adenylate cyclase — or cyclic AMP receptor protein-deficient mutants of *Escherichia coli, J. Bacteriol.*, 136, 812, 1978.
182. Mallick, U. and Herrlich, P., Regulation of synthesis of a major outer membrane protein: cyclic AMP represses *Escherichia coli* protein III synthesis, *Proc. Natl. Acad. Sci. U.S.A.*, 76, 5520, 1979.
183. Green, P. and Inouye, M., unpublished results, 1982.
184. Movva, R., Nakamura, K., and Inouye, M., Regulatory region of the gene for the ompA protein, a major outer membrane protein of *E. coli, Proc. Natl. Acad. Sci. U.S.A.*, 77, 3845, 1980.
185. Cole, S. T., Bremer, E., Hindennach, I., and Henning, U., Characterization of the promoters for the *ompA* gene which encodes a major outer membrane protein of *Escherichia coli, Mol. Gen. Genet.*, 188, 472, 1982.
186. Green, P. and Inouye, M., Roles of the 5′ leader region of the *ompA* mRNA, *J. Mol. Biol.*, 176, 431, 1984.
187. Napoli, C., Gold, L., and Singer, B. S., Translational reinitiation in the rIIB cistron of bacteriophage T4, *J. Mol. Biol.* 149, 433, 1981.
188. Schumperli, D., McKenney, K., Sobieski, D. A., and Rosenberg, M., Translational coupling at an intercistronic boundary of the *Escherichia coli* galactose operon, *Cell*, 30, 865, 1982.
189. Oppeheim, D. S. and Yanofsky, C., Translational coupling during expression of the tryptophan operon of *Escherichia coli, Genetics*, 95, 785, 1980.
190. von Gabain, A., Belasco, J. G., Schottel, J. L., Chang, A. C., and Cohen, S., Decay of mRNA in *Escherichia coli.* Investigation of the fate of specific segments of transcripts, *Proc. Natl. Acad. Sci. U.S.A.*, 80, 653, 1983.
191. Kolter, R. and Yanofsky, C., Attenuation in amino acid biosynthetic operons, *Annu. Rev. Genet.*, 16, 113, 1982.
192. Hahn, J., Grandi, G., Gryczan, T. J., and Dubnau, D., Translational attenuation of ermC: a deletion analysis, *Mol. Gen. Genet.*, 186, 204, 1982.
193. Friden, P., Newman, T., and Freundlich, M., Nucleotide sequence of the *ilv*B promoter-regulatory region. A biosynthetic operon controlled by attenuation and cyclic AMP, *Proc. Natl. Acad. Sci. U.S.A.*, 79, 6156, 1982.
194. Braun, G. and Cole, S., The nucleotide sequence coding for major outermembrane protein OmpA of *Shigella dysenteriae, Nucleic Acid Res.*, 10, 2367, 1982.
195. Cole, S., Sonntag, I., and Henning, U., Cloning and expression in *Escherichia coli* K=12 of the genes for major outer membrane protein OmpA from *Shigella dysenteriae, Enterobacter aeroginaes* and *Serratia marcescens, J. Bacteriol.*, 149, 145, 1982.
196. Beher, M. G. and Schatman, C. A., Regulation of the OmpA outer membrane protein of *E. coli, J. Bacteriol.*, 147, 972, 1981.
197. Aiba, H., Autoregulation of the *Escherichia coli* crp gene: CRP is a transcriptional repressor for its own gene, *Cell*, 32, 141, 1983.
198. Beher, M., Pugsley, A., and Schnaitman, C. A., Correlation between the expression of an *Escherichia coli* cell surface protein and the ability of the protein to bind to lipopolysaccharide, *J. Bacteriol.*, 143, 403, 1980.
199. Rick, P. D., Neumeyer, B. A., and Young, D. A., Effect of altered lipid a synthesis on the synthesis of the OmpA protein in *Salmonella typhimurium, J. Biol. Chem.*, 258, 629, 1983.
200. Mutoh, N., Nagasawa, T., and Mizushima, S., Specialized transducing bacteriophage lambda carrying the structural gene for a major outer membrane matrix protein of *Escherichia coli* K-12, *J. Bacteriol.*, 145, 1085, 1981.
201. Mizuno, T., Chou, M.-Y., and Inouye, M., DNA sequence of the promoter region of the *ompC* gene and the amino acid sequence of the signal peptide of the pro-OmpC protein of *Escherichia coli, FEBS Lett.*, 151, 159, 1983.
202. Mizuno, T. and Inouye, M., unpublished results, 1983.
203. Nikaido, H. Rosenberg, E., and Foulds, J., Porin channels in *Escherichia coli:* studies with β-lactams in intact cells, *J. Bacteriol.*, 153, 232, 1983.
204. Mizuno, T., Chou, M.-Y., and Inouye, M., A comparative study on the genes for three porins of the *Escherichia coli* outer membrane: DNA sequence of the osmoregulated *ompC* gene, *J. Biol. Chem.*, 258, 6932, 1983.
205. Overbeeke, N., Bergmans, H., vanMansfeld, F., and Lugtenberg, B., Complete nucleotide sequence of *phoE*, the structural gene for the phosphate limitation inducible outer membrane pore protein of *Escherichia coli* K-12, *J. Mol. Biol.*, 163, 513, 1983.
206. Bedouelle, H., Schmeissner, U., Hofnung, M., and Rosenberg, M., Promoters of the *malEFG* and the *malK-lamB* operons in *E. coli* K-12, *J. Mol. Biol.*, 161, 519, 1982.
207. Hasegawa, Y., Yamada, H., and Mizushima, S., Interactions of outer membrane proteins 0-8 and 0-9 with peptidoglycan sacculus of *Escherichia coli* K-12, *J. Biochem. (Tokyo)*, 80, 1401, 1976.

208. vanAlphen, W. and Lugtenberg, B., Influence of osmolarity of the growth medium on the outer membrane protein pattern of *Escherichia coli*, *J. Bacteriol.*, 131, 623, 1977.
209. Scott, N. and Harwood, C., Studies on the influence of the cyclic AMP system on major outer membrane proteins of *Escherichia coli*, *FEMS Lett.*, 9, 95, 1980.
210. Osawa, U. and Mizushima, S., Osmoregulation of *E. coli* outer membrane proteins, OmpC and OmpF, *Seikagaku*, 53, 709, 1981.
211. Taylor, R. K., Hall, M. N., and Silhavy, T. J., Isolation of mutations altering expression of the major outer membrane porin proteins using the local anaesthetic procaine, *J. Mol. Biol.*, 166, 273, 1983.
212. Wurtzel, E. T., Chou, M.-Y., and Inouye, M., Osmoregulation of gene expression. I. DNA sequence of the *ompR* gene of the *ompB* operon of *Escherichia coli* and characterization of its gene product, *J. Biol. Chem.*, 257, 13685, 1982.
213. Mizuno, T., Wurtzel, E. T., and Inouye, M., Cloning of the regulatory genes *(ompR* and *envZ)* for the matrix proteins of the *Escherichia coli* outer membrane, *J. Bacteriol.*, 150, 1462, 1982.
214. Mizuno, T. Wurtzel, E. T., and Inouye, M., Osomoregulation of gene expression. II. DNA sequence of the *envZ* gene of the *ompB* operon of *Escherichia coli* and characterization of its gene product, *J. Biol. Chem.*, 157, 13692, 1982.
215. Matthews, B. W., Ohlendorph, D. H., Anderson, W. F., and Takeda, Y., Structure of the DNA-binding region of *lac* repressor inferred from its homology with *cro* repressor, *Proc. Natl. Acad. Sci. U.S.A.*, 79, 1428, 1982.
216. Lundrigan, M. D. and Earhart, C. F., Temperature-dependent regulation of *Escherichia coli* K-12 outer membrane proteins, Abstract K203, *Abstr. Annu. Meet. Am. Soc. Microbiol.*, 210, 1983.
217. DiRienzo, J. M. and Inouye, M., Lipid fluidity-dependent biosynthesis and assembly of the outer membrane proteins of *E. coli*, *Cell*, 17, 155, 1979.
218. Pugsley, A. P. and Schnaitman, C. A., Outer membrane of *Escherichia coli*. VII. Evidence that bacteriophage-directed protein 2 functions as a pore, *J. Bacteriol.*, 133, 1181, 1978.
219. Fralick, J. A. and Diedrich, D. L., Studies on the expression of outer membrane protein 2 in *Escherichia coli*, *Mol. Gen. Genet.*, 188, 139, 1982.
220. Halegoua, S. and Inouye, M., Translocation and assembly of outer membrane proteins of *Escherichia coli*, *J. Mol. Biol.*, 130, 39, 1979.
221. Pages, J.-M. and Lazdunski, C., Transcriptional regulation of *ompF* and *lamB* genetic expression by local anesthetics, *FEBS Lett.*, 15, 153, 1982.
222. Schnaitman, C. A., Outer membrane proteins of *Escherichia coli*. IV. Differences in outer membrane proteins due to strain and cultural differences, *J. Bacteriol.*, 118, 454, 1974.
223. Cavard, D., Regnier, P., and Lazdunski, C., Specific cleavage of colicin A by outer membrane proteases from sensitive and insensitive strains of *Escherichia coli*, *FEBS Lett.*, 14, 283, 1982.
224. Clement, J. M. and Hofnung, M., Gene sequence of the λ receptor, an outer membrane protein of *E. coli*, K-12, *Cell*, 27, 507, 1981.
225. Hatfield, D., Hofnung, M., and Schwartz, M., Genetic analysis of the maltose A region in *Escherichia coli*, *J. Bacteriol.*, 98, 559, 1969.
226. Champon, C., Role of catabolite activator protein in the maltose regulon of *E. coli*, *J. Bacteriol.*, 150, 722, 1982.
227. Bedouelle, H. and Hofnung, M., A DNA sequence containing the control regions of the *malEFG* and *malK-lamB* operon in *Escherichia coli* K-12, *Mol. Gen. Genet.*, 185, 82, 1982.
228. Kikuchi, Y., Yoda, K., Yamasaki, M., and Tomura, G., Nucleotide region of the promoter and the amino-terminal region of alkaline phosphatase structural gene *(phoA)* of *Escherichia coli*, *Nucleic Acids Res.*, 9, 5671, 1981.
229. Inouye, H., Barnes, W., and Beckwith, J., Signal sequence of alkaline phosphatase of *Escherichia coli*, *J. Bacteriol.*, 149, 434, 1982.
230. Gayda, R. C., Henderson, G. W., and Markovitz, A., Neuroactive drugs inhibit trypsin and outer membrane protein processing in *Escherichia coli* K-12, *Proc. Natl. Acad. Sci. U.S.A.*, 76, 2138, 1979.
231. Fiss, E. H., Hollifield, W. C., Jr., and Neilands, J. B., Absence of ferric enterobactin receptor modification activity in mutants of *Escherichia coli* K-12 lacking protein a, *Biochem. Biophys. Res. Commun.*, 91, 29, 1979.
232. Blobel, G. and Dobberstein, B., Transfer of proteins across membranes. I. Presence of proteolytically processed and unprocessed nascent immunoglobulin light chains on membrane-bound ribosomes of murine myeloma, *J. Cell Biol.*, 67, 835, 1975.
233. Redman, C. M., Siekevitz, P., and Palade, G. F., Synthesis and transfer of amylase in pigeon pancreatic microsomes, *J. Biol. Chem.*, 241, 1150, 1966.
234. Ron, E. Z., Kohler, R. E., and Davis, B. D., Polysomes extracted from *Escherichia coli* by freeze-thaw-lysozyme lysis, *Science*, 153, 1119, 1966.
235. Schlessinger, D., Protein synthesis by polyribosomes on protoplast membranes of *B. megaterium*, *J. Mol. Biol.*, 7, 569, 1963.

236. Randall, L. L. and Hardy, S. J. S., Synthesis of exported proteins by membrane-bound polysomes from *Escherichia coli, Eur. J. Biochem.*, 75, 43, 1977.
237. Halegoua, S. Hirashima, A., Sedizawa, J., and Inouye, M., Protein synthesis in toluene-treated *Escherichia coli:* exclusive synthesis of membrane proteins, *Eur. J. Biochem.*, 69, 163, 1976.
238. Halegoua, S., Sedizawa, J., and Inouye, M., A new form of structural lipoprotein of outer membrane of *Escherichia coli, J. Biol. Chem.*, 252, 2324, 1977.
239. Hirashima, A., Wang, S., and Inouye, M., Cell-free synthesis of a specific lipoprotein of the *Escherichia coli* outer membrane directed by purified messenger RNA, *Proc. Natl. Acad. Sci. U.S.A.*, 71, 4149, 1974.
240. Inouye, S., Wang, S., Sedizawa, J., Halegoua, S., and Inouye, M., Amino acid sequence for the peptide extension on the prolipoprotein of the *Escherichia coli* outer membrane, *Proc. Natl. Acad. Sci. U.S.A.*, 74, 1004, 1977.
241. Sedizawa, J., Inouye, S., Halegoua, S., and Inouye, M., Precursors of major outer membrane proteins of *Escherichia coli, Biochem. Biophys. Res. Commun.*, 77, 1126, 1977.
242. Blobel, G. and Dobberstein, B., Transfer of proteins across membranes. II. Reconstitution of functional rough microsomes from heterologous components, *J . Cell. Biol.*, 67, 852, 1975.
243. Milstein, C., Brownlee, G. G., Harrison, T. M., and Mathews, M. B., A possible precursor of immunoglobulin light chains, *Nature (London) New Biol.*, 239, 117, 1972.
244. Wickner, W., The assembly of proteins into biological membranes: the membrane trigger hypothesis, *Annu. Rev. Biochem.*, 48, 23, 1979.
245. Wickner, W., M13 coat protein as a model of membrane assembly, *TIBS*, 8, 90, 1983.
246. Inouye, M., Pirtle, R., Pirtle, I., Sedizawa, J., Nakamura, K., DiRienzo, J., Inouye, S., Wang, S., and Halegoua, S., Mechanism of translocation of the outer membrane proteins of *Escherichia coli* through the cytoplasmic membrane, in *Microbiology 1979*, Vol. 34, Schlessinger, D., Ed., American Society for Microbiology, Washington, D.C., 1979.
247. Smith, W. P., Tai, P.-C., Thompson, R. C., and Davis, B. D., Extracellular labeling of nascent polypeptides traversing the membrane of *Escherichia coli, Proc. Natl. Acad. Sci. U.S.A.*, 74, 2830, 1977.
248. Smith, W. P., Tai, P.-C., and Davis, B. D., Nascent peptide as sole attachment of polysomes to membrane in bacteria, *Proc. Natl. Acad. Sci. U.S.A.*, 75, 814, 1978.
249. Adelman, M. R., Sabatini, D. D., and Blobel, G., Ribosome-membrane interaction, non-destructive disassembly of rat liver rough microsomes into ribosomal membranous components, *J. Cell Biol.*, 56, 206, 1973.
250. Halegoua, S., Hirashima, A., and Inouye, M., Puromycin-resistant biosynthesis of a specific outer membrane lipoprotein of *Escherichia coli, J. Bacteriol.*, 126, 183, 1976.
251. Smith, W. P., Tai, P.-C., and Davis, B. D., Interaction of secreted nascent chains with surrounding membrane in *Bacillus subtilis, Proc. Natl. Acad. Sci. U.S.A.*, 75, 5922, 1978.
252. Smith, W. P., Cotranslational secretion of diphtheria toxin and alkaline phosphatase *in vitro:* involvement of membrane protein(s), *J. Bacteriol.*, 141, 1142, 1980.
253. Randall, L. L., Hardy, S. J. S., and Josefsson, L.-G., Precursors of three exported proteins in *Escherichia coli, Proc. Natl. Acad. Sci. U.S.A.*, 75, 1209, 1978.
254. Maeda, T., Glass, J., and Inouye, M., Accumulation of the prolipoprotein of the *Escherichia coli* outer membrane cause by benzyloxycarbonylalanine chloromethylketone, *J. Biol. Chem.*, 256, 4712, 1981.
255. Davis, B. D. and Tai, P.-C, The mechanism of protein secretion across membranes, *Nature (London)*, 283, 433, 1980.
256. Ito, K., Mondel, G., and Wickner, W., Soluble precursor of an integral membrane protein: synthesis of procoat protein in *Escherichia coli* infected with bacteriophage M13, *Proc. Natl. Acad. Sci. U.S.A.*, 76, 1199, 1979.
257. Zimmermann, R. and Wickner, W., Energetics and intermediates of the assembly of protein OmpA into the outer membrane of *Escherichia coli, J. Biol. Chem.*, 258, 3920, 1983.
258. Sabatini, D. D., Kreibich, G., Morimoto, T., and Adesnik, M., Mechanisms for the incorporation of proteins in membranes and organelles, *J. Cell Biol.*, 92, 1, 1982.
259. Meyer, D., The signal hypothesis — a working model, *TIBS*, 7, 320, 1982.
260. Walter, P. and Blobel, G., Signal recognition particle contains a 7S RNA essential for protein translocation across the endoplasmic reticulum, *Nature (London)*, 299, 691, 1982.
261. Wu, H. C., Hou, C., Lin, J. J. C., and Yem, D. W., Biochemical characterization of a mutant lipoprotein of *Escherichia coli, Proc. Natl. Acad. Sci. U.S.A.*, 74, 1388, 1978.
262. Emr, S. D. and Silhavy, T., Molecular components of the signal sequence that function in the initiation of protein export, *J. Cell Biol.*, 95, 689, 1982.
263. von Heijne, G., Signal sequences are not uniformly hydrophobic, *J. Mol. Biol.*, 159, 537, 1982.

264. Inouye, S., Soberon, X., Franceschini, T., Nakamura, K., Itakura, K., and Inouye, M., Role of positive charge on the amino-terminal region of the signal peptide in protein secretion across the membrane, *Proc. Natl. Acad. Sci. U.S.A.* 79, 3438, 1982.
265. Vlasuk, G. O., Inouye, S., Ito, H., Itakura, K., and Inouye, M., Effects of complete removal of basic amino acid residues from the signal peptide on secretion of lipoprotein in *Escherichia coli, J. Biol. Chem.*, 258, 7141, 1983.
266. Hall, M. N., Gabay, J., and Schwartz, M., Evidence for a coupling of synthesis and export of an outer membrane protein in *Escherichia coli, EMBO,* 2, 15, 1983.
267. Walter, P. and Blobel, G., Translocation of proteins across the endoplasmic reticulum. III. Signal recognition protein causes signal sequence dependent and site specific arrest of chain elongation that is released by microsomal membranes, *J. Cell Biol.*, 91, 557, 1981.
268. Bremer, E., Cole, S. T., Hindennach, I., Henning, U., Beck, E., Kurz, C. and Schaller, H., Export of a protein into the outer membrane of *Escherichia coli* K-12. Stable incorporation of the OmpA protein requires less than 193 amino-terminal amino acid residues, *Eur. J. Biochem.*, 122, 223, 1982.
269. Inukai, M. and Inouye, M., manuscript in preparation.
270. Koshland, D. and Botstein, D., Secretion of beta-lactamase requires the carboxy end of the protein, *Cell*, 20, 749, 1980.
271. Koshland, D. and Botstein, D., Evidence for posttranslational translocation of β-lactamase across the bacterial inner membrane, *Cell*, 30, 893, 1982.
272. Pollitt, S. and Zalkin, H., Role of primary structure and disulfide bond formation in β-lactamase secretion, *J. Bacteriol.*, 153, 27, 1983.
273. Coleman, J., Inukai, M., and Inouye, M., manuscript in preparation.
274. Talmadge, K., Brosius, J. and Gilbert W., An "internal" signal sequence directs secretion and processing of proinsulin in bacteria, *Nature (London)*, 284, 176, 1981.
275. Oliver, D. B. and Beckwith, J., Regulation of a membrane component required for protein secretion in *Escherichia coli, Cell*, 30, 311, 1982.
276. Emr, S. D., Hanely-Way, S., and Silhavy, T. J. Supressor mutations that restore export of a protein with a defective signal sequence, *Cell*, 23, 79, 1981.
277. Shultz, J., Silhavy, T. J., Berman, M. L., Fiil, N., and Emr, S. D., A previously unidentified gene in the *spc* operon of *Escherichia coli* K-12 specifies a component of the protein export machinery, *Cell*, 31, 227, 1982.
278. Daniels, C. J., Bole, D. G., Quay, S. and Oxender, D. L., Role for membrane potential in the secretion of protein into the periplasm of *Escherichia coli, Proc. Natl. Acad. Sci. U.S.A.*, 9, 5396, 1981.
279. Enequist, H. G., Hirst, T. R., Harayama, S., Handy, S. J. S., and Randall, L. L., Energy is required for maturation of exported proteins in *Escherichia coli, Eur. J. Biochem.*, 116, 227, 1981.
280. Date, T., Goodman, J., and Wickner, W., Procoat, the precursor of M13 coat protein, requires an electrochemical potential for membrane insertion, *Proc. Natl. Acad. Sci. U.S.A.*, 8, 4669, 1980.
281. Bayer, M. E., Dolack, M., and Houser, E., Effects of lipid phase transition on the freeze-cleaved envelope of *Escherichia coli, J. Bacteriol.*, 129, 1563, 1977.
282. Haset, C. W. M., Verklej, A. J., DeGier, J., Scheek, R., Ververgaert, P. H. J., and Van Deenen, L. L. M., The effect of lipid transitions on the architecture of bacterial membranes, *Biochim. Biophys. Acta*, 356, 17, 1974.
283. Kleeman, W. and McConnell, H. M., Lateral phase separations in *Escherichia coli* membranes, *Biochim. Biophys. Acta*, 345, 220, 1974.
284. Letellier, L., Moudden, H., and Shechter, E., Lipid and protein segregation in *Escherichia coli* membrane: morphological and structural study of different cytoplasmic membrane fractions, *Proc. Natl. Acad. Sci. U.S.A.*, 74, 452, 1977.
285. Inouye, S., Hsu, C.-P. S., Itakura, K., and Inouye, M., Structural aspects of the *Escherichia coli* lipoprotein peptide cleavage site required for correct processing, *Science*, 221, 59, 1983.
286. Inouye, S., Franceschini, T., Sato, M., Itakura, K., and Inouye, M., Prolipoprotein signal peptidase of *Escherichia coli* requires a cysteine residue at the cleavage site, *EMBO*, 2, 87, 1983.
287. Crowlesmith, I. and Gamon, K., Rate of translation and kinetics of processing of newly synthesized molecules of two major outer membrane proteins, the OmpA and OmpF proteins, of *Escherichia coli* K-12, *Eur. J. Biochem.*, 124, 577, 1982.
288. Green, P. and Inouye, M., *In vivo* and *in vitro* systems for studying bacterial membrane biogenesis, *Meth. Enzymol.*, 96, 74, 1983.
289. Josefsson, L.-G. and Randall, L. L., Different exported proteins in *E. coli* show differences in the temporal mode of processing *in vivo, Cell*, 25, 151, 1981.
290. Wolfe, P. B., Silver, P., and Wickner, W., The isolation of homogeneous leader peptidase from a strain of *Escherichia coli* which overproduces the enzyme, *J. Biol. Chem.*, 257, 7898, 1982.

291. Tokunaga, M., Loranger, J. M., Wolfe, P. B., and Wu, H. C., Prolipoprotein signal peptidase in *Escherichia coli* is distinct from the M13 procoat protein signal peptidase, *J. Biol. Chem.*, 257, 9922, 1982.
292. Zwizinski, C., Date, T., and Wickner, W., Leader peptidase is found in both the inner and outer membranes of *Escherichia coli*, *J. Biol. Chem.*, 256, 3593, 1981.
293. Ito, K., Purification of the precursor form of maltose-binding protein, a periplasmic protein of *Escherichia coli*, *J. Biol. Chem.*, 257, 9895, 1982.
294. Bayer, M. E., The fusion sites between outer membrane and cytoplasmic membrane of bacteria: their role in membrane assembly and virus infection, in *Bacterial Outer Membrane. Biogenesis and Functions*, Inouye, M., Ed., John Wiley & Sons, New York, 1980, 167.
295. Date, T. and Wickner, W., Isolation of the *Escherichia coli* leader peptidase gene and effects of leader peptidase overproduction *in vivo*, *Proc. Natl. Acad. Sci. U.S.A.*, 78, 6106, 1981.
296. Date, T., Demonstration by a novel genetic technique that leader peptidase is an essential enzyme of *Escherichia coli*, *J. Bacteriol.*, 154, 76, 1983.
297. Silver, P. A. and Wickner, W., Genetic mapping of the *E. coli* leader peptidase gene *(lep)*: a new approach for determining the map position of a cloned gene, *J. Bacteriol.*, 154, 569, 1983.
298. Hussain, M., Ichihara, S., and Mizushima, S., Mechanism of signal peptide cleavage in the biosynthesis of the major lipoprotein of the *Escherichia coli* outer membrane, *J. Biol. Chem.*, 257, 5177, 1982.
299. Inukai, M., Takeuchi, M., Shimizu, K., and Arai, M., Mechanism of action of globomycin, *J. Antibiot.*, 31, 1203, 1978.
300. Ichihara, S., Hussain, M., and Mizushima, S., Characterization of new membrane lipoproteins and their precursors of *Escherichia coli*, *J. Biol. Chem.*, 256, 3125, 1981.
301. Mizuno, T., A novel peptidoglycan-associated lipoprotein found in the cell envelope of *Pseudomonas aeruginosa* and *Escherichia coli*, *J. Biochem.*, 86, 991, 1979.
302. Mizuno, T., A novel peptidoglycan-associated (PAL) found in the outer membrane of *Proteus mirabilis* and other gram-negative bacteria, *J. Biochem.*, 89, 1039, 1981.
303. Mizuno, T., Structure of the peptidoglycan-associated lipoprotein (PAL) of the *Proteus mirabilis* outer membrane. II. Sequence of the amino terminal part of the peptidoglycan-associated lipoprotein, *J. Biochem.*, 89, 1059, 1981.
304. Tokunaga, M., Tokunaga, H., and Wu, H. C., Post-translational modification and processing of *Escherichia coli* prolipoprotein *in vitro*, *Proc. Natl. Acad. Sci. U.S.A.*, 79, 2255, 1982.
305. Yamagata, H., Ippolito, C., Inukai, M., and Inouye, M., Temperature-sensitive processing of outer membrane lipoprotein in an *Escherichia coli* mutant, *J. Bacteriol.*, 152, 1163, 1982.
306. Koshland, D., Sauer, R. T., and Botstein, D., Diverse effects of mutations in the signal sequence on the secretion of β-lactamase in *Salmonella typhimurium*, *Cell*, 20, 903, 1982.
307. Boeke, J. D., Russel, M., and Model, P., Processing of filamentous phage pre-coat protein. Effect of sequence variations near the signal peptidase cleavage site, *J. Mol. Biol.*, 144, 103, 1980.
308. Zimmerman, R., Watts, C., and Wickner, W., The biosynthesis of membrane-bound M13 coat protein. Energetics and assembly intermediates, *J. Biol. Chem.*, 257, 6529, 1982.
309. Wu, H., Biosynthesis and assembly of outer membrane lipoprotein in *E. coli*, in *Membrane and Transport*, Vol. 1, Martonosi, A. N., Ed., Plenum Press, New York, 1982, 299.
310. Lai, J.-S., Philbrick, W. M., Hayashi, S., Inukai, M., Arai, M., Hirota, Y., and Wu, H. C., Globomycin sensitivity of *Escherichia coli* and *Salmonella typhimurium*: effects of mutations affecting structures of murein lipoprotein, *J. Bacteriol.*, 145, 657, 1981.
311. Chattopadhyay, P. K. and Wu, H. C., Biosynthesis of the covalently linked diglyceride in murein lipoprotein of *Escherichia coli*, *Proc. Natl. Acad. Sci. U.S.A.*, 74, 5318, 1977.
312. Chattopadhyay, P. K., Engel, R., Tropp, B., and Wu, H., Biosynthesis of murien lipoprotein in *Escherichia coli*. Effects of 3,4-dihydroxybutylphosphonate, *J. Bacteriol.*, 138, 944, 1979.
313. Rotering, H. and Braun, V., Lipid deficiency in a lipoprotein mutant of *Escherichia coli*, *FEBS Lett.*, 83, 41, 1977.
314. Lai, J.-S., Philbrick, W. M., and Wu, H. C., Acyl moieties in phospholipids are precursors for the fatty acids in murine lipoprotein of *Escherichia coli*, *J. Biol. Chem.*, 255, 5384, 1980.
315. Lai, J.-S. and Wu, H. C., Incorporation of acyl moieties of phospholipids into murein lipoprotein intact cells of *Escherichia coli* by phospholipid fusion, *J. Bacteriol.*, 144, 451, 1980.
316. Hussain, M., Ichihara, S., and Mizushima, S., Accumulation of glyceride-containing precursor of the outer membrane lipoprotein in the cytoplasmic membrane of *Escherichia coli* treated with globomycin, *J. Biol. Chem.* 255, 3707, 1980.
317. Zwiebel, L. J., Inukai, M., Nakamura, K., and Inouye, M., Preferential selection of deletion mutations of the outer membrane lipoprotein gene of *Escherichia coli* by globomycin, *J. Bacteriol.*, 145, 654, 1981.

318. Inukai, M., Ghrayeb, J., Nakamura, K., and Inouye, M., Apolipoprotein an intermediate in the processing of the major lipoprotein of the *Escherichia coli* outer membrane, *J. Biol. Chem.* 259, 757, 1984.
319. Wensink, J. and Witholt, B., Conversion of free lipoprotein to the murein-bound form, *Eur. J. Biochem.*, 117, 207, 1981.
320. Tanaka, N., Iseki, M., Miyoshi, T., Aoki, H., and Imanaka, H., Mechanism of action of bicyclomycin, *J. Antibiot.*, 29, 155, 1976.
321. Braun, V. and Bosch, V., *In vivo* biosynthesis of murein-lipoprotein of the outer membrane of *E. coli, FEBS Lett.*, 34, 302, 1973.
322. Braun, V. and Wolff, H., Attachment of lipoprotein and murein (peptidoglycan) of *Escherichia coli* in the presence and absence of penicillin, *J. Bacteriol.*, 123, 888, 1975.
323. Wensink, J., Gilden, N., and Witholt, B., Attachment of lipoprotein to the murein of *Escherichia coli, Eur. J. Biochem.*, 122, 587, 1982.
324. Weigand, R. A., Vinci, D. K., and Rothfield, L. I., Morphogenesis of the bacterial division septum: a new class of septation-defective mutants, *Proc. Natl. Acad. Sci. U.S.A.*, 73, 1882, 1976.
325. Wensink, J. and Witholt, B., Outer-membrane vesicles released by normally growing *Escherichia coli* contain very little lipoprotein, *Eur. J. Biochem.*, 116, 331, 1981.
326. Braun, V. and Sieglin, U., The covalent murein-lipoprotein structure of the *E. coli* cell wall. The attachment site of the lipoprotein on the murein, *Eur. J. Biochem.*, 13, 336, 1970.
327. Garten, W. and Henning, U., Cell envelope and shape of *E. coli* K12. Isolation and preliminary characterization of the major ghost-membrane proteins, *Eur. J. Biochem.*, 47, 343, 1974.
328. Diedrich, D. L. and Schnaitman, C. A., Lysyl-derived aldehydes in outer membrane proteins of *Escherichia coli, Proc. Natl. Acad. Sci. U.S.A.*, 75, 3708, 1978.
329. Chen, R., Kramer, C., Schmidmeyr, W., and Henning, U., Primary structure of major outer membrane protein I of *Escherichia coli* B/r, *Proc. Natl. Acad. Sci. U.S.A.*, 76, 5014, 1979.
330. Chai, T.-J. and Foulds, J., Purification of protein A, and outer membrane component missing in *Escherichia coli* K12 *ompA* mutants, *Biochim. Biophys. Acta*, 493, 210, 1977.
331. Schweizer, M., Hindennach, I., Garten, W., and Henning, U., Major proteins of the *Escherichia coli* outer cell envelope membrane. Interaction of protein II* with lipopolysaccharide, *Eur. J. Biochem.*, 82, 211, 1978.
332. DiRienzo, J. and Inouye, M., Accumulation of a lipopolysaccharide precursor in *Escherichia coli* during conditions of reduced membrane lipid fluidity, *Eur. J. Biochem.*, 135, 351, 1983.
333. Halegoua, S. and Inouye, M., Biosynthesis and assembly of the outer membrane proteins, in *Bacterial Outer Membranes. Biogenesis and Functions*, Inouye, M., Ed., John Wiley & Sons, New York, 1980, 67.
334. Silhavy, T. J., Shuman, H. A., Beckwith, J., and Schwartz, M., Use of gene fusions to study outer membrane protein localization in *Escherichia coli, Proc. Natl. Acad. Sci. U.S.A.*, 74, 5411, 1977.
335. Moreno, F., Fowler, A. V., Hall, M., Silhavy, T. J., Zabin, I., and Schwartz, M., A signal-sequence is not sufficient to lead beta-galactosidase out of the cytoplasm, *Nature (London)*, 286, 356, 1980.
336. Bassford, P. J., Silhavy, T. J., and Beckwith, J. R., Use of gene fusion to study secretion of maltose-binding protein into the *Escherichia coli* periplasm, *J. Bacteriol.*, 139, 19, 1979.
337. Michaelis, S., Guarente, L., and Beckwith, J., *In vitro* construction and characterization of *phoA-lacZ* gene fusions in *Escherichia coli, J. Bacteriol.*, 154, 356, 1983.
338. Blobel, G., Intracellular protein topogenesis, *Proc. Natl. Acad. Sci. U.S.A.*, 77, 1496, 1980.
339. Bayer, M. E., Areas of adhesion between the wall and membrane of *Escherichia coli, J. Gen. Microbiol.*, 53, 395, 1968.
340. Muhlradt, P. F., Menzel, J., Golecki, J. R., and Speth, V., Outer membrane of *Salmonella*. Sites of export of newly synthesized lipopolysaccharide on the bacterial surface, *Eur. J. Biochem.*, 35, 471, 1973.
341. Bayer, M. E. and Thurow, H., Polysaccharide capsule of *Escherichia coli:* microscope study of its size, structure, and sites of synthesis, *J. Bacteriol.*, 130, 911, 1977.
342. Smit, J. and Nikaido, H., Outer membrane of gram-negative bacteria. XVIII. Electron microscopic studies on porin insertion sites and growth of cell surface of *Salmonella typhimurium, J. Bacteriol.*, 135, 687, 1978.
343. Ohki, M., The cell cycle-dependent synthesis of envelope proteins in *Escherichia coli*, in *Bacterial Outer Membrane. Biogenesis and Functions*, Inouye, M., Ed., John Wiley & Sons, New York, 1980, 293.
344. Inukai, M. and Inouye, M., Association of the prolipoprotein accumulated in the presence of globomycin with the outer membrane, *Eur. J. Biochem.*, 130, 27, 1983.
345. Cole, S. T., Chen-Schmeisser, U., Hindennach, I., and Henning, U., Apparent bacteriophage-binding region of an *Escherichia coli*, K-12 outer membrane protein, *J. Bacteriol.*, 153, 581, 1983.
346. Inukai, M., Masui, Y., Vlasuk, G. P. and Inouye, M., Effects of inserting eight amino acid residues into the major lipoprotein on its assembly in the outer membrane of *Escherichia coli, J. Bacteriol.*, 155, 275, 1983.

347. DeMartini, M. and Inouye, M., Interaction between two major outer membrane proteins of *Escherichia coli*: the matrix protein and the lipoprotein, *J. Bacteriol.*, 133, 329, 1978.
348. Yu, F. and Mizushima, S., Stimulation by lipopolysaccharide of the binding of outer membrane proteins 0-8 and 0-9 to the peptidoglycan layer of *Escherichia coli* K12, *Biochem. Biophys. Res. Commun.*, 74, 1397, 1977.
349. Hedgpeth, J., Clement, J.-M., Marchal, C., Perrin, D., and Hofnung, M., DNA sequence encoding the NH_2-terminal peptide involved in transport of λ receptor, an *Escherichia coli* secretory protein, *Proc. Natl. Acad. Sci. U.S.A.*, 77, 2621, 1980.
350. Mutoh, N., Inokuchi, K., Mizushima, S., Amino acid sequence of the signal peptide of *ompF*, a major outer membrane protein of *Escherichia coli*, *FEBS Lett.*, 137, 171, 1982.
351. Jaurin, B. and Grundstrom, T., *ampC* cephalosporinase of *Escherichia coli* K-12 has a different evolutionary origin from that of β-lactamase of the penicillinase type, *Proc. Natl. Acad. Sci. U.S.A.*, 78, 4897, 1981.
352. Sutchliffe, J. G., Nucleotide sequence of the ampicillin resistance gene of *Escherichia coli* plasmid pBR322, *Proc. Natl. Acad. Sci. U.S.A.*, 75, 3737, 1978.
353. Wilson, V. G. and Hogg, R. W., The NH_2-terminal sequence of a precursor form of arabinose binding protein, *J. Biol. Chem.*, 255, 6745, 1980.
354. Higgins, C. F. and Ames, G. F. L., Two periplasmic transport proteins which interact with a common membrane receptor show extensive homology: complete nucleotide sequences, *Proc. Natl. Acad. Sci. U.S.A.*, 78, 6038, 1981.
355. Oxender, D. L., Anderson, J.-J., Daniels, C. J., Landick, R., Gunsalus, R. P., Zurawski, G., and Yanofsky, C., Amino-terminal sequence and processing of the precursors of the leucine specific binding protein, and evidence for conformational differences between the precursor and the mature form, *Proc. Natl. Acad. Sci. U.S.A.*, 77, 2005, 1980.
356. Landick, R. and Oxender, D. L., Bacterial periplasmic binding proteins, in *Membranes and Transport*, Vol. 2, Martonosi, A., Ed., Plenum Press, New York, 1982, 81.
357. Bedouelle, H., Bassford, P. J., Fowler, A. V., Zabin, I., Beckwith, J., and Hofnung, M., Mutations which alter the function of the maltose binding protein of *Escherichia coli*, *Nature (London)*, 285, 78, 1980.
358. Sugimoto, K., Sugisaki, H., Okamoto, T., and Takanami, M., Studies on bacteriophage fd DNA. IV. The sequence of messenger RNA for the major coat protein gene, *J. Mol. Biol.*, 111, 487, 1977.
359. Shalter, H., Beck, E., and Takanami, M., Sequence and regulatory signals of the filamentous phage genome, in *The Single Stranded DNA Phages*, Denhart, D., Dressler, D., and Ray, D., Eds., Cold Spring Harbor Laboratory Press, Cold Spring Harbor, New York, 1979, 139.
360. Dallas, W. S. and Falkow, S., Amino acid sequence homology between cholera toxin and *Escherichia coli* heat-labile toxin, *Nature (London)*, 288, 499, 1980.
361. Spicer, E. K. and Noble, J. A., *Escherichia coli* heat-labile enterotoxin. Nucleotide sequence of the A subunit gene, *J. Biol. Chem.*, 257, 5716, 1982.
362. Koryer, J. and Chang, S., The promoter-proximal region of the *Bacillus licheniformis*, penicillinase gene: nucleotide sequence and predicted leader peptide sequence, *Gene*, 15, 343, 1981.
363. McLaughlin J. R., Murray, C. L., and Rabinowitz, J. C., Unique features in the ribosome binding site sequence of the gram-positive *Staphylococcus aureus* β-lactamase gene, *J. Biol. Chem.*, 256, 11283, 1981.
364. Palva, I., Pettersson, R. F., Kalkkineri, N., Lehtovaara, P., Sarvas, M., Soderlund, H., Takkinen, K., and Kaariainen, L., Nucleotide sequence of the promoter and NH_2-terminal signal peptide region of the α-amylase gene from *Baccillus amyliquefaciens*, *Gene*, 15, 43, 1981.

Chapter 3

FUNCTIONAL ORGANIZATION OF BACTERIAL MEMBRANE PROTEINS

Bijan K. Ghosh and Arati Ghosh

TABLE OF CONTENTS†

I. Introduction 106

II. General Considerations 106
- A. Membrane Bilayer and Lipid Protein Association 106
- B. Membrane Microenvironment and Enzyme Function 108
- C. Intramembranous Movement of Molecules 110

III. Enzymes of Lipid Biosynthesis 110
- A. Phospholipid Biosynthesis and Degradation 110
- B. Phospholipid Turnover 113

IV. Adenosinetriphosphatase (ATPase, 3.6.1.3) 115
- A. Physiological Role and Distribution 115
- B. Purification, Subunit Structure, and Properties 115
- C. Uncoupler *(unc)* Mutant 120
- D. Proton Translocation 122
- E. Organization in the Membrane 124

V. Nitrate Reductase (1.6.6.1; 1.6.6.2.; 1.6.6.3) 125
- A. Physiological Role and Distribution 125
- B. Purification and Properties 126

VI. Membrane-Bound Dehydrogenase 127
- A. Glycerol-3-Phosphate Dehydrogenase (1.1.99.5; 1.1.1.8) 127
- B. L-Lactate Dehydrogenase (1.1.2.3) 128
- C. D-Lactate Dehydrogenase (1.1.2.4) 130
- D. D-Alanine (*), Allohydroxy-D-Proline (*), Choline (1.1.99.1), and Sarcosine Dehydrogenase (1.5.99.1) 130
- E. Malate Dehydrogenase (1.1.1.37.1.1.1.40) 131
- F. Hydrogenase (*) 131
- G. Formate Dehydrogenase (1.2.2.1.; 1.2.1.2) 131
- H. NADH Dehydrogenase (1.6.99.3; 1.6.99.5) 131
- I. Fumarate Reductase (1.3.1.6) 132
- J. Succinic Dehydrogenase (1.3.99.1) 132
- K. Miscellaneous Oxidases, Reductases, and Dehydrogenases (*) 134

VII. Membrane-Bound Hydrolase 134
- A. Peptidase (*) 134

† Numbers in parentheses are enzyme commission numbers for standard nomenclature. Asterisks within parentheses indicate that enzyme commission numbers for the specific enzymes discussed here are not available.

B. Protease (*) 135
C. Alkaline Phosphatase (3.1.3.1) 136
D. Penicillinase (3.5.2.6) 138
E. Miscellaneous Transferase and Esterase (*) 141

VIII. Membrane-Bound Enzymes for Wall Biosynthesis 143

IX. Binding Protein 146
A. Proteins Related to Solute Transport 146
B. Penicillin-Binding Protein 152
C. DNA-Binding Protein 154

Acknowledgment 157

References 157

I. INTRODUCTION

The ratio between plasma membrane surface area and cytoplasmic volume of bacterial cells is much greater than in eukaryotic cells. However, if total cellular membrane surface area is considered, eukaryotic cells may have much greater membrane surface area relative to cytoplasmic volume. In eukaryotic cells the plasma membrane is a specialized structure and retains only a few functions, e.g., permeability barrier and small molecular transport. In bacterial cells, there is generally no other membrane structure besides the plasma membrane; only photosynthetic bacteria house the photosynthetic apparatus in the intracytoplasmic membrane vesicles. It is possible that the plasma membrane in bacteria is an undifferentiated structure that carries out all the physiological functions. These functions are distributed in the plasma and intracellular membrane of eukaryotic cells. It is possible that during the course of cellular evolution the plasma membrane differentiated to give rise to intracellular membranous bodies.[1] Bacterial plasma membrane is a versatile and undifferentiated membrane structure that is the hub of multifarious physiological activities.

The aim of this chapter is to examine the characteristics of the organization of functionally important protein molecules in bacterial plasma membrane. The final goal is to develop the concept that the plasma membrane in bacteria is not only a structural but a functional mosaic, as well. Although the literature on bacterial membrane proteins is substantial, the information presented is a sporadic and unrelated body of data. Hence, this presentation will fall much shorter than the stated goal. One may view this as an attempt to look at a large body of data within a conceptual framework.

II. GENERAL CONSIDERATIONS

A. Membrane Bilayer and Lipid Protein Association

Danielli and Davson in the early 1930s studied the organization of phospholipids obtained from biological sources.[2] They made a suggestion of far reaching and universal implication, that the basic structure of the biological membrane is a phospholipid bilayer. Nearly three decades passed before this idea was seriously incorporated into the thoughts on biological membrane. During this period another landmark observation was made by Robertson by way of electron microscopy, that cells are delimitated by a bilayer structure, which can be interpreted as being composed of a phospholipid bilayer.[3] Although the bilayer concept of membrane passed through a stormy period during the 1950s and early 1960s, skeptical biochemists certainly enriched membrane research by providing a range of experimental data especially on the protein organization of the biological membrane. As the organization of proteins in a functional state is incompatible with the phospholipid bilayer concept, skepticism grew regarding the lipid/protein organization predicted by the bilayer concept. However, a large body of biophysical data made it almost impossible to ignore the bilayer concept. Finally, the fluid mosaic model of Singer and Nicholson[4] became a happy compromise between the static bilayer model and dynamic lipid/protein micelle subunit model.[5] In spite of the skepticism of biochemists, the bilayer model of membrane structure appealed to biologists because it provided a general mechanism for highly effective separation of intra- and extracellular hydrophilic environments. Phospholipid molecules, by a self-assembly process, can generate a hydrophobic barrier between intra- and extracellular environments, restricting free diffusion of material between the above two compartments. As the formation of bilayer occurs by a self-assembly process, a continuous input of energy to maintain the stability of this structure is minimized. However, it is difficult to visualize the organization of membrane proteins into this phospholipid bilayer, which constitutes about 50 to 60% of the dry weight of the membrane. The protein organization must be functional and dynamic, because this is not only a permeability barrier but also the hub of a variety of essential biological activities including the influx of nutrients (sugars, amino acids, etc.) and ions, the efflux of macromolecules (enzymes, binding proteins, etc.) and biosynthesis, as well as the assembly of envelope material.

The structure of the phospholipid bilayer has been studied by a variety of techniques, i.e., light scattering, calorimetry, differential labeling, and electron microscopy (freeze fracture and immunolabeling). A few general characteristics of the plasma membrane have emerged from these studies: (1) the bilayer structure is stable unless degraded by detergents or nonpolar solvents;[6] (2) the hydrophobic middle region of the bilayer has relatively low energy and thus this region is prone to fracture by freeze fracture;[7] (3) there is considerable asymmetry of the protein composition of the two leaflets of the bilayered membrane;[8] (4) the lipid material of the two leaflets of the membrane has considerable fluidity or disorder[9] (this fluid phase changes into a gel phase [ordered structure] at a critical phase transition temperature); and (5) in the lateral plane of the two leaflets of the bilayer, material diffuses freely. In contrast, transbilayer movement of the constituents is restricted and can occur by a special (flip-flop) mechanism.[10]

The supramolecular organization of lipid in the membrane clearly shows the existence of a continuum of stable hydrophobicity. Hence, the proteins existing within the membrane must possess some unique property which will allow them to remain in a stable manner in this hydrophobic environment. Quantitatively, proteins constitute the bulk of the membrane and these associate with the basic bilayer through a variety of bonds. These differently bonded proteins can be removed from the native membrane by a widely differing extraction mechanism. Thus, proteins have been operationally classified on the basis of the characteristics of extractibility from the membrane.[11] These are (1) integral and (2) peripheral proteins.

Integral proteins remain either totally or partially buried within the hydrophobic interior of the membrane. These proteins are usually extracted with amphipathic solvent (i.e., detergent); in the absence of detergent these proteins show poor solubility and tend to aggregate. Besides their hydrophobic association with the membrane these proteins may be covalently bonded to the membrane lipid, requiring lipolysis for their release.[12] Almost 70 to 80% of the proteins located within the membrane are considered integral. They constitute membrane-associated enzymes, antigenic determinents, transport proteins, and drug or hormone receptors.[14]

Peripheral proteins are bound to the membrane through ionic bonding. These can be extracted with high salt or by varying the H^+ ion concentration which influences the ionization characteristics. The ionic bonds may develop between the ionic groups of the phospholipid or the protein. There are many intermediate types of these proteins which may form ionic, hydrophobic, or covalent complexes with integral or other peripheral proteins. Hence, membrane protein classification, based on extractibility from the membrane, may appear theoretically sound, but in reality deviations from the basic types are so wide that the exception becomes the rule. However, this classification provides a theoretical framework for the understanding of the characteristics of the membrane proteins. Differences in the overall character of the integral and soluble proteins have been attributed to the proportion of polar amino acid residues in each type of protein.[15] Overall polarity has been determined from the amino acid composition.[16] Following this criterion soluble proteins appear to have greater polarity (46 ± 6%) than membrane proteins (35 ± 6%); this difference perhaps indicates the native environment of the proteins. Differences in polarity correlate with some empirical properties of membrane proteins, i.e., solubility in an aqueous environment and the ability to reconstitute into bilayers in the presence of lipid.

B. Membrane Microenvironment and Enzyme Function

Questions have been frequently raised as to whether or not the lipid molecules have any regulatory role on the activity, catalytic or otherwise, of the membrane protein molecules. The experimental data in this area are scant because of the difficulty of integral membrane protein extraction and purification. However, studies with several membrane-bound enzymes provide clear data indicating that specific intramembranous microenvironment modulates the activities of membrane proteins.

D-Lactate dehydrogenase has been implicated in the energy-dependent transport of amino acids and sugars across the plasma membrane.[17,18] This 72-kDa monomeric polypeptide containing flavin is purified from *E. coli* plasma membrane. The enzyme has been extracted by chaotropic agent, detergent, or sonication and purified in phospholipid free form.[19] The lipid-free enzyme aggregates, in the absence of the detergent, readily associate with a variety of phospholipids.[20] The delipidated enzyme, exhibiting poor catalytic activity is activated sevenfold after associating with phospholipid. This activation depends on the micellar form of the phospholipid. Thus, ordered lipid structure providing a lipid matrix is essential for the enzyme activity. The membrane-bound enzyme in the native state is surrounded, predominantly by phosphatidylglycerol or phosphatidyl serine. It appears that the lipid microenvironment modulates the catalytic activity of this and a variety of other enzymes.[21]

Pyruvic oxidase, in contrast to D-lactate dehydrogenase, is a peripheral membrane protein of *E. coli*. The pure enzyme contains FAD and has a molecular weight of 60 kDa.[22] The catalytic activity is twenty-five times stimulated after interaction with amphiphiles, namely phospholipids, neutral detergent, and anionic or cationic detergents below their critical miscellar concentrations. The enzyme has high-affinity detergent-binding sites.[23] Binding occurs by hydrophobic association. However, the simultaneous presence of substrate (pyruvate) and co-factor (thiamine pyrophosphate) causes an increase of affinity for the detergent (SDS). It is likely that a ligand-induced conformational change of the pyruvic oxidase causes detergent binding and enzyme activation. The detergents are, in essence, allosteric effectors. In fact, the membrane in vivo may function as an allosteric effector.

If phospholipid and neutral lipid are extracted with 90% cold acetone, D-lactate, NADH, and succinate oxidase activities are lost from *E. coli* plasma membrane.[24] The loss of these enzyme activities by delipidation can be restored after brief incubation at 37°C with coenzyme Q10 and appropriate phospholipids. The three enzyme activities, however, respond differently to this reactivation with different phospholipids. D-Lactate oxidase is most stringent, because it is reactivated only with cardiolipin. On the other hand, succinate oxidase is unspecific; it is best activated by cardiolipin and least by phosphatidylcholine. Phosphatidylserine, which is only a minor phospholipid, is effective in restoring the activity of NADH oxidase. Furthermore, K_m values are significantly altered after reconstitution with these phospholipids. Therefore, these enzyme activities are regulated by phospholipids.

The above are a few examples which show that the functional states of these membrane-bound enzymes are significantly influenced by their association with the membrane. The heterogeneity of membrane lipid plays an important role in maintaining an effective microenvironment around membrane-bound enzyme molecules. This concept has been experimentally substantiated in fatty acid auxtotrophs of *E. coli*.[25,26] In these mutants, the relative proportion of the fatty acids of the membrane phospholipids is changed by a variety of exogenously added fatty acids. The amount of *cis* $\Delta''-$ 18:1 fatty acid is maintained at a critical concentration in control strains by regulating its biosynthesis. In mutants, the concentration of this fatty acid may exceed the amount in control cells by exogenous addition in growth media. This increase slows the growth, increases the permeability to small molecules, and diminishes the membrane-associated NADH oxidase activity.[27] This loss of normal membrane function could be correlated with distinct changes in the physical properties of the phospholipids in the isolated membranes; one aspect of this property is the nature of thermotropic transition of phospholipids.[28] In addition to concentration, the proportion of different acyl chains in a fatty acid molecule has a significant role in conserving the functional state of the membrane. The acyl group structure influences, in addition to the characteristics of thermotropic phase change, the liquid crystalline state of membrane lipid.[29] Data have been generated to show that a high increase of the 18:1 molecular species of fatty acid inhibits NADH oxidase activity whereas an increase of the 16:0 molecular species causes cell lysis. The placement of two different acyl groups on the same phospholipid molecule is important for the weak interactions between phospholipids. The synthesis of phospholipid molecules with the same acyl group is important for adaptation to the changes of environmental temperature through membrane fluidity change.[30] However, an excessive amount of phospholipids with the same acyl groups leads to strong lipid-lipid interactions, causing the formation of quasi-crystalline clusters.[31] This change may alter the annular lipid material surrounding the membrane proteins and may increase the viscosity of bulk lipid decreasing rotational or lateral diffusion of small molecules.[32,33] In order to maintain the growth of *E. coli*, 50% of the membrane lipid must be maintained in a fluid state.[27,34,35] Unsaturation of the fatty acids of the phospholipid is important for the maintenance of this fluidity. Therefore, replacement of the unsaturated fatty acids with saturated ones affects the fluidity and membrane function. However, retention of fluidity with saturated but shorter-chain-length fatty acids may maintain normal cell function.[36] Although the thermal characteristics (i.e., fluidity) of the membrane are governed by the major phospholipid, the minor components (e.g., phosphatidylglycerol) may function by forming a complex with the integral membrane proteins. These proteins form bridges through divalent cations to ionized amino acid residues of the protein molecules. This bridge allows hydrophobic association of the lipid acyl chain with the hydrophobic surfaces of the proteins.[28]

The requirement of fatty acid heterogeneity has been studied in mycoplasma.[37] The results clearly show that neither fatty acid heterogeneity nor a low cooperativity lipid phase transition is essential for proper membrane function in this organism. The low transition cooperativity of mycoplasma membrane lipids is related to the high adaptive nature of this organism. Thus, there is less drastic change in the membrane phase in response to environmental perturbations. Lipid fatty acyl group heterogeneity provides this adaptive capacity to sterol-deficient bacterial membranes excluding mycoplasma. It has been shown in mycoplasma membrane that cholesterol prevents the crystalization of membrane lipid and maintains a state of fluidity required for membrane function.[38]

C. Intramembranous Movement of Molecules

It is well known from the membrane fluidity study that phospholipid molecules exhibit rapid lateral diffusional flow; such flow is restricted across the bilayer of membrane. Transbilayer movement, however, is important for the membrane-membrane exchange reaction. In animal cells this exchange occurs through specific exchange proteins.[39] Such exchange proteins have been suggested in bacterial cells, but have not been demonstrated. However, translocation of phospholipids occur between the outer and the inner membrane of Gram-negative bacteria.[40] This movement is physiologically significant because the biosynthetic enzymes of phospholipid and lipopolysaccharides are located in the inner membrane and, hence, the synthesized materials can be incorporated into the outer membrane only by a transmembrane-translocation reaction.[41-43] Translocation of lipopolysaccharide is irreversible, but the movement of phospholipids may occur in both directions.[43,44] This rapid, reversible, and relatively nonspecific transmembrane translocation of the lipid has been studied in *Salmonella typhimurium*. Exogenously added lipid from phospholipid vesicles incorporates into the outer membrane of intact cells.[45] The material incorporated into the outer membrane is recovered from the inner membrane. This intermembrane movement occurs through the zone of adhesion between the two membranes.[44]

III. ENZYMES OF LIPID BIOSYNTHESIS

A. Phospholipid Biosynthesis and Degradation

Lipid synthesis can be subdivided into two broad categories: fatty acid biosynthesis and phospholipid biosynthesis. In addition, a variety of transferase and kinase activities are also involved.[46,47] The enzymes of fatty acid biosynthesis are cytosolic, but phospholipid biosynthesis enzymes are located in the plasma membrane (Tables 1 and 2).[46] Many of these enzymes have been extensively purified. Experimental approaches of the study fall into three general categories: (1) demonstration of enzyme activities in the crude cell extract which catalyzes specific biochemical reactions; (2) demonstration of intermediates in the biosynthetic pathway by radiolabeling; and (3) studies on the alteration of the lipid moiety in the mutants which have defects in the enzymes of phospholipid biosynthesis.

Starting with the initial acylation reaction of Sn-glycerol-3-phosphate, all the reactions of phospholipid synthesis occur in the plasma membrane. The only exception is phosphatidylserine synthetase, which is tightly bound to ribosomes; however, the enzymes act upon membrane-bound substrates.[48,49] The sequence of reactions is shown in summary form in Figure 1.

The sequence of enzyme reactions described above has been developed from studies using primarily *E. coli* cells. A detailed discussion on the mechanism of action of these enzymes is beyond the scope of this chapter. Other enzymes, in addition to these, have been suggested as being indirectly involved in phospholipid biosynthesis.

The first committed step of phospholipid biosynthesis is Sn-glycerol-3-phosphate acyltransferase activity. However, this enzyme requires activation. Adenylate kinase is a dual-function cytosolic enzyme producing ATP on the one hand and functionally interacting with membrane-bound acyltransferase activity on the other.[50] Purified inner membrane contains several proteins which are precipitated by anti-adenylate kinase antibody. Hence, there may exist two forms of adenylate kinase; the one is an enzyme cytosolic synthesizing ATP and the other activates the membrane-bound acyltransferase. The membrane-bound acyltransferase activity has a number of different substrates

Table 1
DISTRIBUTION OF THE ENZYMES OF PHOSPHOLIPID BIOSYNTHESIS IN *E. COLI*

Enzyme	Total membranes	Specific activity			
		L	M	H	L/H
Glycerol 3-phosphate acyltransferase		54.2	11.3	2.5	21.7
1-Acylglycerol 3-phosphate acyltransferase		100.0	16.4	4.8	20.8
CTP: phosphatidic acid cytidyltransferase	2.5	6.7	3.2	0.73	9.18
Glycerol 3-phosphate: CMP phosphatidyltransferase	23.7	27.2	5.50	2.17	12.5
Cardiolipin synthetase	3.07	7.35	2.00	0.53	13.9
Serine CMP: phosphatidyltransferase	0.123	0.223	0.106	0.026	8.58
Phosphatidylserine decarboxylase	8.53	20.2	4.47	0.97	20.8
Phospholipase A_1	1.93	0.28	5.16	8.88	0.032
Glycerol 3-phosphate dehydrogenase		380.0	70.8	40.5	9.38

Note: Membrane fractions were prepared from *E. coli.* Four membrane fractions were assayed: total membranes, L, M, and H. L represents the cytoplasmic membrane, M unresolved material, and H the outer membrane. Enzyme assays were performed as described in Ref. 41. All specific activities are expressed as nmol/min per mg protein.

From Bell, R. M., Mavis, R. D., Osborn, M. J., and Vagelos, P. R., *Biochim. Biophys. Acta*, 249, 628, 1971. With permission.

and effector binding sites; thus, this enzyme activity may arise from molecules composed of several proteins.[51]

Acyl group asymmetry in phospholipid molecules arises from acyltransferase activity. The activities for the consecutive acylation reactions reside in *E. coli* inner membrane.[52,53] The two enzymes act consecutively, transferring acyl groups to positions 1 and 2; the acylated product of position 1 has been suggested to be the substrate for the other. Acylation specificity is not absolute and depends partly upon the supply of fatty acids. The use of well-defined native acyl-acyl carrier protein shows that the two enzymes differ in K_m values for glycerophosphate.[51] Palmitoyl-acyl carrier protein is the substrate for the position 1, and palmitoleic acid is the substrate for the other.[54] The Sn-glycerol-3-phosphate acyltransferase has been purified from an *E. coli* strain which contains the multicopy plasmid *plsB* gene and overproduces the enzyme.[55] This is an integral membrane protein requiring Triton X-100® for solubilization; 50% of its amino acids are hydrophobic. Catalytic activity of this enzyme, both crude and purified forms, requires reconstitution with phospholipids, preferably homologous.[56]

Cytidine 5′-diphosphate (CDP) diglyceride synthetase has been demonstrated in both Gram-negative and Gram-positive plasma membrane.[47] This enzyme has been extracted from *E. coli* membrane by digitonin extraction and partially purified by DEAE cellulose chromatography. Besides CDP, a substantial amount of deoxy CDP diglyceride is also present. High levels of phosphatidylserine synthetase are present in Gram-negative bacteria. Unlike other enzymes of phospholipid biosynthesis, this enzyme is mainly ribosome associated and only 10 to 30% is present in *E. coli* plasma mem-

Table 2
DISTRIBUTION OF THE ENZYMES OF PHOSPHOLIPID BIOSYNTHESIS IN *S. TYPHIMURIUM*

Enzyme	Total membranes	Specific activity L_1	L_2	M	H	L_1/H	L_2/H
Glycerol 3-phosphate acyltransferase		18.2	12.0	3.8	1.2	15.2	10.0
1-Acylglycerol 3-phosphate acyltransferase		121.0	129.0	25.0	11.1	10.9	11.6
Glycerol 3-phosphate:CMP phosphatidyltransferase	16.9	30.8	28.4	5.68	1.85	16.6	15.4
Cardiolipin synthetase	2.73	3.95	5.02	0.75	0.61	6.48	8.23
Phosphatidylserine decarboxylase	5.07	12.7	12.6	5.00	1.10	11.5	11.5
Phospholipase A_1	2.91	0.40	0.39	3.46	6.13	0.065	0.064
Phospholipase A_1 + lysophospholipase	1.48	0.23	0.43	3.31	3.31	0.069	0.130
Glycerol 3-phosphate dehydrogenase		199.0	133.0	45.0	27.4	7.30	4.85

Note: Membrane fractions were prepared from *S. typhimurium*. Two cytoplasmic membrane fractions L_1 and L_2 were obtained. Other conditions are described in Reference 41.

From Bell, R. M., Mavis, R. D., Osborn, M. J., and Vagelos, P. R., *Biochim. Biophys. Acta*, 249, 628, 1971. With permission.

brane.[47] This minor membrane-associated enzyme of *E. coli* is a peripheral membrane protein,[58] whereas in Gram-positive bacteria this enzyme is an integral membrane protein.[59-61] The enzyme of *Clostridium perfringen* is solubilized by Triton X-100® in the presence of manganese ion. The kinetics (BiBi sequential reaction) of both *B. subtilis* intramembranous enzyme and the minor-membrane-associated enzyme of Gram-negative cells are similar. The major ribosome-associated enzyme of Gram-negative cells follows the BiBi ping-pong kinetic mechanism. In *B. licheniformis*, phosphatidylserine constitutes the amino terminus of the membrane-bound penicillinase. The net amount of phosphatidylserine in *E. coli* membrane is very small because membrane-bound decarboxylase activity converts the phosphatidyl serine to phosphatidylethanolamine. The decarboxylase enzyme can be extracted from *E. coli* inner membrane by Triton X-100®. Purification of this enzyme is achieved by combination of ion-exchange chromatography, gel filtration, and density gradient centrifugation.[62] Catalytic activity in a purified state has an absolute requirement for detergent (Triton X-100®). Phosphatidylglycerophosphate synthetase catalyzes the formation of phosphatidylglycerophosphate from CDP diglyceride and Sn-glycerol-3-phosphate. The enzyme is extracted from *E. coli* plasma membrane and purified by affinity chromatography on CDP-diglyceride modified Sepharose.[63] Phosphatidylglycerophosphate does not accumulate in the cell; this is hydrolyzed by a specific phosphatidylglycerophosphate phosphatase. This enzyme is particulate, but has not been purified to homogeneity.[64] The existence of cardiolipin synthetase has been shown in a variety of bacterial cell extracts but purification has not been achieved.[65-67] Diglyceride kinase has been purified from *E. coli* plasma membrane by Triton X-100® extraction and purification in the presence of a detergent.[68-69]

A large number of phospholipid-degrading enzymes have been detected in *E. coli* membrane as shown in Table 3.[53] Many of these enzymes have been purified to hom-

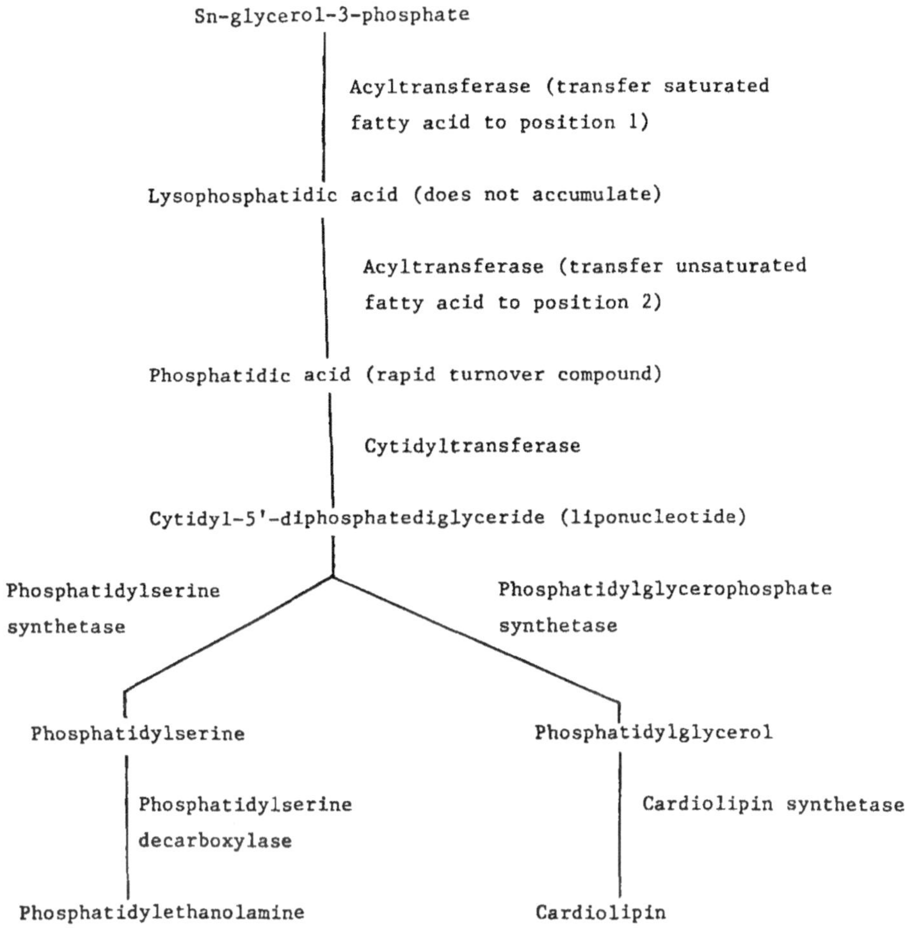

FIGURE 1. Phospholipid biosynthesis in bacteria.

ogeneity; however, the physiological role of these hydrolytic enzymes is not fully understood.

B. Phospholipid Turnover

Phospholipid biosynthesis is regulated in such a way that a rapid membrane growth coincides with cell division by cross-wall formation.[70] It has been suggested that membrane lipids may be involved in the coordination of biosynthetic activities producing crosswalls. The lipid-dependent catalytic activities of cell-wall biosynthesis enzymes may be modulated by the alterations of membrane lipid synthesis.[71,72] In spite of considerable controversary, firm data are now at hand to show that the stimulation of membrane lipid synthesis is related to the cross-wall formation through the activation of cell-wall hydrolase.[70,73,74] Although the protein synthesis continues at a constant rate during cross-wall formation, the lipid synthesis increases, thus causing an increase of the lipid/protein ratio. The composition of phospholipid remains unaltered during this period of increased synthesis. Wide variations of phospholipid composition, however, have no effect on the membrane function.[75] Net hydrophobic environment rather than a specific phospholipid molecule is critical for the membrane enzyme function. In mutants the accumulation of a specific phospholipid exerts a lethal effect by altering the ionic condition, rather than the reduction of a specific molecular species of the phospholipid.

Table 3
PHOSPHOLIPID DEGRADING ENZYME IN *E. COLI*[47]

Enzyme	Location	Known substrates	Bonds cleaved	Purified	Other properties
Lipases					
Phospholipase A	Outer membrane	PE,PG,CL,lyso derivatives	A_1,A_2,L_1,L_4	Yes	Detergent and heat resistant; requires Ca^{2+}
Phospholipase A	Cytoplasm	PG	A_1 or A_2 (or both)	No	Detergent and heat sensitive; requires Ca^{2+}
Lysophospholipase	Inner membrane	Lyso-PE	L_2	No	Heat sensitive; no Ca^{2+} required
Lysophospholipase	Cytoplasm	Lyso-PE,lyso-PG, glyceride	L_1,L_2	Yes	Heat sensitive; no Ca^{2+} required
Phospholipase C	Unknown	PE		No	
Phospholipase D	Cytoplasm	CL		No	Mg^{2+} and possibly ATP are required
Phospholipase D	Cytoplasm	PS		Yes	Associated with phosphatidylserine synthetase
Other hydrolases					
Phosphatidic acid phosphatase	Membrane	Phosphatidic acid		No	Separable from other phosphatases
CDP-diglyceride hydrolase	Inner membrane	CDP-diglyceride	Pyrophosphate bond	Yes	Highly specific, no Ca^{2+} required

Note: Abbreviations: PE, phosphatidylethanolamine; PG, phosphatidylglycerol; CL, cardiolipin; PS, phosphatidylserine; A_1, phospholipase A_1; A_2, phospholipase A_2; L_1, lysophospholipase L_1; L_2, lysophospholipase L_2.

From Raetz, C. R. H., *Microb. Rev.*, 42, 614, 1978. With permission.

The above experimental results suggest that membrane phospholipids undergo both synthesis and degradation to maintain a balanced membrane function during cell growth. A large number of phospholipid-degradative enzymes, present in the membrane, may play an important role in this phospholipid turnover. It has been suggested from detailed pulse chase experiments that in differing regions of the membrane, biosynthesis occurs on one hand and degradation on the other, maintaining a molecular gradient of lipid composition. Such an idea of dynamic membrane compartments may explain a variety of site-specific functions. It has been shown that phosphatidylethanolamine is degraded only in the outer membrane, but phosphatidylgylcerol is hydrolyzed in both the outer and inner membrane. The fatty acids of the hydrolyzed phospholipid, however, are only incorporated in the plasma membrane.[76]

The mechanism of action of specific enzyme reactions has been determined by ESR spectroscopy. The state of diffusion of substrate into the membrane is important for catalytic action; CDP-diglyceride exists in remarkably different locations within the membrane bilayer, depending on the presence or absence of Mg^{2+}. Magnesium ion and CDP-diglyceride form a complex with the membrane surface arresting polyglycerophosphate synthesis. It has been suggested that complex formation with Mg^{2+} does not allow the polar head group to approach the vicinity of the active site of the synthetase. However, phosphatidylethanolamine synthesis is not affected. This suggests that the topological relationship plays an important regulatory role for phospholipid biosynthesis.[77] The importance of topological distribution of enzymes is evident from *de novo* phospholipid synthesis in the plasma membrane vesicles of *E. coli.*[78] In these vesicles, enzymes are membrane associated and have an inverted orientation. Therefore, substrates could interact with the enzyme molecules. The products of biosynthesis are incorporated into the membrane vesicles which maintain intact vesicular structure. Lipid accumulation inside these vesicles causes a lowering of their density. Due to the nonrandom distribution of lipid-synthesizing enzymes, many vesicles remain inactive for lipid synthesis and, hence, do not show any density alteration. Accumulation of lipid in the membrane does not inhibit the *de novo* lipid synthesis. However, such accumulation is not expected because lateral diffusion of enzyme molecules should increase the enzyme substrate mean distance and decrease the enzyme activity. Electron microscopy shows that the enzymes exists in clusters, whereas lipid accumulation occurs in a gel state in the membrane. Genetic modification of the enzymes of membrane lipid biosynthesis has enriched the knowledge on bacterial lipid biosynthesis and control, an aspect which is beyond the scope of this section.

IV. ADENOSINETRIPHOSPHATASE (ATPase)

A. Physiological Role and Distribution

ATP phosphohydrolase (ATPase 3.6.1.3.) is defined as an enzyme participating in two distinct biochemical reactions: biosynthesis of ATP from ADP, and coupling of ATP, generated from substrate-level phosphorylation, to the energization of membrane.[79] The direction of proton translocation is the same for both mitochondria and bacteria.[80] The bacterial system, however, has been suggested to differ from the mitochondrial proton-generating system in establishing proton gradient by the hydrolysis of ATP rather than the synthesis of ATP by the agency of proton gradients.[81]

It has been suggested by Harold that in the bacterial system ATPase plays a role coupling the high energy of ATP to the generation of a proton motive force which supports a diverse group of energy-consuming functions in anaerobic, facultative aerobic, and photosynthetic bacteria.[82] The diagrammatic summary of this coupling function is presented in Figure 2.[82]

The presence of ATPase has been demonstrated in a large number of prokaryotic cells. A list of the occurrence of ATPase has been presented in Tables 4[83] and 5. This list, although not complete, provides a view on the diversity of both occurrence and organization in membrane.

B. Purification, Subunit Structure, and Properties

By definition ATPase is a peripheral-membrane protein. Low salt wash releases membrane-bound ATPase activity.[79,81,83] In fact, Ishikawa and Lehninger first showed the release of ATPase from *Micrococcus lysodeikticus* by "shock wash" using cold water.[97] The subcellular source of ATPase, its extraction, and the method of purification have been shown in Table 6. The ATPase activity capable of energy transfer reactions and H^+ ion exchange in the native membrane can be divided into two basic components, F_1 and F_o. The F_1 activity is insensitive to the inhibitors of energy transfer, oligomycin, and *N-N'*-dicyclohexylcarbodiimide (DCCD). The F_o component, on the other hand, is sensitive to these inhibitors. Broadly, three different treatments release F_1 from the membrane: (1) at low ionic strength after divalent cation depletion, (2) by mild anionic detergent treatment (e.g., SDS or SLS), and (3) by lipid depletion of the membrane with organic solvents. The F_o component (DCCD-sensitive ATPase) is released by nonionic detergent treatment. In addition, the component of ATPase, which is residual in the membrane after removal of the F_1, may restore DCCD sensitivity. The F_o component appears to be anchored into the membrane through hydropho-

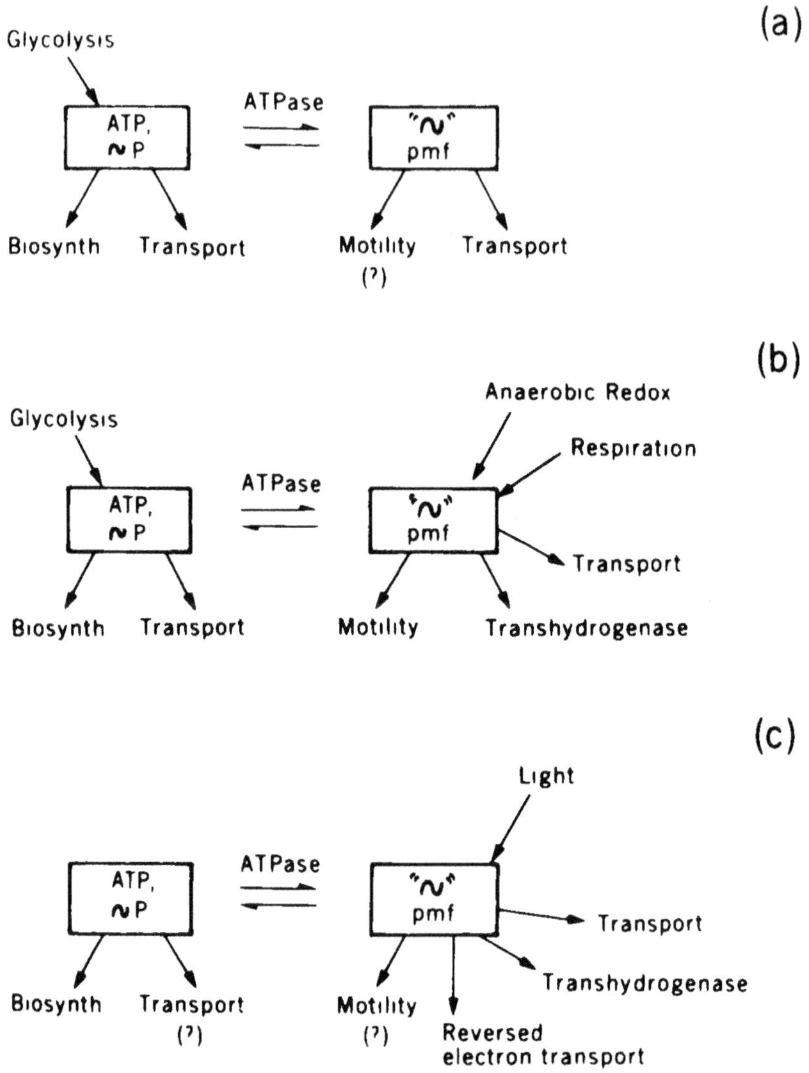

FIGURE 2. Some bacterial energy economics. (a) Fermentative metabolism in organisms such as streptococci or clostridia; (b) metabolism of facultative anaerobes such as *Escherichia coli;* (c) metabolism of photosynthetic bacteria. (From Harold, F. M., *Curr. Top. Bioenerg.*, 6, 83, 1977. With permission.)

bic interaction. Divalent cations are needed to retain F_1 in the membrane. The purification of F_1 is generally achieved by salt precipitation, ion exchange, and gel filtration chromatography (Table 6). The F_o or F_1-F_o complex, however, requires antibody[31] or ADP-affinity chromatography.[85] Functionally, the ATPase activity culminates in oxidative phosphorylation; therefore, sensitivity to oxidative phosphorylation inhibitors is a measure of catalytic activity of the ATPase complex. The DCCD-sensitive ATPase complex is purified from the detergent extracts of the membrane by sucrose density gradient and gel filtration chromatography.[115-118] The maximal ATPase activity and DCCD sensitivity are retained by the addition of phospholipid during purification. If the increase of catalytic activity is considered to be the measure of enzyme purification, a 100- to 500-times purification has been achieved. The increase of catalytic activity,

Table 4
DISTRIBUTION OF MEMBRANE ATPASE IN VARIOUS BACTERIA

Source	Type of preparation	Divalent requirement and activation	Effects of Na^+ and K^+
Gram-Positive			
Bacillus megaterium	Membrane-bound and soluble	Ca^{2+}>Mg^{2+}	None or little
Bacillus stearothermophilus	Membrane-bound and soluble	Mg^{2+}>Ca^{2+}	Not significant
B. subtilis	Membrane-bound	Ca^{2+}>Mg^{2+}	Stimulation at high concentration
Lactobacillus arabinosus	Cell-wall membrane	Mg^{2+}	Unknown
L. fermenti	Membrane-bound	Mg^{2+}>Ca^{2+}	Unknown
Micrococcus lysodeikticus	Membrane-bound and soluble	Ca^{2+}>Mg^{2+}	None; inhibition at high concentration
Staphylococcus aureus	Membrane-bound	Mg^{2+}>Ca^{2+}	Stimulation at high concentration
Streptococcus faecalis	Membrane-bound and soluble	Mg^{2+}	Stimulation at high concentration
Gram-Negative			
Escherichia coli	Membrane-bound and soluble	Mg^{2+}>Ca^{2+}	Slight effects; inhibitory together
Marine pseudomonad	Envelope	Mg^{2+}	No activation
Vibrio parahaemolyticus	Envelope	Mg^{2+}	Stimulation at high concentration
Vitreoscilla sp.	Envelope and soluble	Mg^{2+}	Stimulation at high concentration

From Salton, M. R. J., *Adv. Microb. Physiol.*, 11, 213, 1974. Copyright: Academic Press Inc. (London) Ltd. With permission.

Table 5
DEMONSTRATION OF ATPASE IN DIFFERENT BACTERIA

Organisms	Ref.
Thermophilic bacteria	
Gram-variable bacterium PS3	84
Bacillus acidocaldarius	
Alkalophilic bacillus	
Mesophilic bacteria	
Mycobacterium phlei	85
Streptococcus faecalis (DCCD-resistant)	86
Bacillus subtilis (DNA-dependent ATPase)	87
Vibrio cholerae	88
Acholeplasma laidlawii	89
Clostridia	90
Azotobacter vinelandii	91
Rhodopseudomonas spheroides	92
Proteus	93
Rhodospirillum rubrum	94
Mycoplasma laidlawii	95
Salmonella typhimurium	96

however, is not only caused by the rise of the concentration of ATPase protein relative to the crude extract, but also by the potentiation of latent activity.

ATPase is a multi-subunit enzyme; these subunits are released under differing extraction conditions. It has been shown that DCCD inhibits the cholate- or deoxycho-

Table 6
PURIFICATION OF ATPASE IN DIFFERENT ORGANISMS

Organism	Gram behavior	Subcellular source	Extraction	Purification	Ref.
M. lysodeikticus	Gram-variable	Membrane (sonicated envelope fragment)	Cold water	Density gradient	97
S. faecalis	Gram-positive	Membrane	Mg^{+2}-free buffer	Density gradient	98
M. lysodeikticus	Gram-variable	Membrane	3 m*M* Tris without Mg^{+2}	—	99—102
B. megaterium	Gram-positive	—	—	—	103—105
E. coli	Gram-negative	30,000 × g pellet from 10,000 × g supernatant of sonicated cells	2 m*M* Tris	DEAE column chromatography	106
E. coli	Gram-negative	Spheroplast membrane	Mg^{+2}-free buffer	Ion exchange column chromatography	107
B. stearothermophilus	Gram-positive	—	Dialysis against EDTA buffer and EDTA distilled water	Ion exchange column chromatography	108
E. coli	Gram-negative	Spheroplast membrane ghost, prepared by lysozyme-EDTA	1 m*M* Tris + 0.4% SLS	Ion exchange column chromatography	109
M. phlei (DCCD sensitive)		Sonicated envelope fragments	0.25 *M* sucrose, no ion, 105,000 × g supernatant	Sepharose ADP affinity column chromatography	85
Thermophilic PS-3 (TFI)		Membrane from lysozyme lysate	0.2 *M* EDTA	DEAE cellulose and Sepharose 6B chromatography	110
Thermophilic DCCD sensitive TFO		Membrane from lysozyme lysate	Cholate-washed Triton X-100® extract	Anti TFI Ab-ppt and Ab-solubilization in 6 *M* area → centrifuge → ATPase in pellet	110
Thermophilic PS3 (TFO-F1)		Membrane from lysozyme lysate	Cholate-washed membrane extracted with Triton X-100® + 0.2 *M* Na_2SO_4	DEAE cellulose followed by Sepharose 6B gel filtration in presence of Triton X-100®	111
M. lysodeikticus	Gram-variable	—	*N*-butanol	—	112, 113
E. coli	Gram-negative	—	Chloroform	—	114

E. coli (DCCD sensitive ATPase)	Gram-negative	Membrane	Deoxycholate	Sucrose density	115, 116
E. coli (DCCD sensitive ATPase)	Gram-negative	Membrane	Deoxycholate	Sucrose density and DEAE cellulose chromatography	117
Clostridium pasturianum	Gram-positive	Membrane	Triton X-100®	Sepharose 4B chromatography	118
E. coli	Gram-negative	Membrane	Hexamethyl phosphoric *N*-methyl pyrolidine (aprotic solvent)	DEAE-cellulose, hydroxypropyl Sepharose G-50 chromatography	114
S. faecalis	Gram-positive	Membrane	Sodium deoxycholate and salt precipitation	Two-dimensional gel electrophoresis	115

late-solubilized enzyme as well as the overall membrane-bound activity; but DCCD does not inhibit the activity of low-salt-washed ATPase. ATPase exists in the membrane as an ATPase complex of molecular weight ranging from 250 to 400 kDa. The subunit structure of ATPases isolated from both mesophiles and thermophiles are similar. There are the following five subunits in order of decreasing molecular weights: α, 54 to 60; β, 52 to 56; γ, 32 to 35; δ, 21; and ε, 11 to 13 kDa.[96,119,120]

The purified individual subunits have been reconstituted in various combinations; the subunits are mixed and incubated at 25 to 40°C at pH 6.3 to 7.0. The mixtures containing $\beta + \gamma$ or $\alpha + \beta + \gamma$ regain ATP hydrolytic activity, but no activity is present in $\beta + \delta$ or $\alpha + \beta$ mixtures. Any combination without β subunit is inactive, but $\beta + \delta$ regains activity only when α subunit is added to the mixture. It seems that the active site of the enzyme is located in the β subunit. The ($\alpha + \beta + \delta$)-ATPase is thermolabile and insensitive to sodium azide. In contrast, ($\beta + \gamma$)-ATPase is thermostable and sensitive to sodium azide comparable to native ATPase. The mobility of ($\beta + \gamma$)-ATPase is comparable to native ATPase, but the subunit mixtures without γ show greater mobility than the native ATPase. The ε subunit inhibits the native ATPase activity of *E. coli* but thermophilic ATPase is not inhibited by ε subunit.

Among the five subunits, δ and ε do not have any direct role in catalytic activity; however, the ε subunit has a regulatory role. The δ subunit functions as a membrane-binding agent. If membrane is depleted of this δ subunit (known as nectin) of the ATPase complex, the F_1 ATPase fails to bind to this depleted membrane. The binding, however, has been suggested to be dependent upon the presence of one ATPase receptor which has specific affinity for the δ subunit of the ATPase complex.[121-124]

The ε subunit is a dual-function subunit. In addition to being an inhibitor for oxidative phosphorylation, this subunit functions in binding the ECF_1 ($\alpha + \beta + \gamma$) to the membrane. The following characteristics of this binding have been determined: (1) both δ and ε subunits are required for the ECF_1 binding to the membrane and restoration of oxidative phosphorylation; (2) either δ or ε will bind singly to the ECF_1 but these ternary complexes do not bind to the δ or ε-depleted membrane; (3) ε improves binding of δ to the deficient enzyme; (4) trypsin-treated ECF_1 lacking the γ subunit does not bind to δ or ε subunits; and (5) ε subunit does not bind to the membrane depleted of ECF.[125] The membrane ATPase complex (F_1-F_o) has been purified from both DCCD-sensitive and -resistant *S. faecalis* by salt precipitation and two-dimensional gel electrophoresis.[116] This process eliminates minor impurities as well as a δ-less form of F_1. The F_1 fragment of the purified F_1-F_o complex of both strains has the usual $\alpha,\beta,\gamma,\delta,\varepsilon$ subunits. The F_o fragment has three (27-, 15-, and 6-kDa) components. Although the 27- and 15-kDa components are identical, the 6-kDa component is different in the resistant strain to the extent that DCCD failed to bind to the F_1-F_o complex. The 6-kDa component may regulate membrane organization of the active site of F_1-F_o complex. The 6-kDa component may regulate membrane organization of the active site of F_1-F_o; therefore, a specific alteration of this component does not allow DCCD binding.

C. Uncoupler *(unc)* Mutant

It was suggested in 1971 that mutation in *E. coli* may uncouple oxidative phosphorylation from electron transport.[126] Since this early observation, a large number of uncoupled mutants have been isolated.[79] These have well-defined genetic lesions in different loci of the chromosome and have exhibited uncoupled oxidative phosphorylation and electron transport. The characteristics of uncoupling, however, differed widely, indicating involvement of multiple genes in the regulation of oxidative phosphorylation. The uncoupled mutants (*unc*) are selected by the inability of the facultative anaerobic cells to grow on nonfermentable substances, such as acetate, malate, and succinate. This growth inability may arise from the mutation leading to uncoupled

Table 7
SOME EFFECTS[a] OF MUTATIONS IN VARIOUS *UNC* GENES

Mutant alleles	ATPase activity	Other characteristics	Component of ATPase affected (polypeptide)
unc A401, *unc* A447, *unc* A453	−	Inactive F_1 on membranes and F_1-depleted membranes can be reconstituted with normal F_1	F_1 (α-subunit)
unc A450	−	Low ionic strength washed membranes cannot be reconstituted with normal F_1: mutant α-subunit remains on membranes	F_1 (α-subunit)
unc B402	+	Normal F_1 on membranes; F_1-depleted membranes bind normal F_1 but energy-linked reactions are not reconstituted; ATPase activity in cytoplasm	F_0
unc C424	+	Membrane-bound ATPase activity sensitive to DCCD and bathophenanthroline; cytoplasmic ATPase activity insensitive to DCCD and stimulated by bathophenanthroline	F_1 and/or F_0
unc D405, *unc* D406, *unc* D409, *unc* D427	−	β-subunit with abnormal net charge present on low-ionic-strength washed membranes; these membranes bind differing amounts of normal F_1	F_1 (β-subunit)
unc D412	(−)[b]	F_1 with low ATPase activity on membranes; F_1-depleted membranes can be fully reconstituted with normal F_1	F_1 (β-subunit)
unc E408, *unc* E429	+	ATPase activity found only in cytoplasm; low-ionic-strength washed membranes do not bind normal F_1	F_0

[a] All strains carrying the mutant alleles listed show little or no growth on nonfermentable carbohydrates, have low serobic growth yields, and cannot carry out oxidative phosphorylation.

[b] F_1 prepared from a mutant strain carrying the *unc* D412 allele has about 10% of normal specific activity.

Reproduced with permission from Downie, J. A., Gibson, F., and Cox, G. B., *Annu. Rev. Biochem.*, 48, 103, 1979. Copyright 1979 by Annual Reviews, Inc.

oxidative phosphorylation and electron transport causing the lack of ATP production. These mutants, however, grow on fermentable substrates (glucose, etc.) due to ATP production by substrate-level oxidation. After mutagenesis by *N*-methyl *N'*-nitro-*N*-nitrosoguanidine,[127] hydroxylamine,[128,129] ethylmethyl sulfonate,[130] and ultraviolet irradiation,[131] the mutants are screened for uncoupled phenotype.[127] As *unc* mutants are generally mapped in the *ilv* region of the chromosome, site-specific mutagenesis is also possible by transduction.[128,129]

All the known mutations that uncouple oxidative phosphorylation map at 83 min on the *E. coli* chromosome; 40% of these mutants are co-transducible with its gene, 70% with the *asn* gene, and 80% with the *bglR* gene.[128,129,131-135] In addition, DCCD resistance is closely linked to the *unc* gene.[135,136]

The closely linked characteristic of *unc* genes suggest that *unc* genes form an operon. Polar mutation caused by bacteriophage Mu as mutagen has established the sequential arrangement of these genes to be *unc BEADC*; the *B* gene is located proximal to the *ilv* gene cluster.[137,138] The effects of mutation of the *unc* gene have been summarized in Table 7.[79] Recently, it has been shown that there is an additional gene *ucf* which codes for 18 kDa-protein of the F_o sector.[139] Thus, there appear to be three genes *unc* BFE which code for the subunits of the F_o sector; i.e., 18- , 24- , and 84-kDa proteins. The 84-kDa protein is the DCCD-binding protein.

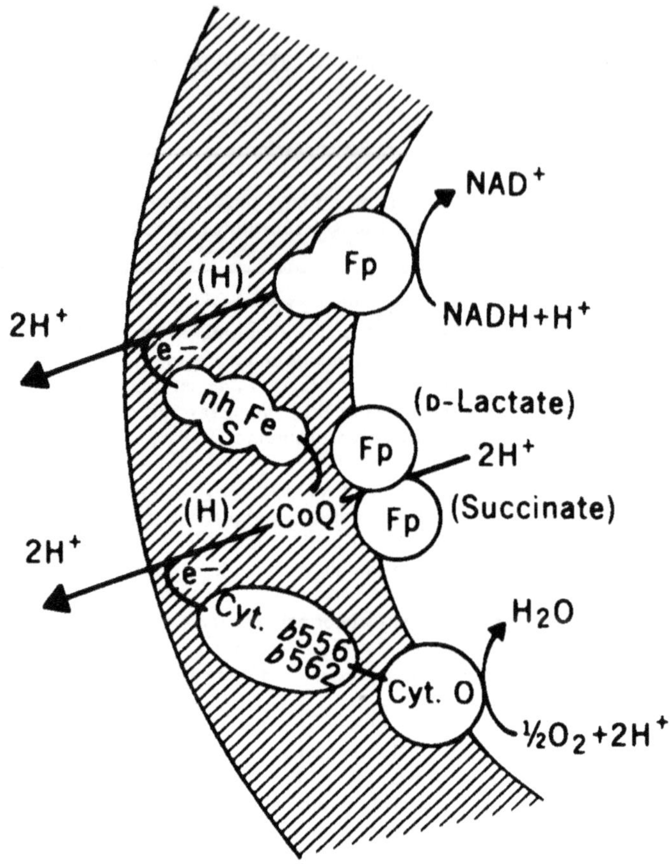

FIGURE 3. A tentative scheme showing two possible respiratory loops in *Escherichia coli*. Oxidation of NADH involves both loops; oxidation of lactate or succinate only the second. (From Harold, F. M., *Curr. Top. Bioenerg.*, 6, 83, 1977. With permission.)

D. Proton Translocation

A clear picture has emerged on the identification of the respiratory chain, the photosynthetic apparatus, and the ATPase complex as pathways for proton translocation. However, the picture on the molecular mechanism of this translocation is still unclear. Particularly in contrast to mitochondria, the knowledge of the bacterial system is poor. There are four clearly defined proton-translocating systems: (1) redox chains, (2) the photosynthetic apparatus, (3) bacteriorhodopsin, and (4) the ATPase and pyrophosphatase systems. The details have been elegantly reviewed by Harold.[82]

The general principle of redox chain operation has been described by Michell, who states that the electron transport is coupled to a proton translocation process.[140] It is suggested that each coupling site translocates two protons outward through a membrane barrier by a proton carrier mechanism, and two electrons flow back by an electron carrier mechanism. Thus, a proton translocation loop is generated. Although the electrogenic proton translocation has been demonstrated in *E. coli*, the constitution of the bacterial electron transport chains is extremely variable;[141] frequently, this variation arises from differing physiological condition. The respiratory loop of *E. coli* is shown in Figure 3.[82] A functional ATPase is not needed for the redox chain coupled to the electrogenic proton translocation.

Proton translocation coupled to photosynthesis has been extensively studied in *Rhodospirillum rubrum* and *Rhodopseudomonas capsulata*. The site of this translocation

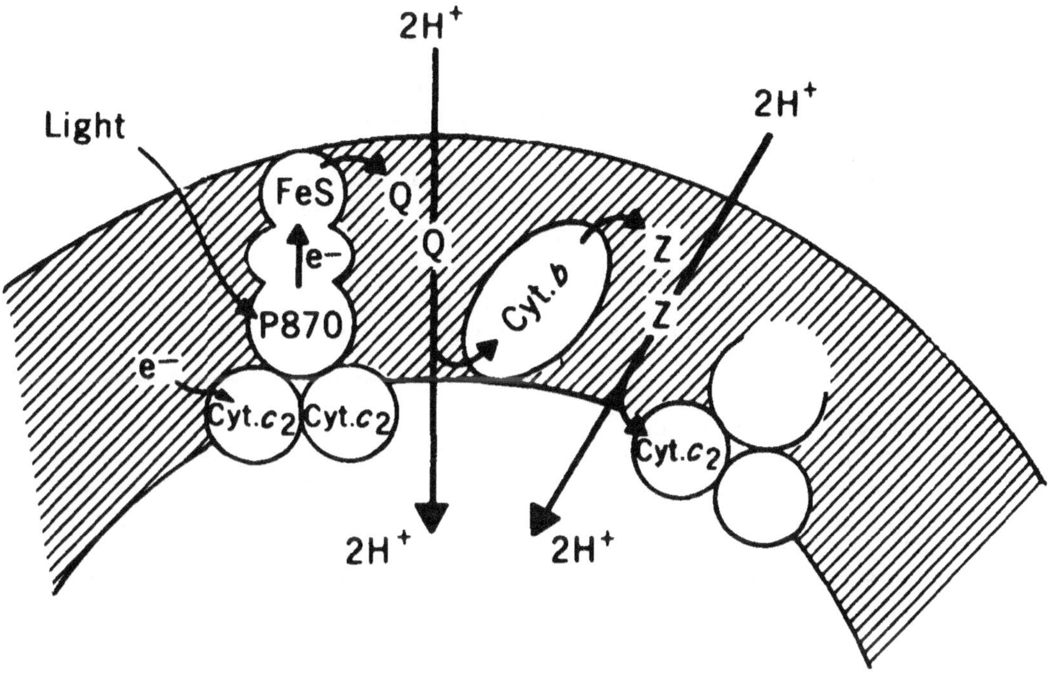

FIGURE 4. Possible pathway for proton translocation during cyclic electron flow in chromatophores of *Rhodopseudomonas spheroides*. (From Harold, F. M., *Curr. Top. Bioenerg.*, 6, 83, 1977. With permission.)

is in the chromatophore membrane which is connected to the plasma membrane. The membrane surface of the chromatophore vesicles is inverted with respect to the plasma membrane surface. Therefore, the translocation of proton across the membrane into chromatophore vesicles can be considered to be equivalent to proton translocation into the extracellular medium. The photosynthetic reaction center is buried in the hydrophobic membrane interior (i.e., chlorophyll) and ejects an electron in the presence of light. This electron is accepted by an acceptor which allows the flow to occur through the cytochrome system. The cytochrome reoxidizes chlorophyll and the light-driven reaction continues.[142] In this manner the membrane becomes energized due to charge separation coupled to proton flow.[143,144] The proton circulation is diagrammatically presented in Figure 4.[82] The electron flow does not require any ATP.

Proton translocation by photoreceptor pigment in *Halobacterium halobium* has been thoroughly investigated because of its simple chemical structure. The bacteriorhodopsin forms a single sheet oriented in such a way that the long axis of the molecule spans the membrane. A hydrophobic pocket is formed by a complex formation of the retinal with the protonated Schiff base to lysine residue of the protein and aromatic amino acid.[144] Studies with reconstituted systems suggested that the proton translocation occurs through a channel mechanism rather than a carrier system.[145,146]

ATP is the "energy currency" of any cell. Hydrolysis of this molecule coupled to proton translocation by ATPase enzyme complex leads to membrane energization essential for a variety of membrane functions. It has been mentioned that the ATPase is a multi-subunit membrane-bound enzyme. These subunits are organized in vivo into a knob structure;[147,148] the head of this knob is oriented towards the cytoplasm. The head corresponds to the F_1 fragment, which is hydrophilic and contains the catalytic center of the enzyme. The base is associated with the membrane and corresponds to the F_o fraction; this is hydrophobic and translocates the proton. The F_o fragment is frequently connected to the head piece by the protein nectin. The base piece functions as a proton translocation channel (ionophore) allowing the proton to be translocated across the hydrophobic barrier to the catalytic site in the interface of the head piece and the membrane;[149] DCCD blocks this F_o ionophore and inhibits both H^+ translocation and ATP hydrolysis. The catalytic site and the head piece, however, are immersed in the cytoplasm, and the proton can only reach the catalytic site through the channel.

It has been suggested that proton translocation can also be effected by the alteration of the ATPase conformation.[150] The binding of ATP to the head piece alters the conformation of the base piece exposing the proton binding site. Further alteration of the conformation caused by ATP hydrolysis releases protons. Thus, a rotational action is generated by ATP binding and hydrolysis; the proton is translocated by this rotational activity.

E. Organization in the Membrane

It has been mentioned earlier that the ATPase has two major components, F_1 and F_o. These components differ in their organization in the membrane as well as in their function; F_1 is generally dissociated from the membrane if there is a low concentration of salt and a metal ion chelating agent in the surrounding medium. This released component (i.e., F_1) catalyzes oxidative phosphorylation; however, a component with identical catalytic activity could be released by *n*-butanol or chloroform extraction. This solvent-extracted F_1 appears to differ significantly in its organization from the F_1 released by salt depletion; in contrast to solvent-extracted F_1, the salt-extracted F_1 fails to reassociate with F_1-depleted membrane. This difference suggests that in vivo functional association of F_1 to the membrane is ionic (divalent ions playing a major role). Nonionic detergent treatment causes dissociation of the F_o component from the membrane. Thus, the F_o component appears to be held to the membrane by hydrophobic bonds. The F_o component functions as ionophore and this channel activity is antagonized by DCCD binding.

The F_1-F_o complex is asymmetrically located in association with the inside surface of the plasma membrane. This asymmetric distribution has been elegantly demonstrated by ^{125}I labeling of the ATPase complex before and after destruction of the permeability barrier. The labeling is done with reagents impermeable to the permeability barrier.[151] Finally, this asymmetric distribution has been fully confirmed by ferritin conjugated anti-ATPase antibody labeling technique.[152] The antibody was prepared against the F_1 fraction of ATPase complex. The ATPase could be labeled only after destruction of the membrane permeability barrier; therefore, the enzyme is bound to the inside half of the plasma membrane bilayer. The membrane, if stained with ammonium molybdate before labeling, shows a large accumulation of particles on the inside surface; these particles have a highly characteristic doughnut shape[153] comparable to any ionophore molecules.[152]

Progress in studies of the organization of the ATPase complex of the mesophilic organism encounters insurmountable difficulty because of the thermolability of the ATPase complex. The ATPase (TF_1-F_o complex) of the thermophilic organism, however, is resistant to such thermal denaturation due to extensive salt bridge formation. A detailed study has been made on the molecular organization of the TF_1-F_o complex from thermophilic bacterium PS3.[154] Purified individual subunits of TF_1-F_o have been used for the stepwise reconstitution of gate, channel, and the energy transformer activity of ATPase complex (Figure 5).[155] Detailed discussion on the gate, channel, and pump activity is beyond the scope of this chapter; however, the definitions of these functional components are (1) gate, a monitor for the passage of solute molecules; (2) channel, a pore which allows the translocation of a specific molecule at the electrochemical gradient of the pore; and (3) pump, a translocator which drives solute mole-

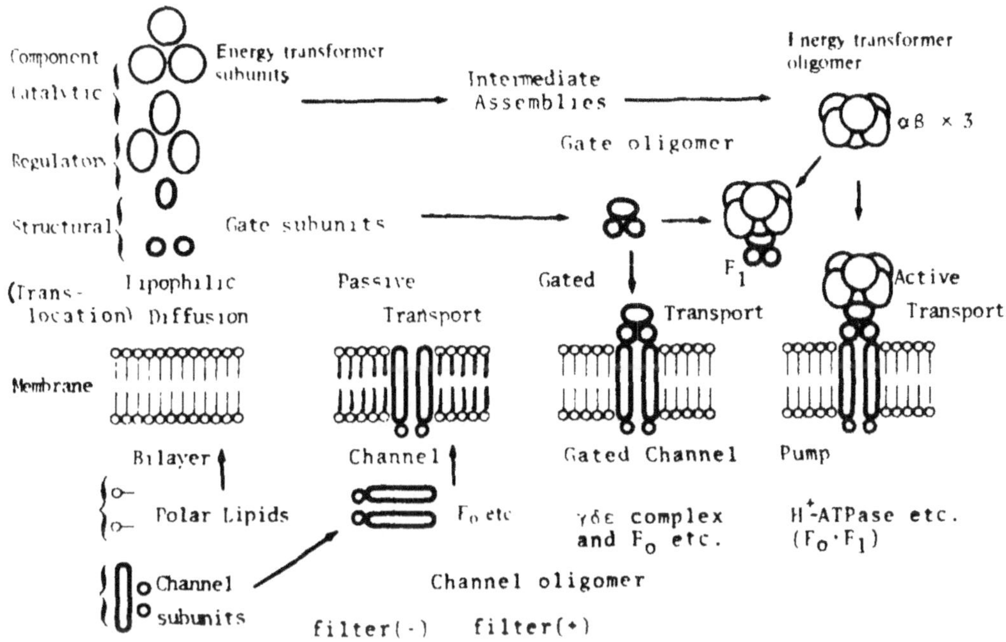

Facilitation	(-)	(+)	(+)	(+)	(+)
Selection	(-)	(-)	(+)	(+)	(+)
Regulation	(-)	(-)	(-)	(+)	(+)
Energy Transformation	( )	(-)	(-)	(-)	(+)

FIGURE 5. Reconstitution of channel, gate and pump. (From Kawawa, Y., *Biochim. Biophys. Acta*, 505, 45, 1978. With permission.)

cules against their electrochemical potential gradient using energy derived from the ATP hydrolysis. This activity, however, is reversible because transport of solutes may generate an electrochemical potential gradient and this energy generated by electrochemical gradient may be used for ATP synthesis.

V. NITRATE REDUCTASE

A. Physiological Role and Distribution

The functions of nitrate reductase in microorganisms are (1) nitrate assimilation and (2) nitrate respiration. Nitrate assimilation is the utilization of nitrate as the sole source of nitrogen for the synthesis of all nitrogen-containing compounds of a cell. Nitrate assimilation can occur under both aerobic and anaerobic conditions of growth. Nitrate respiration, on the contrary, can occur only under anaerobic conditions; in nitrate respiration, nitrate functions as the terminal hydrogen receptor. Prior to utilization, nitrate is reduced to nitrite, which in some microorganisms may further accept hydrogen and be converted into nitrous oxide.[156-159] The nitrate is reduced to nitrite by the enzyme nitrate reductase. The assimilatory nitrate reductase is soluble and is present in the cytoplasm.[160] The respiratory nitrate reductase on the other hand is a membrane-bound enzyme. The molecular properties and organization of the nitrate reductase have been studied extensively in *E. coli.* This enzyme is a membrane-bound multienzyme system which catalyzes the reduction of nitrate with formate as the major hydrogen donor.[161-165]

Table 8
METHODS OF SOLUBILIZATION OF NITRATE REDUCTASE OF DIFFERENT ORGANISMS

Treatment	Organism
Alkali-acetone	*Micrococcus denitrificans*
	M. halodenitrificans
	Escherichia coli
Deoxycholate	*E. coli*
	Klebsiella aerogenes
	Micrococcus denitrificans
	Pseudomonas denitrificans
Sodium dodecylsulfate	*Bacillus stearothermophilus*
Triton X-100®, Triton 144	*E. coli*
Heat (7 min at 60°C)	*E. coli*
t-Amyl alcohol	*Proteus mirabilis*

With permission from Stouthamer, A. H., in *Adv. Microb. Physiol.*, 14, 315, 1976. Copyright: Academic Press, Inc. (London) Ltd.

B. Purification and Properties

The membrane-bound nitrate reductases from different organisms have been solubilized by widely varying extraction procedures (Table 8).[159] Detailed studies comparing the molecular properties of the nitrate reductase released from *E. coli* by different procedures provide clues as to how this enzyme is bound to the membrane. The enzyme can be released by heat treatment (60°C) under alkaline conditions.[166] Suspension of the organism in 50 m*M* phosphate buffer containing 2% Triton X-100® also releases the enzyme from the inner membrane of *E. coli.*[167,168] The enzymes released by both procedures are precipitated with antibody and the polypeptide profiles of the precipitated enzymes are examined by gel electrophoresis after SDS and urea treatment. The enzyme precipitated by antibody contained three subunits, whereas the alkali-heat-solubilized enzyme shows two major subunits and many other degraded products originating from the small subunit. The 142-kDa subunit is the catalytic subunit. The activity of this subunit is destroyed by proteolysis. The small subunit does not affect the catalytic activity by proteolysis, but prevention of the proteolysis of this subunit protects the enzyme from being released from the membrane. The enzyme complex is not held to the membrane through covalently bound phospholipid. It is suggested that the catalytic subunit forms a tetrahedron and the outer faces of the tetrahedron are attached to the B subunit which in turn remains buried in the membrane.[169] The enzyme possibly is exposed on both sides of the membrane. Detergent-extracted nitrate reductase contained a third subunit which has been identified as cytochrome b_1 apoprotein.[170] This cytochrome b_1 has a regulatory function in controlling the amount of enzyme formed as well as the amount of the enzyme inserted into the membrane. It is evident from Table 9 that the nitrate reductase isolated by different procedures from the same organism has different sizes.

These data suggest that the enzyme molecules associate differently under differing conditions, forming aggregates of varying sizes. The heat-released enzyme was purified by single-step gel filtration;[174] this homogeneous protein behaves as an associating-dissociating system and exhibits concentration-dependent sedimentation constants which range from 24S at a high concentration to 10S at a low concentration of protein. The enzyme dissociates into a homogeneous monomer of 200 kDa and may form a

Table 9
EXTRACTION OF NITRATE REDUCTASE FROM *E. COLI* BY DIFFERENT PROCEDURES

Procedure	Mol wt	Ref.
Heat treatment and alkaline extraction	1×10^6	171
Above treatment	0.75×10^6	169
Acetone extraction and deoxycholate treatment	0.32×10^6	172
Only heat treatment	11S	173

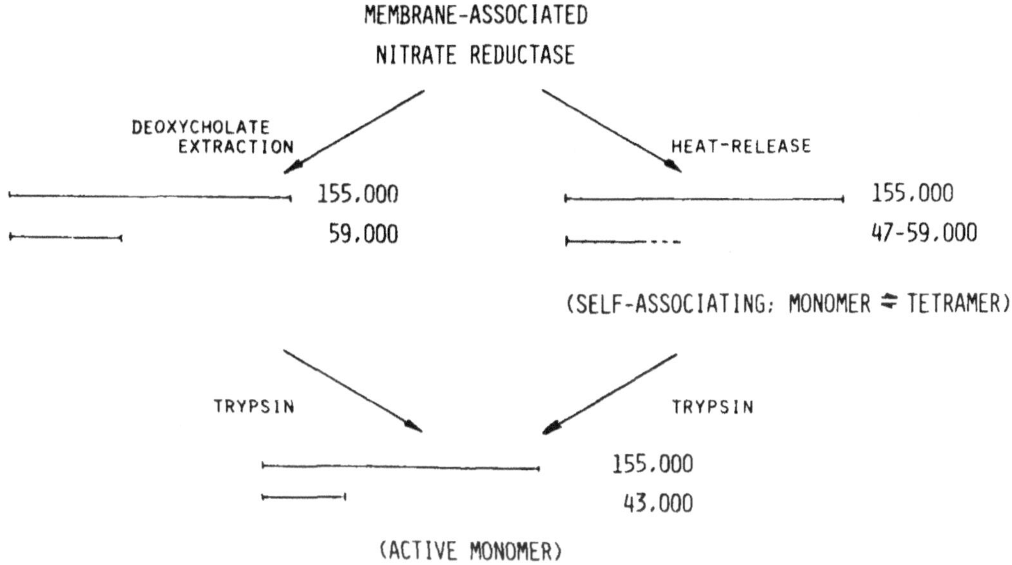

FIGURE 6. Release of membrane bound nitrate reductase. (From DeMoss, J. A., *J. Biol. Chem.*, 252, 1696, 1977. With permission.)

tetramer of 800 kDa. The association-dissociation of the monomer units is reversible. The fractions with the highest specific activity contain 150- and 55-kDa subunits an a ratio of 1:1. A heterogeneous group of enzymes is also released by neutral heat treatment; these enzymes have one common 155-kDa subunit, but the other subunit varies from 47 to 50 kDa. All these enzymes after incubation with trypsin are converted into one form, which loses the property of reversible association-dissociation and contains one 155- and the other 43-kDa subunits.[175] The interrelationship between different forms has been represented diagrammatically in Figure 6.[175] Trypsin treatment removes only a 15-kDa fragment which is responsible for the association-dissociation property of the enzyme. This fragment may also be involved in the binding of the enzyme to the membrane. A multiple molecular form of nitrate reductase having a large catalytic subunit and a small structural subunit is not unique for *E. coli*; comparable subunit composition has also been found in other organisms (Table 10).[159] The nitrate reductase is a metallo-enzyme containing molybdenum. Although the precise role of molybdenum is not known, it appears from electron paramagnetic resonance studies that it may participate in electron transfer reactions.

VI. MEMBRANE-BOUND DEHYDROGENASE

A. Glycerol-3-Phosphate Dehydrogenase

Glycerol-3-phosphate dehydrogenase is an integral membrane protein requiring de-

Table 10
PROPERTIES OF NITRATE REDUCTASE

Property	*Escherichia coli*	*E. coli*	*Klebsiella aerogenes*	*Micrococcus denitrificans*	*M. halodenitrificans*
Molecular weight (daltons)	3.2×10^5	8×10^5	1×10^6	1.6×10^5	1.65×10^5
Subunit structure	Not determined	4 × 142.000 4 × 58.000	4 × 117.000 4 × 57.000 8 × 52.000	Not determined	Not determined
Isoelectric point	4.25			4.2	
Iron content (atoms/mole)	20		48	7.53	2
Molybdenum content (atoms/mole)	1.5	3.2	0.96	0.3	1
Labile sulfide (moles/mole)	18.6		32	10	4
Michaelis constant for nitrate (m*M*)	1.5		0.1	0.25	1.3

tergents or chaotropic agents for solublization.[176] This is a flavin-linked primary dehydrogenase of the cytochrome chain and serves as an essential function for aerobic growth on glycerol.[177,178] In constitutive strains of *E. coli*, 10% of the plasma-membrane protein appears to be composed of this enzyme.[179] The active site of the enzyme faces the cytoplasmic side of the membrane.[180] The purified protein is a dimer and has a subunit molecular weight of 58 kDa. The pure enzyme contains 0.5 mol flavin adenine dinucleotide per mole of 58-kDa protein. The membrane-bound and soluble forms of enzyme have the same broad pH and temperature optimum; however, activation energy of the soluble form is greater than the membrane-bound form. The enzyme activity is inhibited by structural analogues (e.g., dihydroxy acetone phosphate, phosphoenol pyruvate, etc.) and purine nucleotides (ATP and GTP). The phosphorylation of enzyme is not involved in the inhibition because nonhydrolyzable ATP also inhibits the activity.

The detergent can be exchanged for sodium perchlorate to solubilize the enzyme from the membrane. The catalytic activity, however, is lost after detergent removal. This loss of activity is regained by the addition of phospholipid; however, this reactivation does not require any specific phospholipid molecule in the assay medium. Both ATP and GTP appear to be the natural effector molecules in regulating the enzyme activity. The pure solubilized enzyme has been reconstituted, after detergent removal, into artificial lipid vesicles. These reconstituted vesicles stimulate glycerol-3-phosphate-dependent active accumulation of L-proline into vesicles.

B. L-Lactate Dehydrogenase; (LDH)

L-Lactate dehydrogenase is a primary dehydrogenase, which transfers electrons from L-lactate directly to cytochrome in aerobically grown *E. coli*.[181,182] This enzyme is an inducible membrane protein in cells when grown aerobically in DL-lactate (Figure 7).[181] This induction is completed within one-fourth generation. The cells grown in glucose or lactate medium can be induced for LDH after the addition of DL-lactate in the media.[183] The enzyme is located entirely in the plasma membrane and constitutes 3% of the membrane protein. The enzyme is extracted from the washed membrane with 1% cholate or 1% Triton X-100® or 0.1 *M* sodium thiocyanate; however, 1% deoxycholate or 1 *M* sodium chlorate extracts the enzyme with considerable loss of activity.

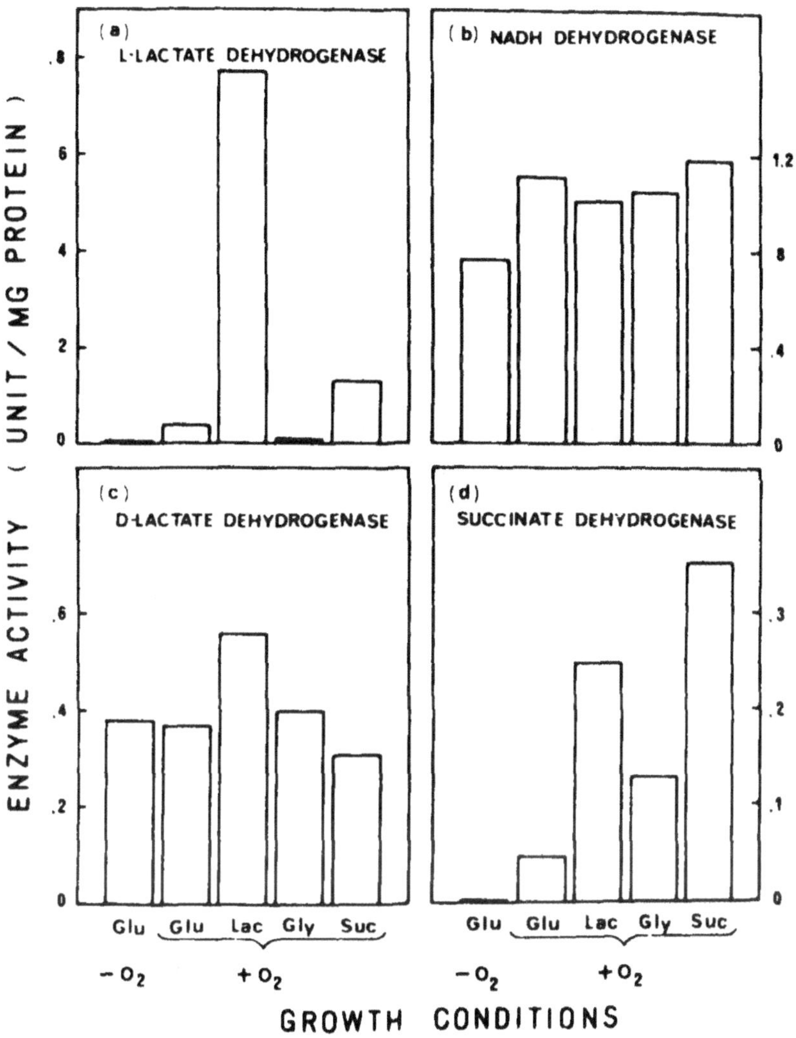

FIGURE 7. Activities of dehydrogenases in membranes of *Escherichia coli* ML-308 to -205 grown under different conditions. Membrane fractions were obtained by lysing cells grown aerobically ($+O_2$) on glucose (Glu), lactate (Lac), glycerol (Gly), or succinate (Suc), or anaerobically ($-O_2$) on glucose. L-Lactate, D-lactate, NADH, and succinate dehydrogenase were assayed and their specific activities are shown. (From Futai, M. and Kimura, H., *J. Biol. Chem.*, 252, 5820, 1977. With permission.)

The enzyme, extracted with cholate and precipitated with ammonium sulfate, is purified by DEAE cellulose or Bio-Gel A-O.5m chromatography in presence of Tween-80.® The specific activity of the pure material is 44 times higher than the crude membrane, but only 3% of the net activity of the crude membrane is recovered in the purified enzyme.

The molecular weight in 0.1% SDS-PAGE is 43 kDa. Comparable to the intrinsic membrane protein in general, L-lactate dehydrogenase forms oligomers in aqueous solution.[184,185] The oligomeric molecular weight after the removal of the detergent is 480 kDa, determined by sucrose density gradient. This oligomeric character is retained in the presence of low concentration of detergents.[185] The enzyme contains 1 mol flavin mononucleotide per 43 kDa pure protein (the optimum activity is between pH 8 to 9 and the isoelectric point is at pH 8.3). The enzyme shows no activity with D-lactate and is inactivated in presence of 10 m*M* pyruvate.

The antibody prepared against purified lactate dehydrogenase is highly specific; however, this antibody shows cross-reaction with D-lactate dehydrogenase. The antibody inhibits 80% of the catalytic activity after interacting with the enzyme; however, the pure enzyme is less sensitive to this inhibition than the enzyme activities in the crude membrane. As this inactivation of the membrane activity is seen only with inverted membrane vesicles prepared by a French press, the active sites of L-lactate dehydrogenase must face the cytoplasmic side of the membrane.[186]

Catalytic activity of this oligomeric enzyme is stimulated by phospholipids. This stimulation has considerable specificity because a mixture of *E. coli* phospholipid, phosphatidyl glycerol, and cardiolipin increases the enzyme activity, whereas phosphatidylethanolamine or dilauryl sulfate has no effect. Thus, acidic phospholipids are important for the activation, which occurs not due to increased accessibility of the dye (electron acceptor) to the enzyme, but due to a change in the conformation of the enzyme molecules.

The amino acid composition shows nonpolar and uncharged polar amino acids to be 80%; hence, a substantial hydrophobic portion is expected to be present in the LDH molecule. The stability of the hydrophobic domain of the molecule is gained by the interaction with phospholipids. Increased accessibility of the antibody molecule to the enzyme oligomer and its increased heat stability after phospholipid treatment suggest the effect of phospholipid on the enzyme conformation. The portion of the LDH molecule interacting with flavin mononucleotide is stabilized by its interaction with phospholipid. The α-helical content of the LDH increases after interacting with the phospholipid. Treatment by low concentration of detergent, which usually destroys α-helical content, does not affect the amount of α-helix after phospholipid treatment. The protection of the melting of the α-helix at different temperatures suggests that the secondary structure of LDH is stabilized by the phospholipid. It appears that the enzyme exists in the lipid bilayer in a monomolecular dispersed state. The conformation of the active site and the secondary structure of the detergent-free oligomeric enzyme is significantly different from the native membrane or phospholipid-treated enzyme.

C. D-Lactate Dehydrogenase

D-Lactate dehydrogenase is a membrane-bound primary dehydrogenase in the respiratory chain of *E. coli.* The enzyme is solubilized from the membrane by detergent treatment and purified.[187,188] The concentrated pure enzyme in the ratio of 1 mg protein to 1 mg Triton X-100® is freed of detergent by Bio-Gel P-100® chromatography. The enzyme is finally purified by extensive dialysis in Tris-HCl buffer, pH 7.5 + 0.05% β-mercaptoethanol, and linear sucrose density gradient.[189] The detergent-free enzyme is a homogeneous single protein consisting of an aggregate of 2 to 3 molecules with an apparent molecular weight ranging from 180 to 200 kDa. The catalytic activity of the aggregate is stimulated with Triton X-100®, phospholipid mixture, cardiolipin, and lecithin; the affinity of the enzyme for phospholipid is greater than the detergent. Phospholipid activation requires substantial preincubation which disperses the enzyme to a monomeric form. The activity of this enzyme, in contrast to the aggregated form, is completely inhibited by a specific antibody. The K_m and maximal velocity of the dispersed enzyme are comparable to the enzyme in the native membrane-bound state. The phospholipid dispersion not only renders the active sites available due to monomer formation, but also alters the conformation of the protein enhancing its intrinsic catalytic activity. It is possible that the amphipathic L-lactate dehydrogenase molecule is oriented in such a way that, compared to cytochrome b5, the catalytic and antigenic sites are exposed to the hydrophilic cytoplasmic and the phospholipid binding sitesembedded in the membrane.[190,192]

D. D-Alanine, Allohydroxy-D-Proline, Choline, and Sarcosine Dehydrogenase

Many substrates supporting bacterial growth are oxidized by membrane-bound dehydrogenases at one or more steps in their catabolism. D-Amino acid dehydrogenases in *E. coli*, *S. typhimurium*, and various *Pseudomonas* species are membrane bound, inducible, and linked to cytochrome.[193-195] Besides their substrate specificities there are very small differences among these enzymes. Unlike constitutive dehydrogenases (e.g., succinate or NADH dehydrogenase) which are integral membrane proteins, many of these dehydrogenases appear to be peripheral membrane proteins.[196,197] D-Alanine dehydrogenase of *P. aeruginosa* is membrane bound; 83% of the enzyme is extracted by 1 *M* Tris-HCl (pH7.0) and 68% by 1 *M* phosphate (pH 7.0). The membrane-bound enzyme is unstable and is released nonspecifically from the membrane when the membrane particles are repeatedly washed in the absence of Mg^{2+}. This enzyme cannot be released from the membrane by anionic detergent extraction (e.g., deoxycholate, cholate, and dodecyl sulfate); nonionic detergents (Triton X-100®, Brij 56®) do not release any enzyme in low concentration (0.1%), but high concentrations release major amounts of the enzyme. The salt-extracted enzyme is truly soluble (proven by equilibrium centrifugation) and is free of any lipid moiety. The apparent molecular weight from sedimentation coefficient is 156 kDa. The inactivation of this dehydrogenase by anionic detergent and tight binding as well as release of enzyme by nonionic detergent (Triton X-100®) suggest that these dehydrogenases may have integral membrane-binding proteins which may function as anchors for those proteins. Magnesium ion may cause the formation of a salt bridge between the enzyme and its binding protein in the membrane.[198]

E. Malate Dehydrogenase

Malate dehydrogenase is purified from *E. coli* membrane particles. A homogeneous protein has been obtained by DEAE cellulose, 5′-AMP-Sepharose, and Sephacryl-200 chromatography.[199] Extensive homology of the amino terminal amino acid sequences between the particle-bound *E. coli* and porcine mitochondrial enzyme suggests evolutionary conservation of the molecule. However, the ctyoplasmic malate dehydrogenase isozyme shows only minor homology with the mitochondrial or *E. coli* enzyme.

F. Hydrogenase

Hydrogenase is a membrane-bound enzyme.[200] Location of the enzyme with respect to the two surfaces of the plasma membrane is examined by radioiodination of the enzyme by membrane-impermeant probes in intact spheroplast, sonicated membrane fragments, and inside-out plasma membrane vesicles. Two membrane-impermeant probes, ^{125}I diazotized-iodosulfanilic acid and lactoperoxidase-catalyzed radioiodination, have been used for labeling, and F_1-ATPase has served as control for monitoring the labeling studies. It can be concluded from the results that hydrogenase is transmembranous, being exposed both to the periplasmic and cytoplasmic surfaces. The hydrogenase gene directs the synthesis of the precursor of the enzyme containing the putative leader peptide.[201]

G. Formate Dehydrogenase

Formate dehydrogenase is a proton translocation enzyme in anaerobically grown *E. coli*.[202-204] The orientation of the enzyme in the membrane is examined by labeling with three membrane-impermeant probes, lactoperoxidase catalyzed iodination, diazotized (^{125}I) diiodosulfanilic acid iodination, and diazobenzene ^{35}S sulfonate labeling; F_1-ATPase labeling has been monitored as a control. It seems that both subunits of the enzyme are transmembranous polypeptides. Immunological methods to separate inside-out and right-side-out vesicles facilitated this localization work.

H. NADH Dehydrogenase

NADH ubiquinone oxidoreductase is identified as respiratory NADH dehydrogenase because this enzyme activity is absent from NADH mutants.[205] In fact, such mutants have no NADH oxidase activity and show the presence of an inactive form of the NADH dehydrogenase complex.[206] This enzyme has been genetically amplified 50 to 100 times more than the wild type. The enzyme purified from potassium cholate-extracted membrane has a 45-kDa single subunit. The enzyme contains 1 mol ubiquinone and 1 mol flavin adenine dinucleotide per mole of enzyme subunit and 70% w/w of phospholipid (predominantly phosphatidylethanol amine). NADH oxidase activity can be reconstituted into vesicles prepared from NADH-negative mutants.[207,208]

NADH dehydrogenase has been purified from *Micrococcus lysodeikicus* membrane.[209] The enzyme has been isolated as an aggregate from the washed membrane with phosphate buffer and EDTA; Triton X-100® extraction releases the enzyme from the aggregate. Further purification has been achieved by DEAE cellulose chromatography and iso-electrofocusing. The apparent molecular weight of this protein is 76 kDa.

I. Fumarate Reductase

Fumarate reductase is a membrane-bound enzyme in *E. coli.* The purified enzyme exhibits a substantial modification of properties after interacting with a broad range of anionic compounds.[210] The properties sensitive to this anionic modification are maximum velocity, K_m for reverse reaction (i.e., succinic dehydrogenase activity), thermal stability, and sensitivity to thiol reagent. In contrast to the anionic modification of purified fumarate reductase, the membrane-bound activity cannot be changed by interaction with anions. In fact, the activities of the purified anion-modified enzyme and the membrane-bound native enzyme are similar. It has been suggested that the enzyme exists in an anionic domain in the membrane.

J. Succinic Dehydrogenase

Succinic dehydrogenase (SDH) is the only membrane-bound enzyme of the Krebs cycle in both bacteria and mitochondria. It is a member of both the Krebs cycle and the respiratory chain. It catalyzes the oxidation of succinate to fumarate and transfers the electrons directly to the respiratory chain. Fumarate reductase and SDH catalyze the same biochemical reaction; however, the equilibriums favor succinate oxidation and fumarate reduction respectively. Both the enzymes have a distinct physiological role; SDH is repressed during anaerobic and fumarate reductase during aerobic growth of *E. coli.*[211-213]

The purification and properties of SDH has been extensively studied in *Rhodospirillum rubrum.*[214] The enzyme is extracted from the chromatophore membrane by sodium perchlorate (a chaotropic ion) and precipitated with ammonium sulfate; both succinate and dithiothreotol are used as protective agents. The water-soluble purified SDH contains covalently bound flavin adenine dinucleotide, nonheme iron, and acid labile sulfur. The molecular weight is 100 kDa. There are two unequal subunits, i.e., FP (60 kDa) and IP (25 kDa). After dissociation by repeated freezing and thawing in the presence of trichloroacetate, the FP subunit forms insoluble aggregate and IP remains in solution. The nonheme iron and acid labile sulfur render the enzyme sensitive to oxygen inactivation; therefore, the enzyme should be kept under a reducing environment. It can be suggested from the studies on *R. rubrum* and *R. spheroides* enzymes that FP subunits contain two iron sulfur clusters (FeS_2^x), and IP subunits with one such cluster (FeS_2^x and FES_4^x).[215-217] These clusters are usually referred to as S_1, S_2, and S_3, respectively. The substrate (i.e., succinate) binding site is in FP subunit.[218] A variety of substances inhibit electron transport by binding with specific functional sites of the subunits. Carboxin binds to S_3 inhibiting electron transport from S_3 to quinone, but S_3 reduction by succinate is not inhibited.[219] It has been suggested that this binding requires of membrane component besides SDH.[220,221]

A detailed discussion on the genetics of SDH is beyond the scope of this paper. One important point is that the regulations of the membrane binding components of the SDH and the SDH subunits may be interrelated. The mutants of SDH are usually selected in a minimal medium containing fumarate as the sole carbon source. The mutants fail to grow if the minimal medium contains succinate instead of fumarate.[212,220,223] In Gram-positive bacteria (e.g., *B. subtilis*), SDH mutants can be selected for by acid accumulation inside the mutagenized cells or by defective sporulation. In the latter case, the colonies after mutagenization are usually pigmented. The mutation occurs at the *citF* gene located at 255 min. It appears from examination of large numbers of mutants that the *citF* locus contains the structural genes for the subunits of the SDH complex, consisting of $cytb_{558}$, FP, and IP.[225] It will be discussed later that $cytb_{558}$ is the anchoring protein in the membrane for SDH.

The catalytic site of SDH is possibly oriented towards the cytoplasm; fumarate produced by the action of SDH is utilized in the cytoplasm. A variety of techniques have been used to determine the orientation of SDH in the membrane. Some of these are antibody adsorption, loss of activity by proteolysis, determination of the catalytic activity using a membrane-impermeable acceptor, and labeling of SDH by enzymatic iodination.[226-228] The data clearly show the location of the enzyme to be on the cytoplasmic side of the plasma membrane.

Detergents are used to solubilize the membrane SDH; however, after detergent removal SDH remains soluble and shows a polarity index above 40%.[214] The purified SDH does not appear to be an integral membrane protein. It is possible that SDH is bound nonhydrophobically to an anchoring receptor protein which in turn is hydrophobically bonded to the membrane. Therefore, SDH, when dissociated from this anchoring protein, does not exhibit the properties typical for a hydrophobic protein. *B. subtilis* heme mutants have been used to study the nature of membrane binding of SDH.[229,230] Auxotrophs for 5-aminolevulenic acid (5 ala) cannot synthetize cytochrome 1 or c, in the absence of 5 ala. In strict aerobic conditions 5 ala auxotrophs of *B. subtilis* continue to grow for a limited period because of the excess cytochrome in the membrane; however, membrane-bound active SDH is not synthesized during this period and soluble cytoplasmic FP and IP subunits which lack catalytic activity continue to accumulate. After resumption of heme synthesis in the presence of 5 ala, FP and IP subunits bind to the membrane at a fast rate. The process may even continue if protein synthesis is arrested by chloramphenicol. It is possible that a limited number of specific binding sites for SDH exist in the membrane. Cytochrome may constitute this binding site at least in part. Triton X-100® -solubilized SDH contains a stoichiometric amount of cytochrome b_{558}; the cells which lack this protein contain inactive FP and IP subunits in the cytoplasm. It has been suggested that apocytochrome is inserted into the membrane co-translationally.[231] This apocytochrome b exposes a binding site for the FP-IP dimer only after binding with protoheme. Covalent binding of FAD is important for the FP-IP dimer binding to the cytochrome, but the role of the iron sulfur center is unknown. This binding process may vary in different organisms, because in *R. rubrum* soluble FP has not been demonstrated. The tight binding of SDH to cytochrome which is an integral protein of the bacterial membrane is a common property for many bacterial cells; this direct structural and functional coupling of SDH and cytochrome is significantly different from that of the eukaryotic system.[211] The SDH of pro- and eukaryotic organisms have comparable structure and amino acid composition. The binding and organization of this protein, on the other hand, vary significantly among different organisms. Thus, the basic protein molecule shows considerable evolutionary conservation but the subcellular organization of this molecule depends upon the physiological and morphological characteristic of the organism.

K. Miscellaneous Oxidases, Reductases, and Dehydrogeneses

A variety of specific oxidases or reductases have been reported to be present in many bacterial cells. Generally, cytochromes are tightly associated to these membrane-bound oxidases.[232,233] Some of these oxido-reductases are associated with ammonia oxidation or nitrogenase activities.[234,235] NADH oxidase has been extracted from *Acholeplasma laidlawii* by extraction with buffer containing 3% Triton X-100®.[236] A homogeneous protein obtained by chromatographic purification is a copper-containing iron-sulfur flavoprotein and consists of a large number of hydrophobic amino acids. There are three subunits in the enzyme: 65, 45, and 19 kDa. Various dehydrogenases, reductases, and electron transfer proteins have been characterized in *Desulfovibrio gigas.*[237] Fumarate reductase, nitrate reductase, menaquinone, and cytochrome b are localized in the plasma membrane of this organism. Fumarate reductase is located on the cytoplasmic surface of the plasma membrane whereas nitrate reductase is transmembranous.

VII. MEMBRANE-BOUND HYDROLASE

Hydrolases have a broad physiological role. They are important for the generation of nutrients from complex substrates, the degradation of metabolites for their turnover, and the biosynthesis of pericellular polymeric substances. The membrane-bound nature of these hydrolytic enzymes is significant because (1) the action on polymeric substrates must be site specific; and (2) the accessibility of the active sites of the enzymes must be restricted to prevent indiscriminate hydrolysis of cellular macromolecules. The following is a brief discussion on the characteristics of membrane binding and the properties of purified hydrolases and their organization in the membrane. The literature on these material is sporadic; hence, it will not be possible to present a comprehensive review.

A. Peptidase

It is known from early studies on microbial nutrition that peptides are a major nitrogen source for bacterial growth. It is unnecessary to mention Koch's nutrient broth, consisting of peptones and meat-extract concentrate which is needed to grow the "disease-causing bacteria." The peptides are hydrolyzed by specific enzymes and the resulting free amino acids are used to support protein synthesis and the growth of bacterial cells. Several mechanisms have been suggested for this peptide utilization:[238] (1) hydrolysis of peptides by wall-bound or periplasmic peptidase and the transport of free amino acids into the cell; (2) concomitant hydrolysis of peptides and transport of amino acids by plasma membrane-bound peptidase and coupled amino acid transport system; and (3) transport of intact peptides followed by intracellular hydrolysis producing free amino acids. Wall-bound, periplasmic, or plasma membrane-associated peptidases are rare. Transport of free amino acids are not favored over peptide transport. Therefore, intact peptide transport may be the most significant among the abovementioned three mechanisms for peptide utilization.[238,239] Studies on *Pseudomonas aeruginosa* suggest tight coupling of the peptide transport and hydrolysis.[240,242] In this case the peptidase is associated with the inside surface of the membrane. It has been shown in a methionine auxotroph that neither the sequence of amino acids of the peptide nor the free carboxyterminal is needed for the transport, but free amino terminal end is needed. Membrane-bound *DD*-carboxypeptidase has been found in *Pseudomonas putida,*[242] *B. subtilis,*[243] and *E. coli.*[244] *Mycoplasma laidlawii* contains peptidase associated with the outside surface of the plasma membrane.[245,246] Highly specific peptidases are needed for cell wall biosynthesis; these are membrane-bound α-D-glutamyl-(L)-meso-diaminopimelate endopeptidase and *DD*-carboxypeptidase.[247] A membrane-bound transpeptidase has been reported to interact with the peptidoglycan cross-linking enzyme system in *Streptomyces rimosus*.[248]

Pyrolidone carboxylpeptidase attacks a peptide bond in which a pyrolidone carboxyl residue in the amino terminal residue is involved; furthermore, the nature of penultimate amino acids determines the rate of hydrolysis. This enzyme, which is a membrane-bound enzyme, is important in the milk and cheese industries.[249] Results of detailed study on membrane association indicate that conformation of this enzyme molecule and the state of membrane lipid determines the membrane association of the enzyme. As membrane lipid composition may vary with growth conditions, the membrane association as well as the activity of this enzyme may change in different growth phases. These studies have a general implication that no conclusion should be drawn regarding the *in situ* location of enzymes. Changes (even minor) in the ionic, osmolarity, or hydrophobic conditions during cell fractionation may drastically alter the membrane binding and activity of the enzymes.

B. Protease

Proteases have important physiological roles; e.g., protein turnover, selective proteolysis for the modification of protein molecules leading to their activation, inactivation, or secretion. There is poor knowledge of the mechanism of proteolysis and the characteristics of related enzymes.[250] Intracellular proteolysis proceeds by repeated endoproteolytic actions; ATP is required for the initial cleavages of the substrate. Membrane fragments of *E. coli*, containing an ATP-stimulated protease, degrade proteins to large peptides which are then hydrolyzed into amino acids by ATP-independent soluble protease. These ATP-stimulated membrane-bound proteases carry out a variety of functions; e.g., catabolism of normal cell protein, processsing of the precursor of secreted proteins, and selective inactivation of specific proteins (repressor molecules). About eight distinct soluble proteases have been detected in *E. coli*.[251] None of these serine proteases are membrane bound; however, one of these is periplasmic and the others are both cytoplasmic and periplasmic. In *B. subtilis* 80% of the cell-bound serine protease is cytoplasmic. The other 20% of this protease is membrane-bound integral protein which is extracted with nonionic detergent.[252] Both the cytoplasmic and membrane-bound proteases have been purified; their properties are identical. Membrane-bound and secreted proteases are also same. Two novel membrane-bound proteolytic enzymes (protease IV and V) have been purified and characterized from *E. coli*.[253]

During the last few years highly specific membrane-bound endoproteases have been reported in bacterial cells. These enzymes act only on the precursor of secretory proteins. Bacteriophage f1 pre-coat protein is synthesized in a precursor form with an amino-terminal signal sequence.[254] During mRNA directed translation of f1 coat protein, if the inverted *E. coli* membrane vesicles are present, the nascent protein molecules are cleaved into two fragments, one a mature f1 coat protein and the other an amino-terminal signal peptide fragment. This endoproteolytic activity is stimulated if the incubation mixture containing plasma membrane vesicles is treated with nonionic detergent Nikkol.® This result indicates that this signal peptidase is an integral membrane protein. Coliphase M_{13} coat protein is an integral plasma membrane protein. The precursor of M_{13} coat protein is synthesized by DNA-programmed in vitro synthesis. The amino-terminal leader peptide of this precursor is cleaved by an endoprotease known as leader peptidase. This enzyme is distributed in both the plasma and outer membrane; the activity of plasma membrane is stimulated by a detergent.[255] This leader peptidase has been extensively purified from the plasma and outer membrane and the properties of purified material from both sources are identical.[256] The leader peptidase gene has been subcloned and transferred into a multicopy plasmid PBR322 yielding a new plasmid PTD 101.[157] *E. coli* with PTD 101 produces 30 times more leader peptidase than the control cells and converts the pre-coat into mature coat at an accelerated rate.

The precursor of *B. licheniformis* 749 penicillinase is synthesized in vitro by translation of mRNA in *E. coli* cell-free systems or in vivo by phenyl-ethyl alcohol-treated whole cells. This precursor is a 35-kDa protein. The in vitro translated product, while treated with *B. licheniformis* membrane fraction, loses 26 amino acids of the pre-penicillinase, producing membrane-bound 31-kDa penicillinase. This form has an amino-terminal cystine residue which anchors this protein to the outside layer of the plasma membrane.[258]

In *E. coli* outer membrane, lipoprotein is synthesized as a precursor form. This precursor requires two types of proteases for processing; one is the signal peptidase which cleaves the signal peptide from the pre-lipoprotein, and the other is signal peptide peptidase which hydrolyzes the cleaved signal peptide into amino acids.[259] These two peptidases have different properties: signal peptidase is heat resistant, has a broad pH optimum, and is highly sensitive to detergent. Signal peptide peptidase, on the other hand, is heat sensitive.

C. Alkaline Phosphatase

Alkaline phosphatase is universally present in animal, plant, and microbial cells. In general, it is a nonspecific enzyme which hydrolyzes monophosphoesterases in alkaline pH liberating inorganic phosphate. A variety of isoenzymes and specific enzymes have also been reported. In spite of extensive studies, the precise physiological role of this enzyme has not been established. However, widely differing specific functions have been reported. It is possible to suggest a broad function of this enzyme in phosphate metabolism. In bacterial cells this enzyme is largely inducible in inorganic phosphate-deficient growth media. It is frequently coregulated with a phosphate-binding protein; hence, the enzyme may function as a component of the inorganic phosphate transport system.

Alkaline phosphatase is located in the cell envelope of both Gram-positive and Gram-negative cells. The distribution in the cell envelope of Gram-negative cells is significantly different from that of Gram-positive cells. In Gram-negative cells the enzyme is located almost entirely in the periplasm and a small amount is outer-membrane bound. In Gram-positive cells a significant amount is plasma-membrane bound and the rest remains in the cell wall and periplasm. Early observations using the classical approach of osmotic shock show that the *E. coli* alkaline phosphatase is located in the periplasm.[260] This periplasmic location has been confirmed almost concomitantly using ultrastructural cytochemistry.[261] Both experimental procedures have the potential for generating artifacts: osmotic shock may have caused redistribution of enzyme molecules from the native sites, due to drastic ionic and physical changes of the cell surface. Reaction product cytochemistry may exhibit sites far from the native location of enzyme molecules. In spite of these possibilities, the enzymes obtained from Gram-negative cells continued to be identified as periplasmic proteins. After a considerable period, subcellular location of alkaline phosphatase has been reinvestigated. In this study, data generated from the simultaneous application of reaction product cytochemistry and immunocytochemistry have been compared. In addition to active dimers, inactive monomers have also been localized. Both the monomers and dimers are located in the periplasm. In contrast to wild-type (smooth) strains, the rough strains show the significant amount of enzyme bound to the outer membrane of *E. coli*.[262,263] The outer membrane prevents the periplasmic enzyme from the denaturing effect of fluctuating H^+ ion concentration.[264] The synthesis of alkaline phosphatase occurs in the membrane-bound polysomes. The nascent enzyme chains have been isolated from the membrane-bound polysomes.[265] The amount of these membrane-bound polysomes decreases in alkaline-phosphatase-repressed cells.[266] Comparable to *E. coli*, biochemical analysis shows 97% of alkaline phosphatase to be present in the periplasm of a marine pseudomonad.[267] Reaction product cytochemistry confirms this observation.[268] By spheroplast formation of *Pseudomonas aeruginosa*, alkaline phosphatase has been shown to be located in the periplasm of this organism.[269] In addition to periplasm, some amount of enzyme is found in the outer membrane and the extracellular fluid.[270] Analysis of activation energy and relative hydrolytic activities on monophosphoesters show that the pure enzyme is different from the in vivo cell-bound enzyme. This difference, however, is eliminated if purified enzyme is bound to LPS in vitro. Thus, the microenvironment around the catalytic site of alkaline phosphatase is maintained by lipopolysaccharide.

Application of comparable technology for the subcellular localization of alkaline phosphatase in Gram-positive cells shows the enzyme to be present in the plasma membrane, cell wall, and periplasm. Subcellular distribution has been extensively studied in *Bacillus*. In *B. subtilis* 90% of the enzyme is plasma-membrane bound.[271] This plasma-membrane-bound enzyme is not extractable with nonpolar solvents but with high salt. The nature of salt influences this extraction procedure; salts of magnesium and nitrate appear to be very effective. The salt extraction is rapid and occurs without any cell damage. In another strain of *B. subtilis* SB15, the enzyme has been localized by ultrastructural cytochemistry.[272] The enzyme is bound to the inside surface of the plasma membrane as aggregates. There are only a few such membrane-bound aggregates present in an individual cell. A few plasma-membrane-bound small and discrete deposits are found in repressed cells, but both number and sizes of these membrane-bound sites increased remarkably after derepression.[273] It has been suggested that in *B. subtilis* plasma membrane there are specific regions which function both as sites of synthesis and membrane incorporation for alkaline phosphatase. In *B. licheniformis* 749/C, reaction product cytochemistry shows discrete aggregates of enzyme bound to the inside surface of the membrane.[274] Immunocytochemistry using colloidal gold confirms this observation.[275] It could be suggested from reaction product cytochemistry data that a substantial amount of enzyme might be released during protoplast formation. Periplasmic enzyme determined by biochemical localization may be largely artifactually generated by translocation from original sites. The membrane-bound enzyme has been labeled with membrane-impermeable ^{125}I-diazobenzine. High labeling after the destruction of the permeability barrier confirms the location of the enzyme on the cytoplasmic surface of the plasma membrane.[274] In *B. licheniformis* MC14 the reaction product cytochemistry shows the enzyme to be bound to the inside surface of the plasma membrane in a few discrete aggregates.[276] Lactoperoxidase ^{125}I-labeling experiments confirm this distribution.[277] However, 30% of the enzyme appears to be bound to the outside surface of the plasma membrane which may have leaked out during protoplast formation. Biochemical localization experiments show that during the growing phase the enzyme is predominantly plasma membrane bound but in the stationary phase, an appreciable amount is present in the cytoplasm.[278] The cells from all the growth phases show periplasmic enzyme, i.e., the enzyme released during protoplast formation. Reaction product cytochemistry shows the enzyme to be located between the cell wall and plasma membrane.[279] Plasma-membrane binding sites for alkaline phosphatase have been suggested to be a specific membrane apparatus.[280] *B. anthracis* also shows plasma-membrane-associated discrete enzyme aggregates.[281] In *Staphylococcus aureus* both acid and alkaline phosphatases are membrane associated.[282] In *Myxococcus xanthus*, reaction product cytochemistry shows alkaline phosphatase to be in the periplasm but associated with both the outside plasma membrane surface and the cell wall.[283] Both reaction product cytochemistry and cell fractionation show the plasma membrane location of the alkaline phosphatase in mycoplasma cells.[284,285] In sporulating cells of *B. subtilis* 168, unlike vegetative cells, alkaline phosphatase is derepressed in the presence of high inorganic phosphate.[286] Reaction product cytochemistry shows this enzyme to be bound to the prespore membrane. There is a strong correlation between an increasing number of electron-dense deposits due to increased phosphatase activity and the formation of spore septum (i.e., Stage II of sporulation).[287]

It is evident from the above discussion that in bacilli a significant amount of alkaline phosphatase is bound to the cytoplasmic surface of the plasma membrane. In addition, the enzyme may be found in the periplasm or cell wall material and secreted into the extracellular medium. It is not known whether the plasma-membrane-bound enzyme and the enzyme in other subcellular locations have any precursor product relationship. Ultrastructural study shows a unique characteristic of the plasma membrane binding. The enzyme molecules appear to form an aggregate and bind to some specific sites suggesting the existence of specific receptors. The enzyme from the membrane has been purified to homogeneity. One common property of the alkaline phosphatase purified from the membrane of bacilli is the requirement for a high magnesium salt concentration (1 to 2 *M*) for solubilization.[288,292] This suggests that the enzyme is bound to the membrane by ionic interaction. The enzyme has been purified by repeated precipitation and dialysis against magnesium salt-free buffer followed by ion exchange, gel filtration chromatography, and/or sucrose density gradient centrifugation. The pure enzyme requires about 0.2 *M* magnesium salt for solubility; this is precipitated by dialysis against magnesium salt-free buffer and crystalized after prolonged incubation in cold.[288] It has been assumed that the enzyme contains a minor polypeptide moiety which acts an anchoring structure to the membrane. However, the pure enzyme in magnesium salt, if heat shocked, loses the magnesium salt requirement for solubility.[292] Although a variety of suggestions have been made, the reason for this salt requirement is still unresolved.[293] Recently, both membrane-bound and extracellular alkaline phosphatase have been purified from *B. licheniformis* 749/C.[294] The pure extracellular enzyme does not require any salt for solubility. Although both these enzymes have the same dimer structure and the same molecular weight, the K_i for inorganic phosphate and the distribution of antigenic determinants is significantly different. Similar differences have been reported for heat-solubilized and salt-dependent enzymes from *B. licheniformis* MC14. It is possible that membrane binding and solubility are interrelated properties which may depend upon conformation of the enzyme molecule. Bacillus enzymes are dimers of mol wt 100 to 120 kDa, contain Zn^{2+}, CO^{2+}, or Mg^{2+}, enriched with basic amino acids (Table 11).[290] A monomeric constitutive alkaline phosphatase has been reported from *B. subtilis* 6160-Bc6 mutant strain.[295] This 45-kDa protein contains both alkaline phosphatase and phosphodiesterase activities. A small 15.5-kDa alkaline phosphatase has been isolated from *Halobacterium*.[296] The catalytic activity of this enzyme is 10 times stimulated with Mn^{2+} but not with Ca^{2+} or Mg^{2+}.

D. Penicillinase

Penicillinase is a hydrolytic enzyme which acts on the β-lactam ring of penicillin causing the loss of inhibitory action of this antibiotic on cell wall biosynthesis. Since its discovery by Abraham and Chain[297] nearly half a century ago, extensive studies have been conducted because of its biomedical significance.[298] Although initially suspected of being a highly specific enzyme, its distribution is broad across the entire microbial kingdom (Table 12).[298] This broad distribution is significant in relation to the evolution of the protein molecule. Penicillinase synthesis in microorganisms has been enchanced following extensive use of penicillins as antibacterial agents. It has been suggested that this protein was present in cells prior to the penicillin enhancement

Table 11
COMPARISON OF AMINO ACID COMPOSITIONS OF ALKALINE PHOSPHATASE OF DIFFERENT ORGANISMS

	Relative micromoles						
Amino acid	Phosphatase^{Glc}	Phosphatase^{Lac}	*B. subtilis* (19)	*B. subtilis* (23)	*B. licheniformis* (6)	*M. sodonensis* (12)	*E. coli* (8)
Lysine	92	104	142	130	118	36	58
Histidine	18	17	31	23	16	27	19
Arginine	17	20	25	28	31	54	27
Aspartic acid	173	142	198	169	106	115	108
Threonine	83	67	85	95	64	67	49
Serine	91	70	71	67	74	50	88
Glutamic acid	100	100	110	102	100	79	102
Proline	22	25	30	32	41	51	45
Glycine	100	100	100	100	100	100	100
Alanine	121	112	126	126	97	74	143
Half-cystine				2		5	9
Valine	61	61	56	61	98	62	50
Methionine	10	19	30	28	30	8	17
Isoleucine	64	73	67	66	45	27	32
Leucine	54	48	63	65	66	51	87
Tyrosine	27	31	43	37	30	23	23
Phenylalanine	40	38	43	38	31	35	18

Note: The values represent the quantities present in the enzyme solutions after a 24-hr hydrolysis. The numbers indicate micromoles of amino acids relative to glycine as 100.

From Baldassare, J. J., Brenckle, G. M., Haffman, M., and Silbert, D. F., *J. Biol. Chem.*, 252, 8797, 1977. With permission.

Table 12
PENICILLINASE DISTRIBUTION IN DIFFERENT MICROORGANISMS

Organism	Comments
	Gram-positive bacteria
Bacillus "subtilis-mesentericus" group	Extracellular
B. anthracis	Extracellular
Staphylococcus spp.	"Penicillin-destroying agent"
S. albus	"Intracellular"
S. aureus	Inducible
Micrococcus lysodeikticus	"Penicillin-destroying agent"
	Actinomycetes[a]
Actinomyces sp.	Destroy penicillin
Streptomyces sp. (100 strains)	Most of the strains produce a thermolabile, diffusable factor which destroys penicillin
Nocardia sp.	Constitutive cell-bound β-lactamase
Mycobacterium spp.	Intracellular; released on autolysis
Myco. tuberculosis strain BCG	"Penicillin-destroying factor"; extracellular β-lactamase; inducible; and constitutive
Myco. phlei	Constitutive
Myco. smegmatis	Intracellular; released on autolysis
	Gram-negative bacteria
Escherichia coli	"Mainly intracellular"
	"Extracellular"
	In Berkefeld filtrates; formed in alkaline medium only; cephalosporins hydrolyzed
"Paracolon group"	Intracellular
	"Extracellular"
Aerobacter-Klebsiella group	Intracellular
	Cephalosporins hydrolyzed
Serratia sp.	Cephalosporins hydrolyzed
Hafnia sp.	Cephalosporins hydrolyzed
Shigella sp.	Cephalosporins hydrolyzed
Salmonella sp.	Cephalothin hydrolyzed
Proteus sp.	Intracellular
	Cephalosporins hydrolyzed
Pseudomonas sp.	Intracellular
	Cephalosporins hydrolyzed
Pasteurella sp.	"Penicillinase activity"
Herellea sp.	Cephalothin hydrolyzed

in very small amounts or in a different form as an ancestral gene product.[299] This product, of course, would be different from the present penicillinase molecule. It is possible that this enzyme is a natural constitutive protein which, in the soil, acts on antibiotics produced by the soil microorganisms. It is also likely that penicillinase participated in a metabolic reaction, which is the target for the antibiotic action. After extensive use of penicillin, a point mutation and selection pressure altered the catalytic activity of the enzyme which now functions as a potent hydrolase detoxifying penicillin. Thus, microevolution of this protein molecule and macroevolution of the whole organism interacted for superior adaptation.

In Gram-positive organisms about 50% of penicillinase is cell bound and 50% is

Table 12 (continued)
PENICILLINASE DISTRIBUTION IN DIFFERENT MICROORGANISMS

Organism	Comments
	Fungi
Papulospora sp.	These fungi destroy penicillin but no β-lactamase found
Mycoderma valida	
Debaryomyces quilliermondii	
Aspergillus oryzae	Takadiastase and clarase destroy penicillin
	Takadiastase and clarase activity due to contaminating bacteria
	Animal
Tissues and serum	No β-lactamase activity was found but other penicillin-destroying mechanisms present in serum and tissues

[a] Presently grouped with the Gram-positive bacteria.

From Citri, N. and Pollock, M. R., *Adv. Enzymol.*, 28, 237, 1966. By copyright permission of John Wiley & Sons.

extracellular.[300] In contrast, in Gram-negative bacteria the enzyme is entirely cell bound; no significant secretion has been demonstrated.[298] This cell-bound enzyme is present predominantly in the periplasm; however, a very small amount of plasma membrane-bound salt-extractable enzyme has been detected.[301,303] It has been discussed in detail in the following chapter[304] that penicillinase is primarily associated with the outside surface of the plasma membrane of *B. licheniformis* through a trypsin-sensitive covalent bond.[305] A small amount is also bound to the cytoplasmic surface of the plasma membrane. Both biochemical and immunocytochemical observation confirm this distribution. However, the cytosol also contains enzyme, which can be detected only by immunological technique; furthermore, this cytosolic material is poorly active.[305] Recently, the colloidal gold labeling technique has shown that the distribution of penicillinase in the plasma membrane is random and there are multiple secretion sites on the outside surface of the wall.[306]

Penicillinase has been purified from a variety of sources (Table 13).[298] In Gram-positive cells, penicillinase is a 29- to 31-kDa protein. The molecular weight and the activity of the Gram-negative enzyme is 50% and 90% less, respectively, than the Gram-positive enzyme. The most important observation from recent studies is the lipoprotein nature of the membrane-bound penicillinase of Gram-positive cells.[304,307,308] The lipophilic modification of the membrane-bound enzyme occurs at the amino terminal, which anchors the enzyme to the membrane. This modification consists of a diacyl glyceryl group in the ether linkage to an amino-terminal cysteine residue and a long chain fatty acid (e.g., palmitate) in the amide linkage. This blocks the amino terminus to Edman degradation and sequencing.[307] This modification resembles *E. coli* outer-membrane-bound lipoprotein.[309] The hydrophobic penicillinase is converted into hydrophilic form by trypsin digestion. It seems this proteolytic post-translational processing is the final reaction to free the enzyme to the extracellular medium.

E. Miscellaneous Transferase and Esterase

Membrane-bound phosphotransferase system catalyzes sugar phosphate transphosphorylation.[310,311] The phosphotransferase system consists of a membrane-bound enzyme (enzyme II) and a soluble enzyme (enzyme I).[312] The phosphoenol pyruvate glu-

Table 13
PROPERTIES OF PENICILLINASES FROM DIFFERENT ORGANISMS

Preparation	Mol wt	Spec. activity[a]	Mol. activity[b]
B. cereus 569	31,500	2.02×10^4	1.60×10^5
B. cereus 569/H	30,800	1.95×10^6	1.53×10^5
B. cereus 5/B	35,200	1.53×10^6	1.48×10^5
B. licheniformis 6346/C	23,200	3.9×10^5	2.10×10^4
B. licheniformis 749/C(c)	22,900	2.18×10^6	1.18×10^5
B. licheniformis 749/C(Sn)	28,100	2.35×10^6	1.08×10^5
Staph. aureus			
Type A	29,600	2.5×10^5	1.97×10^4
Type B	29,600	3.5×10^4	2.76×10^3
Type C	29,600	2.3×10^5	1.82×10^4
E. coli (TEM)	16,700	5.3×10^5	2.0×10^4

[a] Units per milligram N.
[b] Moles benzylpenicillin hydrolyzed per mole of enzyme per minute at 30° at pH 7.0 (for the bacillary enzymes) or pH 5.9 (for the staphylococcal enzymes).

From Citri, N. and Pollack, M. R., *Adv. Enzymol.*, 28, 237, 1966. By copyright permission of John Wiley & Sons.

cose phosphotransferase system catalyzes a transmembrane transphosphorylation reaction known as exchange group translocation reaction. It has been shown by in vitro studies using *E. coli* membrane vesicles that phosphate groups from intravesicular sugar phosphates are transferred across the membrane to phosphorylate extracellular sugars which are translocated inside the vesicle and concomitantly the intravesicular dephosphorylated material is expelled. This mechanism may be physiologically related to facilitated diffusion.

Phosphoenolpyruvate-glucose phosphotransferase is a membrane-bound enzyme. During phosphorylation and uptake of methyl-α-D-glucopyranoside the enzyme becomes sensitive to inactivation by thiol-inactivating reagents, *N*-methyl malamide, or fluoro-dinitrobenzene.[313] The target for this thiol reagent is enzyme II.[314] The enzyme may exist in two different conformations (in the membrane). One is energized and the other deenergized; the energized conformation is sensitive to the thiol-inactivating-reagent sensitive.[315] Functional assay shows presence of phosphotransferase II in the membrane vesicles of *E. coli.* This enzyme exists in the membrane in two forms; one catalyzes vectorial and the other nonvectorial transphosphorylation.[316] The phosphotransferase system of *Rhodopseudomonas spheroides* consists of a membrane-associated soluble factor (SF) which is purified by ion exchange, gel filtration, and hydrophobic chromatography.[317] The protein is an 85-kDa monomer which diamerizes to a stable form in the presence of Mg^{2+} and phosphoenolpyruvate. The enzyme catalyzes the conversion of pyruvate from phosphoenolpyruvate; the phosphoryl group is transferred sequentially to the enzyme and fructose.

Glucosyltransferase is the terminal enzyme for the synthesis of glucan which may be soluble or insoluble. The enzyme for the synthesis of the insoluble form is present as a membrane-associated aggregate form.[318]

Two membrane-bound esterases of *E. coli,* active on alanine and phenylalanine, respectively, are extracted with detergent, suggesting their integral nature.[319] The amounts of enzyme extracted are increased with the addition of salts to the detergent although high salt alone fails to extract the enzyme. Topographical distribution of these enzymes is examined by impermeable substrate accessibility in vesicles of known polarity. The data suggest that one of these two esterases (alanine activity) is located on the outside surface of the plasma membrane, whereas the other (phenylalanine activity) is deeply buried inside the outer membrane. The phenylalanine activity, unlike alanine activity, is located in both the outer and inner membranes of *E. coli.*

Esterases are enzymes for the biosynthesis of fatty acid chains of lipids. They primarily act as a regulatory enzyme controlling the level of free CoA or acyl-CoA, which are feedback inhibitors of lipid biosynthesis.[320] The enzyme is usually found in the cytoplasm.[321] In mycoplasma, however, the activity of this enzyme has been found in osmotically lysed and washed mycoplasma membrane. The enzyme is sensitive to pronase digestion but the pronase does not act when the permeability barrier is intact. The enzyme is bound to the cytoplasmic surface of the plasma membrane. The enzyme cannot be extracted with high salt, deionized water washing, buffer variation, or EDTA. The detergent treatment, however, inactivates the enzyme. These results suggests that the enzyme is not only buried in the hydrophobic environment of the membrane but also requires the hydrophobic environment for catalytic activity.

Staphylocoagulase of *Staphylococcus aureus* is synthesized in oxygen and Mg^{2+}-limited conditions.[322] The major amount of this enzyme (81%) is integral (Triton X-100® extractable) and the rest is periplasmic. The membrane-bound enzyme is likely to be a lipid intermediate precursor for the secreted enzyme; a protease hydrolyzes the precursor causing the ultimate secretion of the enzyme.

A nucleotidase activity is found to be associated with the membrane of marine bacteria.[323] The nucleotidase, however, possesses a variety of specificities, e.g., 5′ nucleoside mono- or diphosphate sugar hydrolases,[324,325] 3′ nucleotidase, 3′ monophosphate phosphohydrolase, cyclic phosphodiesterase,[326,327] and phosphotransferase to form 2′ to 3′ nucleotides. These enzymes are distributed both in the plasma membrane and periplasm. A membrane-bound nucleotidase is released from the membrane particles by trypsin treatment, suggesting a trypsin-sensitive covalent bond binding the enzyme to the membrane surface. The purified enzyme exhibits both 5′ and 3′ nucleotidase activities and migrates as a double in PAGE. It is possible that two enzymes are bound to each other through a trypsin-sensitive bond. Comparison of properties during purification suggests that both activities arise from one enzyme molecule.

A phosphoprotein phosphatase has been demonstrated in *E. coli*; this enzyme functions as a protein kinase.[320] This enzyme has a broad substrate profile; it phosphorylates membrane fragments, histone, etc. This enzyme is extracted with 0.5 *M* KCl and is located largely in the plasma membrane. The enzyme forms a 250-kDa aggregate.

B. amyloliquifaciens can be grown in a medium containing amylaceous polysaccharides as carbon source.[331] The membrane of these cells contains α-glucosidase. This enzyme cannot be extracted with the salt of chelating agents, but detergent with salt effectively extracts the enzyme. The purified enzyme has a molecular weight of 27 ± 1 kDa and an optimum pH of 6.8.

VIII. MEMBRANE-BOUND ENZYMES FOR WALL BIOSYNTHESIS

Prokaryotic cells, whether found on the ocean floor or in the human gut, maintain a modified extracellular environment outside the plasma membrane. Cell surface structures outside the plasma membrane (e.g., cell wall, fibrillar polysaccharide layer, glycocalyx) support this modified extracellular environment. There is a considerable amount of space between the cell wall and plasma membrane (periplasm) within the matrix of the cell wall and inside the glycocalyx. Although in a classical sense the cell wall is considered to provide rigid support to osmotically fragile bacterial protoplasts, recent observations show that the cell wall may function in a cooperative fashion with the plasma membrane.[332] In fact, in Gram-negative cells the wall and the plasma membrane are structurally associated (adhesion zones); these regions perform a variety of wall membrane cooperative functions, e.g., initiation of DNA replication.[333] The structure, biosynthesis, and function of cell wall have been thoroughly investigated;[334,335] discussion on this topic is out of place in this chapter. Our aim is to examine the organization of membrane proteins related to wall synthesis or wall function. Sites of biosynthetic reactions of peptidoglycan, a major wall component, is schematically presented in Figure 8.[336]

Turnover of cell wall is very high. Although intuitively we may assume a direct role of membrane in this turnover phenomenon no clear data are available. Cell wall material (peptidoglycan and teichoic acid) is synthesized in discrete blocks. These nascent blocks do not mix with the ones which are being synthesized.[337] It is possible that after synthesis these blocks are incorporated into the plasma membrane at some specific site which transfers these blocks into specific growth zones at the wall.

Teichoic acid polymer synthesis requires both cell wall and membrane fragments.[338,339] In *Micrococcus luteus* wall-membrane preparation catalyzes the incorporation of ^{14}C-glucose from UDP^{14}C-glucose into teichuronic acid. The teichuronic acid of both the wall and the plasma membrane requires glucose and *N*-acetyl mannosaminuronic acid, but in addition, membrane requires *N*-acetyl glucosamine. The wall-bound teichuronic acid is cross-linked to peptidoglycan. Protoplasts of *B. subtilis* synthesizes ribitol-teichoic acid from a nucleotide precursor in the surrounding medium; if cytidine diphosphate ribitol is present, polyribitol phosphate attached to lipoteichoic acid carrier is synthesized. Enzymes bound to the external surface of the membrane react with the exogenous substrates for the biosynthetic reactions. The glycerol phosphate polymer is synthesized by an external elongation system. The glycerol phosphate repeating units are transferred from phosphatidylglycerol and become linked to the glycerol phosphate through a phosphodiester bond.[340,341] Membrane teichoic acid of *Lactobacillus casei* is a lipid containing polyglycerol phosphate esterified with D-alanine. This synthesis requires membrane fragments, D-alanine-activating enzyme, D-alanine membrane acceptor, ligase, and ATP. Thus, a short segment of polymeric material attaches to the membrane lipid.

In vitro biosynthesis of peptidoglycan has been reported;[342-344] however, due to D-alanine carboxy peptidase activity it is difficult to identify the products of synthesis.[342,345] Membrane isolated for *B. subtilis* 168 incorporates *N*-acetyl glucosamine from UDP-*N*-acetyl glucosamine directly onto undecaprenyl phosphate via transphosphorylation and subsequent transglucosylation. Ether permeabilized *E. coli* has been used to study in vitro peptidoglycan synthesis using a variety of precursors.[346] This indicates that peptidoglycan synthesis does not require an intact membrane structure. Cell wall lytic enzyme, *N*-acetyl muramyl-alanine-amidase is solubilized by nonionic detergent.[347] This enzyme is located at the external surface of the membrane. Thus, it can exert limited lytic activity for the incorporation of newly synthesized wall material.

The peptidoglycan cross-linking enzyme system in streptomyces strains is membrane bound.[348] Transpeptidase activity is fully membrane bound, whereas DD-carboxypeptidase is exocellular and found in two different locations, namely, plasma membrane and periplasm (i.e., released by lysozyme treatment). It is possible that these three forms of enzymes in the three different locations represent molecules in the pathway of secretion. In streptomyces the same cross-linking enzyme system is present but the concentrations vary significantly.

Streptococcus faecalis contains six membrane-bound penicillin-binding proteins; one of these proteins is 43 kDa and shows strong DD-carboxypeptidase activity.[349] Washed membrane suspension (10 mg protein) is extracted with 1% Genapol X-100® in phosphate buffer, pH 8.0. This treatment solubilizes 90 to 97% of the above enzyme activity and 25 to 44% of the membrane protein. DD-carboxypeptidase has been purified by ampicillin-Sepharose affinity chromatography; the final purification step required

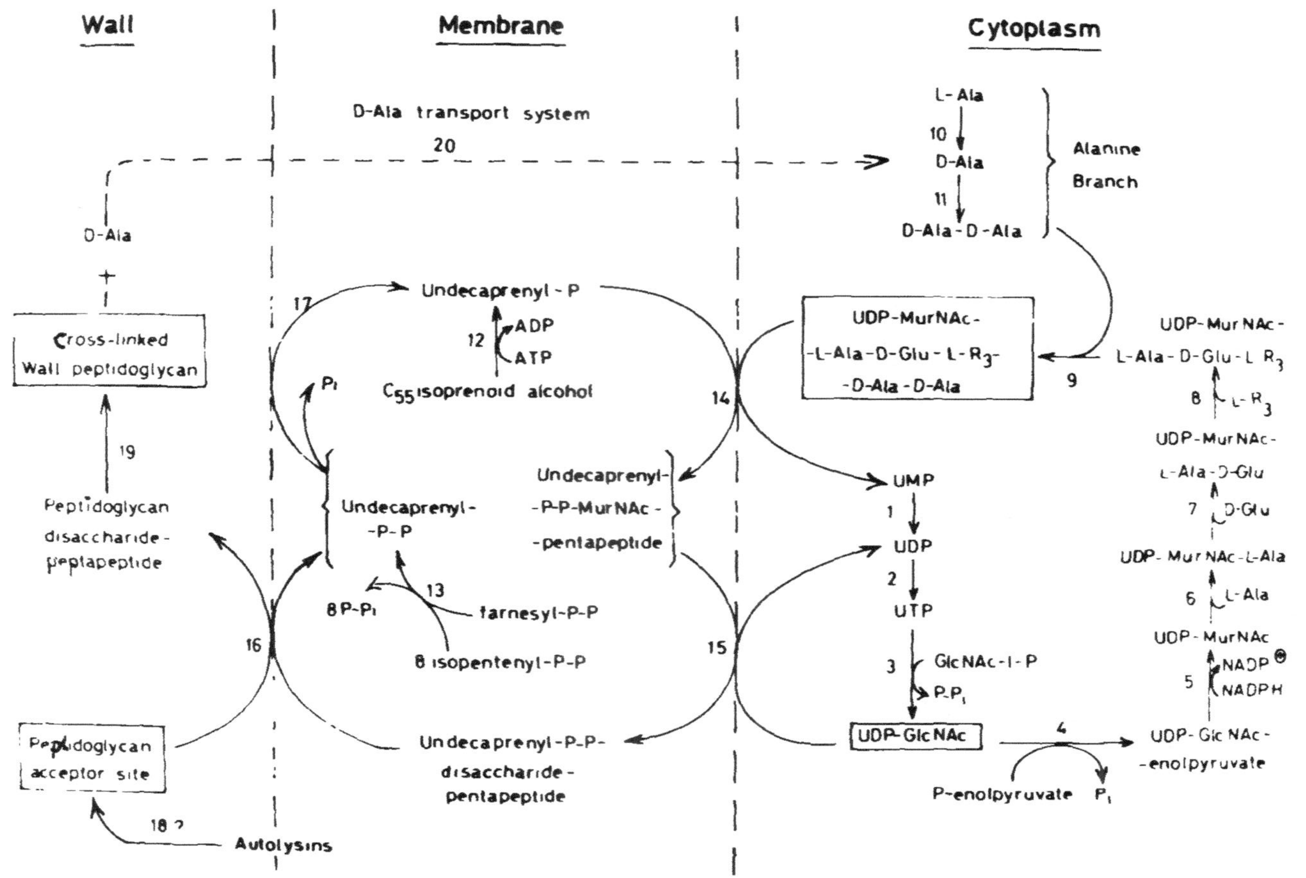

FIGURE 8. Schematic representation of the biosynthesis of a peptidoglycan of chemotype I. The three stages — cytoplasmic, membrane-bound, and wall-bound — are separated by the dashed vertical lines. All the reactions are shown in the anabolic sense. GlcNAc = *N*-acetylglucosamine; MurNAc = *N*-acetylmuramic acid. (From Ghuysen, J. M. and Shockman, G. D., in *Bacterial Membranes and Walls*, Vol. 1, Leive, L., Ed., Marcel Dekker, New York, 1973, 37. By courtesy of the publisher.)

preparative gel electrophoresis (PAGE) containing 0.1% Genapol X-100®. The pure enzyme forms an inactive complex with penicillin; this enzyme is the lethal target for penicillin in *S. faecalis.* Multiple forms of the enzyme differ in their sensitivity to penicillin binding. The 54-kDa exocellular form of the enzyme is highly sensitive, whereas the 40-kDa exocellular form is only slightly sensitive to penicillin binding.[350] *DD*-carboxypeptidase of *Mycobacterium smegmatis* is also membrane bound.[351] This enzyme activity is sensitive to penicillin; however, the organism is resistant to this antibiotic. The same enzyme protein may function as carboxypeptidase and transpeptidase depending on the microenvironment of the enzyme reaction.[352]

D-Alanine carboxy peptidase has been purified from *B. subtilis* membrane.[353,354] This enzyme is a penicillin-sensitive intrinsic protein. This protein has an amino-terminal hydrophilic catalytic domain and a membranous carboxy-terminal consisting of 20 to 30 hydrophobic amino acids. This protein does not contain significant cytoplasmic domain and is anchored to the membrane differently compared to other transmembrane proteins.

Proteus mirabilis L-form protoplasts contain DD-carboxypeptidase, which can be solubilized with detergent. Although phospholipase c or acetone-ammonium hydroxide extraction removes 90% of membrane lipid, DD-carboxypeptidase still remains cell bound. It is possible that the enzyme forms a complex with the hydrophobic proteins; hence, lipid-protein micelle formation may remove the enzyme from the L-form membrane.

An extremely hydrophobic protein, C55-isoprenoid alcohol phosphokinase, has been purified from *Staphylococcus aureus* membrane.[355] The enzyme causes ATP-dependent phosphorylation of C55-isoprenoid alcohol which is required for peptidoglycan and lipopolysaccharide synthesis.[356] The enzyme is extracted with *n*-butanol at pH 4.2 and is stable in organic solvent. The enzyme requires phospholipid as a co-factor, but the phospholipid can be removed during purification. The enzyme in butanol extract is precipitated by cooling to −20°C; the enzyme is extracted from the precipitate by fractional extraction with a methanol and butanol mixture. Successive chromatography on DEAE cellulose and hydroxyphosphorylated Sephadex G50 in organic solvents yielded a pure 17-kDa protein. Nonpolar amino acids constitute 58% (highest reported to date); lipid requirements are nonelectrostatic in nature.

IX. BINDING PROTEIN

Binding protein can be defined as a protein which exhibits high specific affinity for any substance (both high and low molecular weight) without exhibiting any catalytic activity. This specific binding has a wide range of functions. Many of these are related to the transport of small molecules and macromolecules (secretion). As these proteins are devoid of any catalytic activity, their detection, quantitative assay, and purification requires complex biochemical, immunochemical, and genetic technology. Therefore, the studies on the binding proteins have progressed vigorously only recently, after the development of specific technology. Classical research in this area has developed in relation to the transport of nutrients (mainly amino acids and sugars) from the extracellular environments to the cell interior.[357]

A. Proteins Related to Solute Transport

The literature in the general area of amino acid and sugar transport is vast and large numbers of reviews and monographs provide in-depth analyses of the subject. Cohen and Monod first originated the concept that the nutrient transport across the permeability barrier of bacterial membrane proceeds through a stereospecific catalytic interaction with permease, which does not play any role in the metabolism of the compound.[358] It is implied in this classic concept that membrane contains protein molecules which recognize a specific nutrient to be transported inside a cell. Highly specific interaction between a membrane component and the nutrient in question is assumed to be an important step for the transport process. Furthermore, the genetic regulation constitutes a part of the regulatory mechanism for the overall utilization of the nutrient. It was first demonstrated by Fox and Kennedy that lac-Y protein is involved in the active transport of β-galactosidase.[359] For more than a quarter of a century, research continued on nutrient transport (e.g., sugar, ions, amino acids, etc.) following the above-mentioned basic theme. However, information on the molecular organization of the transport apparatus has started to accumulate only recently because there has been a considerable lag in the understanding of the functional organization of the membrane. A quote from Kaback's recent publication is most appropriate: "Hopefully, within the near future, these apparently conflicting observations will be accounted for within one conceptual framework. At the present time, however, this is not possible without ignoring one important observation or another."[360] Two different types of molecules generally participate in the transport activity: (1) carrier molecules which are intrinsic membrane proteins; and (2) binding proteins which are periplasmic proteins. Carrier molecules are primarily intrinsic hydrophobic proteins. These form complexes with transportable material and translocate this material across the membrane into the intracellular environment. Binding proteins are periplasmic proteins which usually form complexes with carrier molecules. A model proposed by Anraku summarizes the possible role of binding protein in the transport activity (Figure 9).[357]

Solute molecules are transported across the membrane by a variety of mechanisms including (1) passive diffusion, (2) facilitated diffusion, (3) active transport, and (4) group translocation. Among these mechanisms both facilitated and active transport require carrier molecules which form specific complexes with solute molecules. Oscillation of the carrier between two surfaces of the plasma membrane transfers material from one side of the membrane to the other; on one side it forms a complex with the solute and releases the solute to the other. This does not require any metabolic energy and does not concentrate material against electrochemical or osmotic gradients. The carrier molecule in active transport forms a complex with solute, but this complex formation is coupled to respiratory chain or proton translocation. In either case metabolic energy input or electrochemical gradient is needed for solute translocation against electrochemical or osmotic gradient.[360,361]

The presence of carrier protein in the membrane has been demonstrated through indirect experiments, e.g., kinetics of solute uptake, binding with a chemical probe molecule.[357,360,361] There are two different advances in the field; one is the production of membrane vesicles of known polarity which actively transports solutes, and the other is the use of probes, which can penetrate the hydrophobic membrane interior and produce fluorescence. The active transport mechanism has been carefully analyzed to show the characteristics of coupling to the oxidation of D-lactate or the artificial electron donor phenazine methosulfate.[362] Another advance in the field is the use of solute covalently bound to the reporter group, e.g., the β-galactoside system.[363] Dansyl galactoside binds in an energy-dependent way to the lac carrier protein present in *E. coli* vesicles.[364] This binding is not followed by transport but causes a large increase in fluorescence and a shift in the emission maximum to a lower wavelength. This unique property makes it possible to titrate the amount of lac carrier proteins which become accessible upon energizing the membrane. The basic assumption is that upon energization of a right-side-out vesicle the lac carrier moves to the outside surface of the membrane of the vesicle and this lac carrier becomes accessible to dansyl galactoside. There is a striking parallelism between the temperature dependence of the initial rate of dansyl galactoside binding and lactose transport. Thus, the amount of lac carrier

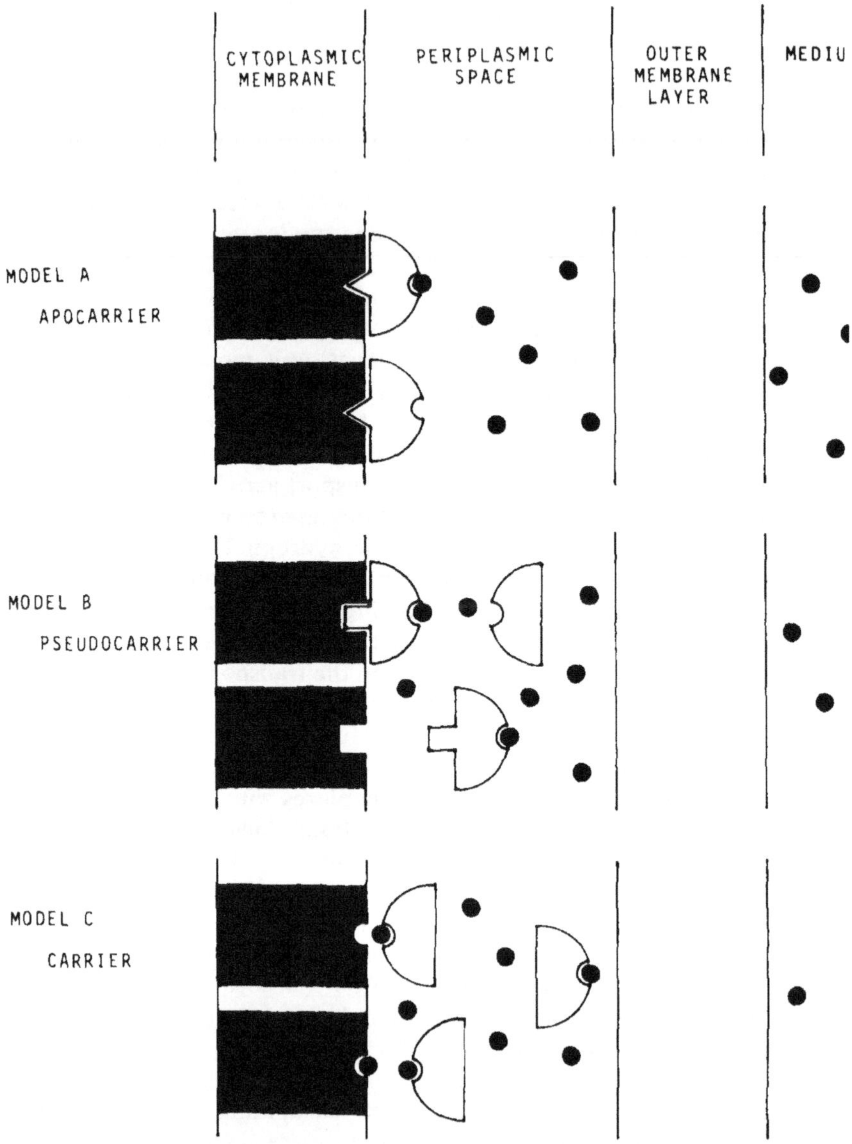

FIGURE 9. Molecular apparatus of active transport and possible roles of binding proteins. Black squares indicate membrane-bound transport proteins and semicircles indicate periplasmic binding proteins having different properties. The black squares with triangular and square indentations are apo-carriers (Model A) and pseudocarriers (Model B), respectively. These indentations represent interaction sites (accepting) for binding proteins. The black squares with semicircular indentations are carrier (Model C), and the indentation represents a binding site for a substrate. Binding proteins also have a binding site for a substrate (circular indentations), and closed circles represent a substrate. (From Anraku, Y., in *Bacterial Transport*, Rosen, B. P., Ed., Marcel Dekker, New York, 1978, 171. By courtesy of the publisher.)

was determined by functional assay. However, the isolation of the membrane-bound active product of the lac operon gene is difficult.[395] Essential cystine residue of this protein is labeled with *N*-ethylene malemide, and the inactive labeled protein is extracted with Triton X-100® or dodecyl sulfate.[365,366] Aprotic solvents (e.g., dimethylsulfoxide, *N*-methyl-pyrolidone, tetramethyl urea, or hexamethyl phosphoric triamide) cause reversible inhibition of β-D-galactoside transport.[367] These solvents in high con-

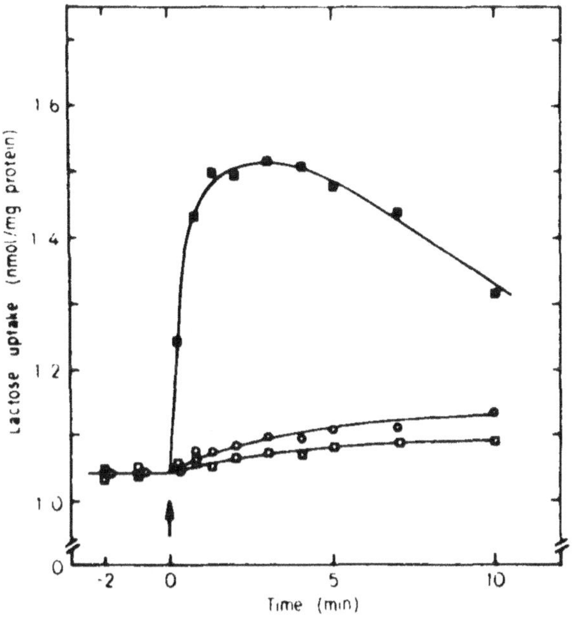

FIGURE 10. Lactose transport in reconstituted ML-35 vesicles in response to an artificial membrane potential. *E. coli* ML-35 vesicles were treated with R extract. K^+-loaded vesicles were suspended in a mixture of 100 m*M* Tris-maleate, pH 7.0, 0.013 *M* $MgSO_4$, and 0.28 *M* sucrose at 0.84 mg protein per milliliter. To measure uptake, suspensions (1 m*l*) were supplemented with 0.5 m*M* (final concentration) of [1-^{14}C]lactose (0.5 μCi/m*l*). After 5 min at 25°C, uptake was initiated (arrow) by addition of valinomycin (1 μg dissolved in 1 μ*l* ethanol); the control tube received 1 μ*l* ethanol only at this point. Aliquots of the suspensions (100 μ*l*) were filtered and washed with 0.4 *M* sucrose, 10 m*M* $MgSO_4$. (—□—□—□—) Ethanol only; (■—■—■) ethanol plus valinomycin; (—O—O—O—) ML-35 vesicles without R extract treatment, plus valinomycin. (From Altendorf, K., Muller, C. R., and Sandermann, H., Jr., *Eur. J. Biochem.*, 73, 545, 1977. With permission.)

centration solubilize 50 to 80% of the β-galactoside transport activity from the membrane of *E. coli*. This extract, if reconstituted into transport-negative *E. coli* membrane vesicles by sonication, restores the active transport of β-galactoside (Figure 10).[367] Protein fractionation of this extract suggests the presence of a lac carrier protein.[368]

A proline transport carrier has been extracted from *E. coli* membrane.[369] Acidic *n*-butanol extraction comparable to that used for the isolation of C55-isoprenoid alcohol phosphokinase removes the carrier activity. A variety of other detergents (e.g., Triton X-100®, Brij® 58, laurylsulfonate, deoxycholate) or organic solvents fail to extract significant amounts of proline carrier activity. Vesicles are prepared with *E. coli* phospholipid and *n*-butanol extract of the plasma membrane. These vesicles preloaded with K^+ show uphill transport in the presence of valinomycin; this transport shows characteristic inhibition by 3-4-dihydroproline, SH-reagent, and energy uncouplers. The crude activity is further fractionated by Sephadex LH-20 chromatography. The stereospecific uphill proline uptake is highly sensitive to the pH of the medium, suggesting the modulation of proline-binding activity by a primary proton binding with the carrier.[370] The simple situation where proline carrier is located in the plasma membrane is further complicated by the observation that the active transport in vesicles containing both outer- and inner-membrane material is many times higher than in plasma membrane vesicles alone.[371] It is possible that periplasmic binding proteins, conserved by the outer membrane, potentiate the membrane carrier function. Genetic analysis further confirmed the existence of carrier proteins in the plasma membrane of *E. coli*.[372] A proline auxotrophic strain has been used to construct mutants with specific defects in the proline transport. Several of these transport-negative mutants show diminished or a lack of proline binding ability. A highly specific nature of the defect has been suggested from intact proline oxidase activity in these strains. It is implied that the energy-coupling ability of the transport activity remains intact;[363] hence, the carrier protein itself becomes defective in the mutant (a function of the regulatory or structural gene).

Alanine carrier has been purified following the same principle of extracting material from the membrane by disrupting hydrophobic bonds and then performing ion exchange chromatography in the presence of detergent. The carrier protein has been extracted by cholate-deoxycholate mixture from the thermophilic bacterium PS3.[374,375] This protein has been purified by subsequent chromatography in DEAE cellulose and CM cellulose in the presence of Triton X-100® or Triton X-100® and urea mixture. The characteristic uphill transport has been shown in proteoliposomes which are either loaded with K^+ or contain cytochrome oxidase.[376,377] The proton conductance in proteoliposomes containing ATPase, caused by TF_o, uncouples the alanine transport.

It has been mentioned earlier in this chapter that the presence of a carrier and a carrier-coupled energy-generation system in the membrane are essential components for active transport. However, the presence of binding proteins showing a narrow range of substrate specificity increases the efficiency as well as specificity of the carrier-mediated transport system. Binding proteins may also perform a general function by retaining substrates in the periplasmic space. Purification of a variety of such binding proteins has been achieved. Data generated from varied experimental approaches establish the presence of binding proteins. First, transports of amino acids, carbohydrates, and phosphate ions are shock sensitive, but the loss of this activity can be restored by the addition of shock material to the shocked vesicles.[378,379] Second, the repression of transport activity concomitantly represses binding protein synthesis.[380-382] Third, the mutants synthesizing a high amount of binding proteins show high transport activity.[383,384] Fourth, isoleucine uptake has been shown to improve by the addition of LIVT protein but not by LS protein, because only the former has the specificity to bind isoleucine. Binding proteins of differing specificity have also been isolated[385] and classified (Table 14).[357] These classifications are arbitrary and only explain a few facts.

Histidine transport in *Salmonella typhimurium* is a thoroughly investigated system. The genetics of the histidine transport mechanism has been elegantly developed. The histidine transport operon directs the synthesis of two products related to pore and binding activity.[386] The *hisP* gene product has been designated as membrane pore protein,[387] and the *hisJ* gene product is periplasmic histidine-binding protein. *HisJ* protein has been purified by hydrophobic chromatography or ion-exchange and gel filtration chromatography.[388,389] The *hisP* gene product is a 24-kDa basic membrane protein which functions as pores of broad specificity.[387,390] Recent results suggest that both *hisP* and *Q* gene products are membrane-bound proteins; these proteins interact to form specific pores for the passage of L-histidine.[391] However, the specificity is ascertained by the periplasmic J protein. It has been suggested that substrate binding alters the conformation of J protein which in turn binds to the P-Q proteins and activates the pore for the transport of L-histidine.

Glutamine transport requires two proteins; one of them is membrane bound and the other is periplasmic glutamine-binding protein.[392,393] Both of these proteins have independent genetic regulatory mechanisms and are unrelated to the metabolism of glutamine.

Table 14
ROLE OF BINDING PROTEIN IN THE TRANSPORT OF BRANCHED CHAIN AMINO ACIDS IN *E. COLI*

Transport system	Substrate[a]	K_t[b]	Energy	Osmotic shock	Inhibitor[c] Thr	Norleu	Others	Repression by leucine[d]	Proposed roles of binding protein
CLASSIFICATION A									
LIV-I	LIVTA	0.2	—[e]	Sensitive	Yes	No	Alanine	LIVT-BP,[f] repressed	LIVT-BP, essential
LIV-II	LIV	2	Δp	Resistant	No	Yes	—	Activity, not repressed	LIVT- and Ls-BPs, not required
L	L	0.2	—	Sensitive	No	No	Trifluoroleucine	Ls-BP,[f] repressed	Ls-BP, essential
CLASSIFICATION B									
LIV-1	LIV	0.4	ATP	Sensitive	No	No	—	Activity, not repressed	LIVT- and Ls-BPs, obligatorily stimulative
LIV-2	LIV	6	ATP	Resistant	No	Yes	—	Activity, repressed	LIVT- and Ls-BPs, additionally stimulative
LIV-3	LIV	6	Δp	Resistant	No	Yes	—	Activity, not repressed	LIVT- and Ls-BPs, additionally stimulative

[a] Abbreviations used are L, leucine; I, isoleucine; V, valine; T, threonine; A, alanine.
[b] K_t value of entry (in μM) for leucine (Classification A) and for isoleucine (Classification B).
[c] Abbreviations used are Thr, threonine; Norleu, norleucine.
[d] Repression of transport activity and synthesis of binding proteins by leucine during the growth of cells.
[e] Not determined.
[f] Abbreviations used are LIVT-BP, LIVT binding protein; Ls-BP, Ls binding protein.

From Anraku, Y., in *Bacterial Transport,* Rosen, B. P., Ed., Marcel Dekker, New York, 1978, 171. Reprinted by courtesy of the publisher.

Maltose and maltodextrins in chemically unmodified form (up to seven glucose units) can be transported against a concentration gradient (1:10) in *E. coli.*[394] The outer membrane has a barrier to the low concentration of the (0.1 m*M*) substrate; this barrier, however, is removed if lambda receptor protein (*malB* gene product) is present in the outer membrane.[395,396] The diffusion barrier is partially overcome by Ca^{+2} treatment. Although Ca^{+2} treatment appears to generate nonspecific outer-membrane disruptions allowing entry of high-molecular-weight substances (e.g., amylopectin, anti-maltose binding protein antibody, etc.) from extracellular medium, many large-molecular-weight periplasmic proteins (e.g., alkaline phosphatase and maltose binding protein) cannot leak out. The active transport of maltose requires five proteins; three of these are plasma-membrane bound, one is water-soluble periplasmic, and another one is outer-membrane-bound lambda receptor, the *MalB* gene product.[397] The plasma membrane components have been identified by *lacZ* gene fusion technique.[398] One of these proteins is an integral membrane protein (i.e., *malF*) and the other is a peripheral membrane protein *malK* gene product extractable with high salt. The *malK* product is a 40-kDa protein which is bound to the inside surface of the plasma membrane through *malG* protein.[399] The periplasmic maltose binding protein is essential for the transport because the plasma membrane substrate-binding site recognizes the substrate only after complexing with the binding protein. It has been postulated that the plasma membrane channels allow unidirectional movement of substrates when energized with ATP. These channels may allow movement of acetylated maltose in maltose metabolism-blocked mutant; however, this movement is energized by electrochemical gradient.[394]

B. Penicillin-Binding Protein

In a large number of bacteria, penicillin binding occurs in the membrane;[400] studies with *Neisseria gonorrhoeae* show that the binding is inversely related to the minimum inhibitory concentration (MIC) of penicillin for different strains.[401] Binding with whole cells does not provide any information on localization of the binding protein; therefore, isolated cell fractions, (i.e., membrane or wall) are used for the study. The membrane of a susceptible cell binds five to ten times more penicillin than the resistant cell. Pretreatment of membrane with the unlabeled penicillin inhibits ^{14}C antibiotic binding. This binding is unrelated to the presence of β-lactamase activity. In susceptible cells, the binding in many organisms is biphasic; 80 to 90% of net binding occurs at low antibiotic concentration. In the resistant cells, however, the binding is proportional to the concentration of penicillin. The binding appears to be heat labile and pH (4.4 to 7.6) dependent.[402] Early observations showed the possibility that penicillin is concentrated against concentration gradient by sensitive cells.[403,404] It has been shown in *B. subtilis* that penicillin binds to two particulate enzymes related to the cell wall metabolism; one is D-alanine carboxypeptidase which is the site of lethal action, and the other transpeptidase which parallels the susceptibility. After ^{14}C penicillin binding, eight radioactive protein bands were detected in *N. gonorrhoeae.*[401] In *B. subtilis* five proteins have been detected which can bind penicillin.[405]

There are many reports on the variety of proteins which bind penicillin through covalent bonds.[405-408] The view is very strong that penicillin inhibits peptidoglycan transpeptidase activity.[409] Four penicillin binding proteins have been isolated from *B. megaterium* and *B. licheniformis* membrane.[410] The washed membrane material from these organisms is extracted with nonionic detergent Genapol X-100®, of pH 8.0 containing varying amounts of NaCl. The extract, free of insoluble material, has been passed through a DEAE-Sepharose CL-6B ion exchange column followed by an ampicillin affinose column.[411] *B. megaterium* membrane extract contains a 123-kDa protein (protein 1) which is 95% pure; however, there are other contaminating materials binding penicillin. Protein 1 has a very high affinity for benzylpenicillin, consistent

Table 15
PENICILLIN-BINDING PROTEIN OF
B. LICHENIFORMIS

Protein	Apparent mol wt	Binding of [^{14}C]benzylpenicillin (% of total)	Affinity for benzylpenicillin (μ/mℓ)
1	123,000	18	0.03
2	89,000	12	0.3
3	83,000	3	0.2
4	46,000	68	3.0

Note: The apparent molecular weights of the proteins were determined by comparing their mobilities on a dodecylsulfate polyacrylamide slab gel with proteins of known molecular weight. The affinities of the penicillin-binding proteins for benzylpenicillin were determined by incubating 50 $\mu\ell$ of membranes (0.5 mg protein) with increasing concentrations of [^{14}C]benzylpenicillin for 15 min at 35° C. The affinities are expressed as the concentrations of [^{14}C]benzylpenicillin that resulted in 50% of the maximum binding to each protein.

From Chase, H. A., Reynolds, P. E., and Ward, J. B., *Eur. J. Biochem.*, 88, 275, 1978. With permission.

with the view that this protein is the lethal target for penicillin.[412] *B. licheniformis* 94, a penicillinase- and protease-negative mutant, contains penicillin-binding protein comparable to *B. megaterium* protein (Table 15);[410] however, there is one exception in that the protein 1 penicillin complex of *B. megaterium* degrades very rapidly. This property has been exploited to purify protein 1 from ^{14}C penicillin prelabeled cells by affinity chromatography.

The similarity of the kinetics of inhibition by benzylpenicillin of the natural peptidoglycan transpeptidase and the kinetics of protein 1 penicillin binding suggests the possibility of protein 1 being peptidoglycan transpeptidase.[412] Purified *DD*-carboxypeptidase-transpeptidase forms a 1:1 complex with penicillin.[413] Many modified penicillins have been synthesized, based on the rationale that these compounds which have both high affinities for penicillin-binding proteins and resistance to penicillinase will be superior bactericidal agents. Cephamycin (Cs-1170), an antibiotic with a methoxy group at the α position of the cephalosporin skeleton, satisfies the above basic assumptions.[414] This compound (1) shows high affinity for high-molecular-weight penicillin-binding proteins (PBP1) of *E. coli* or *Proteus* species; (2) causes strong inhibition of the in vitro assay of peptidoglycan cross-linking reaction releasing alanine: (3) shows high affinity for PBP3 which is related to *E. coli* septum formation;[416,417] (4) causes filamentous cell formation of *E. coli*; and (5) shows high affinity for the low-molecular-weight penicillin-binding proteins (PBPs). These PBPs are assumed to be identical with D-alanine carboxypeptidase, an additional peptidoglycan cross-linking enzyme.[415] Purified penicillin-binding proteins fail to exhibit any cell-wall-synthesizing enzyme activities. A purified high-molecular-weight PBP shows enzymatic acylation reaction for the cell-wall-related substrates; but the significance of this reaction is unknown.[418] Protein 3 of *E. coli* shows a direct effect on the septum-forming reaction sequence.[419] A large number of mutants of *Pseudomonas aeruginosa* have been selected after ethylmethyl sulfonate mutation and tricarboxycillin selection.[420] The PBPs of these mutants have low affinity for penicillin; frequently, some of these mutants produce large quantities of these proteins. It appears that two closely linked separate genes control the binding affinity and the amount of protein formed. In *Clostridium perfringens*, there are six penicillin-binding proteins; their molecular weight ranges from 42 to 100 kDa.[421] The binding affinities for PBP 3 and 4 match with the MIC of the antibiotics, although filament formation is unrelated to binding affinity of MIC. Discrepancies between binding affinities with PBPs and inhibitory action are common for different penicillins. Mecillinum has very high affinity for PBP 5 but inhibits growth at a concentration which does not saturate PBP 4. Alterations in the chemical structures of penicillins may not cause appreciable changes in their biochemical properties, but may cause substantial changes in their antibacterial actions.[422] Ceftizoxime is a much superior antibacterial agent compared to its stereoisomer, but their permeability through the outer membrane of *E. coli* or cephalosporinase stability is identical.[423] The binding affinity of PBP 1B for these compounds differed significantly. Although insensitivity of the antibiotics against penicillinase action differs, their antibacterial activity cannot be correlated to the penicillinase production. Changes of the intensity of action on the same target may be related to the alteration of chemical structure of the molecule. Specificity of the localization of PBP has been shown in *R. spheroides* phototropic and chemoautotropic growth conditions; the major PBP (comparable to *E. coli* PBP 5) is present in the plasma membrane in very high amounts, but only a small amount is present in the outer of intracytoplasmic membrane.[424]

The basic biological implications of PBP and β-lactamase have been a matter of considerable discussion;[425] it can be argued that β-lactamase might have evolved by a divergent evolution from a common precursor molecule or by a convergent manner from a large number of protein molecules. However, these proteins should have a common property, i.e., affinity for β-lactam antibiotics. This line of investigation will be fruitful both from the standpoint of cell wall biosynthesis and evolution of the protein molecule.

C. DNA-Binding Protein

Prokaryotic cells lack nuclei delimited by membrane; hence, DNA remains dispersed in the cytoplasm. Fibrillar DNA and its association with the plasma membrane has been demonstrated by the application of differing techniques of electron microscopy. Results of electron microscopic studies, however, are controversial.[420] Orderly replication of chromosomes by DNA biosynthesis and synchronization of the completion of replication with cell division require a specific attachment of the chromosome to a solid surface (i.e., the plasma membrane in bacteria). This concept has evolved from the replicon model of Jacob et al.[427] The most important aspect of this model is that the replication of DNA and the segregation of chromosome require a membrane attachment site. In spite of the complex technological bottleneck, research in this area has progressed in several directions:

1. It has been shown by morphological, biochemical, and genetic technology that chromosome and membrane are associated.[428-430]
2. The chromosome membrane attachment is specific.[431-430]
3. The modification of DNA membrane attachment significantly affects the physiology of cell.
4. It is possible to identify one or a complex of several molecules which may serve as the anchor point of the DNA molecule as well as functioning as a regulator substance for chromosome replication and cell division.

A valuable review in this area is available.[437] Experimental approaches to show the specificity of chromosome membrane attachment are diverse. Irradiation with gamma rays causes dissociation of 85% DNA from the DNA-membrane complex.[438] However, the radio-resistant association of membrane-DNA increases during the repair of irradiated cells. In gently lysed *E. coli* cells, 90% of the membrane-associated DNA is released by single-stranded specific endonuclease.[439] The nuclease-resistant regions are chemically different and appear to be enriched near the origin of chromosome replication. A marker, *purA*, is close to the replication origin of *B. subtilis* chromosome.[440] In *B. subtilis* cell lysate this *purA* marker region has two forms: the M-complex, tightly associated to the membrane, and the S-complex, loosely associated to the membrane containing salt-extractable (peripheral) membrane proteins. The M-complex DNA fragment after purification is homogeneous, containing origin-labeled DNA. Site-specific restriction endonuclease treatment shows that the DNA originates from a particular region of the chromosome. This complex appears to be a *purA*-DNA-complex. The DNA-membrane complex has affinity for Mg-sarkosyl crystals.[441] Nearly 80% of the cellular DNA can be isolated by this binding; this DNA contains the origin of a replication region closely associated to the membrane. Interruption of protein or DNA synthesis does not dissociate the DNA-membrane complex. Analysis of a variety of markers in *B. subtilis* shows that the most recently replicated marker is attached to the membrane.[435,436,442] It is possible that the replication terminus is attached to the membrane.[443]

Both the origin and the terminus regions of the chromosome of *B. subtilis*[444-448] and *E. coli*[449-451] are membrane bound; although this binding is stable, it is not permanent.[452] The DNA-membrane attachment site has a regulatory role; some component in this region controls cell division which depends on the continuation of protein synthesis.[453-455] Furthermore, a protein (termination protein) has been postulated which causes the release of chromosome from the plasma membrane following the completion of the replication; this release, however, requires continuation of the protein synthesis.[453,456]

An intricate study with *E. coli* shows the presence of a unique DNA-membrane complex consisting of origin of replication. The use of a mutant CT28 made this study possible.[457] This mutant at nonpermissive temperature completes ongoing replication. A shift to a permissive temperature allows replication in all cells and this replicated DNA is labeled; a shift to a nonpermissive temperature arrests all the replication. The lableled cells in Tris-EDTA buffer are lysed by repeated freezing and thawing and the lysate is sheared by repeated passage through a needle having a small (1.59-cm diameter) hole. The lysate free of cell debris is fractionated in a sucrose density gradient. The labeled replicating origin region is associated to the plasma membrane fragments. The DNA-membrane complex is easily disrupted by 0.12 *M* NaCl, pronase, or trypsin treatment. Comparable sensitivity has been described earlier.[458] The origin specific DNA-membrane binding persists after termination of chromosomal replication.

In Gram-positive bacterial cells mesosomes are frequently found to be associated to fibrillar DNA. It has been assumed that this association plays a role in the replication and segregation of chromosomes.[1] It has also been suggested that transforming DNA penetrates through mesosomes.[459-461] Comparison of isolated mesosome and cytoplasmic membrane shows that both the initial rate of binding and the final level of the DNA binding is higher in the mesosome than in the plasma membrane.[462] Further studies with *B. subtilis* showed that cell wall, in contrast to the plasma membrane, binds a very small amount of DNA. Homologous and heterologous DNA binds to the plasma membrane differently, suggesting that the plasma membrane contains various types of DNA receptors.[463] In Gram-negative cells, however, the DNA receptor is also present in the outer membrane.[464] Proteinaceous receptors for plasmid DNA appear to be present in the outer membrane of *E. coli* because the binding is lost after trypsin treatment of the whole cell.[465] This binding is divalent cation (Ca^{+2})-dependent. The plasmid DNA after binding to the outer membrane becomes DNAse resistant because the plasmid is translocated into the periplasm. However, the translocation through the plasma membrane is a rate-limiting activity; therefore, the frequency of transformation is much lower than the amount of DNA uptake by the outer membrane.

The above discussion clearly shows that DNA binds both specifically and nonspecifically with the membrane of the Gram-positive and Gram-negative bacteria. Attempts have also been made to isolate the specific membrane protein(s) which may function as membrane receptor(s) for DNA. Four protein fractions (peaks I, IIa, IIb, and III) have been isolated from *E. coli* membrane by DEAE cellulose chromatography. These have affinity for DNA; their molecular weights vary between 8 and 12 kDa. Peak III protein has specific affinity for single-stranded DNA. The binding activity of the peak IIb is inhibited by polyuridylic acid. Peak I and II proteins are hydrophobic.

Membrane fragments attached to the chromosome have specific affinity for the crystals of Mg-sarkosyl.[466] This membrane-chromosome complex containing the chromosomal replicative fork has three membrane proteins of molecular weights 125, 57, and 42 kDa. Comparison of these proteins of *E. coli* wild-type and temperature-sensitive DNA initiation mutant at restrictive temperature shows that the 57-kDa protein is altered. This change is expressed in restrictive temperature by the accumulation of 57-kDa protein, 3 to 4 hours after the cessation of DNA synthesis. When the culture is returned to normal temperature this accumulation disappears. A 30-kDa protein, common to both whole cell plasma membrane and M-band membrane, disappears from the mutant at a restrictive temperature.

Membrane preparations from radiolabeled male and female strains of *E. coli* K-12 have been analyzed by SDS-PAGE and autoradiography.[467,468] The expression of the transferred operon of the conjugated plasmid leads to the formation of a new protein in the membrane of their hosts. This protein has a molecular weight of 24 kDa and appears to be distributed in both the outer and plasma membrane. This protein may play a significant role in the conjugation process.

The origin, terminus, and replication forks are enriched with the DNA fraction bound to the membrane. This DNA-membrane fragment can be further fractionated and enriched in a sucrose-cesium chloride double gradient[469] and further purified by free-flow electrophoresis.[470,471] Repeated electrophoresis of the DNA-membrane complex obtained by density gradient centrifugation gives a highly purified complex.[472] This complex contains tightly bound 29-, 27-, and 24-kDa proteins. The DNA-membrane complex isolated from *Streptococcus pneumoniae* contains three DNA-polymerases I, II, and III; polymerase III is the most functionally predominant enzyme in the DNA-membrane complex.[473] The modified method of purification gives three DNA-protein-phospholipid subcomplexes containing polymerase activity related to DNA replication.

Bacteriophages have been shown to replicate their DNA in a membrane-bound configuration.[474] Membrane binding is a necessary condition for T_4 DNA replication in *E. coli*.[475] The ϕ 29 phage genes 2 and 3 function to maintain association of viral chromosomes with the cell membrane; cistron 2 encodes 67-kDa protein P2, and cistron 3 encodes 27-kDa protein P3.[476] The P2 protein is a linker protein associating DNA to the membrane, whereas P3 protein is needed for initiation of polymerization of viral DNA. Lambda DNA cosediments with the host (*E. coli*) membrane; this is known as rapid sedimenting complex or RSC.[477] This method is refined by fractionating in $CsCl_2$ gradient yielding DNA-membrane complex in one fraction.[478] However, it has been calculated that there are one to three binding sites for the lambda DNA with the membrane; this result is inconsistent with the 60 to 80 binding sites reported earlier.[478] It has been suggested from detailed genetic analysis that phage T4 gene 32 protein functions in recombination by forming a complex consisting of DNA, recombination enzymes, and membrane components.[479] However, this conclusion is only indirect; no specific role of protein in the membrane association of DNA has been demonstrated.

ACKNOWLEDGMENT

We thank Gloria Trechock, Lori Morlock, and Ranjan Ghosh who gave their untiring help to prepare this manuscript. The work in our laboratory which has been included in this chapter has been funded by Grants NSF 81-10613, NSF PCM 81-08665, and DOE, DE-ASO5-80 ER 10702.

REFERENCES

1. Ghosh, B. K., The mesosome — a clue to the evolution of the plasma membrane, *Sub-Cell. Biochem.*, 3, 311, 1974.
2. Danielli, J. F. and Davson, H., A contribution to the theory of permeability of thin films, *J. Cell Physiol.*, 5, 495, 1935.
3. Robertson, J. D., The ultrastructure of cell membranes and their derivatives, *Biochem. Soc. Symp.*, 16, 3, 1959.
4. Singer, S. J. and Nicholson, G. L., The fluid mosaic model of the structure of cell membranes, *Science*, 173, 720, 1972.
5. Hendler, R. W., Cell membrane structure, in *Protein Biosynthesis and Membrane Structure*, John Wiley & Sons, Inc., New York, 1968, 202.
6. Singer, S. J., The molecular organization of biological membranes, in *Structure and Function of Biological Membranes*, Rothfield, L. I., Ed., Academic Press, New York, 1971, 145.
7. Bronton, D., Fracture faces of frozen membrane, *Proc. Natl. Acad. Sci. U.S.A.*, 55, 1048, 1966.
8. Salton, M. J. J. and Owen, P., Bacterial membrane structure, *Ann. Rev. Microbiol.*, 30, 451, 1976.
9. Edidin, M., Two-dimensional diffusion in membranes., *Symp. Soc. Exp. Biol.*, 28, 1, 1974.
10. Rothman, J. E. and Lenard, J., Membrane asymmetry, *Science*, 195, 743, 1977.
11. Tanner, M. J. A., Isolation of integral membrane proteins and criteria for identifying carrier protein, *Curr. Top. Membr. Transp.*, 12, 1, 1979.
12. Tokes, Z. A. and Chambers, S. M., Proteolytic activity associated with human erythrocyte membranes: self digestion of isolated human erythrocyte membrane, *Biochim. Biophys. Acta*, 389, 325, 1975.
13. Altendorf, K., Lukas, M., Kohl, B., Muller, C. R., and Sandermann, H., Isolation and purification of bacterial membrane proteins by the use of organic solvents: the lactose permease and the carbodiimide protein of the ATPase complex of *E. coli*, *J. Supramol. Struct.*, 6, 229, 1977.
14. Fox, C. F., The structure of cell membrane, *Sci. Am.*, 226, 30, 1972.
15. Segrest, J. P. and Feldman, L. J., Membrane proteins: amino acid sequence and membrane penetration, *J. Mol. Biol.*, 87, 853, 1974.
16. Vanderkooi, G. and Capaldi, R. A., A comparative study of the amino acid compositions of membrane proteins and other proteins, *Ann. N. Y. Acad. Sci.*, 195, 135, 1972.
17. Handler, R. W., Respiration and energy transduction in *E. coli*, in *The Enzymes of Biological Membranes*, Martonosi, A., Ed., John Wiley & Sons, New York, 1976, 75.
18. Kaback, H. R., Transport studies in bacterial membrane vesicles, *Science*, 186, 882, 1974.
19. Futai, M., Membrane D-lactate dehydrogenase from *E. coli:* purification and properties, *Biochemistry*, 12, 2468, 1973.
20. Kovatchev, S., Vaz, W. L. C., and Eibl, H., Lipid dependence of membrane-bound D-lactate dehydrogenase of *E. coli*, *J. Biol. Chem.*, 256, 10369, 1981.
21. Fleischer, S., Bock, H. G., and Gazzotti, P., Studies on phospholipid-protein interaction in D-β-hydroxybutyrate dehydrogenase, a lecithin requiring enzyme, in *Membrane Proteins in Transport and Phosphorylation*, Klingenberg, M. and Azzone, G. F., Eds., North-Holland, Amsterdam, 1974, 125.
22. Raj, T., Russell, P., Flygare, W. H., and Gennis, R. B., Quasielastic light scattering studies on pyruvate oxidase, *Biochim. Biophys. Acta*, 481, 42, 1977.
23. Schrock, H. and Gennis, R. B., High affinity lipid binding sites on the peripheral membrane enzyme pyruvic oxidase: specific ligand effect on detergent binding, *J. Biol. Chem.*, 252, 5990, 1977.
24. Esfahani, M., Rudkin, B. B., Cutler, C. J., and Waldron, P. E., Lipid-protein interactins in membranes: interactins of phospholipids with respiratory enzymes of *E. coli* membranes, *J. Biol. Chem.*, 252, 3194, 1977.

25. Davis, M. B. and Silbert, D. F., Changes in cell permeability following a marked reductin of structural fatty acid content of *E. coli* K-12, *Biochim. Biophys. Acta*, 373, 224, 1974.
26. Baldassare, J. J., Rhinehart, K. B., and Silbert, D. F., Modification of membrane lipid: physical properties in relation to fatty acid structure, *Biochemistry*, 15, 2986, 1976.
27. Jackson, M. B., Cronan, J. E., Jr., An estimate of fluid lipid required for the growth of *E. coli*, *Biochim. Biophys. Acta*, 512, 472, 1978.
28. Takeuchi, Y., Ohnishi, S-I., Ishinga, M., and Kito, M., Spin labelling of *E. coli* membrane by enzymatic synthesis of phosphotidylglycerol and divalent cation-induced interactin of phosphotidylglycerol with membrane proteins, *Biochim. Biophys. Acta*, 506, 54, 1978.
29. Baldassare, J. J., Brenckle, G. M., Haffman, M., and Silbert, D. F., Modification of membrane lipid: functional properties of membrane in relation to fatty acid structure, *J. Biol. Chem.*, 252, 8797, 1977.
30. Nishihara, M., Ishinaga, M., Kato, M., and Kito, M., Temperature sensitive formatin of the phospholipid molecular species in *Escherichia coli* membranes, *Biochim. Biophys. Acta*, 431, 54, 1976.
31. Lee, A. G., Birdsall, N. J. M., Metcalfe, J. C., Toon, P. A., and Warren, G. F., Clusters in lipid bilayers and the interpretation of thermal effects in biological membranes, *Biochemistry*, 13, 3699, 1974.
32. Warren, G. B., Toon, P. A., Birdsall, M. J. M., Lee, A. G., and Metcalfe, J. C., Reconstitution of a calcium pump using defined membrane components, *Proc. Natl. Acad. Sci. U.S.A.*, 71, 622, 1974.
33. Nakamura, M. and Ohnishi, S., Organization of lipids in sarcoplasmic reticulum membrane and Ca^{+2}-dependent ATPase activity, *J. Biochem. Tokyo*, 78, 1039, 1975.
34. Thilo, L. and Overrath, P., Randomization of membrane lipids in relation to transport system assembly in *Escherichia coli*, *Biochemistry*, 15, 328, 1976.
35. Uehara, K., Akutsu, H., Kyogoku, Y., and Akamatsu, Y., Phase transitions of phospholipid bilayers from an unsaturated fatty acid auxotroph of *Escherichia coli*, *Biochim. Biophys. Acta*, 466, 393, 1977.
36. Pluschke, G. and Overath, P., Function of phospholipids in *Escherichia coli:* influence of changes in polar head group composition on the lipid phase transition and characterization of a mutant containing only saturated phospholipid acyl chains, *J. Biol. Chem.*, 256, 3207, 1981.
37. Silvius, J. R. and McElhaney, R. N., Growth and membrane lipid properties of *Acholeplasma laidlawii* B lacking fatty acid heterogeneity, *Nature*, 272, 645, 1978.
38. Rottem, S., Cirillo, V. P., DeKruyff, B., Shinitzky, M., and Razin, S., Cholesterol in mycoplasma membranes: correlation of enzymic and transport activities with physical state of lipids in membranes of *Mycoplasma mycoides* var. *capri* adapted to grow with low cholesterol concentrations, *Biochim. Biophys. Acta*, 323, 509, 1973.
39. Wirtz, K. W. A., Transfer of phospholipids between membranes, *Biochim. Biophys. Acta*, 344, 95, 1971.
40. Jones, N. C. and Osborn, M. J., Translocation of phospholipids between the outer and inner membranes of *Salmonella typhimurium*, *J. Biol. Chem.*, 252, 7405, 1977.
41. Bell, R. M., Mavis, R. D., Osborn, M. J., and Vagelos, P. R., Enzymes of phospholipids metabolism: localization in the cytoplasmic and outer membrane of the cell envelope of *Escherichia coli* and *Salmonella typhimurium*, *Biochim. Biophys. Acta*, 249, 628, 1971.
42. White, D. A., Albright, F. R., Lennarz, W. J., and Schnaitman, C. A., Distribution of phospholipid-synthesizing enzymes in the wall and membrane subfractions of the envelope of *Escherichia coli*, *Biochim. Biophys. Acta*, 249, 636, 1971.
43. Osborn, J. J., Gander, J. E., and Parisi, E., Mechanism of assembly of the outer membrane of *Salmonella typhimurium:* site of synthesis of lipopolysaccharide, *J. Biol. Chem.*, 247, 3973, 1972.
44. Muhlradt, P. F., Menzel, J., Golecki, J. R., and Speth, V., Outer membrane of *Salmonella:* sites of export of newly synthesized lipopolysaccharide on the bacterial surface, *Eur. J. Biochem.*, 35, 471, 1973.
45. Jones, N. C. and Osborn, M. J., Interaction of *Salmonella typhimurium* with phospholipid vesicles: incorporation of exogenous lipids into intact cells, *J. Biol. Chem.*, 252, 7398, 1977.
46. Bell, R. M., Mavis, R. D., Osborn, M. J., and Vagelos, P., Enzymes of phospholipid metabolism: localization in the cytoplasmic and outer membrane of the cell envelope of *Escherichia coli* and *Salmonella typhimurium*, *Biochim. Biophys. Acta*, 249, 628, 1971.
47. Raetz, C. R. H., Enzymology, genetics and regulation of membrane phospholipid synthesis in *Escherichia coli*, *Microb. Rev.*, 42, 614, 1978.
48. Dutt, A. and Dowhan, W., Intracellular distribution of enzymes of phospholipid metabolism in several gram-negative bacteria, *J. Bacteriol.*, 132, 159, 1977.
49. Raetz, C. R. H. and Kennedy, E. P., The association of phosphatidylserine synthetase with ribosomes in extracts of *Escherichia coli*, *J. Biol. Chem.*, 247, 2008, 1972.

50. Golez, S. E. and Cronan, J. E., Adenylate kinase of *Escherichia coli:* evidence for a functional interaction in phospholipid synthesis, *Biochemistry,* 21, 189, 1982.
51. Rock, C. O., Golez, S. E., and Cronan, J. E., Jr., Phospholipid synthesis in *Escherichia coli:* characteristics of fatty acid transfer from acyl-acyl carrier protein to sn-glycerol-3-phosphate, *J. Biol. Chem.,* 256, 736, 1981.
52. Cronan, J. E., Jr. and Gelman, E. P., Physical properties of membrane lipids: biological relevance and regulation, *Bacteriol. Rev.,* 39, 232, 1975.
53. Raetz, C. R. H., Enzymology, genetics and regulation of membrane phospholipid synthesis in *Escherichia coli, Microbiol. Rev.,* 42, 614, 1978.
54. Cronan, J. E., Jr. and Vagelos, P. R., Metabolism and function of the membrane phospholipids of *Escherichia coli, Biochim. Biophys. Acta,* 265, 25, 1972.
55. Lightner, V. A., Larson, T. J., Tailleur, P., Kantor, G. D., Raetz, C. R. H., Bell, R. M., and Modrich, P., Membrane phospholipid synthesis in *Escherichia coli:* cloning of a structural gene (plsB) of the sn-glycerol-3-phosphate acyltransferase, *J. Biol. Chem.,* 255, 9413, 1980.
56. Green, P. R., Merrill, A. H., and Bell, R. M., Membrane phospholipid synthesis in *Escherichia coli:* purification, reconstitution and characterization of sn-glycerol-3-phosphate acyltransferase, *J. Biol. Chem.,* 256, 11151, 1981.
57. Carman, G. M., Zaniewski, R. L., and Cousminer, J. J., CDP-diacylglycerol synthase activity in *Clostridium perfringens, Appl. Env. Microbiol.,* 43, 81, 1982.
58. Larson, T. J. and Dowhan, W., Ribosomal-associated phosphatidylserine synthetase from *Escherichia coli:* purification by substrate-specific elution from phosphocellulose using cytidine 5′-diphospho-1,2-diacyl-sn-glycerol, *Biochemistry,* 15, 5212, 1976.
59. Carman, G. M. and Wieczorek, D. S., Phosphatidylglycerophosphate synthase and phosphatidylserine synthase activities in *Clostridium perfringens, J. Bacteriol.,* 142, 262, 1980.
60. Langley, K. E., Yaffe, M. P., and Kennedy, E. P., Biosynthesis of phospholipids in *Bacillus megaterium, J. Bacteriol.,* 140, 996, 1979.
61. Dutt, A. and Dowhan, W., Characterization of a membrane associated cytidine diphosphate-diacylglycerol-dependent phosphatidylserine synthase in bacilli, *J. Bacteriol.,* 147, 535, 1981.
62. Dowhan, W., Wickner, W. T., and Kennedy, E. P., Purification and properties of phosphatidylserine decarboxylase from *E. coli, J. Biol. Chem.,* 249, 3079, 1974.
63. Hirabayashi, T., Larson, T. J., and Dowhan, W., Membrane-associated phosphatidylglycerophosphate synthetase from *Escherichia coli:* purification by substrate affinity chromatography on cytidine 5′-diphospho-1,2-diacyl-sn-glycerol Sepharose, *Biochemistry,* 15, 5202, 1976.
64. Chang, Y. Y. and Kennedy, E. P., Phosphatidyl glycerophosphate phosphatase, *J. Lipid Res.,* 8, 456, 1967.
65. Finnerty, W. R. and Makula, R. A., Microbial lipid metabolism, *CRC Crit. Rev. Microbiol.,* 4, 1, 1975.
66. Short, S. A. and White, D. C., Biosynthesis of cardiolipin from phosphatidylglycerol in *Staphylococcus aureus, J. Bacteriol.,* 109, 820, 1972.
67. Stanacev, N. Z., Chang, Y. Y., and Kennedy, E. P., Biosynthesis of cardiolipin in *Escherichia coli, J. Biol. Chem.,* 242, 3018, 1967.
68. Schneider, E. G. and Kennedy, E. P., Phosphorylation of ceramide by diglyceride kinase preparations from *Escherichia coli, J. Biol. Chem.,* 248, 3739, 1973.
69. Konig, B., Bohnenberger, E., and Sandermann, H., Jr., Hexamethylphosphoric triamide as a solubilizing agent: purification and reactivation of diglyceride kinase, *FEBS Lett.,* 129, 301, 1981.
70. Carty, C. E. and Ingram, L. O., Lipid synthesis during the *Escherichia coli* cell cycle, *J. Biol. Chem.,* 145, 472, 1981.
71. DiRienzo, J. M. and Inouye, M., Lipid fluidity-dependent biosynthesis and assembly of the outer membrane proteins of *E. coli, Cell,* 17, 155, 1979.
72. DiRienzo, J. M., Nakamura, K., and Inouye, M., The outer membrane proteins of gram-negative bacteria: biosynthesis, assembly, and function, *Annu. Rev. Biochem.,* 47, 481, 1978.
73. Shulman, H. and Kennedy, E. P., Relation of turnover of membrane phospholipids to synthesis of membrane-derived oligosaccharides of *Escherichia coli, J. Biol. Chem.,* 252, 4250, 1977.
74. Gennis, R. B. and Strominger, J. L., Activation of C55-isoprenoid alcohol phosphokinase from *Staphylococcus aureus.* I. Activation by phospholipids and fatty acids, *J. Biol. Chem.,* 251, 1264, 1976.
75. Hawrot, E. and Kennedy, E. P., Phospholipid composition and membrane function in phosphatidylserine decarboxylase mutants of *Escherichia coli, J. Biol. Chem.,* 253, 8213, 1978.
76. Joseleau-Petit, D. and Kepes, A., Degraded and stable phosphatidylglycerol in *Escherichia coli* inner and outer membranes, and recycling of fatty acyl residues, *Biochim. Biophys. Acta,* 711, 1, 1982.
77. Takeuchi, Y., Ohnishi, S. I., Ishinaga, M., and Kito, M., Interaction of CDP-diglyceride spin-label with *Escherichia coli* B membrane fractions and its relationship with phospholipid synthesis, *Biochem. Biophys. Res. Commun.,* 99, 1251, 1981.

78. Demant, E. J. F., Incorporation of large quantities of phospholipid by net de novo synthesis in cytoplasmic membrane vesicles of *Escherichia coli, Biochim. Biophys. Acta*, 664, 229, 1981.
79. Downie, J. A., Gibson, F., and Cox, G. B., Membrane adenosine triphosphatases of prokaryotic cells, *Annu. Rev. Biochem.*, 48, 103, 1979.
80. Singh, A. P. and Bragg, P. D., ATP-dependent proton translocation and quenching of 9-aminoacridine fluorescence in inside-out membrane vesicles of a cytochrome-deficient mutant of *Escherichia coli, Biochim. Biophys. Acta*, 164, 562, 1977.
81. Harold, F. M. and Papinean, D., Cation transport and electrogenesis by *S. faecalis*. II. Proton and sodium extrusion, *J. Memb. Biol.*, 8, 45, 1972.
82. Harold, F. M., Membranes and energy transduction in bacteria, *Curr. Top. Bioeng.*, 6, 83, 1977.
83. Salton, M. R. J., Membrane associated enzymes in bacteria, *Adv. Microbial. Physiol.*, 11, 213, 1974.
84. Kagawa, Y., Protonophoric action of coupling factor ATPase — its reconstitution from purified subunits, reconstitution into ATP synthesizing vesicles and image reconstruction, in *Energy Conservation in Biological Membrane*, Schafer, G. and Klingenberg, M., Eds., Springer-Verlag, Berlin, 195, 1978.
85. Kalra, V. K., Lee, S. H., Ritz, C. J., and Brodie, A. F., Purification and properties of membrane-bound coupling factor-latent ATPase from *Mycobacterium phlei, J. Supramol. Struct.*, 3, 231, 1975.
86. Leimgruier, R. M., Jensen, C., and Abrams, A., Purification and characterization of the membrane adenosine triphosphatase complex from the wild-type and N.N'-dicyclohexylcarbodiimide-resistant strains of *Streptococcus faecalis, J. Bacteriol.*, 147, 363, 1981.
87. Mazza, G. and Riva, S., DNA dependent ATPase from *Bacillus subtillis, J. Bacteriol.*, 146, 305, 1981.
88. Ghosh, S., Chatterjee, K., Haldar, S., Banerjee, D., Tiwari, R., and Chatterjee, G. C., Studies on membrane ATPase activity and lipid level in *vibrio el tor* under normal and nitrofurantoin-resistant conditions, *Acta Microbiol. Pol.*, 30, 231, 1981.
89. Bevers, E. M., Snoek, G. T., Op Den Kamp, J. A. F., and Van Deenen, L. L. M., Phospholipid requirement of the membrane bound magnesium ion dependent ATPase in *Acholeplasma laidlawii, Biochim. Biophys. Acta*, 467, 346, 1977.
90. Riebeling, V. and Jungermann, K., Properties and function of *Clostridial* membrane ATPase EC-3.6.1.3., *Biochim. Biophys. Acta*, 430, 434, 1976.
91. Pandit-Hoven Kamp, H. G., Microbial phosphorylating preparations: *Azotobacter, Meth. Enzymol.*, 10, 152, 1967.
92. Melandri, B. A., Baccarini-Melandri, A., and Fabbri, E., Energy transduction in photosynthetic bacteria. IV. Light-dependent ATPase in photosynthetic membranes from *Rhodopseudomonas capsulata, Biochim. Biophys. Acta*, 275, 383, 1972.
93. Monteil, H., Schoun, J., and Guinard, M., A Na^+,K^+-activated Mg^{2+}-dependent ATPase released from *Proteus* L-form membrane, *Eur. J. Biochem.*, 41, 525, 1974.
94. Lucke, F. K. and Klemme, J. H., Coupling factor adenosine 5'-triphosphatase from *Rhodospirillum rubrum* — simple and rapid procedure for its purification, *Z. Naturforsch.*, 31, 272, 1976.
95. Neeman, Z., Kahane, I., and Rozin, S., Characterization of the Mycoplasma membrane proteins. II. Solubilization and enzymatic activities of *Acholeplasma laidlawii* membrane proteins, *Biochim. Biophys. Acta*, 249, 169, 1981.
96. Bragg, P. D. and Hou, C., Subunit composition function and special arrangement in the Ca^{+2} and Mg^{+2} activated adenosine triphosphatase of *E. coli* and *S. typhimurium, Arch. Biochem. Biophys.*, 167, 311, 1975.
97. Ishikawa, S. and Leninger, A. L., Reconstitution of oxidative phosphorylation in preparation from *Micrococcus lysodeikticus, J. Biol. Chem.*, 237, 2401, 1962.
98. Abrams, A., The release of bound adenosine triphosphatase from isolated bacterial membranes and the properties of the solubilized enzyme, *J. Biol. Chem.*, 240, 3679, 1965.
99. Munoz, E., Nachbar, M. S., Schor, M. T., and Salton, M. R. J., Adenosine triphosphatase of *Micrococcus lysodeikticus:* selective release and relationship to membrane structure, *Biochem. Biophys. Res. Commun.*, 32, 539, 1968.
100. Munoz, E., Freer, J. H., Ellar, D. J., and Salton, M. R. J., Membrane associated ATPase activity from *Micrococcus lysodeikticus, Biochim. Biophys. Acta*, 150, 531, 1968.
101. Munoz, E., Salton, M. R. J., Ng, M. H., and Schor, M. T., Membrane adenosine triphosphatase of *Micrococcus lysodeikticus:* purification, properties of the soluble enzyme, and properties of the membrane bound enzyme, *Eur. J. Biochem.*, 7, 490, 1969.
102. Nachbar, M. S. and Salton, M. R. J., Characteristics of a lipid rich NADH dehydrogenase-containing particulate fraction obtained from *Micrococcus lysodeikticus* membrane, *Biochim. Biophys. Acta*, 223, 309, 1970.
103. Ishida, M. and Mizushima, S., Membrane ATPase of *Bacillus megaterium*. I. Properties of membrane ATPase and its solubilized form, *J. Biochem. Tokyo*, 66, 33, 1969.

104. Ishida, M. and Mizushima, S., The membrane ATPase of *Bacillus megaterium*. II. Purification of membrane ATPase and their recombination with membrane, *J. Biochem. Tokyo*, 66, 133, 1969.
105. Mirsky, R. and Barlow, V., Purification and properties of ATPase from the cytoplasmic membrane of *Bacillus megaterium* KM, *Biochim. Biophys. Acta*, 241, 835, 1971.
106. Kobayashi, H. and Anraku, Y., Membrane-bound adenosine triphosphatase of *Escherichia coli*. I. Partial purification and properties, *J. Biochem.*, 71, 387, 1972.
107. Davis, P. L. and Bragg, P. D., Properties of a soluble Ca^{+2} and Mg^{+2} activated ATPase released from *Escherichia coli* membranes, *Biochim. Biophys. Acta*, 266, 273, 1972.
108. Hachimori, A., Muramatsu, N., and Nosoh, Y., Studies on an ATPase of thermophilic bacteria. I. Purification and properties, *Biochim. Biophys. Acta*, 206, 426, 1970.
109. Evans, D. J., Jr., Membrane adenosine triphosphatase of *Escherichia coli:* activation by calcium ion and inhibition by monovalent cations, *J. Bacteriol.*, 100, 914, 1969.
110. Yoshida, M., Sone, N., Hirata, H., and Kagowa, Y., A highly stable adenosine triphosphatase from a thermophilic bacterium: purification, properties and reconstitution, *J. Biol. Chem.*, 250, 7910, 1975.
111. Sone, N., Yoshida, M., Hirata, H., and Kagawa, Y., Purification and properties of a dicyclohexylcarbodiimide-sensitive adenosine triphosphatase from a thermophilic bacterium, *J. Biol. Chem.*, 250, 7919, 1975.
112. Salton, M. R. J. and Schor, M. T., Subunit structure and properties of two forms of adenosine triphosphatase released from *Micrococcus lysodeikticus* membranes, *Biochem. Biophys. Res. Commun.*, 49, 350, 1972.
113. Salton, M. R. J. and Schor, M. T., Release and purification of *Micrococcus lysodeikticus* ATPase from membrane extracted with n-butanol, *Biochim. Biophys. Acta*, 345, 74, 1974.
114. Cox, G. B., Downie, J. A., Fayle, D. R. H., Gibson, F., and Radik, J., Inhibition, by a protease inhibitor, of the solubilization of the F1-portion of the Mg^{+2}-stimulated adenosine triphosphatase of *Escherichia coli, J. Bacteriol.*, 133, 287, 1978.
115. Nieuwenhuis, F. J. R. M., Thomas, A. A. M., and Van Dam, K., Solubilization by cholate or deoxycholate of a dicyclohexylcarbodi-imide-sensitive adenosine triphosphatase complex from *Escherichia coli, Biochem. Soc. Trans.*, 2, 512, 1974.
116. Hare, J. F., Purification and characterization of a dicyclohexylcarbodiimide-sensitive adenosine triphosphatase complex from membranes of *Escherichia coli, Biochem. Biophys. Res. Commun.*, 66, 1329, 1975.
117. Foster, D. L. and Fillingame, R. H., Purification of a dicyclohexylcarbodiimide-sensitive, energy-transducing adenosine triphosphatase complex from *Escherichia coli, Fed. Proc.*, 37, 1520, 1978.
118. Clarke, D. J. and Morris, J. G., Partial purification of a dicyclohexylcarbodi-imide-sensitive membrane adenosine triphosphatase complex from the obligately anaerobic bacterium *Clostridium pasteurianum, Biochem. J.*, 154, 725, 1976.
119. Hanson, R. L. and Kennedy, E. P., Energy-transducing adenosine triphosphatase from *Escherichia coli:* purification, properties and inhibition by antibody, *J. Bacteriol.*, 114, 772, 1973.
120. Vogel, G. and Steinhart, R., ATPase of *Escherichia coli:* purification, dissociation and reconstitution of the active complex from the isolated subunits, *Biochemistry*, 15, 208, 1976.
121. Baron, C. and Abrams, A., Isolation of a bacterial membrane protein, Nectin, essential for the attachment of adenosine triphosphatase, *J. Biol. Chem.*, 246, 1542, 1971.
122. Smith, J. B. and Sternweis, P. C., Purification of membrane attachment and inhibitory subunits of the proton translocating adenosine triphosphatase from *Escherichia coli, Biochemistry*, 16, 306, 1977.
123. Smith, J. B. and Sternweis, P. C., Restoration of coupling factor activity to *Escherichia coli* ATPase missing the delta subunit, *Biochem. Biophys. Res. Commun.*, 62, 764, 1975.
124. Sternweis, P. C. and Smith, J. B., Characterization of the purified membrane attachment (δ) subunit of the proton translocating adenosine triphosphatase from *Escherichia coli, Biochemistry*, 16, 4021, 1977.
125. Sternweis, P. C., The e subunit of *Escherichia coli* coupling factor 1 is required for its binding to the cytoplasmic membrane, *J. Biol. Chem.*, 253, 3123, 1978.
126. Butlin, J. D., Cox, G. B., and Gibson, F., Oxidative phosphorylation in *Escherichia coli* K12: mutations affecting magnesium-ion or calcium ion-stimulated adenosine triphosphatase, *Biochem. J.*, 124, 75, 1971.
127. Cox, G. B. and Downie, J. A., Isolation and characterization of mutants of *Escherichia coli* K12 affected in oxidative phosphorylatin or quinone biosynthesis, *Meth. Enzymol.*, 56, 106, 1979.
128. Thipayathasana, P., Isolation and properties of *Escherichia coli* ATPase mutants with altered divalent metal specificity for ATP hydrolysis, *Biochim. Biophys. Acta*, 408, 47, 1975.
129. Simoni, R. D. and Shandell, A., Energy transduction in *Escherichia coli* genetic alteration of a membrane polypeptide of the (Ca^{+2}, Mg^{+2}) ATPase complex, *J. Biol. Chem.*, 250, 9421, 1975.
130. Simoni, R. D. and Shallenberger, M. K., Coupling of energy to active transport of amino acids in *E. coli, Proc. Natl. Acad. Sci. U.S.A.*, 69, 2663, 1972.

131. Daniel, J., Roisin, M. P., Burstein, C., and Kepes, A., Mutants of *Escherichia coli* X12 unable to grow on nonfermentable carbon substrates, *Biochim. Biophys. Acta*, 376, 195, 1975.
132. Butlin, J. D., Cox, G. B., and Gibson, F., Oxidative phosphorylation in *Escherichia coli* K12: the genetic and biochemical characterization of a strain carrying a mutation in the uncB gene, *Biochim. Biophys. Acta*, 292, 366, 1973.
133. Schairer, H. U. and Gruber, D., Mutants of *Escherichia coli* K12 defective in exidative phosphorylation, *Eur. J. Biochem.*, 37, 282, 1973.
134. Schairer, H. U., Friedl, P., Schmid, B. I., and Vogel, G., The use of several energy-coupling reactions in characterizing mutants of *Escherichia coli* K12 defective in oxidative phosphorylation, *Eur. J. Biochem.*, 66, 257, 1976.
135. Friedl, P., Schmid, B. I., and Schairer, H. U., A mutant ATP synthetase of *Escherichia coli* with an altered sensitivity to N, N′-dicyclohexylcarbodiimide: characterization in native membranes and reconstituted proteoliposomes, *Eur. J. Biochem.*, 73, 461, 1977.
136. Fillingame, R. H., Identification of the dicyclohexylcarbodiimide-reactive protein component of the adenosine 5′-triphosphate energy transducing system of *Escherichia coli*, *J. Bacteriol.*, 124, 870, 1975.
137. Gibson, F., Downie, J. A., Cox, G. B., and Radik, J., Mu-induced polarity in the unc operon of *Escherichia coli*, *J. Bacteriol.*, 134, 728, 1978.
138. Downie, J. A., Senior, A. E., Gibson, F., and Cox, G. B., A fifth gene (uncE) in the operon concerned with oxidative phosphorylation in *Escherichia coli*, *J. Bacteriol.*, 137, 711, 1979.
139. Downie, J. A., Cox, G. B., Langman, L., Ash, G., Becker, M., and Gibson, F., Three genes coding for subunits of the membrane sector (fo) of the *Escherichia coli* adenosine triphosphatase complex, *J. Bacteriol.*, 145, 200, 1981.
140. Michell, P., Chemiosmotic coupling in oxidative and photosynthetic phosphorylation, *Biol. Rev. Cambridge Philos. Soc.*, 41, 445, 1966.
141. Jones, C. W., Brice, J. M., Downs, A. J., and Drozd, J. W., Bacterial respiration — linked proton translocation and its relationship to respiratory-chain composition *Eur. J. Biochem.*, 52, 265, 1975.
142. Crofts, A. R., Evans, E. H., and Cogdell, R. J., The relation between H^+ uptake and electron flow in chromatophores from photosynthetic bacteria, *Ann. N.Y. Acad. Sci.*, 227, 227, 1974.
143. Drachev, L. A., Kondrashin, A. A., Samuilov, V. D., and Skulachev, V. P., Generation of electric potential by reaction center complexes from *Rhodospirillum rubrum*, *FEBS Lett.*, 50, 219, 1975.
144. Lewis, A. J., Spoonhower, R. A., Bogomolni, R. A., Lozier, R. H., and Stoeckenius, W., Tunable laser resonance Raman spectroscopy of bacteriorhodopsin, *Proc. Natl. Acad. Sci. U.S.A.*, 71, 4462, 1974.
145. Racker, E. and Hinkle, P. C., Effect of temperature on the function of a proton pump, *J. Membr. Biol.*, 17, 181, 1974.
146. Racker, E. and Stoeckenius, W., Reconstitution of purple membrane vesicles catalyzing light-driven proton uptake and adenosine triphosphate formation, *J. Biol. Chem.*, 249, 662, 1974.
147. Abrams, A. and Smith, J. B., Bacterial membrane ATPase, in *The Enzymes*, Boyer, P., Ed., Vol. 10, Academic Press, New York, 1974, 395.
148. Pedersen, P. L., Mitochondrial adenosine triphosphatase, *J. Bioenerg.*, 6, 243, 1975.
149. Mitchell, P., Cation-translocating adenosine triphosphatase models: how direct is the participation of adenosine triphosphate and its hydrolysis products in cation translocation?, *FEBS Lett.*, 33, 267, 1973.
150. Skulachev, V. P., in: MTP International Review of Science, *Biochemistry*, Ser. 1, Vol. 3, Energy Transducing Mechanisms, Racker, E., Ed., University Park Press, Baltimore, Maryland, 1975, 31.
151. Salton, M. R. G., Schor, M. T., and Ng, M. H., Internal localization of *Micrococcus lysodeikticus* membrane ATPase by iodination with 125-I, *Biochim. Biophys. Acta*, 290, 408, 1972.
152. Oppenheim, J. D. and Salton, M. R. J., Localization and distribution of *Micrococcus lysodeikticus* membrane ATPase determined by ferritin labeling, *Biochim. Biophys. Acta*, 298, 297, 1973.
153. Gomez-Puyon, A. and Gomez-LoJero, L., in *Current Topics in Bioenergetics*, Vol. 6, Sanaki, D. R., Ed., Academic Press, New York, 1977, 221.
154. Kagawa, Y., Sone, N., Hirata, H., and Yoshida, M., Reconstitution of ATPase and carriers from thermophilic biomembranes, *TIBS*, (Feb.), p. 31, 1979.
155. Kagawa, Y., Reconstitution of the energy transformer, gate and channel subunit reassembly, crystalline ATPase and ATP synthesis, *Biochim. Biophys. Acta*, 505, 45, 1978.
156. Payne, W. J., Reduction of nitrogenous oxides by microorganisms, *Bacteriol. Rev.*, 37, 409, 1973.
157. Pichinoty, F., Bacterial reduction of oxygenated inorganic compounds of nitrogen (Fr), *Bull. Inst. Pasteur Paris*, 71, 317, 1973.
158. Brown, C. M., MacDonald-Brown, D. S., and Meers, J. L., Physiological aspects of microbial inorganic nitrogen metabolism, *Adv. Microb. Physiol.*, 11, 1, 1974.
159. Stouthamer, A. H., Biochemistry and genetics of nitrate reductase in bacteria, in *Adv. Microbial Physiol.*, Rose, A. H. and Tempest, D. W., Eds., Academic Press, New York, 14, 315, 1976.

160. Pichinoty, F. and Piechaud, Detection and characterization of bacterial nitrate reductase, *Ann. Del'Inst. Past. Paris,* 114, 77, 1968.
161. Iida, K. and Taniguchi, S., Studies on nitrate reductase system of *Escherichia coli.* I. Particulate electron transport system to nitrate and its solubilization, *J. Biochem.,* 46, 1041, 1959.
162. Itagaki, E., Fugita, T., and Sato, R., Cytochrome b1-nitrate reductase interaction in a solubilized system from *Escherichia coli, Biochim. Biophys. Acta,* 51, 390, 1961.
163. Itagaki, E., Fugita, T., and Sato, R., Solubilization and properties of formate dehydrogenase and cytochrome b1 from *Escherichia coli, J. Biochem,* 52, 131, 1962.
164. Ruiz-Herrera, J., Showe, M. K., and DeMoss, J. A., Nitrate reductase complex of *Escherichia coli* K12 — isolation and characterization of mutants unable to reduce nitrate, *J. Bacteriol.,* 97, 1291, 1969.
165. Ruiz-Herrera, J. and DeMoss, J. A., Nitrate reductase complex of *Escherichia coli* K-12 — participation of specific formate dehydrogenase and cytochrome B-1 components in nitrate reduction, *J. Bacteriol.,* 99, 720, 1969.
166. MacGregor, C. H. and Schnaitman, C. A., Nitrate reductase in *E. coli:* properties of the enzyme and *in vitro* reconstitution from enzyme deficient mutants, *J. Supramol. Struct.,* 2, 726, 1975.
167. MacGregor, C. H. and C. A. Schnaitman, Alterations in the cytoplasmic membrane proteins of various chlorate resistant mutants of *Escherichia coli, J. Bacteriol.,* 108, 564, 1971.
168. Schnaitman, C. A., Solubilization of the cytoplasmic membrane of *Escherichia coli* by Triton X-100®, *J. Bacteriol.,* 108, 545, 1971.
169. MacGregor, C. H., Schnaitman, C. A., Normansell, D. E., and Hodgins, M. G., Purification and properties of nitrate reductase from *Escherichia coli* K-12, *J. Biol. Chem.,* 249, 5321, 1974.
170. MacGregor, C. H., Anaerobic cytochrome b1 in *Escherichia coli:* association with and regulation of nitrate reductase, *J. Bacteriol.,* 121, 1111, 1975.
171. Taniguchi, S. and Itagaki, E., Nitrate reductase of nitrate respiration type from *E. coli.* I. Solubilization and purification from the particulate system with molecular characterization as a metalloprotein, *Biochim. Biophys. Acta,* 44, 263, 1960.
172. Forget, P., Bacterial nitrate reductases — solubilization, purification and properties of enzyme A of *Escherichia coli* K-12, *Eur. J. Biochem.,* 42, 325, 1974.
173. Showe, M. K. and DeMoss, J. A., Localization and regulation of synthesis of nitrate reductase in *Escherichia coli, J. Bacteriol.,* 95, 1305, 1968.
174. Lund, K. and DeMoss, J. A., Association-dissociation behavior and subunit structure of heat-related nitrate reductase from *Escherichia coli, J. Biol. Chem.,* 251, 2207, 1976.
175. DeMoss, J. A., Limited proteolysis of nitrate reductase purified from membranes of *Escherichia coli, J. Biol. Chem.,* 252, 1696, 1977.
176. Schryvers, A., Lohmeier, E., and Weiner, J. H., Chemical and functional properties of the native and reconstituted forms of the membrane-bound, aerobic glycerol-3-phosphate dehydrogenase of *Escherichia coli, J. Biol. Chem.,* 253, 783, 1978.
177. Lin, E. C. C., Kohn, J. P., Chused, T. M., and Jorgensen, S. E., Utilization of L-alpha-glycerophosphate by *Escherichia coli* without hydrolysis, *Proc. Natl. Acad. Sci. U.S.A.,* 48, 2145, 1962.
178. Kistler, W. S., Hirsch, C. A., Cozzarelli, N. R., and Lin, E. C. C., Second pyridine nucleotide-independent L-alpha-glycerophosphate dehydrogen12e in *Escherichia coli* K-12, *J. Bacteriol.,* 100, 1133, 1969.
179. Weiner, J. H., and Heppel, L. A., Purification of the membrane bound and pyridine nucleotide-independent L-glycerol-3-phosphate dehydrogenase from *Escherichia coli, Biochim. Biophys. Res. Commun.,* 47, 1360, 1972.
180. Weiner, J. H., The localization of glycerol-3-phosphate dehydrogenase in *Escherichia coli, J. Membr. Biol.,* 15, 1, 1974.
181. Futai, M. and Kimura, H., Inducible membrane-bound L-lactate dehydrogenase from *Escherichia coli, J. Biol. Chem.,* 252, 5820, 1977.
182. Pratt, E. A., Fung, L. W. M., and Flowers, J. A., Studies of membrane bound D-lactate dehydrogenase of *Escherichia coli, Fed. Proc.,* 36, 896, 1977.
183. Kline, E. S. and Mahler, H. R., The lactate dehydrogenase of *E. coli, Ann. N.Y. Acad. Aci.,* 119, 905, 1965.
184. Singer, S. J., The molecular organization of membranes, *Annu. Rev. Biochem.,* 43, 805, 1974.
185. Kimura, H. and Futai, M., Effects of phospholipids on L-lactate dehydrogenase from membranes of *Escherichia coli:* activation and stabilization of the enzyme with phospholipids, *J. Biol. Chem.,* 253, 1095, 1978.
186. Futai, M., Orientation of membrane vesicles from *Escherichia coli* prepared by different procedures, *J. Membr. Biol.,* 15, 15, 1974.
187. Futai, M., Membrane D-lactate dehydrogenase from *Escherichia coli,* purification and properties, *Biochemistry,* 12, 2468, 1973.

188. Kohn, L. D. and Kaback, H. R., Mechanisms of active transport in isolated bacterial membrane vesicles. XV. Purification and properties of the membrane-bound D-lactate dehydrogenase from *Escherichia coli, J. Biol. Chem.*, 248, 7012, 1973.
189. Tanaka, Y., Anraku, Y., and Futai, M., *Escherichia coli* membrane D-lactate dehydrogenase: isolation of the enzyme in aggregated form and its activation by Triton X-100® and phospholipids, *J. Biol. Chem.*, 80, 821, 1976.
190. Ito, A. and Sato, R., Purification by means of detergents and properties of cytochrome b5 from liver microsomes, *J. Biol. Chem.*, 243, 4922, 1968.
191. Spatz, L. and Strittmatter, P., A form of cytochrome b5 that contains an additional hydrophobic sequence of 40 amino acid residues, *Proc. Natl. Acad. Sci. U.S.A.*, 68, 1024, 1971.
192. Robinson, N. C. and Tanford, C., The binding of deoxycholate, Triton X-100®, sodium dodecyl sulfate and phosphatidylcholine vesicles to cytochrome b5, *Biochemistry*, 14, 369, 1975.
193. Franklin, F. C. H. and Venables, W. A., Biochemical, genetic, and regulatory studies of alanine catabolism in *Escherichia coli* K-12, *Mol. Gen. Genet.*, 149, 229, 1976.
194. Pioli, D., Venables, W. A., and Franklin, F. C. H., D-alanine dehydrogenase: its role in utilization of alanine isomers as growth substrates by *Pseudomonas aeruginosa* PA01, *Arch. Microbiol.*, 110, 287, 1976.
195. Bater, A. J., Venables, W. A., and Thomas, S., Allohydroxy-D-proline dehydrogenase: an inducible membrane-bound enzyme in *Pseudomonas aeruginosa* PA01, *Arch. Microbiol.*, 112, 287, 1977.
196. Singer, S. J., The molecular organization of membranes, *Annu. Rev. Biochem.*, 43, 805, 1974.
197. Kung, H. F. and Henning, U., Limiting availability of binding sites for dehydrogenases on the cell membrane of *Escherichia coli, Proc. Natl. Acad. Sci. U.S.A.*, 69, 925, 1972.
198. Bater, A. J. and Venables, W. A., The characterization of inducible dehydrogenases specific for the oxidation of D-alanine, allohydroxy-D-proline, choline and sarcosine as peripheral membrane proteins in *Pseudomonas aeruginosa, Biochim. Biophys. Acta*, 468, 209, 1977.
199. Fernley, R. T., Lentz, S. R., and Bradshaw, R. A., Malate dehydrogenase: isolation from *E. coli* and comparison with the eukaryotic mitochondrial and cytoplasmic forms, *Biosci. Rep.*, 1, 497, 1981.
200. Graham, A., The organization of hydrogenase in the cytoplasmic membrane of *Escherichia coli, Biochem. J.*, 197, 283, 1981.
201. Glick, B. R., Zeisler, J., Banaszuk, A. M., Friesen, J. D., and Martin, W. G., The identification and partial characterization of a plasmid containing the gene for the membrane-associated hydrogenase from *E. coli, Gene*, 15, 201, 1981.
202. Graham, A. and Boxer, D. H., The organization of formate dehydrogenase in the cytoplasmic membrane of *Escherichia coli, Biochem. J.*, 195, 627, 1981.
203. Haddock, B. A., Mandrand-Berthelot, M. A., Graham, A., and Boxer, D., Formate dehydrogenase EC-1.2.1.2 from *Escherichia coli*, in *Microbial Growth on C1 Compounds*, Vol. 13, Proc. 3rd Int. Symp., Dalton, H., Ed., Heyden and Son, London, 1980, 270.
204. Cox, J. C., Edwards, E. S., and DeMoss, J. A., Resolution of distinct selenium containing formate dehydrogenases from *Escherichia coli, J. Bacteriol.*, 145, 1317, 1981.
205. Jaworowski, A., Campbell, H. D., Poulis, M. I., and Young, I. G., Genetic identification and purification of the respiratory NADH dehydrogenase of *Escherichia coli, Biochemistry*, 20, 2041, 1981.
206. Young, I. C. and Wallace, B. J., Mutations affecting reduced nicotinamide adenine dinucleotide dehydrogenase complex of *E. coli, Biochim. Biophys. Acta*, 449, 376, 1976.
207. Jaworowski, A., Mayo, G., Shaw, D. C., Campbell, H. D., and Young, I. G., Characterization of the respiratory NADH dehydrogenase of *Escherichia coli* and reconstitution of NADH oxidase in NDH mutant membrane vesicles, *Biochemistry*, 20, 3621, 1981.
208. Thomson, J. W. and Shapiro, B. M., The respiratory chain NADH dehydrogenase of *Escherichia coli* isolation of an NADH: quinone oxidoreductase from membranes and comparison with the membrane-bound NADH: dichlorophenolindophenol oxidoreductase, *J. Biol. Chem.*, 256, 3077, 1981.
209. Zhukova, I. G., Shaposhnikov, G. L., and Ostrovskii, D. N., Proteins of bacterial membranes purification of NADH dehydrogenase from *Micrococcus lysodeikticus* by electrofocusing, *Biokhimiya*, 41, 1840, 1977.
210. Robinson, J. J. and Weiner, J. H., The effects of anions on fumarate reductase isolated from the cytoplasmic membrane of *Escherichia coli, Biochem. J.*, 199, 473, 1981.
211. Hederstedt, L. and Rutberg, L., Succinate dehydrogenase — a comparative review, *Microbiol. Rev.*, 45, 542, 1981.
212. Hirsch, C. A., Rasminsky, M., Davis, B. D., and Lin, E. C. C., A fumarate reductase in *Escherichia coli* distinct from succinate dehydrogenase, *J. Biol. Chem.*, 238, 3770, 1963.
213. Spencer, M. E. and Guest, J. R., Isolation and properties of fumarate reductase mutants of *Escherichia coli, J. Bacteriol.*, 114, 563, 1973.
214. Davis, K. A., Hatefi, Y., Crawford, I. P., and Baltscheffsky, H., Purification, molecular properties, and amino acid composition of the subunits of *Rhodospirillum rubrum* succinate dehydrogenase, *Arch. Biochem. Biophys.*, 180, 459, 1977.

215. Carithers, R. P., Yoch, D. C., and Arnon, D. I., Isolation and characterization of bound iron-sulfur proteins from bacterial photosynthetic membranes. II. Succinate dehydrogenase from *Rhodospirillum rubrum* chromatophores, *J. Biol. Chem.*, 252, 7453, 1977.
216. Hatefi, Y., Davis, K. A., Baltscheffsky, H., Baltscheffsky, M., and Johansson, B. C., Isolation and properties of succinate dehydrogenase from *Rhodospirillum rubrum*, *Arch. Biochem. Biophys.*, 152, 613, 1972.
217. Ingledew, W. J. and Prince, R. C., Thermodynamic resolution of the iron-sulfur centers of the succinic dehydrogenase of *Rhodopseudomonas sphaeroides*, *Arch. Biochem. Biophys.*, 178, 303, 1977.
218. Kennedy, W. C., Mowery, P. C., Seng, R. L., and Singer, T. P., Localization of the substrate and oxalacetate binding site of succinic dehydrogenase, *J. Biol. Chem.*, 251, 2369, 1979.
219. Mowery, P. C., Steenkamp, D. J., Ackrell, B. A. C., Singer, T. P., and White, G. A., Inhibition of mammalian succinate dehydrogenase by carboxins, *Arch. Biochem. Biophys.*, 178, 495, 1977.
220. Coles, C. J., Singer, T. P., White, G. A., and Thorn, G. D., Studies on the binding of carboxin analogs to succinate dehydrogenase, *J. Biol. Chem.*, 253, 5573, 1978.
221. Hatefi, Y. and Galante, Y. M., Isolation of cytochrome b560 from complex II (succinate-ubiquinone oxidoreductase) and its reconstitution with succinate dehydrogenase, *J. Biol. Chem.*, 225, 5530, 1980.
222. Spencer, M. E. and Guest, J. R., Proteins of the inner membrane of *Escherichia coli*: identification of succinate dehydrogenase by polyacrylamide gel electrophoresis with sdh amber mutants, *J. Bacteriol.*, 117, 947, 1974.
223. Creaghan, I. T. and Guest, J. R., Amber mutants of the alphaketoglutarate dehydrogenase gene of *E. coli* K-12, *J. Gen. Microbiol.*, 71, 207, 1972.
224. Chern, C. K., Ando, A., Kusaka, I., and Fukui, S., A succinate dehydrogenase-deficient mutant of *Agrobacterium tumefaciens*, *Agric. Biol. Chem.*, 40, 779, 1976.
225. Ohne, M., Rutberg, B., and Hoch, J. A., Genetic and biochemical characterization of mutants of *Bacillus subtilis* defective in succinate dehydrogenase, *J. Bacteriol.*, 115, 738, 1973.
226. Owen, P. and Salton, M. R. J., Membrane asymmetry and expression of cell surface antigens of *M. lysodeikticus* established by immunoelectrophoresis, *J. Bacteriol.*, 132, 974, 1977.
227. Oelze, J., Proteins exposed at the surface of chromatophores of *Rhodospirillum rubrum*: the orientation of isolated chromatophores, *Biochim. Biophys. Acta*, 509, 450, 1978.
228. Konings, W. N., Localization of membrane proteins in membrane vesicles of *Bacillus subtilis*, *Arch. Biochem. Biophys.*, 167, 570, 1975.
229. Hederstedt, L. and Rutberg, L., Biosynthesis and membrane binding of succinate dehydrogenase in *Bacillus subtilis*, *J. Bacteriol.*, 144, 941, 1980.
230. Holmgren, E., Hederstedt, L., and Rutberg, L., Role of heme in synthesis and membrane binding of succinic dehydrogenase in *Bacillus subtilis*, *J. Bacteriol.*, 138, 377, 1979.
231. Lascelles, J., Heme-deficient mutants of *Staphylococcus aureus*, *Meth. Enzymol.*, 56, 172, 1979.
232. Yu, E. K., and Devoe, I. W., L-cysteine oxidase activity in the membrane of *Neisseria meningitidis*, *J. Bacteriol.*, 145, 280, 1981.
233. Yang, T., Tetramethyl-P-phenylenediamine oxidase of *Pseudomonas aeruginosa*, *Eur. J. Biochem.*, 121, 335, 1982.
234. Tokuyama, T. and Asano, K., Oxidation of ammonia with cell fragments of *Nitrosomonas europaea*, *Hakko Kogaku Zasshi*, 56, 725, 1979.
235. Wetegrove, R. L., Wyss, O., and Jurtshuk, P. Jr., Membrane bound oxido-reductase associated with the nitrogenase complex in *Azotobacter vinelandii*, *Abstr. Annu. Meet. Am. Soc. Microbiol.*, 77, 206, 1977.
236. Reinards, R., Kubicki, J., and Ohlenbusch, H. D., Purification and characterization of NADH oxidase from membranes of *Acholeplasma laidlawii*, a copper-containing iron-sulfur flavoprotein, *Eur. J. Biochem.*, 120, 329, 1981.
237. Odom, J. M. and Peck, J. D., Jr., Localization of dehydrogenase, reductase, and electron transfer components in the sulfate-reducing bacterium *Desulfovibrio gigas*, *J. Bacteriol.*, 147, 161, 1981.
238. Payne, J. W., Peptides of microorganism, *Adv. Microb. Physiol.*, 13, 55, 1976.
239. Haas, M. W., Becker, J. M., and Miller, R. V., Peptidase activity in the inner membrane of *Pseudomonas aeruginosa*, *Biochim. Biophys. Acta*, 643, 256, 1981.
240. Miller, R. V. and Becker, J. M., Peptide utilization in *Pseudomonas aeruginosa*: evidence for membrane-associated peptidase, *J. Bacteriol.*, 133, 165, 1978.
241. Barak, Z. and Gilvarg, C., Peptide transport, in *Biomembranes*, Vol. 7, Eisenberg, H., Katchalski-Katzir, E., and Manson, L. A., Eds., Plenum Press, New York, 1975, 167.
242. Cascieri, T., Jr. and Mallette, M. F., Peptide utilization by *Pseudomonas putida* and *Pseudomonas maltophilia*, *J. Gen. Microbiol.*, 92, 283, 1976.
243. Blumberg, P. M. and Strominger, J. L., Inactivation of D-alaine carboxypeptidase by cephalosporins is not lethal to *Bacillus subtilis*, *Proc. Natl. Acad. Sci. U.S.A.*, 68, 2814, 1971.
244. Presslitz, J. E. and Ray, V. A., DD-carboxypeptidase and peptidoglycan transpeptidase from *Pseudomonas aeruginosa*, *Antimicrob. Agents Chemother.*, 7, 578, 1975.

245. Choules, G. L. and Gray, W. R., Peptidase activity in the membranes of *Mycoplasma laidlawii*, *Biochim. Biophys. Res. Commun.*, 45, 849, 1971.
246. Pecht, M., Giberman, E., Keysary, A., Yariv, J., and Katchalshi, E., Hydrolysis of alanine oligopeptides by an enzyme located in the membrane of *Mycoplasma laidlawii*, *Biochim. Biophys. Acta*, 290, 267, 1972.
247. Arminjon, F., Guinand, M., Vacheron, M. J., and Michel, G., Specificity profiles of the membrane bound gamma-D-glytamyl-L-meso-diamino pimelate endo peptidase and LD carboxypeptidase from *Bacillus sphaericus*, 9602, *Eur. J. Biochem.*, 73, 557, 1977.
248. Dusart, J., Leyh-Bouille, M., and Ghuysen, J. M., The peptidoglycan cross-linking enzyme system in Streptomyces strain R-61, strains K-15 and *Streptomyces rimosus* kinetic coefficients involved in the interactions of the membrane bound transpeptidase with peptide substrates and beta lactam antibiotics, *Eur. J. Biochem.*, 81, 33, 1978.
249. Exterkate, F. A., Pyrrollidone carboxylylpeptidase in *Streptococcus cremoris* dependence on an interaction with membrane components, *J. Bacteriol.*, 129, 1281, 1977.
250. Voellmy, R. W. and Goldberg, A. L., ATP stimulated endoprotease is associated with the cell membrane of *E. coli*, *Nature*, 290, 419, 1981.
251. Swamy, K. H. and Goldberg, A. L., Subcellular distribution of various proteases in *Escherichia coli*, *J. Bacteriol.*, 149, 1027, 1982.
252. Strongin, A. Y., Izotova, L. S., Abramov, Z. T., Ermakova, L. M., Gorodetsky, D. I., and Stepanov, V. M., On the appearance of *Bacillus subtilis* intracellular serine protease in the cell membrane and culture medium. Comparison of the enzyme and other *Bacillus subtilis* serine protease, *Arch. Microbiol.*, 119, 287, 1978.
253. Pacaud, M., Purification and characterization of two novel proteolytic enzymes in membranes of *Escherichia coli* protease IV and protease V, *J. Biol. Chem.*, 257, 4333, 1982.
254. Chang, C. N., Blobel, G., and Model, P., Detection of prokaryotic signal peptidase in an *Escherichia coli* membrane fraction endo proteolytic cleavage of nascent phage F-1 precoat protein, *Proc. Natl. Acad. Sci. U.S.A.*, 75, 360, 1978.
255. Mandel, G. and Wickner, W., Translational and post-translational cleavage of M13 procoat protein: extracts of both the cytoplasmic and outer membranes of *Escherichia coli* contain leader peptidase activity, *Proc. Natl. Acad. Sci. U.S.A.*, 76, 236, 1979.
256. Zwizinski, C., Date, T., and Wickner, W., Leader peptidase is found in both the inner and outer membranes of *Escherichia coli*, *J. Biol. Chem.*, 256, 3593, 1981.
257. Date, T. and Wickner, W., Isolation of the *Escherichia coli* leader peptidase gene and effects of leader peptidase overproduction *in vivo*, *Proc. Natl. Acad. Sci. U.S.A.*, 78, 6106, 1981.
258. Chang, C. N., Nielsen, J. B., Izui, K., Blobel, G., and Lampen, J. O., Identification of the signal peptidase cleavage site in *Bacillus licheniformis* prepenicillinase, *J. Biol. Chem.*, 257, 4340, 1982.
259. Hussain, M., Ichihara, S., and Mizushima, S., Mechanism of signal peptide cleavage in the biosynthesis of the major lipoprotein of the *Escherichia coli* outer membrane, *J. Biol. Chem.*, 257, 5177, 1982.
260. Neu, H. C. and Heppel, L. A., On the surface localization of enzymes in *E. coli*, *Biochim. Biophys. Res. Commun.*, 14, 215, 1964.
261. Done, J. C., Shorey, D., Loke, J. P., and Pollak, J. K., The cytochemical localization of alkaline phosphatase in *E. coli* at the E. M. level, *Biochem. J.*, 96, 27c, 1965.
262. MacAlister, T. J., Irvin, R. T., and Costerton, J. W., Cell surface localized alkaline phosphatase of *E. coli* as visualized by reaction product deposition and ferritin-labeled antibodies, *J. Bacteriol.*, 130, 318, 1977.
263. MacAlister, T. J., Irvin, R. T., and Costerton, J. W., Immunocytological investigation of protein synthesis in *E. coli*, *J. Bacteriol.*, 130, 329, 1977.
264. MacAlister, T. J., Irvin, R. T., and Costerton, J. W., Cell envelope protection of alkaline phosphatase against acid denaturation in *E. coli*, *J. Bacteriol.*, 130, 339, 1977.
265. Varenne, S., Piovant, M., Pages, J. M., and Lazdunski, C., Evidence for synthesis of alkaline phosphatase on membrane-bound polysomes in *E. coli*, *Eur. J. Biochem.*, 86, 603, 1978.
266. Mikhaleva, N. I., Gulevskaia, S. A., Nesmeianova, M. A., and Fikhte, B. A., Structural and biochemical characteristics of *E. coli* cells with various levels of secretable alkaline phosphatase, *Mikrobiologiya*, 52, 131, 1983.
267. Thompson, L. M. and Macleod, R. A., Biochemical localization of alkaline phosphatase in the cell wall of a marine pseudomonad, *J. Bacteriol.*, 117, 819, 1974.
268. MacAlister, T. J., Costerton, J. W., Thompson, L., Thompson, J., and Ingram, J. M., Distribution of alkaline phosphatase within the periplasmic space of gram-negative bacteria, *J. Bacteriol.*, 111, 827, 1972.
269. Bhatti, A. R., Do Voe, I. W., and Ingram, J. M., The release and characterization of some periplasm-located enzymes of *Pseudomonas aeruginosa*, *Can. J. Microbiol.*, 22, 1425, 1976.

270. Day, D. F. and Ingram, J. M., *In vitro* studies of an alkaline phosphatase-cell wall complex from *Pseudomonas aeruginosa, Can. J. Microbiol.*, 21, 9, 1975.
271. Wood, D. A. W. and Tristram, H., Localization in the cell and extraction of alkaline phosphatase from *Bacillus subtilis, J. Bacteriol.*, 104, 1045, 1970.
272. Ghosh, B. K., Wouters, J. T. M., and Lampen, J. O., Distribution of the sites of alkaline phosphatase(s) activity in vegetative cells of *Bacillus subtilis, J. Bacteriol.*, 108, 928, 1971.
273. Ghosh, B. K. and Ghosh, A., Quantitation of alkaline phosphatase sites in repressed and derepressed *Bacillus subtilis* by electron microscopic histochemistry, *Bacteriol. Proc.*, G46, 1971.
274. Ghosh, A., Vallespir, S., and Ghosh, B. K., Specificity of subcellular distribution of alkaline phosphatase in *Bacillus licheniformis* 749/C, *Can. J. Microbiol.*, 30, 113, 1984.
275. Ghosh, B. K., Guan, T., and Ghosh, A., Immunoelectron microscopic localization of a secretory protein in bacteria, *J. Cell Biochem.*, 7B, 353(1428), 1983.
276. McNicholas, J. M. and Hulett, F. M., Electron microscopic histochemical localization of alkaline phosphatase(s) in *B. licheniformis, J. Bacteriol.*, 129, 501, 1977.
277. Spencer, D. B. and Hulett, F. M., Lactoperoxidase-125I localization of salt-extractable alkaline phosphatases on the cytoplasmic membrane of *Bacillus licheniformis, J. Bacteriol.*, 145, 934, 1981.
278. Glynn, J. A., Schaffel, S. D., McNicholas, J. M., and Hulett, F. M., Biochemical localization of the alkaline phosphatase of *Bacillus licheniformis* as a function of culture age, *J. Bacteriol.*, 129, 1010, 1977.
279. Hansa, J. G., Laporta, M., Kuna, M. A., Reimschuessel, R., and Hulett, F. M., A soluble alkaline phosphatase from *Bacillus lucheniformis* MC14. Histochemical localization, purification, characterization and comparison with the membrane-associated alkaline phosphatase, *Biochim. Biophys. Acta*, 657, 390, 1981.
280. Korinteli, V. I. and Kushnarev, V. M., Alkaline phosphatase binding by the membrane apparatus of *B. subtilis* 83, *Zh. Mikrobiol. Epidermiol. Immunobiol.*, 3, 34, 1976.
281. Moluar, J., Electron microscopic localization of alkaline phosphatase in adenine dependent mutant of *B. anthrasis, Acta Microbiol. Acad. Sci. Hung.*, 19, 7, 1972.
282. Okabayashi, K., Futai, M., and Mizuno, D., Localization of acid and alkaline phosphatases in *Staphylococcus aureus, Jpn. J. Microbiol.*, 18, 287, 1974.
283. Voelz, H. and Ortigoza, R. O., Cytochemistry of phosphatases in *Myxococcus xanthus, J. Bacteriol.*, 96, 1357, 1968.
284. Munkres, M. and Wachtel, A., Histochemical localization of phosphatase in *Mycoplasma gallisepticum, J. Bacteriol.*, 93, 1096, 1967.
285. Pollack, J. D., Fractionation of Mycoplasma cells for enzyme localization, *Life Sci.*, 4, 973, 1965.
286. Glenn, A. R. and Mandelstam, J., Sporulation in *Bacillus subtilis* 168. Comparison of alkaline phosphatase from sporulating and vegetative cells, *Biochem. J.*, 123, 129, 1971.
287. Glenn, A. R. and Coote, J. G., Cytochemical studies on alkaline phosphatase production during sporulation in *Bacillus subtilis, Biochem. J.*, 152, 85, 1975.
288. Takeda, K. and Tsugita, A., Phosphoesterases of *Bacillus subtilis*. Crystallization and properties of alkaline phosphatase, *J. Biochem. Tokyo*, 61, 231, 1967.
289. Yoshizumi, F. K. and Coleman, J. E., Metalloalkaline phosphatase from *Bacillus subtilis*. Physicochemical and enzymatic properties *Arch. Biochem. Biophys.*, 160, 255, 1974.
290. Ghosh, R., Ghosh, A., and Ghosh, B. K., Properties of the membrane bound alkaline phosphatase from glucose and lactate grown cells of *Bacillus subtilis* SB15, *J. Biol. Chem.*, 252, 6813, 1977.
291. LeHegarat, J. C. and Anagnostopoulos, C., Purification, subunit structure and properties of two repressible phosphohydrolases of *Bacillus subtilis, Eur. J. Biochem.*, 39, 525, 1973.
292. Hulett-Cowling, F. M. and Campbell, L. L., Purification and properties of an alkaline phosphatase of *Bacillus licheniformis, Biochemistry*, 10, 1364, 1971.
293. Schaffel, S. D. and Hulett, F. M., Alkaline phosphatase from *Bacillus licheniformis*. Solubility dependent on magnesium, purification and characterization, *Biochim. Biophys. Acta*, 526, 457, 1978.
294. Ghosh, A., Vallespir, S., Menegus, M., Nelson, R., Kumar, R., and Ghosh, B. K., Membrane bound and extracellular alkaline phosphatases of *Bacillus licheniformis* 749/C, *Eur. J. Biochem.*, (communicated), 1983.
295. Yamane, K. and Maruo, B., Purification and characterization of extracellular soluble and membrane-bound insoluble alkaline phosphatase possessing phosphodiesterase activities in *Bacillus subtilis, J. Bacteriol.*, 134, 100, 1978.
296. Fitt, P. S. and Peterkin, P. I., Isolation and properties of a small manganese-iron-stimulated bacterial alkaline phosphatase, *Biochem. J.*, 157, 161, 1976.
297. Abraham, E. P. and Chain, E., An enzyme from bacteria able to destroy penicillin, *Nature*, 146, 837, 1940.
298. Citri, N. and Pollock, M. R., The biochemistry and function of β-lactamase (Penicillinase), *Adv. Enzymol.*, 28, 237, 1966.

299. Ambler, R. P., The structure of β-lactamases, *Philos. Trans. R. Soc. London Ser. B*, 286, 321, 1980.
300. Pollock, M. R., Exoenzymes, in *The Bacteria*, Vol. 4, Gunsalus, I. C. and Stanier, R. Y., Eds., Academic Press, New York, 1962, 121.
301. Mantsala, P. and Suominen, I., Comparison of periplasmic and membrane associated β-lactamase, *Acta Chem. Scand. Ser. B.*, 35, 567, 1981.
302. Choma, C. T. and Yamazaki, H., Simple assay and extraction of periplasmic penicillinase in *E. coli*, *Can. J. Microbiol.*, 27, 547, 1981.
303. Sutcliffe, J. G., Nucleotide sequence after ampicillin resistance gene of *E. coli* plasmid PBR 322, *Proc. Natl. Acad. Sci. U.S.A.*, 75, 3737, 1978.
304. Ghosh, B. K. and Ghosh, A., Protein secretion in bacteria, in *Organization of Prokaryotic Cell Membrane*, Vol. 3, Ghosh, B. K., Ed., CRC Press, Boca Raton, Fla., 1984.
305. Ghosh, A. and Ghosh, B. K., Immunoelectron microscopic localization of penicillinase in *Bacillus licheniformis*, *J. Bacteriol.*, 137, 1374, 1979.
306. Guan, T., Ghosh, A., and Ghosh, B. K., Collidal gold double labeling of two secretory proteins in bacterial cell, *Proc. Annu. Mtg. Electron Microscopic Soc. Am.*, p. 614, 1983.
307. Nielsen, J. B. K. and Lampen, J. O., Membrane-bound penicillinases in gram-positive bacteria, *J. Biol. Chem.*, 257, 4490, 1982.
308. Nielsen, J. B. K., Caulfield, M. P., and Lampen, J. O., Lipoprotein nature of *Bacillus licheniformis* membrane penicillinase, *Proc. Natl. Acad. Sci. U.S.A.*, 78, 3511, 1981.
309. Braun, V. and Hantke, K., Bacterial cell surface receptor, in *Organization of Prokaryotic Cell Membrane*, Vol. 2, Ghosh, B. K., Ed., CRC Press, Boca Raton, Fla., 1981, 1.
310. Saier, M. H., Jr., Newman, M. J., and Rephaeli, A. W., Properties of a phosphoenolpyruvate: mannitol phosphotransferase system in *Spirochaeta aurantia*, *J. Biol. Chem.*, 252, 8890, 1977.
311. Saier, M. H., Jr., Feucht, B. U., and Mora, W. K., Sugar phosphate:sugar transphosphorylation and exchange group translocation catalyzed by the enzyme II complexes of the bacterial phosphoenolpyruvate: sugar phosphotransferase system, *J. Biol. Chem.*, 252, 8899, 1977.
312. Saier, M. H., Jr., Cox, D. F., and Moczydlowski, E. G., Sugar phosphate:sugar transphosphorylation coupled to exchange group translocation catalyzed by the enzyme II complexes of the phosphoenolpyruvate:sugar phosphotransferase system in membrane vesicles of *Escherichia coli*, *J. Biol. Chem.*, 252, 8908, 1977.
313. Haguenauer-Tsapis, R. and Kepes, A., Unmasking of an essential thiol during function of the membrane bound enzyme II of the phosphoenolpyruvate glucose phosphotransferase system of *Escherichia coli*, *Biochim. Biophys. Acta*, 465, 118, 1977.
314. Haguenauer-Tsapis, R. and Kepes, A., Changes in accessibility of the membrane bound transport enzyme glucose phosphotransferase of *E. coli* to protein group reagents in presence of substrate or absence of energy source, *Biochem. Biophys. Res. Commun.*, 54, 1335, 1973.
315. Haguenauer-Tsapis, R. and Kepes, A., The role of enzyme I in the unmasking of an essential thiol of the membrane-bound enzyme II of the phosphoenolpyruvate-glucose phosphotransferase system of *Escherichia coli*, *Biochim. Biophys. Acta*, 469, 211, 1977.
316. Saier, M. H., Jr. and Schmidt, M. R., Vectorial and nonvectorial transphosphorylation catalyzed by enzyme II of the bacterial phosphotransferase system, *J. Bacteriol.*, 145, 391, 1981.
317. Brouwer, M., Elferink, M. G., and Robillard, G. T., Phosphoenolpyruvate-dependent fructose phosphotransferase system of *Rhodopseudomonas sphaeroides:* purification and physicochemical and immunochemical characterization of a membrane-associated enzyme I, *Biochemistry*, 21, 82, 1982.
318. Bozzula, J. L., Kuramitsu, H. K., and Maynard, M. T., Localization of streptococcus mutants GS-5 glucosyltransferases and intracellular invertase, *Infect. Immunol.*, 32, 830, 1981.
319. Pacaul, M., Identification and localization of two membrane-bound esterases from *Escherichia coli*, *J. Bacteriol.*, 149, 6, 1982.
320. Rottem, S., Trotter, S. L., and Barile, M. F., Membrane-bound thioesterase activity in Mycoplasmas, *J. Bacteriol.*, 129, 707, 1977.
321. Vance, D. E., Esders, T. E., and Bloch, K., On the role of a palmityl thioesterase in fatty acid elongation, *J. Biol. Chem.*, 248, 2310, 1973.
322. Engels, W., Kamps, M. A., Secretion of staphylocoagulase by *Staphylococcus aureus:* the role of a cell-bound intermediate, *Antonie van Leeuwenhoek; J. Microbiol. Ser.*, 47, 509, 1981.
323. Simonis, M., Rade, S., and Ahlers, J., Studies on a membrane-bound nucleotidase from the marine bacterium MP-22. Comparison of a photometric and luminometric method for kinetic investigations, *Int. J. Biochem.*, 13, 733, 1981.
324. Mauck, J. and Glaser, L., Periplasmic nucleoside diphosphate sugar hydrolase from *Bacillus subtilis*, *Biochemistry*, 9, 1140, 1970.
325. Neu, H. C., The 5′-nucleotidases (uridine diphosphate sugar hydrolases) of the *Enterobacteriaceae*, *Biochemistry*, 7, 3766, 1968.
326. Anraku, Y., A new cyclic phosphodiesterase having a 3′-nucleotidase activity from *Escherichia coli* B. I. Purification and some properties of the enzyme, *J. Biol. Chem.*, 239, 3412, 1964.

327. Neu, H. C., The cyclic phosphodiesterases (3′-nucleotidases) of the *Enterobacteriaceae, Biochemistry*, 7, 3774, 1968.
328. Brunngraber, E. F. and Chargaff, E., Transferase from *Escherichia coli* effecting low-energy phosphate transfer to nucleosides and nucleotides, *Proc. Natl. Acad. Sci. U.S.A.*, 67, 107, 1970.
329. Brunngraber, E. F. and Chargaff, E., A nucleotide phosphotransferase from *Escherichia coli.* Purification and properties, *Biochemistry*, 12, 3005, 1973.
330. Tuttle, R. C., and Saier, M. H., Jr., Phosphoprotein phosphatase activity associated with the cytoplasmic membrane of *Salmonella typhimurium* and *Escherichia coli, J. Biol. Chem.*, 252, 4416, 1977.
331. Urlaub, H. and Wobber, G., Alpha glucosidase, a membrane bound enzyme of alpha glucan metabolism in *Bacillus amyloliquefaciens* purification and partial characterization, *Biochim. Biophys. Acta*, 522, 161, 1978.
332. Bayer, M. E., Structural and functional evidence of cooperativity between membranes and cell wall in bacteria, *Int. Rev. Cytol.*, 12(Suppl.), 39, 1981.
333. Sandler, N. and Keynan, A., Cell wall synthesis and initiation of deoxyribonucleic acid replication in *Bacillus subtilis, J. Bacteriol.*, 148, 443, 1981.
334. Ghuysen, J. M., Strominger, J. L., and Tipper, D. J., Peptidoglycan biosynthesis, in *Comprehensive Biochemistry*, Vol. 26A, Florkin, M. and Stolz, E. H. Eds., American Elsevier, New York, 1968, 53.
335. Beveridge, T. J., Ultrastructure, chemistry, and function of the bacterial wall, *Int. Rev. Cytol.*, 72, 229, 1981.
336. Ghuysen, J. M. and Shockman, G. D., Biosynthesis of peptidoglycan, in *Bacterial Membranes and Walls*, Vol. 1, Leive, L., Ed., Marcel Dekker, New York, 1973, 37.
337. Schlaeppi, J. M., Pooley, H. M., and Karamata, D., Identification of cell wall subunits in *Bacillus subtilis* and analysis of their segregation during growth, *J. Bacteriol.*, 149, 329, 1982.
338. Ward, J. B., Teichoic and teichuronic acids: biosynthesis, assembly and location, *Microbiol. Rev.*, 45, 211, 1981.
339. Kalomiris, E., Bardin, C., and Neumaus, F. C., Biosynthesis of peptidoglycan in *Gaffkya homari:* reactivation of membranes by freeze-thawing in the presence and absence of walls, *J. Bacteriol.*, 150, 535, 1982.
340. Murazumi, N., Sasaki, Y., Okada, J., Araki, Y., and Ito, E., Biosynthesis of glycerol teichoic acid in *Bacillus cereus:* formation of linkage unit disaccharide on a lipid intermediate, *Biochem. Biophys. Res. Commun.*, 99, 504, 1981.
341. Cabacungan, E. and Pieringer, R. A., Mode of elongation of the glycerol phosphate polymer of membrane lipoteichoic acid of *Streptococcus faecium* ATCC 9790, *J. Bacteriol.*, 147, 75, 1981.
342. Bettinger, G. E., Chatterjee, A. N., and Young, F. E., Incorporation of *N*-acetyl-D-glucosamine from UDP-*N*-acetyl-D-glucosamine by isolated membranes of *Bacillus subtilis:* identification of undecaprenyl poly(*N*-acetylglucosaminyl pyrophosphate, *J. Biol. Chem.*, 252, 4118, 1977.
343. Chase, H. A., Shepherd, S. T., and Reynolds, P. E., Studies on the penicillin-binding components of *Bacillus megaterium, FEBS Lett.*, 76, 199, 1977.
344. Blumberg, P. M. and Strominger, J. L., Interaction of penicillin with bacterial cell: penicillin binding proteins and penicillin-sensitive enzymes, *Bacteriol. Rev.*, 38, 291, 1974.
345. Ghuysen, J. M., The bacterial DD-carboxypeptidase-transpeptidase enzyme system: a new insight into the mode of action of penicillin, in; *E. R. Squibb Lectures on Chemistry of Microbial Products*, Brown, W. E., Ed., University of Tokyo Press, Tokyo, 1977.
346. Maass, D. and Pelzer, H., Murein biosynthesis in ether permeabilized *Escherichia coli* starting from early peptidoglycan precursors, *Arch. Microbiol.*, 130, 301, 1981.
347. Forsberg, C. W. and Ward, J. B., *N*-acetylmuramyl-L-alanine amidase of *Bacillus licheniformis* and its L-form, *J. Bacteriol.*, 110, 878, 1972.
348. Leyh-Bouille, M., Dusart, J., Nguyen-Disteche, M., and Ghuysen, J. M., The peptidoglycan cross-linking enzyme system in *Streptomyces* strains R61, K15 and rimosus: exocellular, lysozyme-releasable and membrane-bound enzymes, *Eur. J. Biochem.*, 81, 19, 1977.
349. Coyette, J., Ghuysen, J. M., and Fontana, R., Solubilization and isolation of the membrane-bound DD-carboxypeptidase of *Streptococcus faecalis* ATCC 9790: properties of the purified enzyme, *Eur. J. Biochem.*, 88, 297, 1978.
350. Leyh-Bouille, M., Nguyen-Disteche, M., and Ghuysen, J. M., On the DD-carboxypeptidase enzyme system of *Streptomyces* strain K15, *Eur. J. Biochem.*, 115, 579, 1981.
351. Eun, H. M., Yapo, A., and Petit, J. P., DD-carboxypeptidase activity of membrane fragments of *Mycobacterium smegmatis.* Enzymatic properties and sensitivity to beta-lactam antibiotics, *Eur. J. Biochem.*, 86, 97, 1978.
352. Frere, J. M., Leyh-Bouille, M., Ghuysen, J. M., Neito, M., and Perkins, H. R., Exocellular DD-carboxypeptidase-transpeptidases from *Streptomyces, Meth. Enzymol.*, 45, 610, 1976.
353. Waxman, D. J. and Strominger, J. L., Primary structure of the COOH-terminal membranous segment of a penicillin-sensitive enzyme purified from two bacilli, *J. Biol. Chem.*, 256, 2067, 1981.

354. Martin, H. H., Kroll, H. P., and Gmeiner, J., Properties of lipids in membranes of the stable protoplast L-form of *Proteus mirabilis, Experientia,* 33, 136, 1977.
355. Sandermann, J., Jr. and Strominger, J. L., C55-Iso-prenoid alcohol phosphokinase: an extremely hydrophobic protein from the bacterial membrane, *Proc. Natl. Acad. Sci. U.S.A.,* 68, 2443, 1971.
356. Rothfield, L. and Romeo, D., Role of lipids in biosynthesis of bacterial cell envelope, *Bacteriol. Rev.,* 35, 14, 1971.
357. Anraku, Y., Active transport of amino acids, in *Bacterial Transport,* Rosen, B. P., Ed., Marcel Dekker, New York, 1978, 171.
358. Cohen, G. N. and Monod, J., Bacterial permeases, *Bacteriol. Rev.,* 21, 169, 1957.
359. Kennedy, E. P., The lactose permease system of *Escherichia coli,* in *The Lactose Operon,* Beckwith, J. R. and Zipser, D., Eds., Cold Spring Harbor Laboratory, Cold Spring Harbor, New York, 1970, 49.
360. Kaback, H. R., Transport studies in bacterial membrane vesicles: cytoplasmic membrane vesicles devoid of soluble constituents catalyze the transport of many metabolites, *Science,* 186, 882, 1974.
361. Kaback, H. R., Molecular biology and energetics of membrane transport, *J. Cell Physiol.,* 89, 575, 1976.
362. Schuldiner, S., Rudnick, G., Weil, R., and Kaback, H. R., Mechanism of beta-galactoside transport in *Escherichia coli* membrane vesicles, *T.I.B.S.,* 1, 41, 1976.
363. Reeves, J. P., Shechter, E., Weil, R., and Kaback, H. R., Mechanisms of active-transport in isolated bacterial membrane vesicles. XVI. dansyl-galactoside, a fluorescent probe of active-transport in bacterial membrane vesicles, *Proc. Natl. Acad. Sci. U.S.A.,* 70, 2722, 1973.
364. Therisod, H., Letellier, L., Weil, R., and Shechter, E., Functional lac carrier proteins in cytoplasmic membrane vesicles isolated from *Escherichia coli.* I. Temperature dependence of dansyl galactoside binding and β-galactoside transport, *Biochemistry,* 16, 3772, 1977.
365. Fox, C. F. and Kennedy, E. P., Specific labeling and partial purification of M protein, a component of beta galactoside transport system of *Escherichia coli, Proc. Natl. Acad. Sci. U.S.A.,* 54, 891, 1965.
366. Jones, T. H. D. and Kennedy, E. P., Characterization of membrane protein component of lactose transport system of *Escherichia coli, J. Biol. Chem.,* 244, 5981, 1969.
367. Altendorf, K., Muller, C. R., and Sandermann, H. Jr., β-D-galactoside transport in *Escherichia coli:* reversible inhibition by aprotic solvents and its reconstitution in transport negative membrane vesicles, *Eur. J. Biochem.,* 73, 545, 1977.
368. Kohl, B. and Sandermann, H., Solubilization of *Escherichia coli* membrane proteins by aprotic solvents, *FEBS Lett.,* 80, 408, 1977.
369. Amanuma, H., Motojima, K., Yamaguchi, A., and Anraku, Y., Solubilization of a functionally active proline carrier from membranes of *Escherichia coli* with an organic solvent, *Biochim. Biophys. Res. Commun.,* 74, 366, 1977.
370. Amanuma, H., Itoh, J., and Anraku, Y., Proton-dependent binding of proline to carrier in *Escherichia coli* membrane, *FEBS Lett.,* 78, 173, 1977.
371. Heerikhuizen, H. V., Boekhout, M., and Witholt, B., Proline transport activity in *Escherichia coli* membrane vesicles of different buoyant densities, *Biochim. Biophys. Acta,* 470, 543, 1977.
372. Motojima, K., Yamato, I., and Anraku, Y., Proline transport carrier-defective mutants of *Escherichia coli* K-12: properties and mapping, *J. Bacteriol.,* 136, 5, 1978.
363. Frank, L. and Ranhand, B., Proline metabolism in *Escherichia coli, Arch. Biochem. Biophys.,* 167, 325, 1964.
374. Hirata, H., Sone, N., Yoshida, M., and Kagawa, Y., Solubilization and partial purification of alanine carrier from membranes of a thermophilic bacterium and its reconstitution into functional vesicles, *Biochem. Biophys. Res. Commun.,* 69, 665, 1976.
375. Hirata, H., Sone, N., Yoshida, M., and Kagawa, Y., Isolation of the alanine carrier from the membranes of a thermophilic bacterium and its reconstitution into vesicles capable of transport, *J. Supramol. Struct.,* 6, 77, 1977.
376. Hirata, H., Solubilization and purification of alanine carrier from thermophilic bacteria and reconstitution into vesicles capable of transport, *Meth. Enzymol.,* 56, 430, 1979.
377. Hinkle, P. C., Kim, J. H., and Racker, E., Ion transport and respiratory control in vesicles formed from cytochrome-oxidase and phospholipids, *J. Biol. Chem.,* 247, 1338, 1972.
378. Wilson, O. H. and Holden, J. T., Stimulation of arginine transport in osmotically shocked *Escherichia coli* W cells by purified arginine binding protein fractions, *J. Biol. Chem.,* 244, 2743, 1969.
379. Anraku, Y., Kobayashi, H., Amanuma, H., and Yamaguchi, A., Transport of sugars and amino acids in bacteria. VII. Characterization of reactions of restoration of active transport mediated by binding protein, *J. Biochem.,* 74, 1249, 1973.
380. Anraku, Y., Transport of sugars and amino acids in bacteria. III. Studies on restoration of active transport, *J. Biol. Chem.,* 243, 3128, 1968.

381. Anraku, Y., Naraki, T., and Kanzaki, S., Transport of sugars and amino acids in bacteria. VI. Changes induced by valine in branched-chain amino acid transport systems of *Escherichia coli*, *J. Biochem.*, 73, 1149, 1973.
382. Penrose, W. R. and Zand, R., Reversible conformational changes in a leucine binding protein from *Escherichia coli*, *J. Biol. Chem.*, 245, 1432, 1970.
383. Ames, G. F. L. and Lever, J., Components of histidine transport-histidine binding proteins and HISP proteins, *Proc. Natl. Acad. Sci. U.S.A.*, 66, 1096, 1970.
384. Rahmanian, M. and Oxender, D. L., Derepressed leucine transport activity in *Escherichia coli*, *J. Supramol. Struct.*, 1, 55, 1972.
385. Amanuma, H. and Anraku, Y., Transport of sugars and amino acids in bacteria. XII. Substrate specificities of the branched chain amino acid-binding proteins of *Escherichia coli*, *J. Biochem.*, 76, 1165, 1974.
386. Ames, G. F. L., Noel, K. D., Taber, H., Negri Spudich, E., Nikaido, K., Afong, J., and Ardeshir, F., Fine structure Map of the histidine transport genes in *Salmonella typhimurium*, *J. Bacteriol.*, 129, 1289, 1977.
387. Ames, G. F. L. and Nikaido, K., Identification of a membrane protein as a histidine transport component in *Salmonella typhimurium*, *Proc. Natl. Acad. Sci. U.S.A.*, 75, 5447, 1978.
388. Shaltiel, S., Ames, G. F. L., and Noel, K. D., Hydrophobic chromatography in the purification of the histidine-binding protein j from *Salmonella typhimurium*, *Arch. Biochem. Biophys.*, 159, 174, 1973.
389. Lever, J. E., Purification and properties of a component of histidine transport in *Salmonella typhimurium*. The histidine binding protein, *J. Biol. Chem.*, 247, 4317, 1972.
390. Ames, G. F. L., Noel, K. D., Taber, H., Spudich, E. N., Nikaido, K., Afong, J., and Ardeshir, F., Find structure map of the histidine transport genes in *Salmonella typhimurium*, *J. Bacteriol.*, 129, 1289, 1977.
391. Ho, C., Giza, Y., Takahashi, S., Ugen, K. E., Cottam, P. F., and Dowd, S. R., A proton nuclear magnetic resonance investigation of histidine binding protein j of *Salmonella typhimurium:* a model for transport of L-histidine across cytoplasmic membrane, *J. Supramol. Struct.*, 13, 131, 1980.
392. Masters, P. W. and Hong, J., Reconstitution of binding protein dependent active transport of glutamine in spheroplasts of *Escherichia coli*, *Biochemistry*, 20, 4900, 1981.
393. Masters, P. S. and Hong, J. S., Genetics of the glutamate transport system in *Escherichia coli*, *J. Bacteriol.*, 147, 805, 1981.
394. Boos, W., Aspects of maltose transport in *Escherichia coli:* established facts and educated guesses, *Ann. Microbiol., Paris*, 133A, 145, 1982.
395. Drass, J. M., Boos, W., and Hengge, R., Reconstitution of maltose transport in malB mutants of *Escherichia coli* through calcium-induced disruption of the outer membrane, *J. Bacteriol.*, 146, 10, 1981.
396. Brass, J. M., Reconstitution of maltose transport in malB and malA mutants of *Escherichia coli*, *Ann. Microbiol. Paris*, 133A, 171, 1982.
397. Shuman, H. A., Active transport of maltose in *Escherichia coli* K-12. Role of the periplasmic maltose-binding protein and evidence for a substrate recognition site in the cytoplasmic membrane, *J. Biol. Chem.*, 257, 5455, 1982.
398. Shuman, H. A., The maltose-maltodextrin transport system of *Escherichia coli*, *Ann. Microbiol. Paris*, 133A, 153, 1982.
399. Shuman, H. A. and Silhavy, T. J., Identification of the malK gene product. A peripheral membrane component of the *Escherichia coli* maltose transport system, *J. Biol. Chem.*, 256, 560, 1981.
400. Suginaka, H., Blumberg, P., and Strominger, J., Multiple penicillin-binding components in *Bacillus subtilis*, *Bacillus cereus*, *Staphylococcus aureus*, and *Escherichia coli*, *J. Biol. Chem.*, 247, 5279, 1972.
401. Rodriguez, W. J. and Saz, A. K., Differential binding of penicillin-susceptible and -resistant Gonococci, *Antimicrob. Agents Chemother.*, 13, 589, 1978.
402. Cooper, P. D., The site of action of penicillin: some properties of the penicillin binding components of *S. aureus*, *J. Gen. Microbiol.*, 12, 100, 1958.
403. Eagle, H., The binding of penicillin in relation to its cytotoxic action. I. Correlation between the penicillin sensitivity and combining activity of intact bacteria and cell free extract, *J. Exp. Med.*, 99, 207, 1954.
404. Eagle, H., The binding of penicillin in relation to its cytotoxic action. II. The reactivity with penicillin of resistant variants, streptococcus, pneumococcus and diphylococcus, *J. Exp. Med.*, 100, 103, 1954.
405. Blumberg, P. M. and Strominger, J. L., Isolation by covalent affinity chromatography of the penicillin-binding components from membranes of *Bacillus subtilis*, *Proc. Natl. Acad. Sci. U.S.A.*, 69, 3751, 1972.
406. Spratt, B. J., Properties of the penicillin-binding proteins of *Escherichia coli* K-12., *Eur. J. Biochem.*, 72, 341, 1977.

407. Shepherd, S. T., Chase, H. A., and Reynolds, P. E., The separation and properties of two penicillin-binding proteins from *Salmonella typhimurium, Eur. J. Biochem.*, 78, 521, 1977.
408. Chase, H. A., Shepherd, S. T., and Reynolds, P. E., Studies on the penicillin-binding components of *Bacillus megaterium, FEBS Lett.*, 76, 199, 1977.
409. Ghuysen, J. M., The concept of the penicillin target from 1965 until today, *J. Gen. Microbiol.*, 101, 13, 1977.
410. Chase, H. A., Reynolds, P. E., and Ward, J. B., Purification and characterization of the penicillin-binding protein that is the lethal target of penicillin in *Bacillus megaterium* and *Bacillus licheniformis, Eur. J. Biochem.*, 88, 275, 1978.
411. Pollock, J. J., Nguyen-Disteche, M., Ghuysen, J. M., Coyette, J., Linder, R., Salton, M. R. J., Kim, K. S., Perkins, H. R., and Reynolds, P. E., Fractionation of the DD-carboxypeptidase-transpeptidase activities solubilized from membranes of *Escherichia coli* K-12, strain 44, *Eur. J. Biochem.*, 41, 439, 1974.
412. Reynolds, P. E., Shepherd, S. T., and Chase, H. A., Identification of the binding protein which may be the target of penicillin action in *Bacillus megaterium, Nature (London)*, 271, 568, 1978.
413. Frere, J. M., Ghuysen, J. M., Reynolds, P. E., Moreno, R., and Perkins, H. R., Binding of β-lactam antibiotics to the exocellular DD-carboxypeptidase-transpeptidase of Streptomyces R39, *Biochem. J.*, 143, 241, 1974.
414. Ohya, S., Yamazaki, M., Sugawara, S., Tamaki, S., and Matsuhashi, M., New cephamycin antibiotics, CS-1170: binding affinity to penicillin-binding proteins and inhibition of peptidoglycan cross-linking reactions in *Escherichia coli, Antimicrob. Agents Chemother.*, 14, 780, 1978.
415. Tamaki, S., Nakajima, S., and Matsuhashi, M., Thermosensitive mutation in *Escherichia coli* simultaneously causing defects in penicillin-binding protein 1Bs and in enzyme activity for peptidoglycan synthesis *in vitro, Proc. Natl. Acad. Sci. U.S.A.*, 74, 5472, 1977.
416. Spratt, B. G., Distinct penicillin-binding proteins involved in the division, elongation, and shape of *Escherichia coli* K-12, *Proc. Natl. Acad. Sci. U.S.A.*, 72, 2999, 1975.
417. Spratt, B. G., Temperature-sensitive cell division mutants of *Escherichia coli* with thermolabile penicillin-binding protein, *J. Bacteriol.*, 131, 293, 1977.
418. Waxman, D. J., Lindgren, D. M., and Strominger, J. L., High molecular weight penicillin-binding protein from membranes of Bacilli, *J. Bacteriol.*, 148, 950, 1981.
419. Ishino, F. and Matsuhashi, M., Peptidoglycan synthetic enzyme activities of highly purified penicillin-binding protein 3 in *Escherichia coli:* a septum-forming reaction sequence, *Biochem. Biophys. Res. Commun.*, 101, 905, 1981.
420. Godfrey, A. J. and Bryan, L. E., Mutation of *Pseudomonas aeruginosa* specifying reduced affinity for penicillin G. *Antimicrob. Agents Chemother.*, 21, 216, 1982.
421. Murphy, T. F., Barza, M., and Park, J. T., Penicillin binding proteins in *Clostridium perfringens, Antimicrob. Agents Chemother.*, 20, 809, 1981.
422. Wright, R. B., Makover, S. D., and Telep, E., RO 13-9904: affinity for penicillin-binding protein and effect on cell wall synthesis, *J. Antibiot. (Tokyo)*, 34, 590, 1981.
423. Shigi, Y., Kojo, H., Wakasugi, M., and Nishida, M., Differences between ceftizoxime and its stereoisomer in antibacterial activity and affinity for penicillin-binding proteins, *Antimicrob. Agents Chemother.*, 19, 393, 1981.
424. Shepherd, W. D., Kaplan, S., and Park, J. T., Penicillin-binding proteins of *Rhodopseudomonas sphaeroides* and their membrane localization, *J. Bacteriol.*, 147, 354, 1981.
425. Horikawa, S. and Ogawara, H., Membrane-bound penicillin-binding proteins of *Escherichia coli.* Comparison of a strain carrying an R-factor and the parent strain, *J. Antibiot. (Tokyo)*, 31, 1283, 1978.
426. Jacob, F., Ryter, A., and Cuzin, F., On the association between DNA and membrane in bacteria. *Proc. R. Soc. London Ser. B*, 164, 267, 1966.
427. Jacob, F., Brenner, S., and Cuzin, F., On the regulation of DNA replication in bacteria. *Cold Spring Harbor Symp. Quant. Biol.*, 28, 329, 1963.
428. Dworsky, P. and Schaechter, M., Effects of rifampin on the structure and membrane attachment of the nucleoid of *Escherichia coli, J. Bacteriol.*, 116, 1364, 1973.
429. Ryter, A. and Jocob, F., Etude au microscope electronique de la liaison entre noyau et mesosome chez *Bacillus subtilis, Ann. Inst. Pasteur Paris*, 107, 384, 1964.
430. Tremblay, G. Y., Daniels, M. J., and Schaechter, M., Isolation of a cell membrane DNA-nascent RNA complex from bacteria, *J. Mol. Biol.*, 40, 65, 1969.
431. Ganesan, A. T. and Lederberg, J., A cell membrane bound fraction of bacterial DNA, *Biochem. Biophys. Res. Commun.*, 18, 824, 1965.
432. Lundquist-Parker, D. and Glaser, D. A., Chromosomal sites of DNA-membrane attachment in *Escherichia coli, J. Mol. Biol.*, 87, 153, 1974.
433. Smith, D. W. and Hanawalt, P. C., Properties of the growing point region in the bacterial chromosome, *Biochim. Biophys. Acta*, 149, 519, 1967.

434. O'Sullivan, M. A. and Sueoka, N., Membrane attachment of the replication orgins of a multifork (dichotomous) chromosome in *Bacillus subtilis, J. Mol. Biol.,* 69, 237, 1972.
435. Snyder, R. W. and Young, F. E., Association between the chromosome and the cytoplasmic membrane in *Bacillus subtilis, Biochem. Biophys. Res. Commun.,* 35, 354, 1969.
436. Sueoka, N. and Quinn, W. G., Membrane attachment of the chromosome replication origin in *Bacillus subtilis, Cold Spring Harbor Symp. Quant. Biol.,* 33, 695, 1968.
437. Leibowitz, P. J. and Schaechter, M., The attachment of the bacterial chromosome to the cell membrane, *Int. Rev. Cytol.,* 41, 1, 1975.
438. Murozov, I. I., Sulimova, T. V., Ryabchenko, N. I., and Myasnik, M. N., Sensitivity and repair of DNA membrane complex of *Escherichia coli* strain B-R and *Escherichia coli* strain B-S-1 irradiated with gamma-rays, *Radiobiologiva,* 17, 178, 1977.
439. Abe, M., Brown, C., Hendrickson, W. G., Boyd, D. H., Clifford, P., Cote, R. H., and Schaechter, M., Release of *Escherichia coli* RNA from membrane complexes by single-strand endonucleases, *Proc. Natl. Acad. Sci. U.S.A.,* 74, 2756, 1977.
440. Kazuo, Y. and Yoshikawa, H., Chromosome-membrane association in *Bacillus subtilis.* III. Isolation and characterization of a DNA-protein complex carrying replication origin markers, *J. Mol. Biol.,* 110, 219, 1977.
441. Harmon, J. M. and Taber, H. W., Some properties of a membrane DNA complex isolated from *Bacillus subtilis, J. Bacteriol.,* 129, 789, 1977.
442. Ivarie, R. D. and Pene, J. J., Association of the *Bacillus subtilis* chromosome with the cell membrane: resolution of free and bound deoxyribonucleic acid on renografin gradients, *J. Bacteriol.,* 104, 839, 1970.
443. Hye, R. J., O'Sullivan, M. A., Howard, K., and Sueoka, N., Membrane association of origin terminus and replication fork in *Bacillus subtilis,* in *Microbiology,* Schlessinger, D., Ed., *American Society for Microbiology,* Washington, D.C., 7, 83, 1976.
444. Ganesan, A. T. and Lederberg, J., A cell-membrane bound fraction of bacterial deoxyribonucleic acid, *Biochem. Biophys. Res. Commun.,* 18, 824, 1965.
445. Ivarie, R. and Pene, J., Association of many regions of the *Bacillus subtilis* chromosome with the cell membrane. *J. Bacteriol.,* 114, 571, 1973.
446. Snyder, R. W. and Young, F. E., Association between the chromosome and the cytoplasmic membrane in *B. subtilis, Biochem. Biophys. Res. Commun.,* 35, 354, 1969.
447. Sueoka, N. and Quinn, W., Membrane attachment of the chromosome replication origin in *B subtilis, Cold Spring Harbor Symp. Quant. Biol.,* 33, 695, 1968.
448. Yamaguchi, K. and Yoshikawa, H., Topography of chromosome membrane junction in *B subtilis, Nature (London) New Biol.,* 244, 204, 1973.
449. Fielding, P. and Fox, C. F., Evidence for stable attachment of DNA to membrane at the replication origin of *E. coli, Biochem. Biophys. Res. Commun.,* 41, 157, 1970.
450. Parker, D. C. and Glaser, D. A., Chromosomal sites of DNA-membrane attachment in *E. coli, J. Mol. Biol.,* 87, 153, 1974.
451. Smith, D. W. and Hanawalt, P. C., Properties of the growing point region in the bacterial chromosome, *Biochim. Biophys. Acta,* 149, 519, 1967.
452. Matsushita, T., Winston, S. E., and Shotola, A., The replicative role of DNA membrane attachment in *Bacillus subtilis, Abstr. Am. Soc. Microbiol.,* 77, 153, 1977.
453. Winston, S. and Matsushita, T., Protein synthesis and the release of the replication terminus from the cell membrane in *Bacillus subtilis,* in *Microbiology,* Schlessinger, D., Ed., Soc. Microbiol., Washington, D.C., U.S.A., 7, 123, 1976.
454. Jones, N. C. and Donachie, W. D., Chromosome replication, transcription and control of cell division in *E. coli, Nature (London) New Biol.,* 243, 100, 1973.
455. Jones, N. C. and Donachie, W. D., Protein synthesis and the release of the replicated chromosome from the cell membrane, *Nature (London),* 251, 252, 1974.
456. Donachie, W. D., Jones, N. C., and Teather, R. T., The bacterial cell cycle, in *Microbial Differentiation,* J. M. Ashworth and J. E. Smith Society of General Microbiology Symposium, Cambridge University Press, New York, 23, 9, 1973.
457. Sakakibara, Y. and Tomizawa, J. I., Gene N and membrane association of lambda DNA, in *The Bacteriophage Lambda,* Hershey, A. D., Ed., Cold Spring Harbor Laboratory, Cold Spring Harbor, New York, 1971, 691.
458. Olsen, W. L., Heidrich, H. G., Hanning, K., and Hofschneider, P. H., Deoxyribonucleic acid-envelope complexes isolated from *Escherichia coli* by free flow electrophoresis: biochemical and electron microscope characterization, *J. Bacteriol.,* 118, 646, 1974.
459. Tichy, P. and Landman, O. E., Transformation in quasi-spheroplasts of *Bacillus subtilis, J. Bacteriol.,* 97, 42, 1969.

460. Verneulen, C. A. and Venema, G., Electron microscope and autoradiographic study of ultrastructural aspects of competence and deoxyribonucleic acid absorption in *Bacillus subtilis:* localization of uptake and of transport of transforming deoxyribonucleic acid in competent cells, *J. Bacteriol.*, 118, 342, 1974.
461. Wolstenholme, D. R., Vermeulen, C. A., and Venema, G., Evidence for the involvement of membranous bodies in the processes leading to genetic transformation in *Bacillus subtilis, J. Bacteriol.*, 92, 1111, 1966.
462. Hampton, M. L. and Archer, L. J., Binding of exogenous DNA to membrane fractions from *Bacillus subtilis*, in *Microbiology-1976*, Schlessinger, D., Ed., American Society for Microbiology, Washington, D.C., 1976, 101.
463. Lopez, P., Perez-Urena, M. T., Garcia, E., Alda, M. T., and Espinosa, M., Detection of different types of receptors for deoxyribonucleic acid in competent *Bacillus subtilis, Zentralbl.-Bakteriol. (Naturwiss.)*, 136, 524, 1981.
464. Weisshaupl, V. and Dworsky, P., Studies on the deoxyribonucleic acid-bearing portion of the cell envelope of *Escherichia coli, Z. Allg. Mikrobiol.*, 18, 681, 1978.
465. Weston, A., Brown, M. G., Perkins, H. R., Saunders, J. R., and Humphreys, G. O., Transformation of *Escherichia coli* with plasmid deoxyribonucleic acids: calcium-induced binding of deoxyribonucleic acid to whole cells and to isolated membrane fractions, *J. Bacteriol.*, 145, 780, 1981.
466. Trembly, G. Y., Daniels, M. J., and Schaechter, M., Isolation of a cell membrane-DNA-nascent-RNA complex from bacteria, *J. Mol. Biol.*, 40, 67, 1969.
467. Minkley, E. G., Jr. and Ippen-Ihler, K., Identification of a membrane protein associated with expression of the surface exclusion region of the F transfer operon, *J. Bacteriol.*, 129, 1613, 1977.
468. Harmon, J. M. and Taber, H. W., Relationship of membrane proteins to chromosomal replication to *Bacillus subtilis*, in *Microbiology*-1976, Schlessinger, D., Ed., American Society for Microbiology, Washington, D.C., 1976, 105.
469. Imada, S., Carroll, L. E., and Sueoka, N., DNA membrane complex in *Bacillus subtilis*, in *Microbiology*, Schlessinger, D., Ed., American Society for Microbiology, Washington, D.C., 1976, 116.
470. Heidrich, H. G. and Olsen, W. L., Deoxyribonucleic acid-envelope complexes from *Escherichia coli:* a complex-specific protein and its possible function for the stability of the complex, *J. Cell Biol.*, 67, 444, 1975.
471. Yamaguchi, K. and Yoshikawa, H., Chromosome-membrane association in *Bacillus subtilis*. III. Isolation and characterization of a DNA-protein complex carrying replication origin marker, *J. Mol. Biol.*, 110, 219, 1977.
472. Toyada, H., Yamaguchi, K., and Yoshikawa, H., Chromosome-membrane association in *Bacillus subtilis*. IV. Further purification of DNA membrane complex by using a combination of centrifugation and electrophoresis, *J. Bacteriol.*, 136, 799, 1978.
473. Firshein, W., Gelman, I. W., Enrichment of DNA polymerase III activity in a DNA membrane complex purified from pneumococcus: the possible existence of subcomplexes, *Mol. Gen. Genet.*, 182, 87, 1981.
474. Siegel, P. J. and Schaechter, M., The role of the host cell membrane in the replication and morphogenesis of bacteriophages, *Annu. Rev. Microbiol.*, 27, 261, 1973.
475. Boikov, P. Y., Shevchenko, N. A., Localization of membrane sites for binding and initiation of replication of bacteriophase T-4 DNA in cells of *Escherichia coli, Dokl. Biochem.*, 229, 323, 1976.
476. McGuire, J. C., Gilpatrick, M. W., and Pene, J. J., DNA replication of bacteriophage ϕ29. Effect of two viral genes on the association of phage chromosomes with the host cell membrane, *Virology*, 78, 234, 1977.
477. Hallick, L. M. and Echols, H., Genetic and biochemical properties of an association complex between host components and lambda DNA, *Virology*, 52, 105, 1973.
478. Miller, R. C., Association of replicative T4 deoxyribonucleic acid and bacterial membranes, *J. Virol.*, 10, 920, 1972.
479. Mosig, G., Berguist, W., and Bock, S., Multiple interactions of a DNA-binding protein *in vivo*. III. Phage T4 gene-32 mutations differentially affect insertion-type recombination and membrane properties, *Genetics*, 86, 5, 1977.

Chapter 4

PROTEIN SECRETION IN BACTERIA*

Bijan K. Ghosh and Arati Ghosh

TABLE OF CONTENTS

I. Introduction 176

II. General Considerations 176
A. Definition of Secretion 176
B. Physiological Role of Secretion 177
C. Basic Nature of the Secretory Process 177
1. Biosynthesis 177
2. Translocation 177
D. Envelope Organization Relevant to Secretion 179

III. Physiological Regulation 187

IV. Molecular Mechanism in Gram-Negative Bacteria 197
A. Periplasmic Protein 197
1. Maltose-Binding Protein (MBP) 197
2. Leucine-Specific Binding Protein (L-SBP) 199
3. Alkaline Phosphatase 200
4. Arabinose-Binding Protein (ABP) 204
B. Plasma Membrane Protein 206
1. M13 Coat Protein 206
2. f1 Coat Protein 207
C. Outer-Membrane Protein 209
1. λ Phage Receptor 209
2. Lipoprotein 212
3. Omp Protein 214

V. Molecular Mechanism in Gram-Positive Cells 214
A. Penicillinase 214
B. α-Amylase 219
C. Alkaline Phosphatase 220

VI. Role of Protein Sequence 220

VII. Secretory Apparatus 223
A. Physiological Regulation 223
B. Morphological Demonstration 224
C. Genetic Regulation 228
D. Biochemical Mechanism 232

VIII. Relation to Pathogenicity and Toxin Production 239

* This chapter is dedicated to Professor J. O. Lampen, a pioneer in this area of research. He talked about microbial secretion at a time when most microbiologists were not even aware of the secretory process as a significant biological activity of microorganisms..

IX. Concluding Remarks on Industrial Applications 243

Acknowledgments 244

References 245

> *"The most important discoveries of the laws, methods and progress of Nature have nearly always sprung from the examination of the smallest objects which she contains."*
>
> J. B. Lamarck

I. INTRODUCTION

Studies on secretion apparently contradicted this comment of the great naturalist. Progress in the understanding of the basic mechanism of secretion has largely depended upon animal cells as models. In the last 10 years, however, studies on the mechanism of secretion using bacteria as the model have been very vigorously pursued. The data do not suggest any major deviation from the general mechanism proposed in eukaryotes. The investigations in bacteria have primarily focused on the molecular biology of secretion. This line of investigation is rapidly opening a new area of inquiry on the poorly understood genetic regulation of membrane function. Furthermore, parallel studies on bacterial and animal cell models have initiated investigation on the physiological response of bacterial cells to extracellular stimulus exerted on the membrane. This area of bacterial physiology has been poorly developed compared to animal cell physiology.

The body of literature on secretion is growing at a very fast pace. Models are proposed and again are reconsidered to accommodate new data. Hence, as regards to mechanisms, they are multiple and all are tentative. It is not possible to visualize the underlying unity of these mechanisms. The aim of this chapter is to develop a comprehensive conceptual framework rather than an exhaustive account of all the published data. Therefore, there may be some omission of information. This is not because of deliberate design but rather to avoid confusion, triviality, or redundancy. This presentation is aimed at correlating the facts and integrating ideas for developing secretion as a general biological phenomenon; and because of the large number of authoritative reviews available,[1-9] there may still be some redundancy.

II. GENERAL CONSIDERATIONS

A. Definition of Secretion

Secretion is defined as a cellular process consisting of multiple molecular activities that are initiated in the cytosol and continuing as membrane-bound as well as intramembranous activities. Secretion includes synthesis and translocation of protein molecules through the membrane permeability barrier.

Thus, broad secretory activities have three components: (1) biosynthesis, (2) translocation of synthesized molecules through the membrane, and (3) processing of the nascent protein molecules which might occur concomitantly during the synthesis and translocation (i.e., co-translational) or after completion of the synthesis and translocation (i.e., post-translational). In eukaryotic cells the secretory proteins are stored intracellularly and ultimately externalized in response to physiological need by way of stimulus-coupled exocytotic phenomena. This process has not been demonstrated in bacteria.

B. Physiological Role of Secretion

The transport of both small and large molecules across the permeability barrier of plasma membrane is a basic function supporting the viability and homeostatis of cells. Secretion is universally present in cells whether they constitute a component of a multicellular organism or are unicellular free-living or parasitic organisms. Physiological functions of secreted proteins include:

1. Biosynthesis: enzymes are secreted for the biosynthesis of pericellular structures; e.g., cell wall, peptidoglycan, receptors for the adherence to any substrate surface.
2. Digestion: a variety of hydrolytic enzymes are secreted for degradation of extracellular macromolecules in order to obtain nutrients.
3. Extracellular self-assembly: subunits of flagellin and pillin and subunits for regular structure (RS) layer are secreted for self-assembly outside the cell.
4. Binding and transport of nutrients: a variety of specific amino acid or sugar-binding proteins which function as components of transport machinery are secreted.
5. Protection: a variety of enzymes for hydrolysis of toxic materials, receptors for phages, and toxins for the destruction of other cells are secreted.

C. Basic Nature of the Secretory Process[1-9]

Secretion proceeds sequentially through several distinct steps. Different secreted protein molecules exhibit considerable diversity in the characteristics of some of these steps, suggesting that there are multiple mechanisms of protein secretion. Although differing molecular mechanisms have been suggested, no clear understanding of these complex mechanisms has developed. Therefore, any possible underlying unity of these apparently diverse mechanisms cannot be appreciated. The following discussion is a brief outline of present knowledge on the general mechanism of secretion as it occurs in all living cells.

1. Biosynthesis

Transcription of mRNA and formation of polysome occurs in the cytosol. The only distinctive characteristic of secretory proteins in this step is the code in the mRNAs for a short amino-terminal hydrophobic amino acid sequence (signal sequence). This gives the nascent secretory protein molecule an affinity for the membrane of secretory cells (i.e., plasma membrane in bacteria and endoplasmic reticulum in eukaryotic cells).

The ribosome-containing emerging translation product forms a complex by binding to a specific cytosolic particle consisting of several protein molecules and RNA. This binding arrests the translation and ensures the binding of this complex to the membrane.

Membrane-receptor proteins recognize and bind to the ribosome complex. The translation of the peptide in the bound ribosome resumes and the amino-terminal hydrophobic amino acid stretch inserts into the membrane.

2. Translocation

The insertion of this hydrophobic amino acid stretch causes the formation of a proteinacious channel in the membrane. This channel allows continued vectorial movement of the translation product from the membrane-bound ribosome across the membrane.

The amino-terminal hydrophobic amino acid stretch is proteolytically removed by hydrolysis at a specific site when the translation product arrives at the outside of the plasma membrane surface. The hydrophobic amino acid stretch is released within the membrane, from where it may enter the cytosol and be completely hydrolyzed into amino acids.

The translation from the membrane-bound ribosome and vectorial movement of the amino acid chain to the outside membrane surface continues and folding of mature proteins occurs on the outside surface of the membrane. Final localization of this protein on the membrane surface or its dissociation to the extracellular medium depends upon a variety of processing factors which may differ for different protein molecules.

Upon completion of translation, the ribosome dissociates from the membrane into subunits.

This process may continue in a cyclic fashion if there is a supply of mRNA; the ribosomal subunits may be reused and hydrolysis of the hydrophobic amino acid stretch may maintain a pool of hydrophobic amino acids without much depletion. The amount of energy needed for the continuation of the process is unknown; it is strongly felt that the energy supply of protein synthesis is not enough and a constant input of energy is required for the continuation of the process. However, the release of conformation energy (i.e., energy released due to folding of the linear amino acid chain into minimum energy conformation) has the potential for the supply of energy needed for vectorial movement of the amino acid chain through the putative membrane channel.

Folding of the secretory protein molecule on the outside membrane surface has implications that are unique for secretory proteins. These are frequently toxic; hence, their segregation protects the cell from their toxic effect. Their folding outside the cytosol assures their stability in the excellular environment. However, the secreted molecules may be extensively modified (e.g., glycosylated) during translation.

Although a large body of data (discussed later in the chapter) and a variety of theoretical considerations support the above model, strong experimental data have accumulated indicating other mechanisms for the translocation of synthesized molecules, two of which are described below. The biosynthetic process, however, is common for these alternative mechanisms of translocation.

In the first alternative mechanism of translocation:

1. Translation of the polypeptide chain continues without any vectorial movement through the membrane. Thus, the polypeptide chain continues to accumulate in association with the inside surface of the membrane.
2. When the folded polypeptide chain reaches a critical size (80 to 100% of the mature size) an abrupt depolarization of the membrane translocates the polypeptide through the membrane.
3. The amino-terminal stretch is removed by proteolysis at a specific site and the mature secretory protein is released into the extracellular environment.

This process may also continue in a cyclic manner if there is a supply of mRNA available and membrane depolarization repolarization occurs without any interruption.

In the second alternative mechanism:

1. The amino-terminal hydrophobic amino acid stretch is inserted into the membrane; however, the protein molecules accumulate within the membrane without vectorial translocation through the putative membrane channel.

2. This intramembranous precursor of the secretory protein is proteolytically processed forming the true extracellular protein, which is released from the membrane. The process may continue in a cyclic fashion until the supply of mRNA is exhausted.

Finally, proteins may also be translocated across a membrane permeability barrier in a fully mature form by a membrane-receptor-mediated mechanism. This mode of translocation has not been demonstrated in prokaryotic cells.

D. Envelope Organization Relevant to Secretion[10]

The membrane composition and structure of both bacterial and eukaryotic cells are basically similar. In regard to secretion, functional analogies exist between bacterial plasma membrane and eukaryotic endoplasmic reticulum (ER); however, the bacterial plasma membrane is a multifunctional organelle, whereas eukaryotic ER is specialized for synthesis and secretion of proteins. Morphological evidence for the difference is the abundant association of ribosomes to the ER membrane surface whereas the bacterial membrane rarely shows any bound ribosome (Figures 1A through 1C). These differences in organization may lead to differences in the process of secretion. In eukaryotes, the proteins are translocated through ER into the cisternal space and further processed, sorted, and delivered to a variety of subcellular organelles. Frequently, many potently hydrolytic secretory enzymes are stored in inactive forms; activation and release to intracellular locations from this storage depot occurs in response to physiological need regulated by hormonal and neural stimulation. In bacterial cells, however, proteins are translocated through the membrane into wall-periplasmic space. These proteins may be retained in that location, or can be processed and translocated for association with the wall or outer-membrane structure, or diffusion into the extracellular space. Theoretical consideration suggests that due to lack of any specialized secretory membrane, the bacterial secretory mechanism requires a high degree of precision; otherwise, conflicting traffic through the membrane may hinder the secretory activity. Furthermore, because of the lack of any storage, the secretory proteins would require precise coupling between the secretory activity and the physiological need. Thus, compared to higher organisms, the bacterial protein secretory mechanism is expected to be less elaborate but more precise.

The bacterial cell envelope consists of plasma membrane, wall polymer, periplasmic space, outer membrane, and capsular layer. Based on the organization of these components, prokaryotic cells can be divided into two broad groups, i.e., Gram-positive and Gram-negative cells. The organization of these components has been diagrammatically presented in Figures 2 and 3.[11] In both Gram-positive and -negative cells cytoplasm is surrounded by plasma membranes, which are comparable except for minor differences in composition (Figures 4 and 5).[12]

After translocation through the plasma membrane, the proteins encounter widely differing environments in Gram-positive and Gram-negative cells. In both organisms, there are peptidoglycan layers opposed to the plasma membrane which resist the deformation of the plasma membrane arising from the cytoplasmic turgor pressure. Peptidoglycan is a highly cross-linked linear polymer of carbohydrate material and has a variety of ionic groups.[13] Thus, the protein molecules may encounter a sieving action, which may be more elaborate in Gram-positive than in Gram-negative cells. In Gram-positive cells the peptidoglycan layer is bound to teichoic acid, a ribitol-glycerol polyphosphate polymer, which may have an additional effect on the diffusion of secreted proteins. However, it is unknown how much, if any, diffusional barrier is exerted by the peptidoglycan layer.

Beyond the peptidoglycan layer in Gram-negative cells, the secreted proteins may

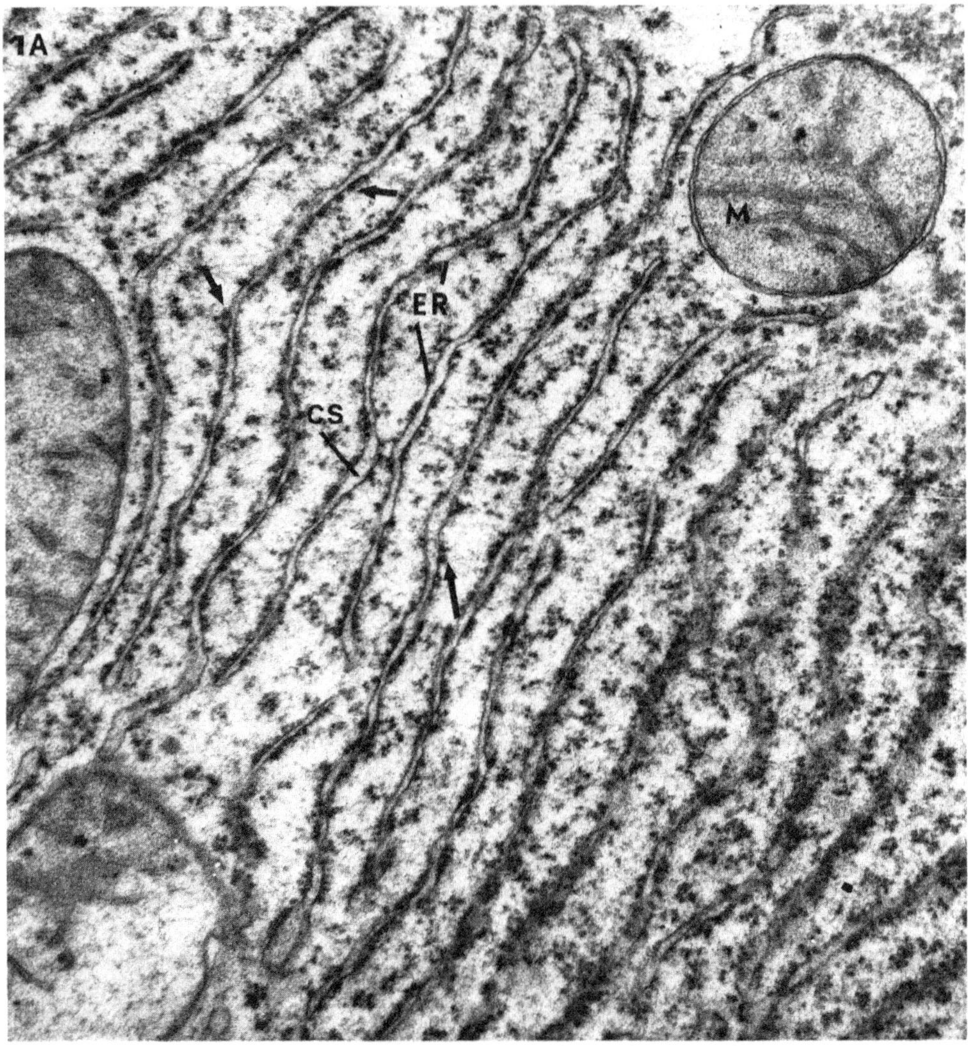

A

FIGURE 1. (A) Endoplasmic reticulum of monkey liver; ribosomes (arrows) are attached to the cytoplasmic side of ER membrane surface; M, mitochondria, ER, endoplasmic reticulum, CS, cisternal space. (Courtesy of Dr. G. Zaki, Department of Pathology, Squibb Institute for Medical Research, Lawrenceville, N.J.) (B) Endoplasmic reticulum of *Trichoderma reesei* (filamentous fungus) shows ribosomes bound to cytoplasmic surface of ER (arrows); note terminal expansion of ER cisternal space (arrowhead). (C) Distribution of ribosome in *Bacillus licheniformis* 749/C. In order to prevent cytoplasmic contraction during dehydration (common for bacterial cells), samples have been prepared by freeze substitution. Ribosomes are distinct (arrows) and the cytoplasm retains a hydrated appearance. Note that the plasma membrane surface maintains a ribosome-free zone (arrowheads), suggesting the lack of extensive ribosome plasma membrane association. (Based on the work of B. K. Ghosh and his associates.)

move into a fluid compartment (i.e., periplasm) of unique composition separated from the extracellular environment by the outer membrane and from the intracellular environment by the plasma membrane (Figure 6).[14] The secreted protein molecules released into this fluid compartment may move by diffusion. In Gram-positive cells, periplasm (defined as above) is not present (Figures 2, 7A, and 7B).[15] However, periplasm has been operationally defined as a hypothetical space from which material is released by osmotic shock or when the peptidoglycan layer is hydrolyzed and the cells are osmotically supported.[16-18] Thus, secreted periplasmic proteins are found in both organisms.

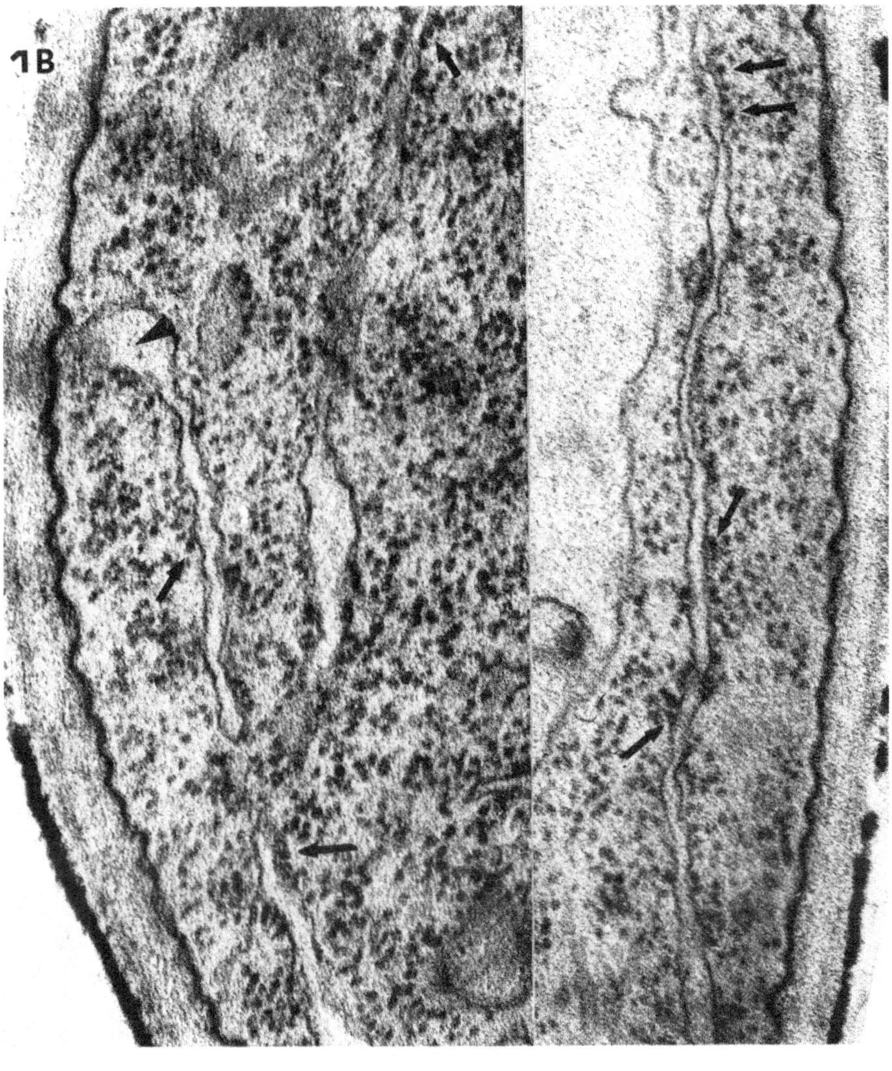

FIGURE 1B

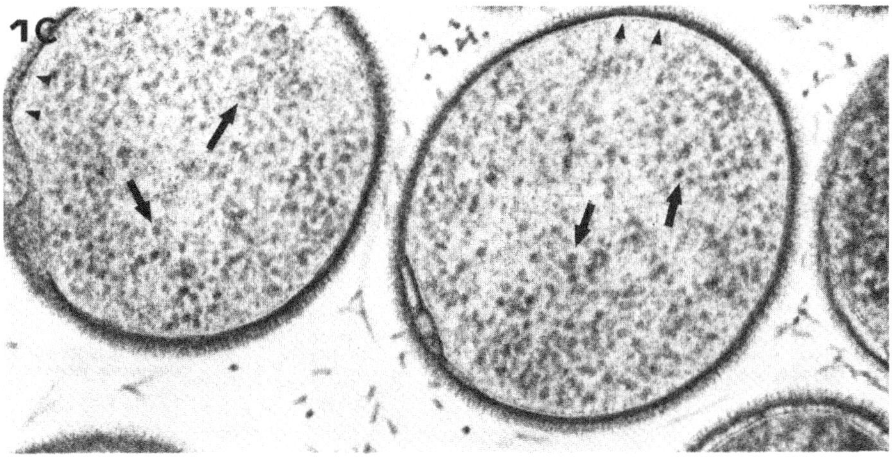

FIGURE 1C

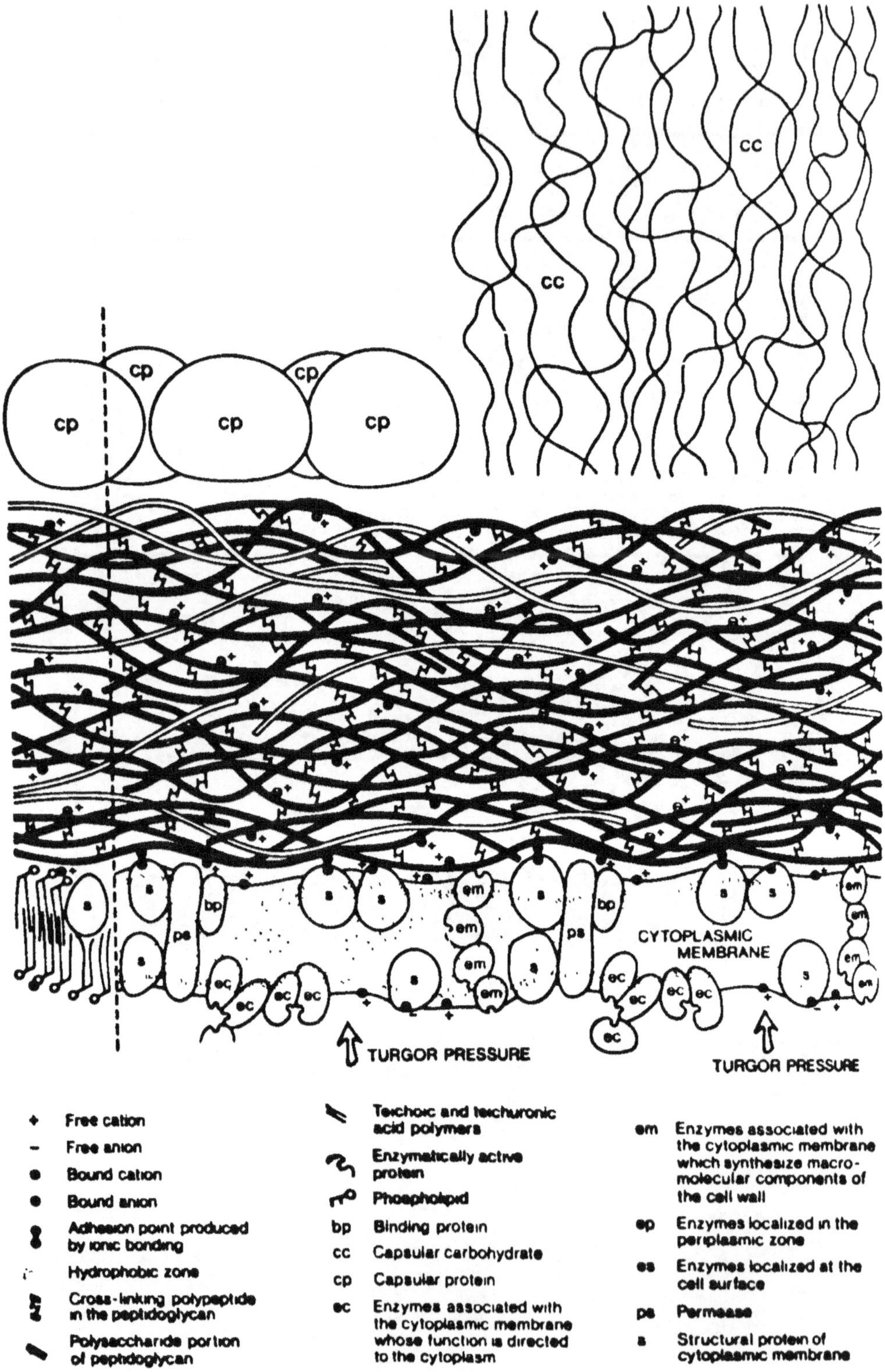

FIGURE 2. Structure of the cell wall of Gram-positive bacteria. (With permission from Costerton, J. W. and Cheng, K. J., *J. Antimicrob. Chemother.*, 1, 363, 1975. Copyright: Academic Press, Inc, (London) Ltd.)

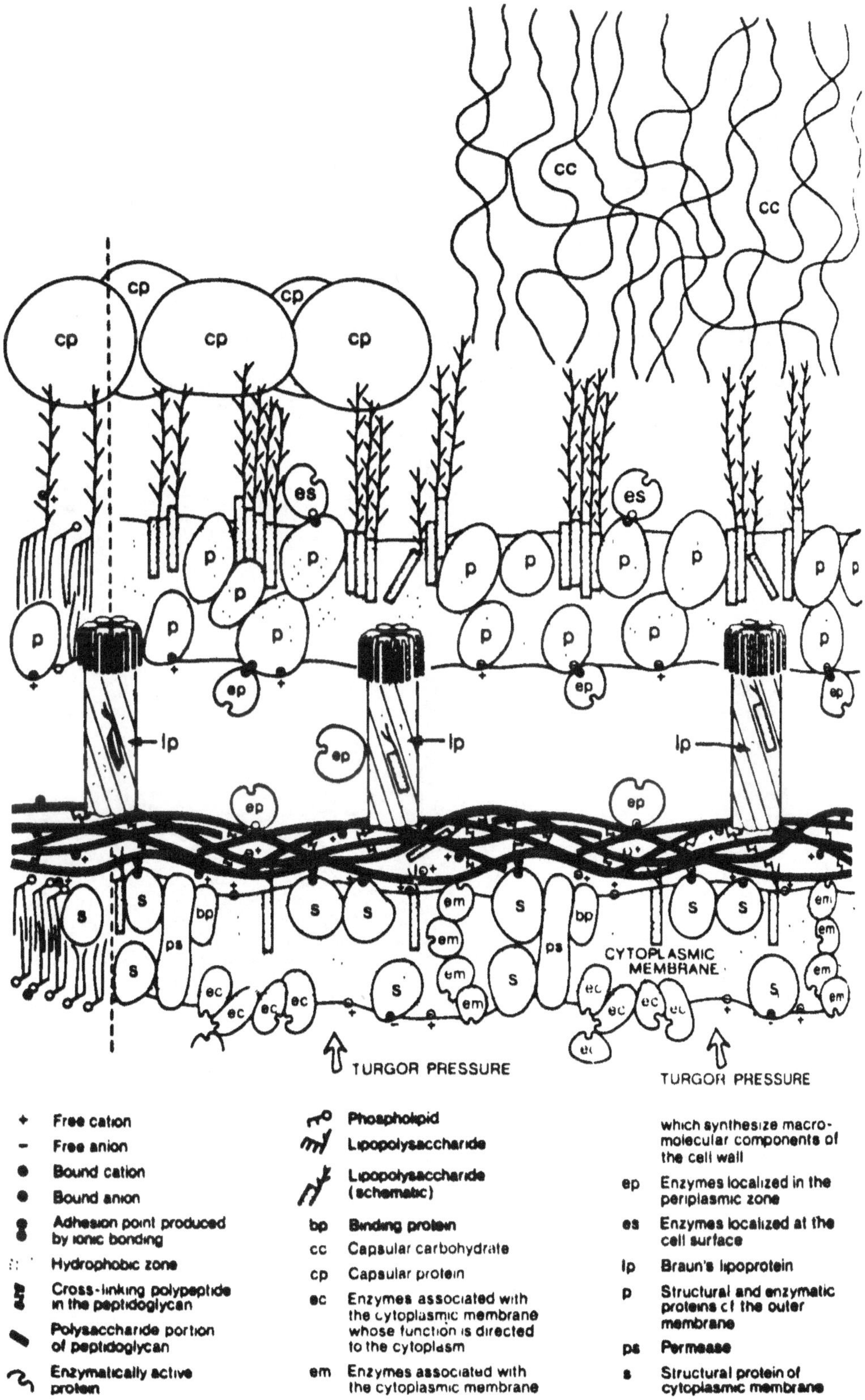

FIGURE 3. Structure of the cell wall of Gram-negative bacteria. (With permission from Costerton, J. W. and Cheng, K. J., *J. Antimicrob. Chemother.*, 1, 363, 1975. Copyright: Academic Press, Inc. (London) Ltd.)

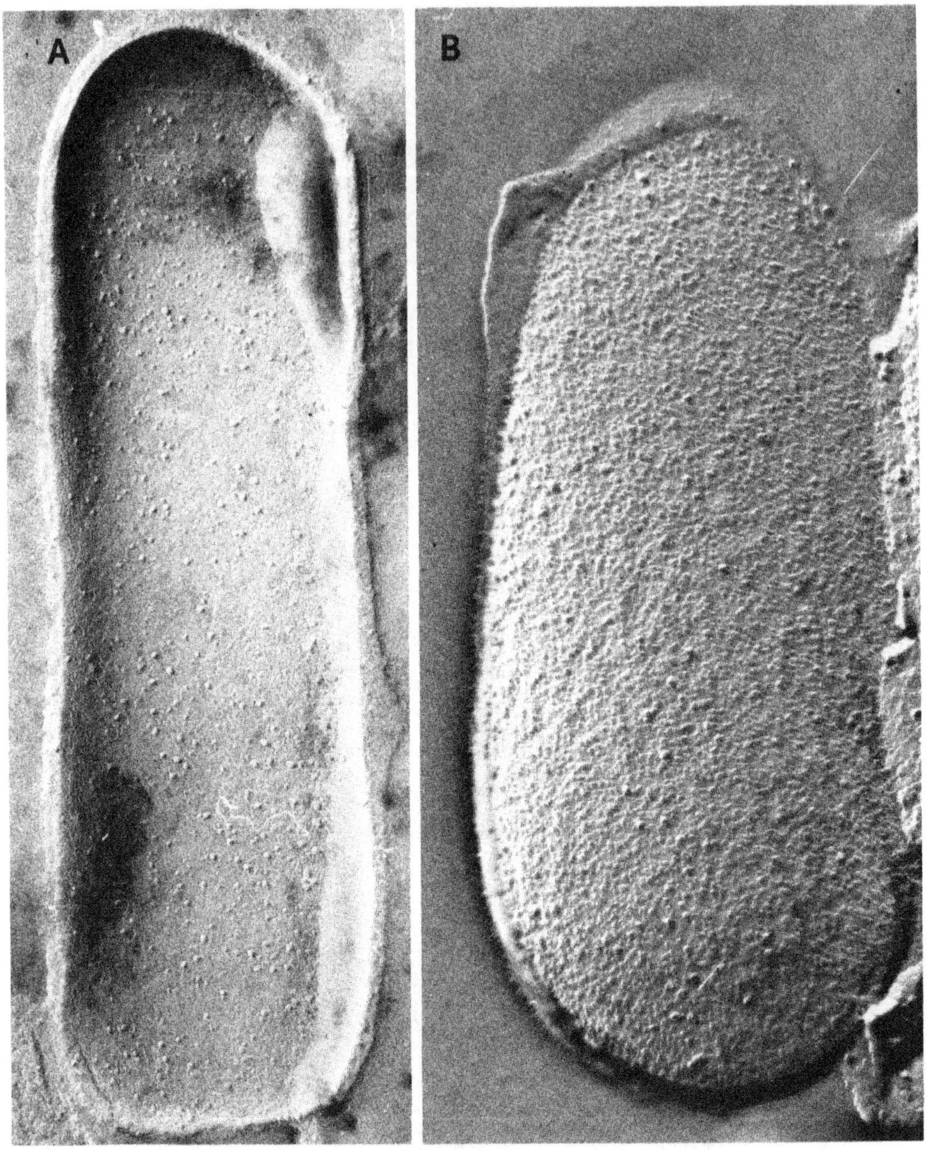

FIGURE 4. Freeze fracture image of Gram-positive (*Bacillus licheniformis* 749/C) bacterial plasma membrane. The samples are unfixed and untreated with glycerol. (A) Concave fracture face shows outside half of the plasma membrane bilayer; (B) Convex fracture face shows the inside half of the plasma membrane bilayer. Note that the intramembranous particles are asymmetrically distributed in the two fracture faces; there is also considerable heterogeneity of membrane particle sizes. (Based on the work of B. K. Ghosh and his associates.)

In Gram-negative organisms the periplasm is surrounded by an outer membrane which is interconnected with the peptidoglycan layer through a covalently bound lipoprotein known as Braun lipoprotein.[19] In addition, there are lipoproteins noncovalently associated to peptidoglycan.[19-21] The outer membrane is an asymmetric bilayered structure consisting of phospholipid in the inside half and lipopolysaccharide in the outside half of the bilayer. Transbilayer pore structures for specific molecules restrict molecular traffic from extracelluler environment into the periplasm.[19] Proteins secreted into the periplasm may be incorporated into either layer of the outer membrane.

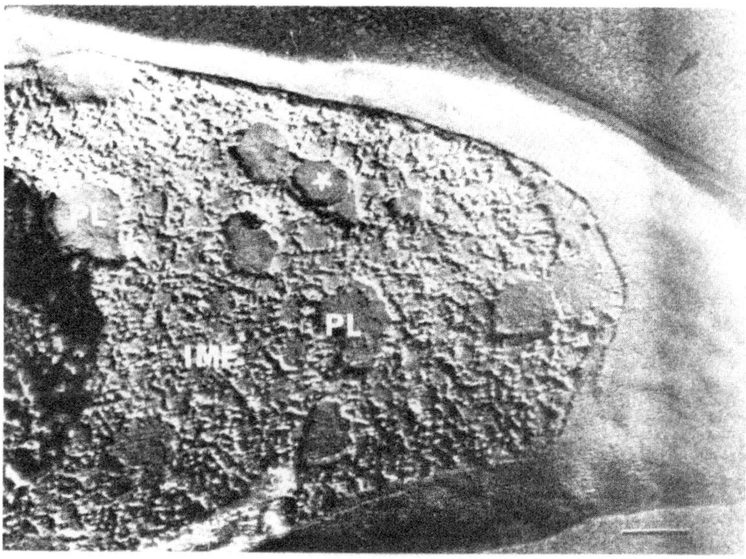

A

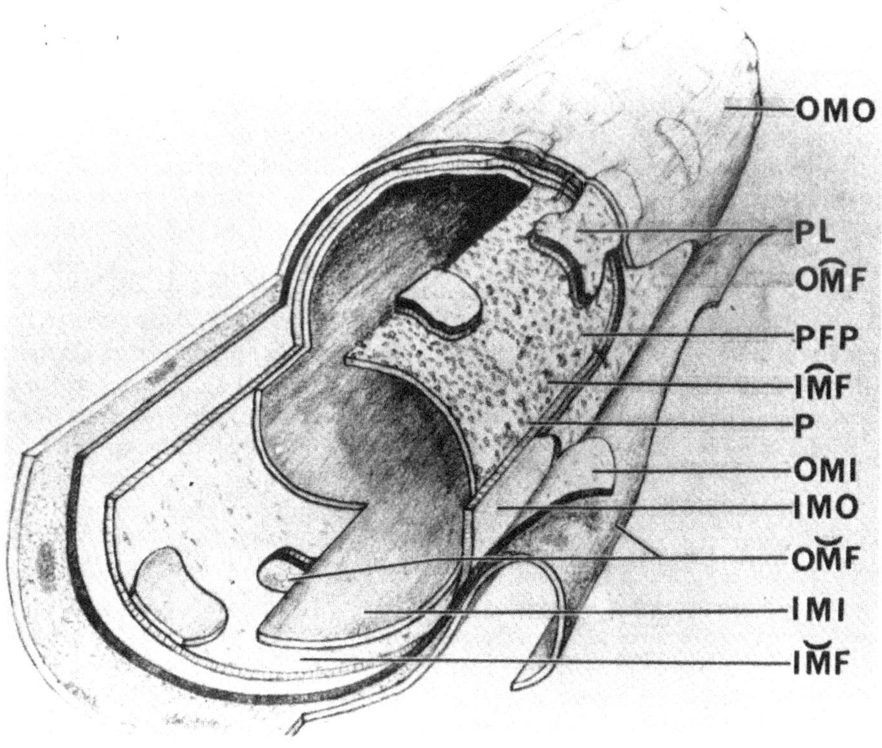

B

FIGURE 5. (A) Freeze fracture of Gram-negative organism; and (B) diagrammatic representation of the various fracture planes of the *E. coli* cell envelope. Symbols: PL, plateaus; OMO, outer face of outer membrane; OMF, fracture plane (convex) of outer membrane; PFP, particle-free patches; IMF, fracture plane (concave) of the inner membrane; P, particles within the fracture plane of IMF; OMI, innermost plane of the outer membrane; IMO, outerface of inner membrane; IMI, innermost face of the inner membrane; ‿, concave membrane fracture faces; ⌒, convex fracture faces; *, a plateau is seen with an additional layer of material on top. (From Bayer, M. E. and Leive, L., *J. Bacteriol.*, 130, 1364, 1977. With permission.)

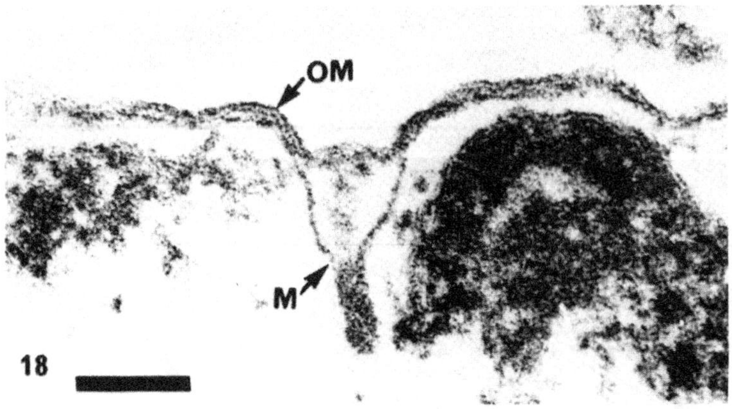

FIGURE 6. Gram-negative wall structure; OM, outer membrane; M, murein. (From Beveridge, T. J., *Int. Rev. Cytol.*, 72, 229, 1981. With permission.)

The outside surfaces of the outer membrane of Gram-negative and the wall layers of Gram-positive cells contain intertwining strands of capsular proteins and polysaccharides.[22] This layer is highly variable and may not be present in cells growing in artificial growth medium; these are extensive in natural habitat. However, it is evident that protein molecules translocated through the plasma membrane cannot freely diffuse into the extracellular medium; this restriction in diffusion is more severe in Gram-negative than in Gram-positive cells. It is likely that the free diffusion of secreted molecules into the growth medium is usually counterproductive. Dilution caused by such diffusion renders the secreted molecules functionally ineffective and causes an energy drain for the cells because synthesis and secretion of proteins is a highly energy-consuming function of cells. Hence, in wild-type cells in natural habitat, secretion is expected to be a slow site-directed function.

The bacterial plasma membrane is a bilayer structure (Figure 8). The core of this bilayer is composed of two symmetrical monolayers of phospholipid generating a hydrophobic core.[23] Both surfaces of this hydrophobic bilayer have ionized phospholipid head groups that interact with protein molecules. These lipid monolayers are in a fluid state and the molecules move by two-dimensional diffusion; however, interbilayer exchanges appear to be restricted.[24] Both amphipathic and hydrophobic proteins interact with the hydrophobic lipid bilayer; however, the former may interact with ionized groups of amphipathic membrane proteins or phospholipid head groups.[25] Transbilayer proteins are large enough to penetrate a 7-nm-thick membrane bilayer. The ionic domains of these molecules should be exposed on both surfaces of the membrane, separated by 7-nm stretches of hydrophobic domains.[26] Such membrane proteins are expected to be rare because of these unique conformational properties. It appears that only a limited number of proteins make direct hydrobic interactions, but a large number of proteins remain associated with the plasma membrane through direct or indirect ionic interaction.[27] As most proteins interact ionically, primary interaction of any secretory protein with the plasma membrane is expected to be ionic. If a protein traverses the plasma membrane by diffusion, transbilayer translocation must take place in a hydrophobic form. Finally, the protein would be released through the ionic layer, presumably interacting ionically. During this process of diffusion the secretory protein must undergo reversible conformational alterations. The mechanism of membrane translocation is one of the most challenging aspects of the study of protein secretion. A variety of suggestions have been made on the translocation mechanism[1-9] based on available data: (1) translocation in a linear form through the proteinacious hydrophilic channel, (2) diffusion or electrophoretic translocation of complete protein is through hydrophilic channels, (3) depolarization and transient destruction of membrane bilayer allowing leakage of complete protein molecules. Translocation by complex formation with carrier molecules, although common for small molecules, has not been suggested for proteins. Bacterial plasma membrane has an asymmetric distribution of proteins in the two halves of the membrane bilayer (Figure 4); hence, the mechanism of entry into membrane may be significantly different from the release through the outside half of the membrane bilayer.

III. PHYSIOLOGICAL REGULATION

Secretion of protein is recognized to be a normal physiological property of the cells of higher organisms. Many cells have differentiated and become specialized for secretory activity in animal and plant cells. In prokaryotic cells secretion appears to be physiologically an energy drain unless the molecules secreted into the extracellular milieu are site-directed and avoid dilution. Secretion of a penicillin-degrading enzyme (i.e., β-lactamase) from bacteria was recognized very early; a classic in the field is a paper published by Duthie in 1944.[28] A systematic investigation by Pollock and his associates since early 1950 has established the studies on β-lactamase as a model for the study of extracellular (or secreted) enzymes in bacteria.[29,30] Secretion of this enzyme is physiologically relevant for the bacterial cell because it degrades a toxic substance which acts at a site in the cell envelope outside the permeability barrier. Thus, the results on β-lactamase secretion have stimulated studies on the secretion of a variety of proteins (enzymes, nonenzymatic protein, toxins, etc.) which perform a multitude of functions supporting the growth and viability of bacteria. This early work, which is primarily aimed at the physiological phenomena, has been superseded recently by studies on the molecular biology of secretion, but many of these early observations point to physiologically significant aspects of the multistep secretory activity. It has been clearly recognized that the synthesis and secretion of proteins are tightly coupled activities, as suggested by Pollock: "The two processes, liberation and formation, are indissolubly connected, i.e., are two separate fractions of the same biochemical event".[29] The major thrust of the study has been to identify and isolate those reactions which are responsible for the physiological regulation of the release of secretory proteins. This final release into the extracellular medium may not directly depend upon the synthesis of secretory proteins. It is possible that an attempt has been made to gather information generally parallel to higher organisms, where release of proteins occurs from storage organelles into the extracellular milieu by stimulus coupled reactions.[31] This final externalization reaction is regulated by a variety of physiological effector molecules and is not directly dependent upon synthesis of these proteins. Recent data show that secretion in bacterial cells is so firmly a cotranslational event that the secretion could be regulated by transcription alone. However, it appears that the cotranslational event may be divided into several distinct sequential steps. Therefore, the earlier data on the reactions of secretory activity independent of biosynthesis are important and may provide clue to future studies on the physiology of secretion.

It was recognized quite early that β-lactamase from Gram-positive cells is released into the extracellular medium, whereas the same enzyme from Gram-negative cells is not externalized (Tables 1 and 2).[30] This difference many have arisen from the differences in the structure of the cell envelope as discussed earlier. As the earlier studies on enzyme secretion depended upon the assays of enzymes into the cell-free culture medium, the studies on secretion have been conducted primarily on Gram-positive cells. It has been shown that *B. licheniformis* 6346 secretes 60% of cell-bound enzyme, while no cellular autolysis has occurred.[29,32,33] However, it has been clearly demonstrated

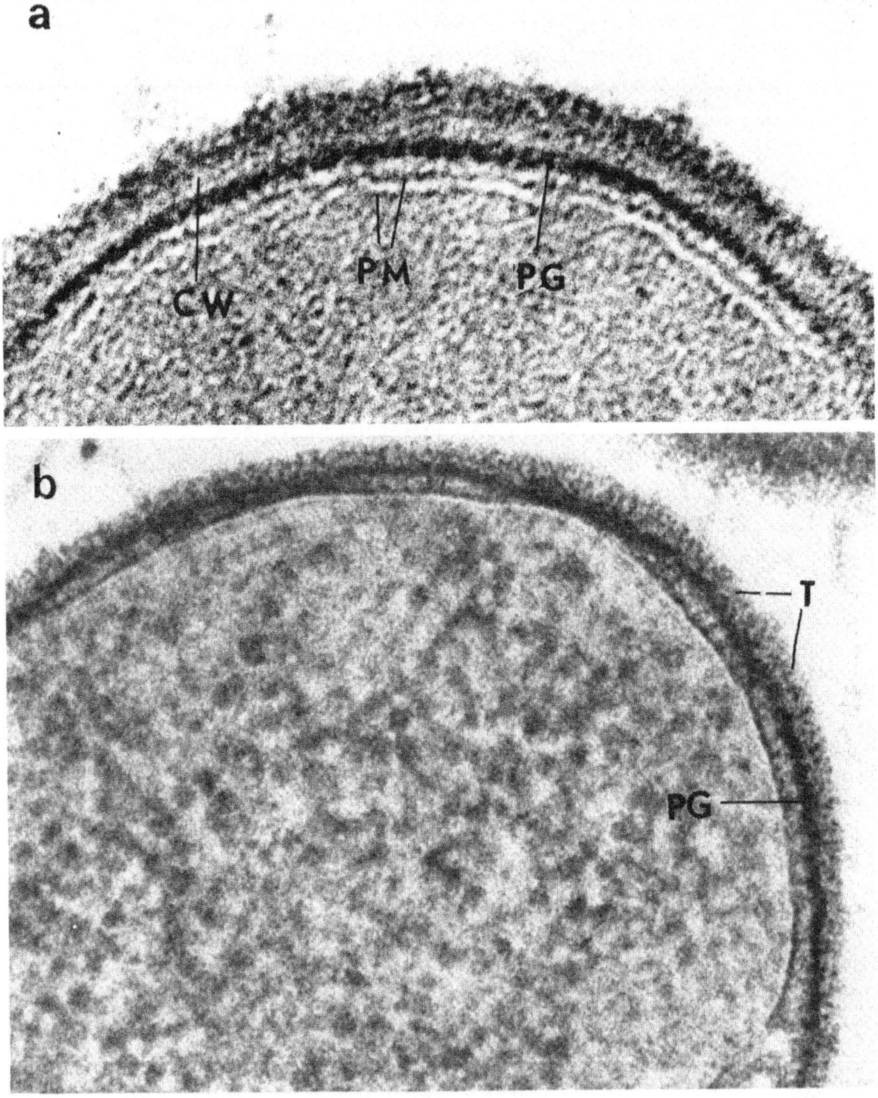

A

FIGURE 7. (A) Gram-positive wall structure. (a) Thin section of *Bacillus licheniformis* 749/C; CW, cell wall; PM, plasma membrane; PG, peptidoglycan. (b) Thin section of *B. licheniformis* 749/C after freeze substitution. Note the thick peptidoglycan layer and thin fibrilar teichoic acid (T) bound to the peptidoglycan (PG). (Based on the work of B. K. Ghosh and his associates.) (B) Arrangements of wall components; (a) and (c), Gram-negative wall; (b), (d), and (e), Gram-positive wall. (From Ghosh, B K., *Subcell. Biochem.*, 3, 311, 1974. With permission.)

that the cell-bound enzyme is located external to the permeability barrier covalently associated with the outside surface of the membrane.[34] Hence, according to the recent definition, the secretion of this β-lactamase has been completed and these studies are truly an examination of enzymatic mechanism liberating the enzyme into the extracellular milieu.

A detailed study in this general direction has been undertaken by Lampen and his associates.[35,36] It has been shown that penicillinase in *B. licheniformis* is distributed in two subcellular membrane fractions: (1) mesosome membrane, a tubule-vesicle frac-

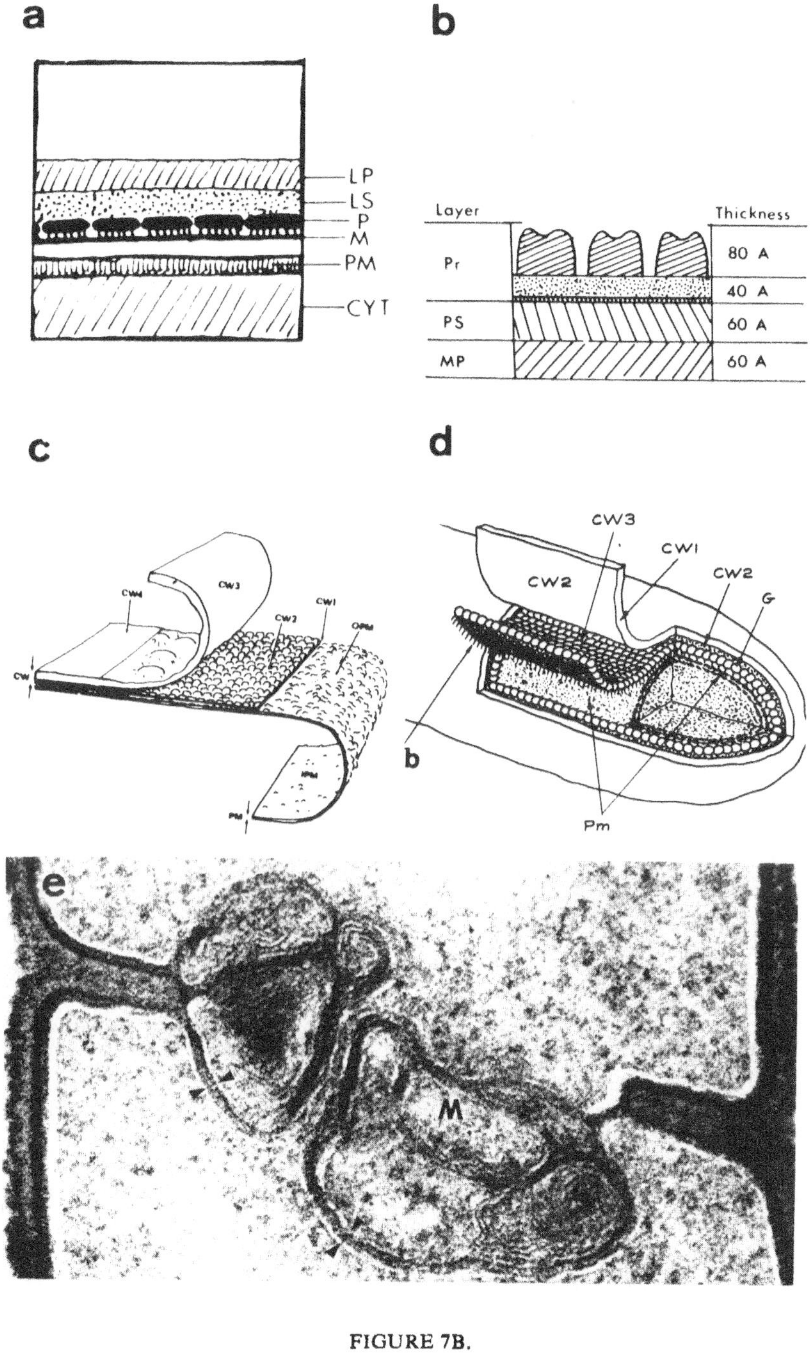

FIGURE 7B.

tion released on protoplast formation and (2) plasma membrane fraction.[37] Penicillinase is concentrated in the tubule vesicle fraction because the specific activity is very high; i.e., 8% of the total protein of this tubule-vesicle membrane fraction is penicillinase.[38] The cell-bound enzyme could be released from nongrowing (chloramphenicol-treated) whole cells, washed protoplasts, and the mesosome fraction in alkaline pH (pH 9.0). This release is fast and could be terminated in acid pH (pH 6.5). The mesosome fraction showed much greater releasing activity than the intact protoplasts. A variety of membrane stabilizing agents (e.g., quinacrine, chlorpromazine and chloro-

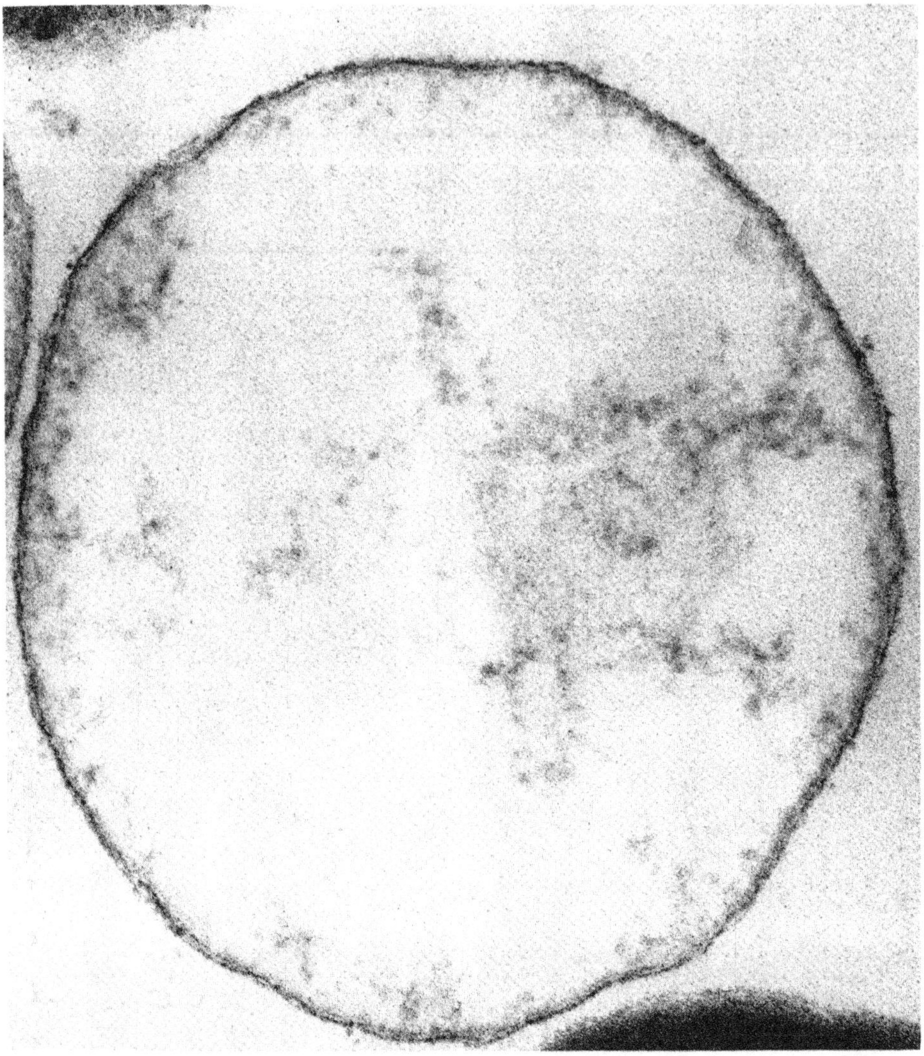

FIGURE 8. Washed membrane of *Bacillus licheniformis* 749/C. Note the clear definition of the bilayer structure. (Based on the work of Ghosh and his associates.)

quine) potently inhibited this release. Although penicillinase secretion is remarkably diminished by chloramphenicol inhibition of protein synthesis, the extent of inhibition is not total in contrast to synthesis.[39] Experiments with protoplasts, which continue synthesizing protein, show that at pH 7.5 almost the entire amount of penicillinase synthesized is released without any apparent incorporation into the plasma membrane. This release, however, could be strongly inhibited by chloramphenicol, suggesting that the secretion is coupled to the protein synthesis. These results confirm the earlier suggestion that there is a rate-limiting step of penicillinase secretion from whole cells which may not require protein synthesis but may depend upon an enzyme activity, as well as H^+ ion concentration and membrane fluidity.[29,32,33,40,41] Furthermore, the synthesized penicillinase appears to be partitioned into a specific compartment of plasma membrane containing a proteinase ("releasing factor") which causes a slow release from the whole cells in absence of synthesis.[39,42]

This question of penicillinase distribution in differing subcellular membrane frac-

Table 1
DISTRIBUTION OF PENICILLINASE IN BACILLI

Strain	Extracellular (%)	Induction ratio[a]	Mean maximum activity[b]
Bacillus cereus 569	80—90	300—500	2000—3000
B. cereus 569/11	80—90	1	2000—3000
B. cereus 5/B	84	1	2000
B. subtilis T98	90	1	—
B. licheniformis 749	40	150	—
B. licheniformis 749/C	ca. 50	1	4000
B. licheniformis 6346/C	ca. 50	1	600
B. anthracis	85	1—2	2000

[a] Ratio of activity of fully induced to noninduced culture.
[b] Micromoles benzylpenicillin hydrolyzed per hour per milligram dry weight of fully induced or consititutive cells.

From Citri, N. and Pollock, M. R., in *Advances in Enzymology*, Vol. 28, Nord, F. F., Ed., John Wiley & Sons, New York, 1966, 237. With permission.

Table 2
DISTRIBUTION OF PENICILLINASE IN GRAM-NEGATIVE CELLS

Organism	Induction ratio	Mean maximum activity	Method of disruption[a]	Crypticity factor[b]	
				BP	CC
E. coli	1	0.1—3.0[c,d]	S	2—2.5	1
				10	
			E	4—10	—
			T	10	—
				5	
Klebsiella spp.	1	3.9[c]	S	1.5—3	1
			E	5	—
Aerobacter aerogenes	2.5	0.5—300[d]	E	3—11	—
A. cloacae	1	50[d]	E	25	—
Pseudomonas	6	16[c]	S	1	1
pyocyanea	440			10	3.9
Proteus morgani	18—44	7.4[c]	S	1—2	1
Pr. vulgaris	17—38	3.4[c]	S	1	1
Pr. rettgeri	1	1.4[c]	S	1	1.6
Pr. mirabilis	3.5—10	4.0[c]	S	2.5	0

[a] S, sonication; E, extrusion (French) press; T, toluenization.
[b] Ratio of activity of disrupted to intact cell preparation on benzylpenicillin (BP) and cephalosporin (CC).
[c] Micromoles benzylpenicillin hydrolyzed per hour per milliliter of fully grown and fully induced or constitutive culture.
[d] Micromoles benzylpenicillin hydrolyzed per hour per 10^9 cells (disrupted).

From Citri, N. and Pollock, M. R., in *Advances in Enzymology*, Vol. 28, Nord, F. F., Ed., John Wiley & Sons, New York, 1966, 237. With permission.

tions in relation to secretion has been analyzed by a specific immunoelectron microscopic method.[43] The membrane-bound penicillinase molecules are asymmetrically distributed on two surfaces of the membrane, i.e., more in association with the inside surface than with the outside surface. In addition, cytosol contains a significant amount of penicillinase which has poor reactivity to antibody, poor catalytic activity, and high sensitivity to proteolysis. It has been suggested that there are a limited number of sites on the inside surface of the membrane at which ribosomes with extruding penicillinase signal sequences may bind. The saturation of these membrane-binding sites may lead to completion of the translation of newly initiated penicillinase in the cytosol. Folding of these molecules in the cytosol may produce molecules with protease sensitive conformation, in contrast to membrane-bound enzyme which is less protease sensitive. The more numerous binding sites on the inside surface compared to the release sites on the outside surfaces of the membrane may indicate that movement of penicillinase molecules does not occur through putative transmembrane pores. As penicillinase is known to be a lipoprotein, the molecule may diffuse in the plasma membrane plane in a hydrophobic conformation by lateral diffusion.[44] It may eventually be released from membrane by a flip-flop mechanism.[24] The above concept of the membrane translocation of penicillinase explains: (1) the extended residence time in the membrane;[40] (2) the partitioning in a specific compartment of plasma membrane, and (3) the inhibition of release by agents affecting membrane fluidity. Finally, the proteolytic activity frees the enzyme from the hydrophobic material (lipid and signal sequence), and the hydrophilic exoenzyme is released.

There are a limited number of binding sites for penicillinase in the outside surface of the plasma membrane.[45] After trypsin removal of penicillinase from these sites on the protoplast membrane surface, only a limited amount of enzyme could be reincorporated into the membrane sites equivalent to those present in the prestriping condition. However, this incorporation may occur in partially folded form. These studies on penicillinase secretion in Gram-positive organisms which spread over a 20-year period, provide evidence that there are two distinct phases of secretion: (1) cotranslational, occurring in association with the inside surface of plasma membrane, and (2) post-translational, occurring within the membrane interior as well as in association with the outside surface of the plasma membrane. It is likely that physiological regulation is exerted on this post-translational reaction.

This physiological line of investigation has been further extended in *B. licheniformis* cells using alkaline phosphatase secretion as a model.[46-49] The rationale for this approach is to examine whether or not a common physiological regulatory mechanism controls the secretion of both the enzymes (alkaline phosphatase and penicillinase). Thus, it will be possible to identify the existence of the specific regulatory mechanism (if any) for the secretion of individual proteins from bacterial cells. The following suggests that secretory mechanisms may differ for these two enzymes.

1. Inhibitors of protein synthesis. The results of the effect of translation inhibitor chloramphenicol (Figure 9A and B)[49] and transcription inhibitor rifampin show that alkaline phosphatase secretion is more tightly coupled to synthesis than penicillinase. Treatments with these inhibitors cause an immediate or total halt of alkaline phosphatase secretion, whereas penicillinase secretion continues at a low rate before coming to a halt. This is consistent with the data that penicillinase molecules, which partition into a specific membrane compartment, continue to be released by an enzymatic-releasing mechanism. However, no such compartment could be demonstrated for alkaline phosphatase.
2. The action of inorganic phosphate. Both deprivation and enrichment of inorganic phosphate (Pi) in the growth medium influence these enzyme secretions differently (Figures 9C and D).[49] The Pi starvation brought about by the suspension of cells, pregrown in casitone (CH/S) medium, in a chemically defined medium, deprived of Pi almost totally stopped penicillinase synthesis but allowed the secretion to continue uninhibited. In the identical condition, alkaline phosphatase synthesis is derepressed but there is no secretion. However, the secretion

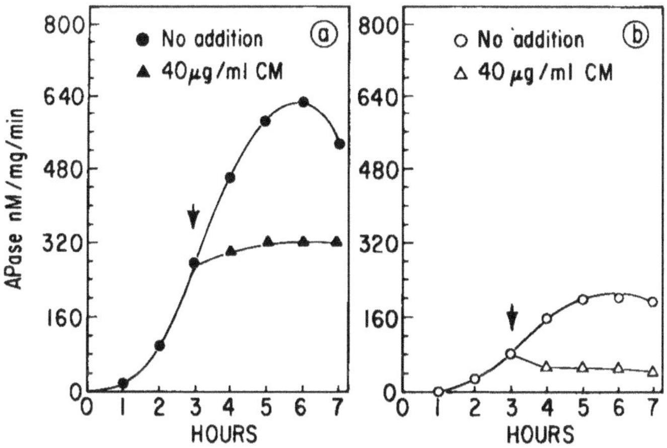

FIGURE 9. (A) Effect of chloramphenicol (CM) on (a) synthesis and (b) secretion of alkaline phosphatase. Note that both the synthesis and secretion are arrested after the addition of CM to the growth media. (Based on the work of Ghosh and his associates. See Reference 49.) (B) Effect of CM on the (a) synthesis and (b) secretion of penicillinase. After 3 hr. growth, the cells were transferred into new media with and without 40 μg/mℓ CM. Note that the synthesis totally halts within 2 hr. after CM addition; however, secretion continues at a slow rate. This resistance to CM inhibition is more pronounced at pH 9.1. Solid line (—), without CM; dashed (- - -), with CM; △, pH 5.5; ○, pH 7.4; □, pH 9.1. (C) Effect of inorganic phosphate (Pi) on the synthesis and secretion of alkaline phosphatase. Note that with only 0.05 m*M* Pi increment the enzyme synthesis is 80% reduced; secretion is also proportionately reduced. A closed symbol indicates synthesis and an open symbol secretion; • and ○, 0.34 m*M* Pi; ■ and □, 0.35 m*M* Pi; ▲ and △, 0.39 m*M* Pi. (D) Effect of Pi on the (a) synthesis and (b) secretion of penicillinase. Cells pregrown in CH/S medium (•) and resuspended in Pi-deficient chemically defined medium (□). a closed symbol indicates synthesis and an open symbol secretion. Note the Pi deficiency arrests synthesis but secretion continues. (E) Effect of varying H^+ ion concentration (pH) on alkaline phosphatase (a) synthesis and (b) secretion. Solid line (—), without CM; dashed line (- - - -), with 40 μg/mℓ CM. Note 50 to 80% inhibition of synthesis and secretion by increasing or decreasing H^+ concentration. □, pH 5.5; ○, pH 7.4; △, pH 9.1. (F) Effect of Mg^{2+} on alkaline phosphatase (a) synthesis and (b) secretion. Note at 100 m*M* concentration, synthesis is 70% inhibited but secretion shows only small change. (G) Effect of Ca^{+2} on alkaline phosphastase (a) synthesis and (b) secretion. Note at 100 m*M* concentration synthesis is 30% stimulated; the secretion, however, is nine times increased. (H) alkaline phosphatase (a) synthesis and (b) secretion in the presence of adenine nucleotides. Note that AMP stimulates synthesis but inhibits secretion; ADP and ATP arrest both synthesis and secretion.

of penicillinase in the absence of synthesis depended upon a high quantity of enzyme retained in the cell before transfer into the Pi-limited medium. Thus, penicillinase secretion, unlike alkaline phosphatase, possibly depended upon the enzyme being partitioned into a specific membrane compartment.

3. The effect of H^+ ion concentration. The changes in H^+ ion concentration also affect synthesis and secretion differently (Figure 9E).[49] Increased H^+ concentration causes inhibition of synthesis and secretion of both the enzymes. However, in alkaline pH a much greater proportion of synthesized penicillinase is secreted compared to alkaline phosphatase. This pH-dependent variation of secretion may occur while protein synthesis is inhibited.
4. The effect of divalent cations. Divalent cations, too, influence the synthesis and secretion of these enzymes differently (Figures 9F and 9G).[49] Calcium is well known as a physiological effector molecule specifically modulating membrane functions.[50] As magnesium ion causes specific dissociation of the alkaline phosphatase-membrane complex, these cation effects have been studied in growth media specifically made magnesium free.[46-49] Calcium and magnesium ions in a con-

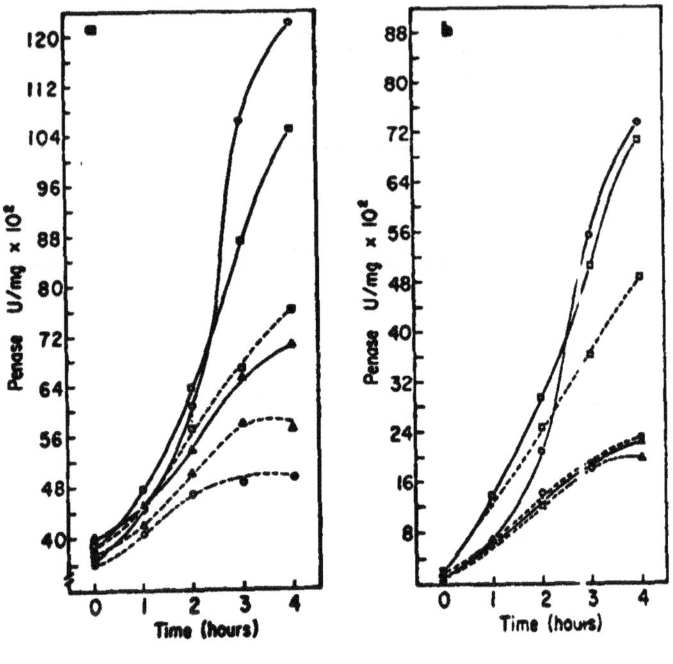

FIGURE 9 B

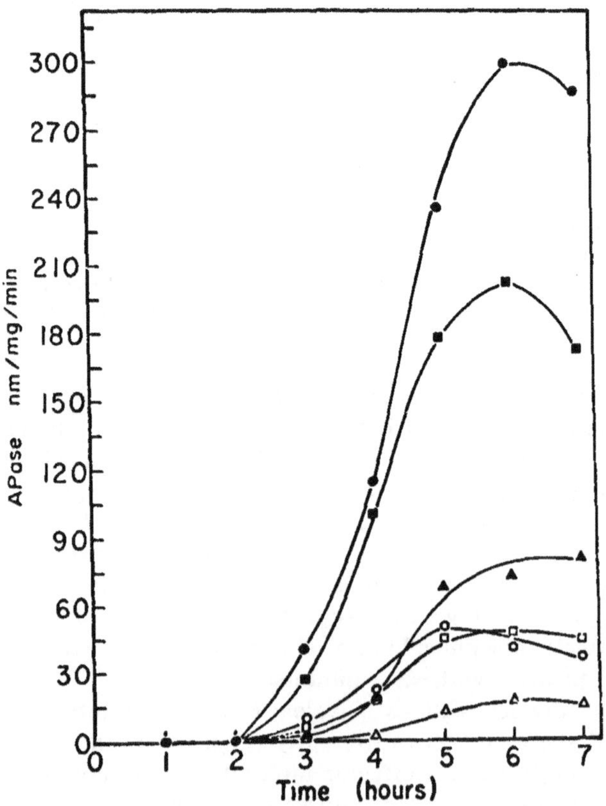

FIGURE 9 C

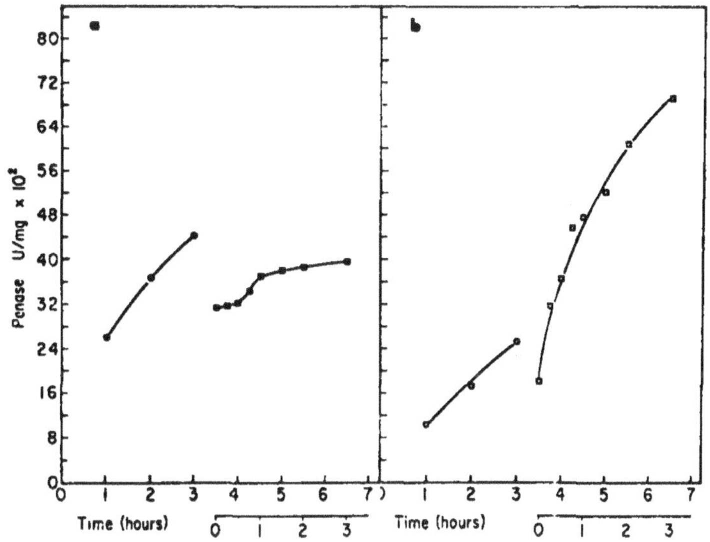

FIGURE 9D

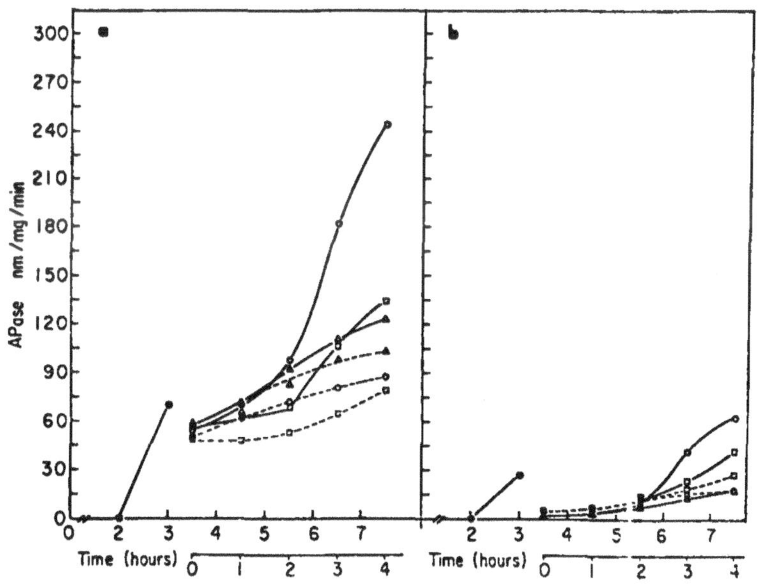

FIGURE 9E

centration range of 0 to 100 m*M* show no effect on the synthesis or secretion of penicillinase. On the other hand, when increasing the magnesium ion concentration (0 to 100 m*M*), a gradual reduction of alkaline phosphatase synthesis has been observed (80% inhibition at 100 m*M* concentration). A comparable increase of calcium ion concentration causes a 37% stimulation. Although the magnesium ion inhibits synthesis, secretion is unaffected. Calcium ion on the other hand enhances secretion, and this enhancement is greater than the stimulation of synthesis. In fact, in the presence of 100 m*M* Ca^{2+}, 100% of the alkaline phosphatase can be secreted. It is possible that calcium ion may specifically stimulate secretory activity.

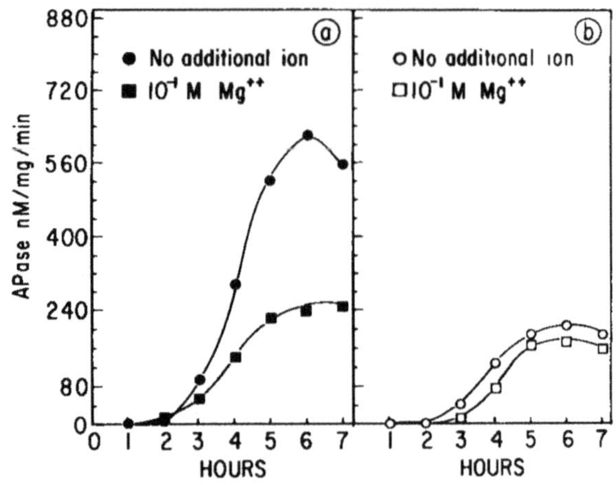

FIGURE 9F

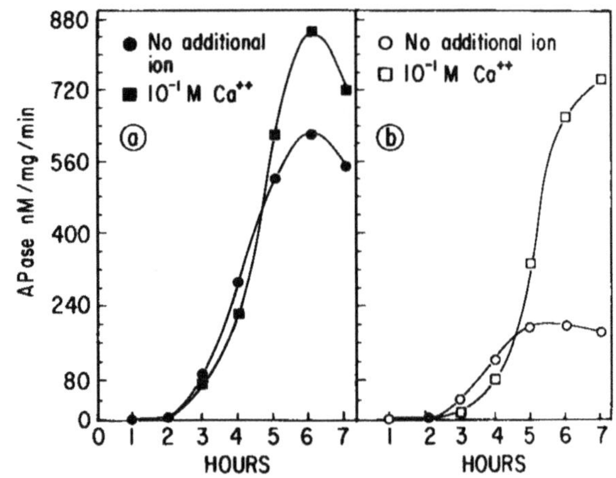

FIGURE 9G

5. The effect of adenine nucleotides. Adenine nucleotides modulate a variety of physiological activities by affecting membrane function.[51] Although cAMP is known for this function, such regulation is not expected in bacilli because of the absence of adenyl cyclase which synthesizes cAMP.[52] The effect of these nucleotides on the synthesis of alkaline phosphatase (Figure 9H)[49] is different for the three compounds: AMP causes 40% stimulation of synthesis; ADP causes 85% inhibition and ATP 30% inhibition of synthesis. Secretion on the other hand is inhibited by all the three nucleotides.

However, this inhibitory effect of AMP is profound because this occurs in spite of stimulation of synthesis. Although an extensive study compared to alkaline phosphatase has not been undertaken in other organisms, remarkable effects of calcium ion in stimulating protease secretion have been reported. The protease secretion from sarcina strain (coccus P) is highly stimulated.[53,54]

These results clearly show that the secretory process is a physiologically regulated function which may have common as well as diverse aspects unique for individual protein molecules. This will be further elucidated while discussing the possible existence of a secretory apparatus in bacterial membrane. Although protein secretion ap-

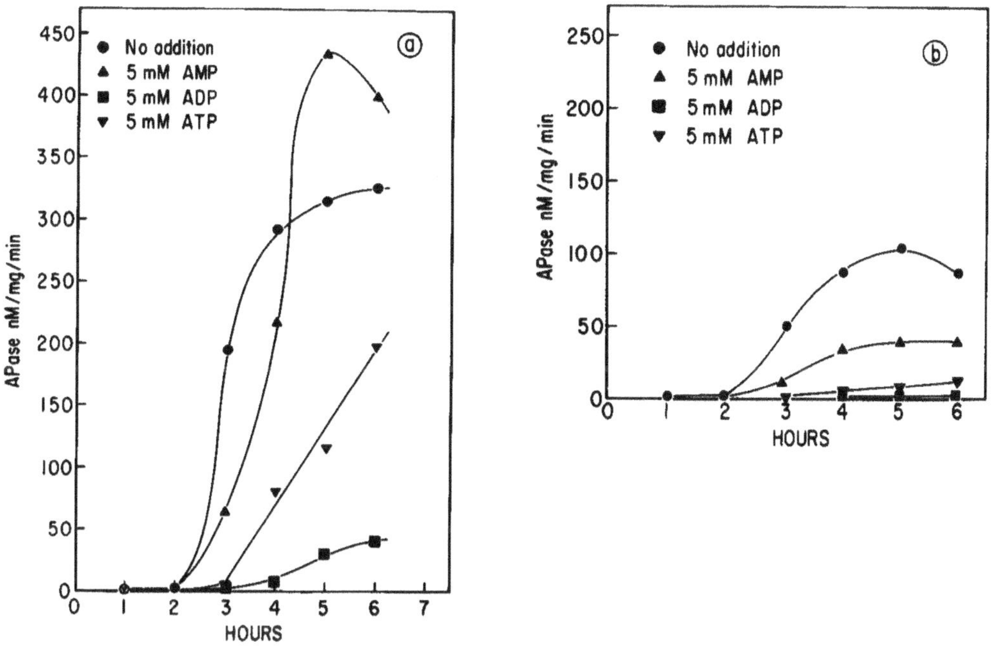

FIGURE 9H

pears to be a multistep process and the nature of the individual steps has been delineated, the occurrence of all these steps is not experimentally verified for any specific protein molecule. In bacterial cells the existence of some of the individual steps is only conjectural. As mentioned earlier, a universal mechanism is expected, and it appears plausible that some aspects of the secretory phenomena of the prokaryotic cells may be unique. Hence, in order that similarities and differences may become evident, molecular mechanims of secretion from the same class of proteins have been discussed together.

IV. MOLECULAR MECHANISM IN GRAM-NEGATIVE BACTERIA

A. Periplasmic Protein

The periplasm is an operationally defined space outside the permeability barrier of the plasma membrane and isolated from the extracellular environment by the limited permeability barrier of outer membrane. This space contains a specific group of proteins having primarily transport or hydrolytic function. These proteins may remain free in the hydrophilic periplasmic environment or may remain anchored to the outside surface of the plasma membrane or inside surface of the outer membrane. Hence, a constant flow of proteins is physiologically important to maintain a unique composition of the periplasm. The following is a discussion on the secretion of some of these proteins.

1. Maltose-Binding Protein (MBP)

Maltose-binding protein is a part of the maltose transport system (Figure 10).[55] The application of the gene fusion technique originated by Casadaban[56] allowed the construction of strains of *E. coli* producing cytoplasmic and secretory hybrid proteins (Figure 11).[57] These studies conducted by Beckwith and his associates played a pivotal role in establishing the function of the signal sequence in bacterial protein secretion.[57,58] Fusion between the *lacZ* gene coding for the cytoplasmic protein β-galactosid-

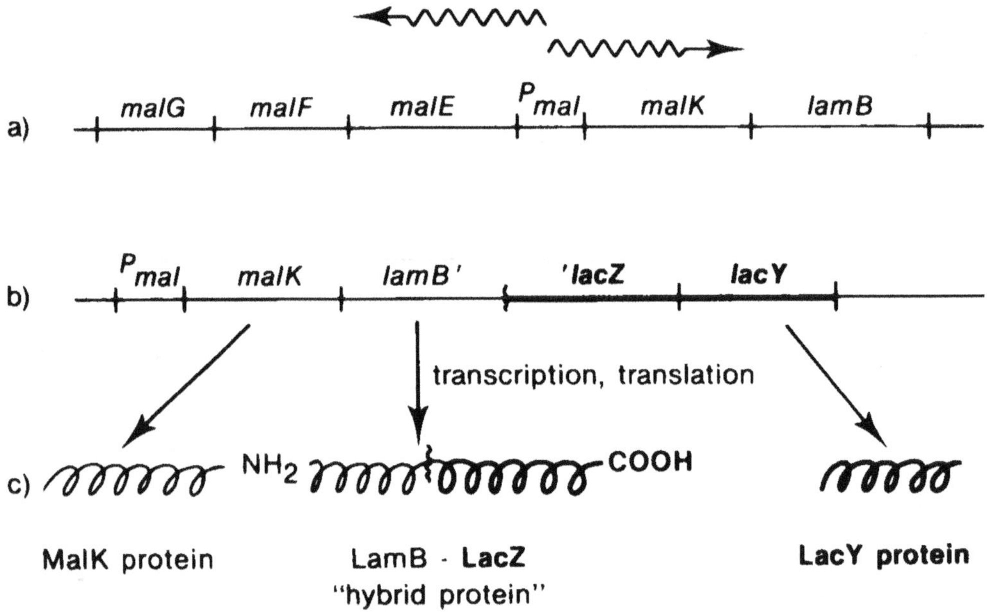

FIGURE 10. (a) The two divergent operons that comprise the *malB* locus in *E. coli*. These two operons specify five proteins, which together make up the active transport system for maltose and maltodextrins. Both operons, and thus the synthesis of all five proteins, are induced by maltose in the growth media. At least three of these proteins are destined to be localized to noncytoplasmic compartments. The *malE* gene specifies the periplasmic MBP, *malF* specifies an inner-membrane protein, and *lamB* specifies the outer-membrane protein λ receptor. The *malK* and *malG* gene products have not been indentifed. Transcription of these two operons is initiated from a central region (Pmal). Direction of transcription is shown by wavy arrows. (b) Genetic structure of a *lamB-lacZ* fusion. The "fusion joint" is designated by a short, vertical, wavy line. Transcription and subsequent translation of the operon shown in (b) result in the production of three proteins. (c) The MalK protein is required for maltose transport. The *LacY* protein is required for lactose transport. The structure of the *LamB-LacZ* hybrid protein is shown. The hybrid protein is comprised of λ receptor sequences at the NH_2- terminus and a major functional portion of β-galactosidase at the COOH-terminus. All of the hybrid proteins discussed here have essentially identical amounts of β-galactosidase sequences at the COOH-terminus. In order for *E. coli* to express a Lac^+ phenotype, strains must have both *LacZ* (β-galactosidase) and *LacY* (lactose transport) protein activities. (From Emr, S. D., Hall, M. N., and Silhavy, T. J., *J. Cell Biol.*, 86, 701, 1980. By copyright permission of The Rockefeller University Press.)

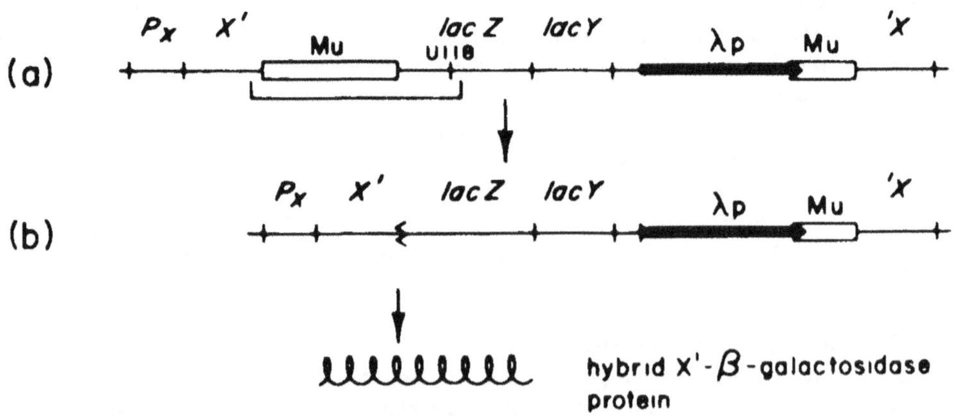

FIGURE 11. Steps leading to the formation of hybrid proteins between β-galactosidase and other bacterial proteins. (From Silhavy, T. J., Bassford, P. J. Jr., and Beckwith J. R., *Bacterial Outer Membranes. Biogenesis and Function*, Inouye, M. Ed., John Wiley & Sons, New York, 1979. With permission.)

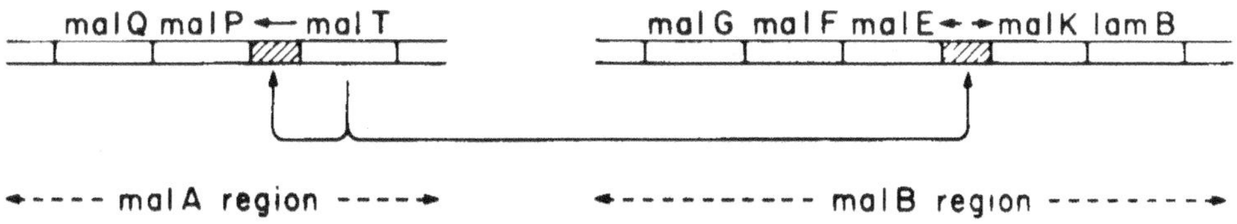

FIGURE 12. The *malA* and *malB* regions of *E. coli*. (From Silhavy, T. J., Bassford, P. J. Jr., and Beckwith, J. R., *Bacterial Outer Membranes. Biogenesis and Function*, Inouye, M. Ed., John Wiley & Sons, New York, 1979. With permission.)

ase and the *malE* gene of the *malB* locus (Figure 12)[57], coding for maltose-binding protein produces hybrid proteins which are located in the plasma membrane.[59] The natural product of the *malE* gene is periplasmic; however, the fused gene product is always cytoplasmic as long as the β-galactosidase catalytic activity is retained. The proportion of the codes of individual genes may vary widely in the fused gene. If a large portion of the fused product is comprised of *malE* gene, the product may be precipitable with anti-MBP antibody. This hybrid protein, in spite of containing a large portion of periplasmic mature MBP, remains bound to the cytoplasmic membrane.[60] One such strain of *E. coli* MM18 appears to be secretion blocked, because it accumulates a precursor form of MBP.[60-63] This strain after induction with maltose has been used to isolate precursors of MBP (pre-MBP) from the lysed cell.[61] The induced cells are converted into spheroplasts and washed to remove any periplasmic MBP. These spheroplasts are ruptured by sonication and the pre-MBP is purified from the cytosol by affinity chromatography and chromatofocusing. The crude cytosol contains mature MBP (40 kDa), pre-MBP (43 kDa), and a fragment of pre-MBP (42 KDa) which has lost eight amino acid residues from the amino terminal.[64] The identical signal sequences of these pre-MBP molecules, determined directly and indirectly from DNA sequencing, proves that purified pre-MBP of secretion-blocked mutant is an authentic precursor protein.[65] Furthermore, purified signal peptidase hydrolyzes the signal peptide from the pre-MBP and the resulting product is indistinguishable from the mature MBP.[66] Hence, it is clear that the secretion-blocked mutant accumulates precursor form of MBP.

2. Leucine-Specific Binding Protein (L-SBP)

Branched chain amino acids are transported by a complex transport mechanism regulated by several genes; *livK* is one of these.[67] This gene has been cloned into a plasmic vector;[68] after recloning this gene has been used for in vitro synthesis and amino acid sequence determination.[69] The cloned gene after restriction endonuclease digestion and ligation was used for a transformation experiment.[70] A coupled transcription-translation system is used for in vitro synthesis of the gene products, which are analyzed by SDS gel electrophoresis and immunoprecipitation. Antiserum against LIV-binding protein cross-reacts with L-SBP. This antibody has been used for the immunoprecipitation of the in vitro synthesized material.[71] Two major in vitro products, 39 and 42 kDa, are found after a 40-min incubation of the in vitro synthesis reaction mixture; however, after prolonged incubation in the presence of DNase, RNase, and membrane fragments, the quantity of the 39-kDa fragment increases, suggesting the presence of a membrane associated processing activity.

The location of the *livK* gene and the direction of transcription is determined by DNA sequence analysis of the fragments and comparison with the amino acid sequence of L-SBP. Furthermore, the *Bgl* II gene fragment coding for the amino-terminal end of the pre-L-SBP has been determined. The data from restriction endonuclease mapping and microsequencing using the Edman degradation matched (Figure 13 and Table 3).[70] The mature form is more resistant to trypsin or chymotrypsin digestion than to

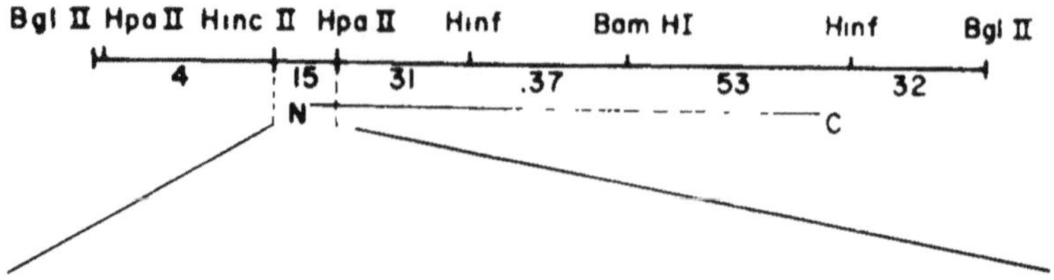


```
CAT AAA ACC AAA GCA ACA CAA CAT CAC GAA TGG GGA
            Initiation
                |      Precursor
               Met  Lys  Ala  Asn  Ala  Lys  Thr  Ile  Ile
TTT TTG ACT ATG AAA GCG AAT GCG AAA ACT ATC ATC

Ala  Gly  Met  Ile  Ala  Leu  Ala  Ile  Ser  His  Thr  Ala
GCA GGG ATG ATT GCA CTG GCA ATT TCA CAC ACC GCT
    Processing
        |     Mature form
Met  Ala  Asp  Asp  Ile  Lys  Val  Ala  Val  Val  Gly  Ala
ATG GCT GAC GAT ATT AAA GTC GCC GTT GTC GGC GCG

Met  Ser  Gly ---
ATG TCC GG ---
```

FIGURE 13. DNA-sequence of a restriction fragment carrying the region corresponding to the amino-terminal segments of the L-SBP. (From Oxender, D. L. et al., *Proc. Natl. Acad. Sci. U.S.A.*, 77, 2005, 1980. With permission.)

papain; in contrast, the precursor form is sensitive to any proteolytic digestion. Based on these data it has been suggested that prior to the cleavage of the signal sequence, the precursor protein conformation is significantly different from the mature protein. This specific conformation of the precursor form may be required for the recognition of the membrane receptor for the secretory apparatus.

3. Alkaline Phosphatase

Alkaline phosphatase is widely distributed in nature. In a variety of Gram-negative organisms the periplasmic location of this enzyme has been shown by both ultrastructural cytochemistry and cell fractionation.[72-77] In *Pseudomonas aeruginosa*, active alkaline phosphatase is released into the extracellular medium due to a pH-dependent outer-membrane permeability change.[75] Multiple forms of alkaline phosphatase are frequently coregulated with other envelope proteins.[78,79] Therefore, secretion of this multifunctional enzyme may be under complex physiological regulation. The following discussion will include data only on translocation through the plasma membrane of *E. coli.*

E. coli alkaline phosphatase has been synthesized in vitro by both a transcription-coupled translation system and an mRNA translation system. In the former case ϕ80d*phoA* DNA was used as a template.[80] Although this study did not address the question of a precursor form of the enzyme, the results point to some important properties of the nascent alkaline phosphatase. All three steps needed for the formation of mature active enzyme occur in vitro (Table 4):[80] (1) dimerization, (2) binding with zinc, and (3) oxidation of the sulfhydryl groups to disulfide. Furthermore, the in vitro synthesized enzyme is as resistant to heat and proanase digestion as the authentic mature

Table 3
EDMAN DEGRADATION OF L-SBP PRECURSOR SYNTHESIZED IN VITRO

Edman cycle	Labeled amino acid released (relative cpm in %)[a]		
	Met	Lys	Thr
1	100	2	15
2	3	52	23
3	2	5	12
4	2	8	14
5	1	4	16
6	1	100	38
7	2	7	100
8	1	3	27
9	3	4	15
10	3	9	23
11	4	—	—
12	19	—	—
13	8	—	—

Note: Deduced sequence: Met-Lys-X-X-X-Lys-Thr-X-X-X-X-Met-X.

[a] cpm in cycle sample as a percentage of the cpm released in the cycle with the highest yield.

From Oxender, D. L. et al., *Proc. Nat. Acad. Sci. U.S.A.*, 77, 2005, 1980. With permission.

enzyme. The synthesis and subtle post-translational reactions needed to form the mature enzyme could be completed in vitro in the absence of the plasma membrane. This suggests that in vivo the enzyme is not processed cotranslationally. This question of post- vs. co-translational processing was further pursued in an in vitro transcription coupled translation system.[81] The size of the immunoprecipitated in vitro synthesized enzyme in SDS-polyacrylamide gel is larger than the authentic mature enzyme. The nascent material dimerizes in presence of Zn^{2+} and becomes catalytically active. This material, in contrast to the authentic mature enzyme, is significantly hydrophobic (it strongly binds to a hydrophobic column). This large, catalytically active, hydrophobic enzyme could be converted to the authentic mature form by incubating with the outer membrane (Figure 14).[81] This observation, though, shows the presence of hydrophobic signal sequence, and generates a series of questions with regard to the in vivo situation: (1) does the strong hydrophobicity of the precursor molecule arise due to a different conformation compared to the mature form; (2) if the dimerization occurs inside the cell can the protein be translocated through the plasma membrane in a linear configuration; and (3) how does the hydrophobic molecule pass through the hydrophilic periplasm to be processed in the outer membrane?

Attempts have been made for the direct demonstration of the synthesis of alkaline phosphatase by membrane-bound polysome using a two-step indirect immunoprecipitation method (Figure 15).[82] Antibody has been produced against the uncoiled alkaline phophatase isolated from a constitutive strain of *E. coli*. Washed spheroplasts are lysed by freeze thawing and homogenization in the presence of RNase inhibitor. The super-

Table 4
IN VITRO SYNTHESIS OF ALKALINE PHOSPHATASE

	Alkaline phosphatase	
	0.D.410 nm	0.D.410 nm/hr × 1000
Experiment 1		
Complete	0.856	130
−S30	0.054	—
−DNA	0.096	—
$-ZnCl_2$[a]	0.431	56
+10 m*M*-EDTA[a]	0.104	1.3
Experiment 2		
Complete	1.272	370
+25 μg/mℓ chloramphenicol	0.206	15
10^{-5} M-ATA	0.188	9.3
−DNA	0.160	—
Experiment 3		
Complete	1.210	370
+rifampicin (40 μg/mℓ)	0.090	—
−DNA	0.108	—

Note: Each tube in Experiment 1 contained 2.25 μg of DNA while Experiments 2 and 3 contained 9 μg of DNA. The enzyme assay was terminated after 6 hr in Experiment 1 and after 3 hr in Experiments 2 and 3. Activity of the enzyme per unit time, shown in column 2 was calculated, subtracting the activity seen in control without DNA.

[a] Omitted or added during "the activation step".

natant, obtained from the lysate after centrifugation at 27,000 × g, is centrifuged in a discontinuous density gradient (1.0 to 2.5 *M* sucrose). The polysome fraction obtained from the interface of the two sucrose concentrations shows no reaction with nonspecific IgG (e.g., anti-myeloma protein antibody). However, antibodies against both uncoiled and coiled alkaline phosphatase reacted; hence, the protein chains on the polysome are likely to be partially folded.

Gene-fusion technique has been employed to show the presence of an amino-terminal signal sequence in the alkaline phosphatase molecule (Figure 16).[83,84] Fused *phoA-lacZ* genes code for hybrid proteins which, if produced in large quantity, become lethal to the cell. As the hybrid protein is under the regulation of *phoA* promotor, inorganic phosphate deprivation derepressed the protein. Furthermore, a plasmid containing the *phoR* mutation, because of constitutive character, causes overproduction of the hybrid protein and cell death. The major amount of the hybrid protein is present in the plasma membrane and a minor amount in the cytoplasm; however, no protein is found in the periplasm. Comparable to the MBP system, the alkaline phosphatase β-galactosidase hybrid protein is catalytically inactive if it is membrane bound. These cells containing membrane-bound hybrid protein are *lac*⁻. On the other hand, in *lac*⁺ cells the hybrid protein is cytoplasmic. Some strains, however, produce active β-galactosidase regulated by inorganic phosphate starvation; but these cells synthesize the enzyme by a translation restart mechanism. Thus, only the presence of β-galactosidase activity signifies *phoA-lacZ* gene fusion.

This study has been further extended to examine the effect of mutation in the *phoA* region of the *phoA-lacZ* fused gene.[84] The rationale is that the hybrid product, coded by the fused gene, is membrane bound, producing Lac⁺ phenotype. If there is a muta-

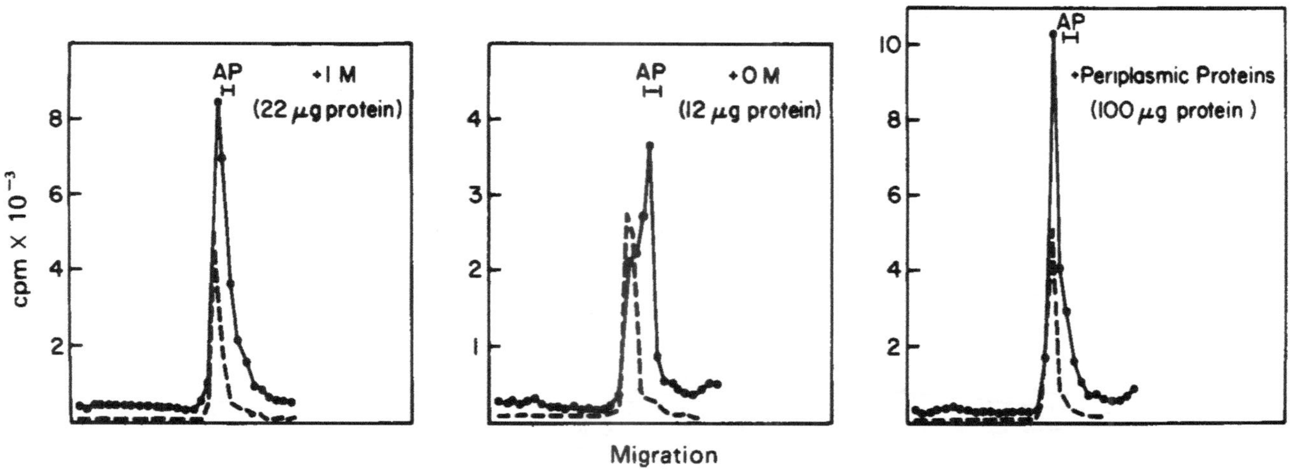

FIGURE 14. Processing of the precursor by purified membrane fractions. (A) Inner membrane, IM; (B) outer membrane, OM; (C) or the periplasmic protein fraction. At the end of the incubation, membrane fractions were dissolved by adding EDTA and Triton X-100®, 20 m*M* and 1% final concentration, respectively, and the samples were subjected to immunoprecipitation. •-•-•, ^{3}H-labeled sample incubated with membrane or periplasmic fraction; - - - -, precursor marker labeled with ^{14}C; AP, position of commercial alkaline phosphatase marker. (From Inouye, H. and Beckwith, J., *Proc. Natl. Acad. Sci. U.S.A.*, 74, 1440, 1977. With permission.)

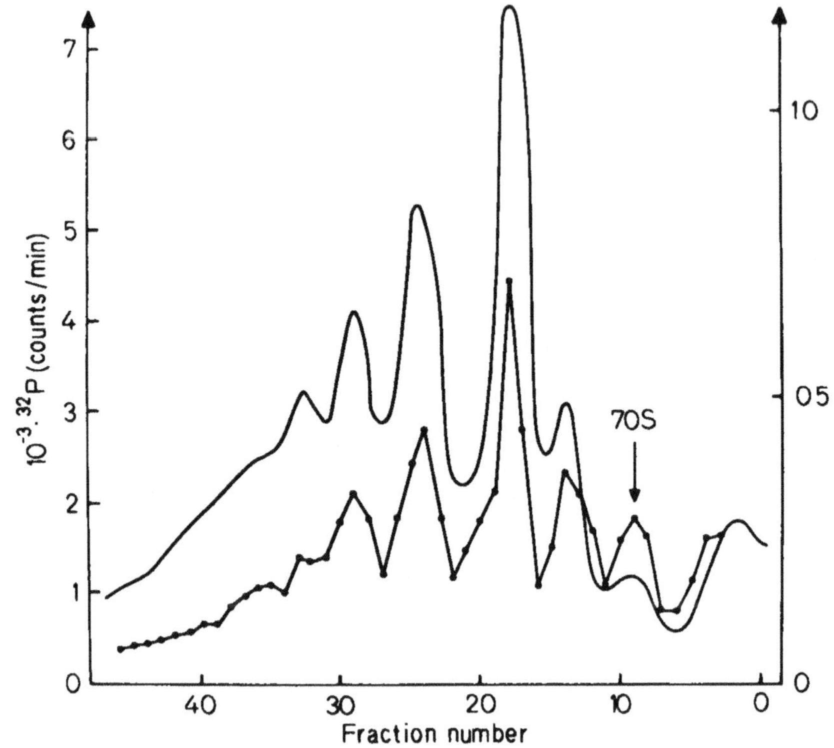

FIGURE 15. Sedimentation profile of membrane-associated polyribosomes. Two A_{260} units of the polysome preparation were layered over a 12.3-m*l* linear sucrose gradient (0.5 to 1.5 *M*) and centrifuged for 120 min at 286,000 × g in a Beckman SW41 rotor. The absorbance at 254 nm (—) was monitored during the collection of the gradient. Samples of the fractions were counted in a scintillation counter (•-•-•). (From Varenne, S. et al., *Eur. J. Biochem.*, 86, 603, 1978. With permission.)

tion in the *phoA* portion of the fused gene, the hybrid product will be internalized due to an alteration in the signal sequence. With the internalized hybrid protein being an active β-galactosidase, the strain will also be Lac^+. Such strains have been constructed and mutations in the *phoA* region have been reclaimed into wild-type *phoA* genes. In these reconstructed strains, alkaline phosphatase is internalized. Finally, genes from two of these strains are cloned and the amino-terminal amino acids are sequenced by DNA sequencing. A specific single amino acid change in the signal sequence appears to have blocked the function of the signal sequence (Figure 17).[84] The specific change in one strain is hydrophobic amino acid leucine into a charged amino acid, arginine, and the other change is hydrophobic leucine into a hydrophobic uncharged amino acid, glutamine.

4. Arabinose-Binding Protein (ABP)

The presence of a precursor form of ABP has been studied by both in vitro and in vivo synthesis systems.[85,86] In the in vitro system after cessation of ^{35}S-methionine incorporation, the incubation mixture has been incubated with puromycin and the released material has been immunoprecipitated and examined by gel electrophoresis and autoradiography. The bands compared to the mature protein show a molecular weight 3 to 4 kDa higher, indicating the synthesis of the precursor of ABP. On comparison, cytoplasmic protein EF-TU does not show the synthesis of any precursor form. In vivo short pulse and chase experiments showed that the large precursor form could be

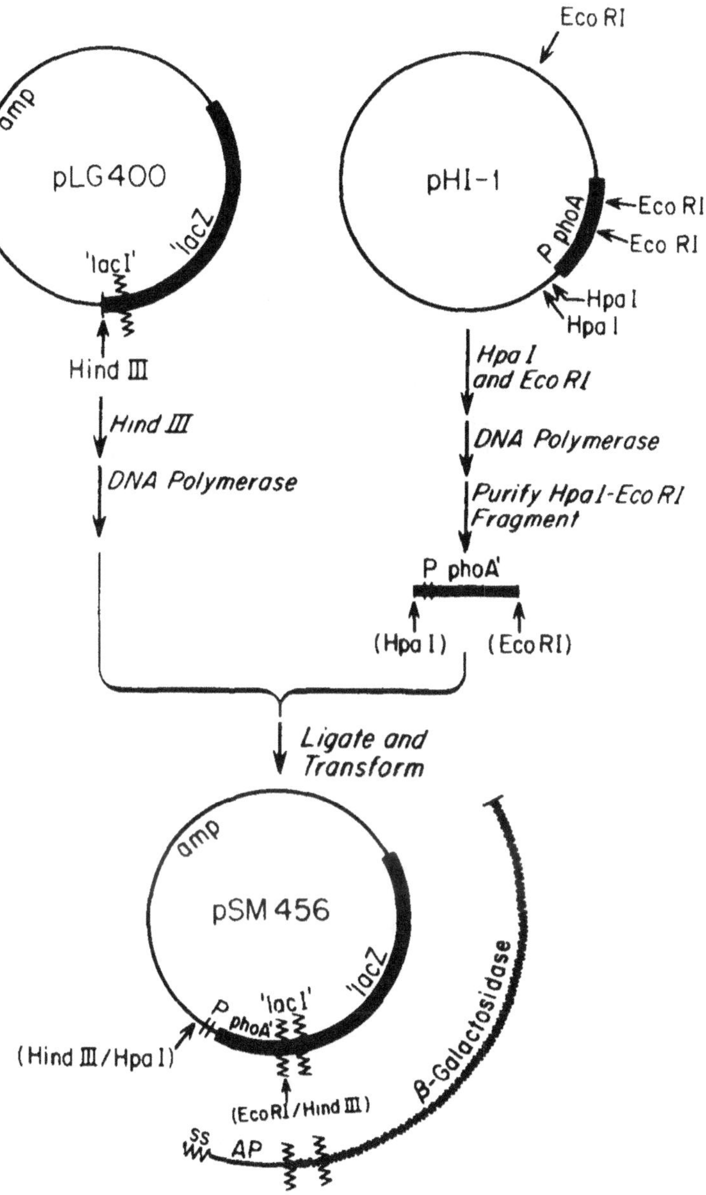

FIGURE 16. Strategy for construction of a *phoA-lacZ* fusion-bearing plasmid. The vector pLG400 (upper left) contains a small portion of *lacI* and nearly all of *lacZ* (heavy line). DNA encoding the amino-terminal 24 residues of β-galactosidase is missing and is replaced by DNA encoding 38 residues internal to the *lacI* gene product (amino acids 318 to 355). DNA at the boundary between *lac'I* and vector DNA contains a unique *Hind* III site (and a *Bam*HI site, not shown). Relevant restriction sites in pH1-1 (upper right) are shown in relation to the position of the *phoA* gene (heavy line). P designates the *phoA* promoter. The pLG400 vector was digested with *Hind* III, and the cohesive ends were filled in with DNA polymerase. pH1-1 was digested with *HpaI* and *Eco* RI, and cohesive ends were filled in with DNA polymerase. The 1.2-kb *HpaI-Eco*RI fragment, which contains a portion of the *phoA* gene, was ligated to the linearized, filled-in pLG-400 vector DNA. The ligation mixture was transformed into MC1000. Among the transformants are those which contain *phoA-lacZ* fusions as shown for pSM456 (bottom). Such fusions are expected to encode a hybrid protein containing a portion of alkaline phosphatase (AP) including its signal sequence (SS) joined to β-galactosidase. (*Hind* III/*HpaI*) and (*Eco*RI/*Hind* III) indicate the positions at which vector and fragment DNA were joined. After the filling in and ligation steps, the original restriction sites are destroyed and can no longer be digested with *Hpa* I, *Eco*RI, or *Hind* III. Only one representative construction is shown here. A similar strategy was employed (with the appropriate modifications) to construct other fusions. (From Michaelis, S., Guarente, L., and Beckwith, J., *J. Bacteriol.*, 154, 356, 1983. With permission.)

```
GTG.AAA.CAA.AGC.ACT.ATT.GCA.CTG.GCA.CTC.TTA.CCG.TTA.CTG.TTT.ACC.CCT.GTG.ACA.AAA.GCC.
Met-Lys-Gln-Ser-Thr-Ile-Ala-Leu-Ala-Leu-Leu-Pro-Leu-Leu-Phe-Thr-Pro-Val-Thr-Lys-Ala
                                   ↓               ↓
                                  CAG             CGG
                                  Gln             Arg
                                  (68)            (61)
```

FIGURE 17. DNA nucleotide sequence and amino acid sequence of *phoA* signal sequences. (From Michaelis, S. et al., *J. Bacteriol.*, 154, 366, 1983. With permission.)

chased into the protein band of the mature form. Limited proteolysis converted the precursor form into a mature form.[86] However, this processing is not comparable to in vivo signal peptidase processing because *S. aureus* v8 protease has been used instead of the processing enzyme of the same organism. The precursor form of ABP has been shown by mRNA-directed in vitro synthesis in the *E. coli* S-30 translation system.[75] The synthesized product has been purified by affinity column chromatography and this pure material is 3 kDa larger than ABP. The amino acid sequence of the cyanogen bromide peptide fragment consisting of 23 amino acids has been determined by the Edman degradation technique. This sequence has the characteristics of signal peptide.

B. Plasma Membrane Protein

Earlier in the chapter, secretion was defined as a series of molecular events starting from the synthesis to the externalization reaction including movement through the permeability barrier. Proteins destined for incorporation into the plasma membrane follow the same course of molecular events excluding the final externalization reaction. Bacteriophage coat proteins are a good model to study this process.

1. M13 Coat Protein

Studies on the filamentous coliphage M13 coat protein have provided significant information on the intramembranous event(s) of protein secretion.[87-89] This small protein consists of only 50 amino acids, residues of 20 to 40 of which are hydrophobic and presumably span the hydrophobic region of membrane.[90-92] This product of virus gene 8 in *E. coli* represents 7% of the total and 30% of the plasma membrane protein synthesis during infection.[93] Although the experimental data are elegant, before suggesting the mechanism to be universal, it should be noted that the study relates to a small protein molecule consisting of a disproportionately high amount of hydrophobic amino acids. In addition, the commitment of the synthesis by the cell is quantitatively unphysiological compared to other constitutive proteins.

M13 protein is synthesized in a precursor form (pre-coat). It has 23 amino acids at the amino terminal which have been defined as a leader sequence and are comparable to the amino-terminal signal sequence. Experimental data have been generated from short pulse and chase experiments to follow the conversion of pre-coat to mature coat protein. This experimental protocol has been successful because in amber 7 M13 phage-infected *E. coli* a substantial pool of pre-coat exists in the cytosol and conversion to coat is slow.[95] The most important outcome of this experimental approach is that CCCP (carbonyl cyanide-*m* chlorophenylhydrazone), an agent dissipating proton motive force (monitored by proton uptake), retarded the conversion of pre-coat to mature coat protein; however, in a spontaneous CCCP-resistant mutant, this block of pre-coat protein conversion is not seen. Arsenate (100 m*M*), in spite of inhibiting both net protein synthesis and oxidative phosphorylation, by diminishing the high-energy phosphate pool, does not affect this pre-coat to coat protein conversion process. A variety of uncouplers (e.g., dinitrophenol, KCN, NaN_3) inhibited both pre-coat to coat conversion and leader peptidase activity. CCCP, however, even at high concentrations does not inhibit leader peptidase activity. The binding of pre-coat to membrane is also not affected. Therefore, CCCP appears to act primarily by dissipating proton motive force.

The results on the effect of limited proteolysis showed the location of pre-coat with respect to the outside and inside surfaces of the plasma membrane during pre-coat to coat conversion.[88] Thus, an attempt has been made to decide whether or not the movement of the pre-coat protein through the plasma membrane occurs in the precursor form. The following data support the view that pre-coat form is translocated through the plasma membrane. After CCCP treatment, (1) if the outer surface of the plasma membrane is rendered accessible to proteolysis by conversion of whole cells into spheroplast, coat protein is digested but the cytoplasmic marker proteins are not; (2) in the whole cells the pre-coat persisted and could not be digested by proteolysis; and (3) following the rupture of the membrane permeability barrier by sonication both the pre-coat and the cytoplasmic marker proteins are proteolyzed. The effect of CCCP is reversible; in the CCCP-inhibited spheroplast samples proteolysis of the pre-coat, the conversion of pre-coat to coat and proteolysis of coat protein is revived after removal of CCCP from the spheroplasts.

This study has been extended to show that the only function of the electrochemical gradient of the plasma membrane is to translocate the pre-coat through the plasma membrane.[89] The experimental model is a mutant M13 am8H1R6 phage, which during infection shows abundant production of R6 pre-coat protein, but the conversion of pre-coat to coat is essentially slow. The amino acids of the signal sequence of R6 pre-coat is altered in the mutant in positions +2 (glu > leu), +6 (pro > ser), +12 (asn > asp); in addition, the R6 pre-coat is a poor substrate for the membrane leader peptidase.[96] The conversion of R6 pre-coat to coat is practically insensitive to CCCP action. Valinomycin treatment in the presence of K^+ (not Na^+) causes a collapse of the membrane potential, but does not affect R6 pre-coat to coat protein conversion. This effect in the mutant is not seen in the wild-type cells. Both R6 pre-coat and coat are sensitive to proteolysis in spheroplasts. Thus, R6 pre-coat appears to be a transmembrane protein unlike the wild type which is bound to the inside surface of the plasma membrane. The slow conversion of R6 pre-coat to coat protein and CCCP insensitivity largely depends upon the net charge of the amino acid in the position +2. The ineffectiveness of CCCP on the mutant does not appear to be due to slow leader peptidase processing on the outside surface of the membrane because the slow processing continues in the presence of CCCP, which completely blocks the conversion in wild type. Furthermore, although the genetic overproduction of leader peptidase stimulates the pre-coat to coat protein conversion, the resistance to CCCP action is retained. Taking these facts together, it appears that in the mutant R6 the pre-coat does not translocate through the membrane; hence, the pre-coat to coat conversion remains insensitive to CCCP action. Finally, the attempts to build arguments on qualitatively derived rate constants of different reactions of the translocation acitvity are at best vaguely suggestive (Figure 18).[89] It should be recognized that the study suffers from over-interpretation of the results obtained from nonquantitative electrophoresis data based solely on visual inspection of gel electrophoresis plates. Hence, a generally applicable conclusion on the mechanism of secretion cannot be drawn. However, the study is important for the clear demonstration of the role of membrane potential on secretory activity.

2. *fl Coat Protein*

In vitro transcription-coupled translocation system has been used to examine the integration of fl coat protein in *E. coli* plasma membrane.[97-98] The experimental design is based on a classical approach used for cotranslational protein transport study. The incubation mixture of in vitro synthesis has been treated with *E. coli* vesicles, which

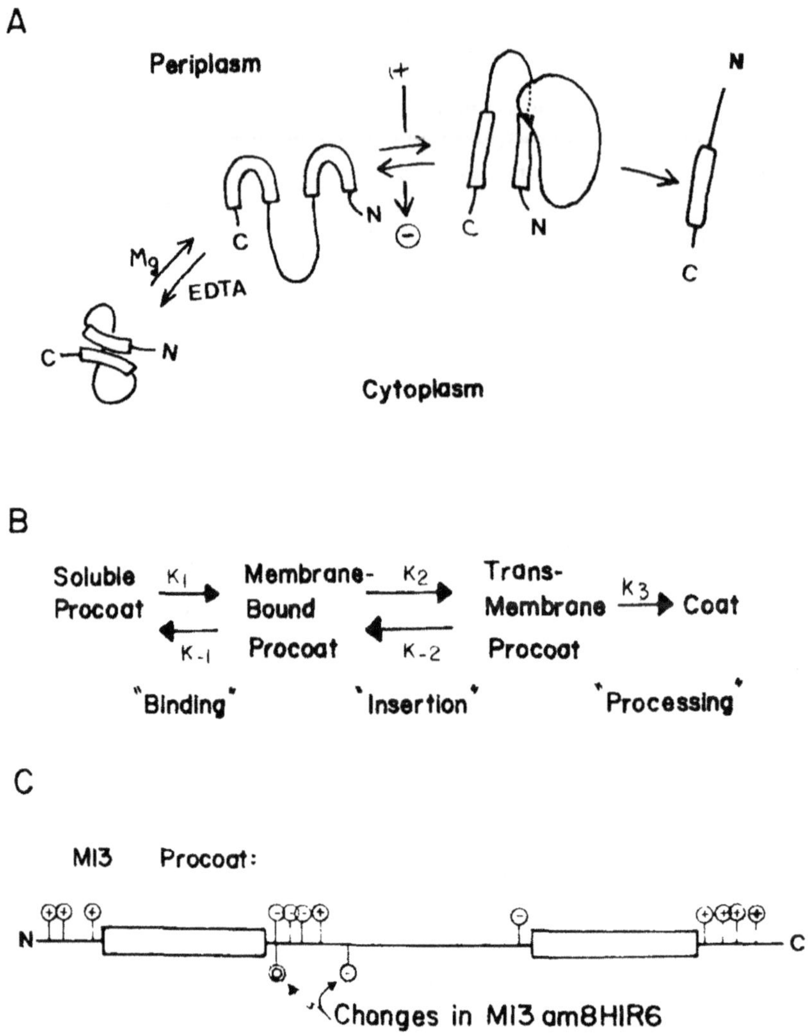

FIGURE 18. A working model of the biogenesis of M13 coat protein. The dark-shaded transmembrane protein in (A) is leader peptidase. The names of the reaction steps and assembly intermediates are shown in (B). The hydrophobic portions of pre-coat are shown as □ in (A) and (C). The charges of M13 precoat are shown above the polypeptide chain in (C), whereas the alterations which shift a charge from residue 2 to residue 12 of coat-R6 are shown below. (From Zimmermann, R., Watts, C., and Wickner, W. T., *J. Biol. Chem.*, 257, 6529, 1982. With permission.)

were claimed to have originated from the plasma (inner) membrane, and these have inverted orientation, i.e., the native cytoplasmic surface of the membrane is exposed to the in vitro incubation mixture. The cells are ruptured in a French press and the vesicles are sedimented by isopycnic centrifugation (20 hr) at 180,000 × g. Therefore, the vesicle sizes must have been very small. If the in vitro synthesis is continued in the presence of these vesicles, the nascent coat protein is fragmented into small and large polypeptide pieces; the addition of the detergent Nikkol® stimulated this fragmentation activity. This fragmentation does not occur if vesicles are added after the completion of in vitro synthesis. Plasma-membrane-associated endoproteolytic activity appears to have caused this fragmentation of nascent peptide. Edman degradation and amino acid sequence analysis of the small fragments show a sequence characteristic for the signal sequence. During the DNA-directed synthesis, if vesicles are present and the incubation mixture is digested with pronase, the molecules incorporated into the vesicles are protected from proteolysis. The material within the vesicle contains an intact amino-terminal sequence because the intravesicular material is precipitated with antibody which is specific for this amino terminal peptide.[99] In a mixture, the amino and carboxy terminal peptides could be separated from each other by high-voltage electrophoresis.[100] Finally, *E. coli* synthesizing in vivo coat protein, if pronase digested, produces immunoprecipitable fragments which contain intact amino-terminal regions. These authors observe that the coat protein binds very strongly with the inverted vesicle membrane. This bound form cannot be dissociated with salt or EDTA and can be fully digested with pronase. This result is interpreted as a negative proof that membrane integration of coat protein could occur post-translationally. However, no pulse chase experiments have been conducted to examine whether or not membrane integration of fl and M13 coat proteins have any comparable mechanisms of plasma membrane integration.

C. Outer-Membrane Protein

The structure and composition of the outer membrane of Gram-negative bacteria has been discussed in Volume 2 of this series.[19] Biosynthesis and assembly of outer-membrane protein has been discussed in detail in an accompanying chapter of this volume.[20] Therefore, the discussion on the secretion of outer-membrane protein will be limited to the discussion of the data which delineate the mechanism of secretion. Extensive study has been made on phage λ receptor, omp proteins, and outer-membrane lipoproteins. The mechanism of the secretion of outer-membrane protein is complex. From the site of biosynthesis in membrane-bound ribosome the protein traverses the plasma membrane, a highly cross-linked peptidoglycan layer, and a fluid-filled periplasmic space (containing multifarious molecules) and finally arrives at the asymmetric bilayered membrane structure consisting of lipopolysaccharide and phospholipid. Thus, one may visualize that extensive post-translational phenomena must exist for the insertion of protein molecules into the outer membrane.

1. λ Phage Receptor

Lambda phage receptor is a *lamB* gene product of *E. coli.* The receptor for λ phage is a protein located in the outer membrane. This *lamB* gene product is a multifunctional protein which has three major functions; it is (1) a receptor for the λ phage,[101] (2) a transport protein for maltose and maltodextrin,[102] and (3) an attractant protein for sugar chemotaxis.[103] The gene-fusion technique described earlier has been used to fuse *lacZ-lamB* genes (Figure 19).[104] The hybrid product of the fused gene consisted of the material coded by the transcribable portions of the genes; hence, in some strains the *lamB* gene product is high and in others low. The hybrid proteins are controlled by maltose regulatory activity; a substantial amount of hybrid protein showing β-galactosidase activity is located in the outer membrane. The *malE-lacZ* fused gene product, on the other hand, fails to be transported through the membrane and remains plasma membrane bound. The organism containing *lacZ-lamB* fused gene shows poor growth rate and a defect in the cell division if the cells are grown in the presence of maltose. This maltose sensitivity may arise from the outer-membrane location of the large hybrid protein without any β-galactosidase activity. However, it is argued that the maltose-sensitive strains could become maltose resistant if the hybrid protein is cytoplasmic and shows high β-galactosidase activity; this would change the phenotype of the cells from lac^- to lac^+. Such strains are selected on the basis of maltose resistance and lac^+ character. Maltose-sensitive cells contain a high amount of membrane-bound inactive β-galactosidase which has a molecular weight 2 to 3 kDa higher than the pro-

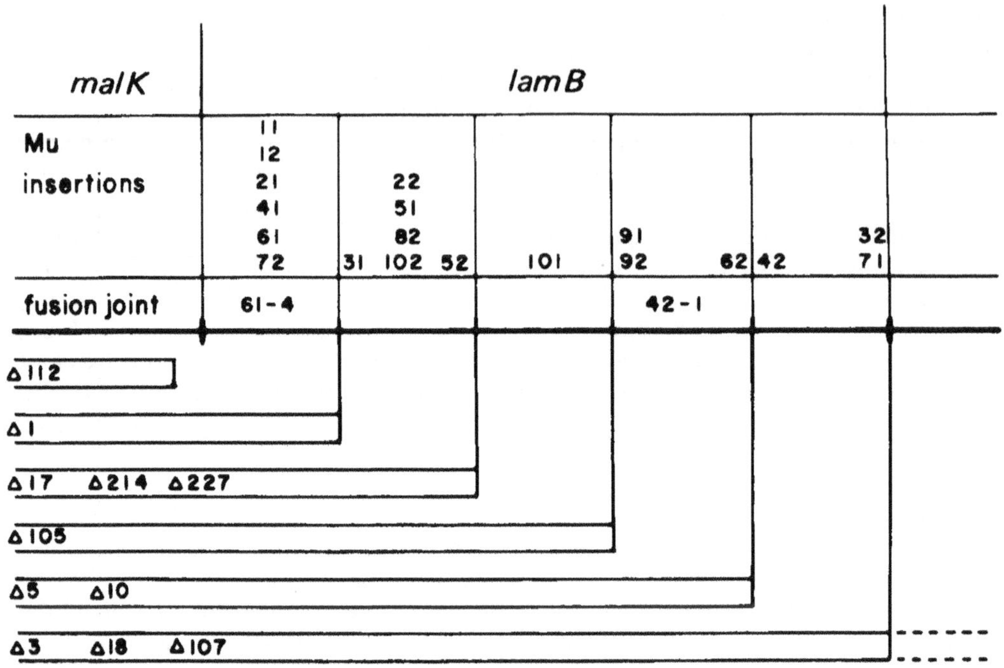

FIGURE 19. Location of the fusion joints in gene *lamB*. The Mu insertions were all isolated during the couse of this work. When they start with the same digit (i.e., 11 and 12, for instance) they were isolated from the same Mu-treated culture and could be identical. Within each interval defined by deletion endings, the mutations have been ordered according to their frequency of recombination with the deletion that ends closest to the left. Only tenfold differences in recombination have been considered. Deletion Δ112 is shown to end within *malK* because it recombines with some known *malK* point mutations. Strain pop 3105 contains fusion number 61-4; strain pop 3186 contains fusion number 42-1. (From Silhavy, T. J. et al., *Proc. Nat. Acad. Sci. U.S.A.*, 74, 5411, 1977. With permission.)

teins of resistant strains containing cytoplasmic hybrid protein. Therefore, a significant expression of *lamB* gene alters the location of cytoplasmic β-galactosidase. This point has been further proven by causing a mutation in the *lamB* region of the fused gene; such mutants can be selected on the basis of maltose sensitivity and lac⁻ character (Figure 20).[105] Two classes of mutations are selected: (1) deletion mutation in the lamB region of the fused gene coding for low-molecular-weight hybrid protein molecules, and (2) point mutation in the lamB region of the hybrid gene coding for unaltered-molecular-weight hybrid protein molecules. If the same *lamB* mutation is introduced in the wild type, the product is a high-molecular-weight cytoplasmic protein.

This overall approach of bringing one gene expression under the influence of another has been further extended.[106] In this study the *lamB* gene has been cloned in a phage such that *lamB* expression can be under lac promoter control, and the hybrid protein can be induced by IPTG (isopropyl-β-thiogalactoside). Thus, *lamB* gene can be expressed independently of the *malB* locus. The IPTG-induced λ receptor is located in the outer membrane. The phage DNA-programmed in vitro product (i.e., pre λ) has a higher molecular weight than the mature protein. The 25 amino acids of the amino terminal are characteristic of the signal sequence. During in vitro translation, this pre-λ protein can be processed with inverted *E. coli* outer-membrane vesicles to form mature λ receptor protein. However, if DNA of the mutants which shows defective *in vivo* transport of λ receptor is used to program in vitro pre λ production, the in vitro product cannot be processed. Therefore, any specific alteration of signal sequence interferes with the processing activity.

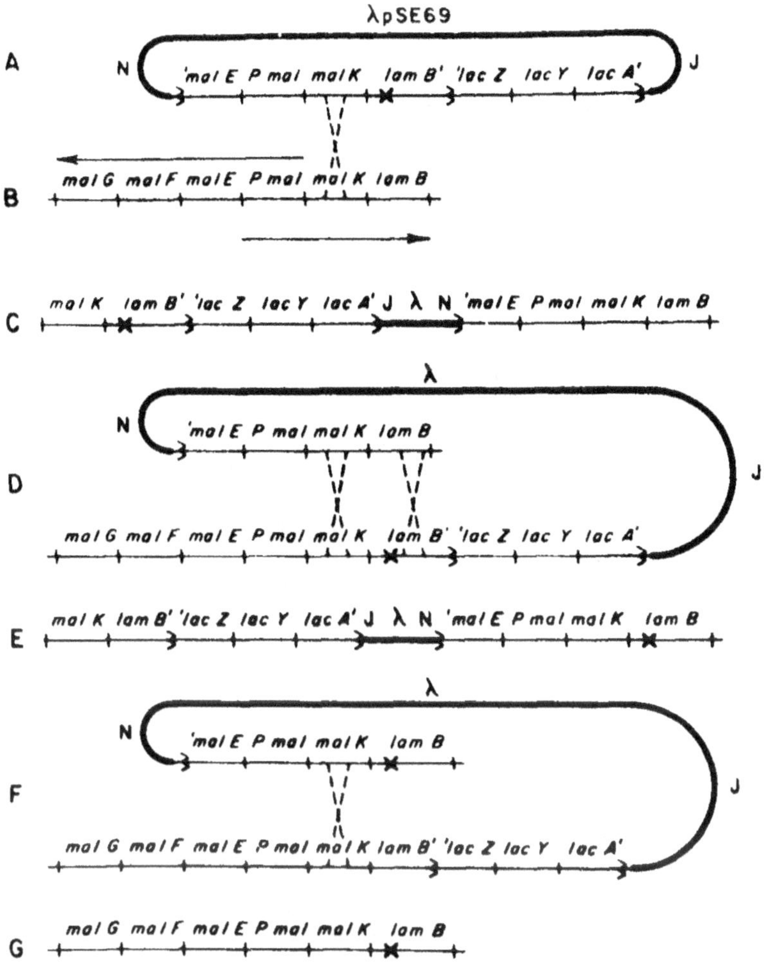

FIGURE 20. Recombination of the *lamB* mutations *S68* and *S69* onto wild-type gene *lamB*. (A) The λp transducing phage that carries the *lamB-lacZ* fusion and the *lamB* mutation S69 (designated here with a ×.) (B) A portion of the chromosome of strain MC4100 depicting the two divergent operons that comprise the *malB* locus. The promoter here is designated "Pmal". Because strain MC4100 is Δ(*lac*) and because the λp transducing phage is Δ(*att*), lysogenization occurs primarily via recombination between homologous *malB* DNA. Insertion of the phage as shown yields a lysogen (C) that has the phenotype $\lambda^{s}Mal^{+}Lac^{+}$. A reciprocal crossing-over as shown in (D) will recombine the *S69* mutation onto the wild type *lamB-lacZ* fusion 42-1. Accordingly, the phenotype of the resulting lysogen (E) is now $\lambda^{R}Mal^{+}Lac^{+}$. Loss of the prophage by a recombinational event as depicted in (F) removes the *lamB-lacZ* fusion and thus eliminates the Mal^{+} phenotype of the lysogen. This can be used as a selection procedure for curing the lysogen of the prophage. Such a curing event yields strain SE2069, which is $Mal^{+\lambda R}$ because the mutation *S69* is now present in the *lamB* gene (G). It is possible to demonstrate that the steps depicted here are easily reversible, i.e., the mutation *S69* can be moved from the hybrid *lamB-lacZ* gene to the wild-type gene *lamB* and vice versa by the recombinational event depicted in (D). This provides evidence that the mutation present in strain SE2069 (G) is identical to the mutation present in λpSE69 (A). Procedures for the construction of strain SE2068 were identical. (From Emr, S. D., Schwartz, M., and Silhavy, T. J., *Proc. Natl. Acad. Sci. U.S.A.*, 75, 5802 1978. With permission.)

The DNA sequence strategy has been used to determine the amino acid sequence of the amino-terminal region of pre λ receptor protein. The amino acid sequence of the beginning of the *lamB* gene derived from the DNA nucleotide sequence matches with the sequence of the primary translation product.[107] The amino terminal sequence contains 25 amino acids; 19 of these amino acids are hydrophobic. However, preceding

```
                      SD          1   2   3   4   5
                              f-MET MET ILE THR LEU
AAAAAAGAAAAGCAATGACTCAGGAGATAGA ATG ATG ATT ACT CTG
TTTTTTCTTTTCGTTACTGAGTCCTCTATCT TAC TAC TAA TGA GAC
                HinfI
-30       -20        -10       -1+1           10

6   7   8   9   10  11  12  13  14  15  16  17  18
ARG LYS LEU PRO LEU ALA VAL ALA VAL ALA ALA GLY VAL
CGC AAA CTT CCT CTG GCG GTT GCC GTC GCA GCG GGC GTA
GCG TTT GAA GGA GAC CGC CAA CGG CAG CGT CGC CCG CAT
HhaI
    20           30            40           50

19  20  21  22  23  24  25
MET SER ALA GLN ALA MET ALA val asp phe his gly tyr
ATG TCT GCT CAG GCA ATG GCT GTT GAT TTC CAC GGC TAT
TAC AGA CGA GTC CGT TAC CGA CAA CTA AAG GTG CCG ATA

     60            70            80            90

ala arg ser gly
GCA CGT TCC GGT
CGT GCA AGG CCA
            hpaII
       100
```

FIGURE 21. DNA and amino acid sequence in the region of the amino terminus of *lamB*. The notation SD (Shine-Dalgarno sequence) indicates a possible ribosome-binding site for initiation of translation of *lamB* mRNA. Amino acids found in the extra NH_2-terminal peptide are given in capital letters; those occurring in the mature protein are given in lower case letters. (From Hedgpeth, J. et al., *Proc. Natl. Acad. Sci. U.S.A.*, 77, 2621, 1980. With permission.)

this hydrophobic stretch there is a basic peptide terminating with lysine (Figure 21),[107] suggesting the possibility of ionic interaction with the plasma membrane surface.

2. *Lipoprotein*

Lipoproteins consititute a major fraction of the outer membrane of Gram-negative bacteria.[19,20,108] These lipoproteins are associated with the peptidoglycan layer and are also incorporated into the outer membrane (Figure 22).[109] One of these lipoproteins (Braun lipoprotein, BLP) exists in two different forms. One is covalently associated with the peptidoglycan layer through carboxy-terminal lysine; the other lipoprotein, peptidoglycan-associated lipotrotein (PAL), is firmly associated through noncovalent bonding. These proteins are generally synthesized in the precursor form in the membrane-bound ribosome and require extensive post-translational processing for binding with the peptidoglycan layer and assembly into the outer membrane.[21,110] This processing can be inhibited by phenethyl alcohol (PEA) and the antibotic globomycin. In the presence of PEA post-translational processing is inhibited causing accumulation of the precursors in the periplasm. The inhibitory effect of PEA depends upon its concentration, the species from which the outer membrane is taken, and the nature of the secreted protein. The inhibitory effect of PEA may vary when differing concentrations are used. However, neither the outer membrane from different species nor the different secretory proteins are equally sensitive to PEA inhibition. This is evident from com-

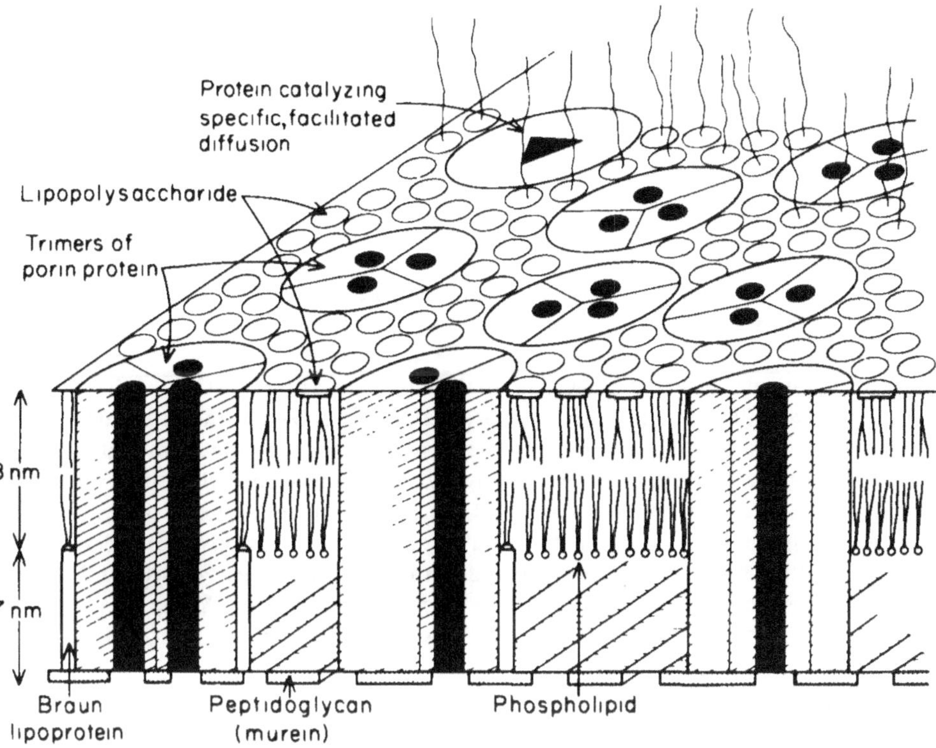

FIGURE 22. Hypothetical model of the cell wall of *E. coli* and *S. typhimurium*. (With permission from Nikaido, H. and Nakae, T., in *Advances in Microb. Physiol.*, 20, 163, 1979. Copyright: Academic Press [London] Ltd.)

paring the inhibiting effect of PEA at identical concentrations on the processing of different proteins: matrix protein > TolG proteins > lipoproteins. The mechanism of action of PEA is not known, but the effect of PEA on increasing membrane fluidity suggests that it may dissociate the coupled biosynthesis and processing activity by disorganizing membrane-associated processing enzyme.

Globomycin treatment causes accumulation of pre-BLP and pre-PAL; both precursors are peptidoglycan associated. Pre-PAL is located solely in the plasma membrane, but Pre-BLP is located in both the plasma membrane and the outer membrane. Both pre-BLP and BLP remain associated with the peptidoglycan through the lysine residue of the carboxy terminus. Thus, the BLP molecule can be transported through the plasma membrane while the hydrophobic signal sequence remains within the membrane. The following information supports the idea that PAL behaves in a manner similar to BLP. The PAL molecule is bound to the peptidoglycan layer through a 1-kDa segment which does not include the amino terminal. Both the pre- and mature PAL are bound to the peptidoglycan; therefore, comparable to the BLP molecule, the PAL molecule is exported through the membrane while the hydrophobic signal sequence remains within the plasma membrane. Finally, pulse chase experiments show that these precursors can be converted into mature forms. Hence, it is possible that the signal sequence, in addiltion to being an anchoring structure within the membrane, may constitute a part of the intramembraneous translocation machinery.

The outer-membrane lipoprotein (a structural protein) is a small (7.2 kDa) molecule. The mRNA of this protein is very stable and contains more nucleotides than required for the translation of the 53 amino acids of the protein molecule. The protein translated from this mRNA contains a 20-amino acid signal sequence at the amino terminal.

At the beginning of this signal sequence there are a few charged amino acids; these presumably bind to the charged group of phospholipids anchoring the translating ribosome to the inside plasma membrane surface; the hydrophobic segment of the signal sequence penetrates the hydrophobic interior of the plasma membrane in the form of a loop.[111] Amino acids at specific locations of the lipoprotein signal sequence play a crucial role because a single substitution of any amino acids of the signal sequence in a mutant blocks the secretion.[112]

3. Omp Protein

Many outer-membrane proteins (known as omp) have been detected in Gram-negative bacteria. These proteins are synthesized in the membrane-bound ribosome and the possible precursor molecules contain signal sequences comparable to that of outer-membrane lipoproteins.[113-116] However, the mechanism of integration of these proteins into the outer membrane is obscure. Theoretically, it can be assumed that after cotranslational translocation through the membrane and signal sequence removal, the molecules undergo post-translational processing for integration into the outer membrane. Therefore, a transient intermediate form of the mature protein is expected to be present in the plasma membrane. Such intermediate forms of protein have not been detected in vivo:[117-119] however, perturbation of membrane fluidity (e.g., growth at 25 instead of 37° C) causes accumulation of transient intermediates of ompA and ompF proteins in the cytoplasm, suggesting that processing and translocation into the outer membrane are not fully coupled to synthesis.[113]

V. MOLECULAR MECHANISM IN GRAM-POSITIVE CELLS

There is a basic difference between the Gram-positive and Gram-negative cell envelope organization which has been discussed earlier. The most significant aspect in relation to secretion is the absence of an outer permeability barrier in Gram-positive cells. This outer permeability barrier is inert biosynthetically; however, it contains porin molecules that control the molecular traffic through it. The movement of protein molecules into the true extracellular environment of Gram-negative cells is restricted and these molecules are likely to be retained within a unique periplasmic environment. In Gram-positive cells, although the rigid layer of the wall (i.e., peptidoglycan and teichoic acid) is more elaborate than the Gram-negative wall, the movements of the molecules which have passed through the plasma membrane may not be severely restricted. However, a variety of ionic groups of the wall matrix may restrict the movement of secreted protein molecules through ionic interaction. This difference in the envelope organization leads to a dual definition of secretion. One is mechanistic, explaining the translocation of molecules from the site of synthesis through the plasma membrane permeability barrier. This mechanism is universal, but does not necessarily lead to release of protein into the extracellular environment. The other is an operational definition, explaining translocation to the true extracellular environment. According to this operational definition, Gram-positive cells are more adapted for secretion and hence have been perceived to have greater industrial impact.[120,121] A large number of exoenzymes are produced by Gram-positive bacteria, expecially organisms of the genus *bacillus* (Table 5).[121] Physiological regulation of synthesis and secretion of these enzymes has been discussed in the earlier part of this chapter. Data on the molecular mechanism of secretion are available for only a few of these and will be discussed below.

A. Penicillinase

The β-lactamase enzymes can be classified according to their size, substrate specific-

Table 5
EXTRACELLULAR ENZYMES OF THE GENUS *BACILLUS*

Enzyme	Species	Comments
Carbohydrases		
Agarase	*Bacillus* sp.	Hydrolyzes the β-1,4 linkage of agarose
α-Amylase	*B. amyloliquefaciens* *B. caldolyticus* *B. coagulans* *B. licheniformis* *B. macerans* *B. stearothermophilus* *B. subtilis* *B. subtilis* var. *amylosacchariticus* Alkalophilic *Bacillus* spp.	Endohydrolysis of the α-1,4-glucosidic linkages in polysaccharides; the different species produce enzymes with different properties
β-Amylase	*B. cereus* *B. megaterium* *B. polymyxa* Alkalophilic *Bacillus* spp.	Exohydrolysis of the α-1,4-glucosidic linkages in polysaccharides yielding β-maltose
Arabinase	*B. subtilis*	Probably an endo α-1,5-arabinase
Cellulase	*B. brevis* *B. firmus* *B. polymyxa* *B. pumilus* *B. subtilis*	Hydrolysis of carboxymethyl cellulose to cellobiose
Chitinase	*B. circulans*	Four enzymes induced by growth on crabshell chitin
Chitosanase	*Bacillus* sp. R-4	Hydrolyzes cell walls of *Rhizopus* spp.
Cyclodextrin glucanotransferase	*B. macerans* *B. megaterium* Alkalophilic *Bacillus* spp.	Synthesizes cyclo-(Schardinger) dextrins from starch
Dextranase	*B. megaterium* *B. subtilis*	Cell-bound enzymes catalyzing the exohydrolysis of dextran to glucose
Galactanase	*B. amyloliquefaciens* *B. subtilis* var. *amylosacchariticus*	Hydrolysis of the β-1,4-galactosidic linkages in soybean arabinogalactan
β-1,3-Glucanase	*B. circulans* *B. polymyxa* *B. subtilis* Alkalophilic *Bacillus* sp.	Endohydrolysis of the β-1,3-glycosidic linkages in laminarin and related glucans
β-1,6-Glucanase	*B. circulans*	Hydrolysis of pustulan and related glucans
Isoamylase	*B. amyloliquefaciens* *B. polymyxa*	Hydrolysis of the α-1,6-glycosidic branch linkages in glycogen, amylopectin, etc.
Levansucrase	*B. subtilis*	
Lichenase	*B. pumilus*	Hydrolysis of the β-1,4-glucosidic linkages of lichenenan
Maltase	*B. subtilis*	Hydrolysis of the α-1,4 linkage of maltose and maltotriose
Mannanase	*B. amyloliquefaciens*	Endohydrolysis of the β-1,4-mannosidic linkages of mannan
Pectate lyase	*B. circulans* *B. polymyxa* *B. pumilus* *B. sphaericus* *B. stearothermophilus* *B. subtilis* Alkalophilic *Bacillus* spp.	Endocleavage of polygalacturonic acid by an eliminative reaction
Phosphomannase	*B. circulans*	Cleaves phosphomannan from yeast cell walls

Table 5 (continued)
EXTRACELLULAR ENZYMES OF THE GENUS *BACILLUS*

Enzyme	Species	Comments
Pullulanase	Alkalophilic *Bacillus* spp.	Endohydrolysis of the α-1,6 linkage of pullulan
Xylanase	*B. amyloliquefaciens* *B. firmus* *B. polymyxa* *B. subtilus* *B. subtilis* var. *amylosaccharititcus*	Hydrolysis of xylans; the specificity of the enzymes has not been studied in detail
Proteases		
Alkalophilic protease	Alkalophilic *Bacillus* spp.	Serine enzymes from alkalophilic species with very high pH optima
Aminopeptidase	*B. licheniformis* *B. subtilis*	
Esterase	*B. subtilis*	Serine enzyme with high esterolytic and low proteolytic activity
Halophilic protease	*Bacillus* sp.	Produced optimally in media containing 1.0 *M* NaCl
Metal protease	*B. amyloliquefaciens* *B. cereus* *B. licheniformis* *B. megaterium* *B. polymyxa* *B. subtilis* *B. subtilis* var. *amylosaccharititcus* *B. thermoproteolyticus* *B. thuringiensis*	Enzymes require Ca^{2+} for stability and Zn^{2+} for activity; pH optimum at or near neutrality
Serine protease	*B. amyloliquefaciens* *B. licheniformis* *B. pumilus* *B. subtilis* *B. subtilis* var. *amylosaccharititcus*	The subtilisins; alkaline pH optima, serine residue at or near the active site
Serine-metal protease	*B. licheniformis* *B. pumilus*	Hybrid enzymes with characteristics of both the serine and metal proteases
Penicillinases		
β-Lactamase	*B. anthracis* *B. cereus* *B. licheniformis* *B. megaterium* *B. subtilis*	Hydrolysis of the amide bond in the β-lactam ring of penicillins and cephalosporins
Penicillin amidase	*B. megaterium*	Hydrolysis of the peptide linkage of penicillin
Nucleases and phosphatases		
Alkaline phosphatase	*B. amyloliquefaciens* *B. cereus* *B. subtilis* Alkalophilic *Bacillus* sp.	Often cell bound, the enzyme is extracellular in these species
Deoxyribonuclease ribonuclease	*B. amyloliquefaciens* *B. cereus* *B. pumilus* *B. subtilis*	A large number of DNases, RNases, and phosphodiesterases with individual properties have been purified
3-Nucleotidase	*B. subtilis*	Active on both ribo- and deoxyribonucleotides
5-Nucleotidase	*B. cereus* *B. megaterium*	Cell-bound enzyme in these species
	B. subtilis	Also possesses nucleoside diphosphate-sugar hydrolase activity

Table 5 (continued)
EXTRACELLULAR ENZYMES OF THE GENUS *BACILLUS*

Enzyme	Species	Comments
	Bacillus spp.	Also possesses RNA phosphodiesterase activ ity
Bacteriolytic enzymes		
Endo-*N*-acetylglucosa-minidase	*B. licheniformis*	
	B. subtilis	
Exo-*N*-acetylglucosamin-idase	*B. subtilis*	
Endo-*N*-acetylmurami-dase	*B. subtilis*	True lysozyme
Exo-*N*-acetylmuramidase	*B. subtilis*	
N-acetyl-muramyl-L-alanine amidase	*B. licheniformis*	A cell-bound enzyme; the major autolysin
	B. subtilis	
Lipase	*B. licheniformis*	Hydrolysis of triacylglycerol to diacylgly-cerol and a fatty acid anion
Phospholipase C	*B. anthracis*	Responsible for the "egg-yoke" reaction
	B. cereus	
	B. thuringiensis	
Thiaminase	*B. thiaminolyticus*	

From Priest, F. G., *Bacteriol. Rev.*, 41, 711, 1977. With permission.

ity, and sequence homology.[121-124] There are three classes: (1) a 23-kDa metallothio protein active on cephalosporins and distributed only in *B. cereus*, (2) a 30-kDa protein active mainly on penicillin and exhibiting sequence homology; this enzyme may have both membrane-bound and secreted molecular forms, and (3) a 39-kDa protein active on cephalosporin, which is a chromosomal protein of Gram-negative cells (e.g., *AmpC* gene product).[124,125] The enzymes of classes (1) and (3) are entirely periplasmic in Gram-negative cells.

The penicillinase of class (2) has several molecular forms in *B. licheniformis*.[44] The location and interrelationship of these multiple molecular forms have been presented schematically in Figure 23. Biosynthesis of penicillinase occurs presumably by membrane-bound ribosome. Synthesis of a precursor form of penicillinase has been achieved using three different procedures: (1) coupled transcription translation system;[126,127] about 3.5 kb *B. licheniformis* DNA fragment containing the structural gene of penicillinase, is transcribed and translated in *E. coli* cell-free system; the labeled immunoprecipitated material (34 kDa) is 3 kDa heavier than the membrane-bound enzyme; (2) translation of *B. licheniformis* mRNA in *E. coli* cell-free system;[128] the immunoprecipitated product of translation is a 35-kDa protein, 2 kDa heavier than the membrane-bound enzyme. Both the membrane-bound enzyme and the product of in vitro synthesis after trypsinization produced exo-small enzymes (Figure 23) of identical molecular weight and amino terminal sequence; (3) in vivo translation by phenethyl alcohol-treated cells,[110,128,129] the immunoprecipitated synthesized material has identical molecular weight to the in-vitro-synthesized material.

The amino terminal sequence of the in vitro product and the sequence derived from the DNA sequence analysis suggest that this precursor molecule has an amino-terminal extension of 42 amino acids relative to the exo-small form of the enzyme (Figure 23). This pre-penicillinase has been transcribed and translated in *E. coli* cells lysogenic for λ vector containing penicillinase gene.[130] The following data support the idea that penicillinase is a lipoprotein. The complete sequence of pre-penicillinase contains leu-Ala-gly-cys in positions 26 to 29 (Figure 24).[130] This hydrophobic section is present in com-

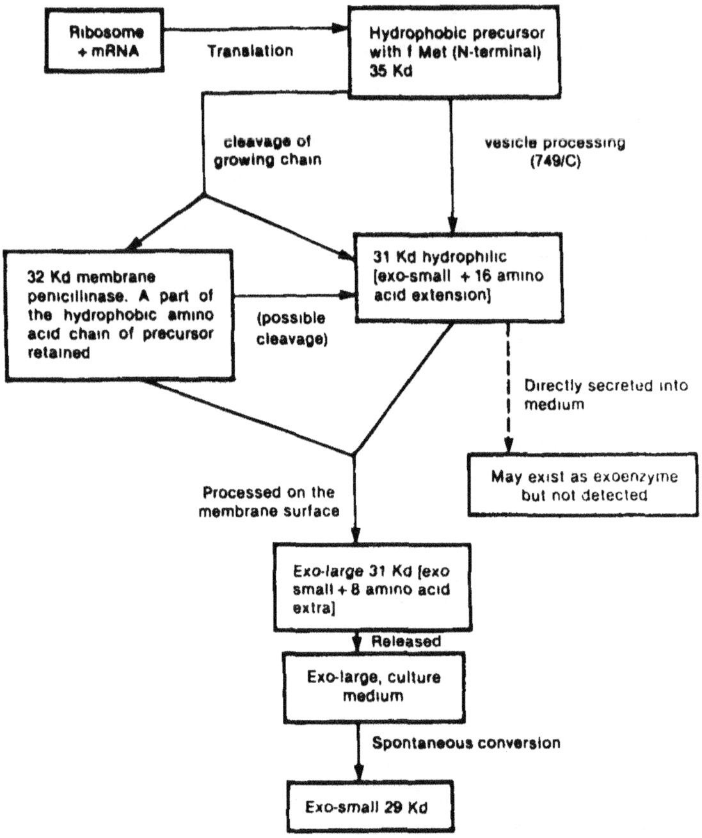

Fig. 23 Scheme of Multiple Forms of Penicillinase

FIGURE 23. Multiple forms of penicillinase of *Bacillus licheniformis* based on the work of J. O. Lampen and his associates (see text.)

parable positions 18 to 21 of a pre-lipoprotein sequence of *E. coli*. The pre-penicillinase incorporates in vivo [^{3}H] glycerol, [^{35}S] cysteine, and [^{3}H] palmitate, but does not incorporate [^{32}P] orthophosphate.[129] In vivo and in vitro labeling of pre-penicillinase correspond and suggest it is a lipoprotein. Incorporation of palmitate into pre-penicillinase is comparable to incorporation into membranes of mammalian cells.[129,131,132] However, caution shouid be applied in interpreting this labeling experiment because firmly bound membrane phospholipid may contaminate the protein fraction.[44] However, the information that membrane penicillinase is lipoprotein is firm because [^{3}H]palmitate could be released from the labeled enzyme as a free fatty acid after 0.1 *M* NaOH treatment. After in vivo synthesis, Braun's lipoprotein, lipopolysaccharide, and pre-penicillinase co-migrated, but only the lipopolysaccharide incorporated ^{32}P label.[130] A glycerol-cysteine moiety could be isolated from the pre-penicillinase double labeled with [^{3}H]glycerol and [^{35}S]cysteine, suggesting the existence of covalently linked glyceride and fatty acid in pre-penicillinase. The amino-terminal tetrapeptide is modified by covalent bonding with thioether comparable to the modification of Braun's lipoprotein in *E. coli*.[116]

Globomycin, which inhibits the processing of *E. coli* pro-lipoprotein, acts by binding noncovalently to the glyceride moiety of the enzyme.[126,133] It appears that a highly specific transferase is involved in the reaction with the tetrapeptide segment.[130] Thus, globomycin affects this transferase activity and interferes with the processing of both

(A) Met–Lys–Ala–Thr–Lys–Leu–Val–Leu–Gly–Ala–Val–Ile–Leu–Gly–Ser– (10)

Thr–Leu–[Leu–Ala–Gly–Cys]–Ser–Ser–Asn–Ala–Lys–Ile–Asp–Gln–Leu– (20, 30)

(B) Met–Lys–Leu–Trp–Phe–Ser–Thr–Leu–Lys–Leu–Lys–Lys–Ala–Ala–Ala– (10)

Val–Leu–Leu–Phe–Ser–Cys–Val–Ala–[Leu–Ala–Gly–Cys]–Ala–Asn–Asn– (20, 30)

Gln–Thr–Asn–Ala–Ser–Gln–Pro–Ala–Glu–Lys–Asn–Glu–Lys–Thr– (40)

↑ exoS (after Ala) ; ↑ exoF (after Glu)

FIGURE 24. Amino acid sequence of the amino-terminal portion of the prolipoprotein in *E. coli* (A) and pre-penicillinase in *B. licheniformis* (B). (From Lai, J. S., *Proc. Natl. Acad. Sci. U.S.A.*, 78, 3506, 1981. With permission.)

pre-lipoprotein and pre-penicillinase.[129,130] This lipoprotein modification is physiologically significant because the altered enzyme may remain anchored to the outside surface of the membrane allowing interaction with the substrates present in the extracellular microenvironment of the bacteria.

The question whether the cysteine-glyceride modification is co- or post-translational has been addressed.[133] The protoplasts of *B. licheniformis* have been labeled with membrane-impermeable ^{125}I-iodosulfanillic acid during penicillinase synthesis. If the labeling is done after puromycin treatment the growing chain could be labeled. This experiment shows that the cysteine-glyceride modification occurs cotranslationally during the pre-penicillinase polypeptide chain growth.[133] However, it should be noted at this stage that the in vitro translated product is presumably processed by a signal peptidase of *B. licheniformis* membrane fraction.[128] Therefore, a dual and possibly sequential cotranslational modification of pre-penicillinase occurs, first, cleavage of signal sequence and second, glyceride-cysteine modification. Physiologically these modifications may direct the enzyme towards two destinations: *one* the extracellular medium (i.e., exoenzyme), and the other, the outside surface of the plasma membrane (i.e., membrane-bound enzyme form).

B. α-Amylase

α-Amylase hydrolyzes the α-β glycosidic bonds in starch and related polysaccharides. This enzyme exists as both membrane-bound and extracellular forms.[134-136] The enzyme has one structural gene *(amyE)* and several regulatory (*amyR*) genes.[137-139] The expression of the *amyE* gene is regulated by the *amyR* gene such that both the quantity and rate of enzyme synthesis are controlled. The distance between *amyR* and *amyE* is smaller than the distances between average genes and they may even be linked. In fact it has been suggested that *amyR* is the promotor for the *amyE.*[140] Mutations affecting synthesis and secretion of amylase lead to pleotropic effects, e.g., synthesis and secretion of another exoenzyme, envelope structure, flagella, and sporulation, etc. A gene for this pleotropic effect called *pap* has been mapped.[141-143]

There are two different forms of amylase. Both are 58 kDa; one is hydrophilic extracellular, and the other a hydrophobic membrane-bound form.[144] *B. subtilis* α-amylase gene has been cloned into pUB110 plasmid and expressed in another strain of *B. subtilis.*[145,146] In vitro transcription-coupled translation of cloned DNA produced a primary secreted enzyme 2 kDa larger than the in vivo form. Partial nucleotide sequencing of the 5′ region of the amylase gene and amino acid sequencing of the in vitro product shows a 31-amino acid signal sequence. This is unusually long but is comparable to pre-penicillinase which is 34 amino acids long. The sequence excluding the signal sequence matches the complete amino acid sequence of the purified exo α-amylase. The signal sequence has a hydrophobic stretch consisting of 15 amino acids preceded by a cluster of charged amino acids. Upstream from the amino terminal is a "Shine Dalgarno" sequence which may be important for mRNA ribosome binding, and further upstream there is a "Pribnow box" sequence which is the binding site for RNA polymerase. Finally, the expression of the cloned gene is 500 times more efficient than the wild type, suggesting the presence of both the full promoter and a multicopy plasmid.

C. Alkaline Phosphatase

Alkaline phosphatase has been shown to be present both as membrane-bound and extracellular forms. Different species as well as different strains of bacilli have both secretor and nonsecretor phenotypes: *B. licheniformis* 749/C secretes 30 to 50% of the synthesized enzyme, whereas strain MC14 does not secrete; *B. subtilis* Marburg strain secretes but SB15 strain does not. Furthermore, the synthesis and secretion are affected by a variety of physiological effector molecules, e.g., Mg^{2+}, Ca^{2+}, adenosine phosphates, etc.[46-49] Alkaline phosphatase is a Zn^{2+}-containing dimer; the subunit molecular weights of both membrane-bound and secreted enzymes are 60 kDa.[147-148] The extracellular enzyme in *B. licheniformis* 749/C has some properties different from the membrane-bound enzyme: (1) poor and unstable catalytic activity in contrast to stable membrane-bound enzyme, (2) high affinity for inorganic phosphate, and (3) the antibody prepared against purified membrane-bound enzyme, when used to compare the reactivity of the extracellular and membrane-bound alkaline phosphatase antigens, evoked nonidentity between these enzyme forms in Ouchterlony plates, suggesting significant differences of the antigenic determinants of the membrane-bound and secreted enzymes. It is possible that the membrane-bound and extracellular enzymes differ in conformation as a result of translocation of enzyme molecules through the membrane. *B. licheniformis* 749/C alkaline phosphatase mRNA has been translated in the S-30 fraction of the same organism.[149] The translated product has been immunoprecipitated; the molecular weight of this nascent product is 5 to 6 kDa higher than the mature form, and corresponds to a 30- to 35-amino acid extension. Hence, the precursor molecule appears to have a long signal sequence; this observation is consistent with results from other Gram-positive organisms.[44,126,128,145] Alkaline phosphatase can be extracted from the membrane with $MgCl_2$. This magnesium extractibility is lost from a secretion blocked mutant of *B. licheniformis* 749/C, which implies a membrane-receptor function in alkaline phosphatase secretion.[150]

VI. ROLE OF PROTEIN SEQUENCE

The above discussion shows that the amino acid sequence of protein molecules plays a pivotal role in the protein secretion from bacterial cells. There are two components of this sequence: signal sequences of the presecretory molecules at the amino-terminal end and the sequences of mature protein molecules. The evidence is overwhelming that a bacterial protein without a signal sequence cannot be secreted. Comparative data on the structure of signal sequence are given in Table 6;[58] sequence data of *B. licheniformis* penicillinase have been given in Figure 24 for further comparison with the Gram-positive signal sequence. Examination of the sequences does not show the conservation of a specific sequence. However, there are few unique properties consitently present in all the signal sequences:

1. There is a charged segment present at the start of all the sequences. The length of this segment varies from two charged amino acids to eight. The role of this charged segment appears to be to establish binding of the presecretory molecule

Table 6
PROKARYOTIC SIGNAL SEQUENCES

Protein	Charges segment	Hydrophobic segment*
Inner-membrane proteins phage fd, major coat protein	MET LYS LYS SER LEU VAL LEU LYS	ALA SER VAL ALA VAL ALA THR LEU VAL PRO MET LEU SER PHE ALA↓ALA GLU GLY
Phage fd, minor coat protein	MET LYS LYS	LEU LEU PHE ALA ILE PRO LEU VAL VAL PRO PHE TYR SER HIS SER↓ALA GLU THR
Periplasmic proteins alkaline phosphatase	MET LYS	GLN SER THR ILE ALA LEU ALA LEU LEU PRO LEU LEU PHE THR PRO VAL THR LYS ALA↓ARG THK PRO
Maltose-binding protein	MET LYS ILE LYS THR GLY ALA ARG	ILE LEU ALA LEU SER ALA LEU THR THR MET MET PHE SER ALA SER ALA LEU ALA↓LYS ILE GLU
Leucine-specific-binding protein	MET LYS ALA ASN ALA LYS	THR ILE ILE ALA GLY MET ILE ALA LEU ALA ILE SER HIS THR ALA MET ALA↓ASP ASP ILE
Leucine-isoleucine-valine-binding protein	MET ASN ILE LYS GLY LYS	ALA LEU LEU ALA GLY CYS ILE ALA LEU ALA PHE SER ASN MET ALA LEU ALA↓
Histidine-binding protein of *Salmonella typhimurium*	MET LYS LYS	LEU ALA LEU SER LEU SER LEU VAL LEU ALA PHE SER SER ALA THR ALA ALA PHE ALA↓ALA ILE PRO
Lysine-arginine-ornithine-binding protein of *S. typhimurium*	MET LYS LYS	THR VAL LEU ALA LEU SER LEU LEU ILE GLY LEU GLY ALA THR ALA ALA SER TYR ALA↓ALA LEU PRO
ampCβ-lactamase	MET PHE LYS	THR THR LEU CYS ALA LEU LEU ILE THR ALA SER CYS SER THR PHE ALA↓ALA PRO GLN
TEM β-lactamase of PBR322	MET SER ILE GLN HIS PHE ARG	VAL ALA LEU ILE PRO PHE PHE ALA ALA PHE CYS LEU PRO VAL PHE ALA↓HIS PRO GLU
Arabinose-binding protein	MET LYS X THR LYS	LEU VAL LEU GLY ALA VAL ILE LEU THR ALA GLY LEU SER X GLY[ALA] X [ALA]↓GLU ASN LEU
Outer-membrane proteins lipoprotein	MET LYS ALA THR LYS	LEU VAL LEU GLY ALA VAL ILE LEU GLY SER THR LEU LEU ALA GLY↓CYS SER SER
LamB	MET MET ILE THR LEU ARG LYS	LEU PRO LEU ALA VAL ALA VAL ALA ALA GLY VAL MET SER ALA GLN ALA MET ALA↓VAL ASP PHE
OmpA	MET LYS LYS	THR ALA ILE ALA ILE ALA VAL ALA LEU ALA GLY PHE ALA THR VAL ALA GLN ALA↓ALA PRO LYS
OmpF	MET MET LYS ARG	ASN ILE LEU ALA VAL ILE VAL PRO ALA LEU LEU VAL ALA GLY THR ALA ASN ALA↓ALA GLU ILE

Table 6 (continued)
PROKARYOTIC SIGNAL SEQUENCES

Protein	Charges segment	Hydrophobic segment[a]
Toxins		
Heat-labile toxin, B-sub unit	MET ASN LYS VAL LYS	CYS TYR VAL LEU PHE THR ALA LEU LEU SER SER LEU TYR ALA HIS GLY↓ALA PRO GLN
Heat-labile toxin, A-sub unit	MET LYS	ASN ILE THR PHE ILE PHE PHE ILE LEU LEU ALA SER PRO LEU TYR ALA↓ASN GLY ASP
Heat-stable toxin, ST1	MET LYS LYS	LEU MET LEU ALA ILE PHE ILE SER VAL LEU SER PHE PRO SER PHE SER GLN SER?THR GLU SER

[a] The site of proteolytic cleavage is denoted by a downward arrow (↓).

with the membrane surface. Inouye suggested in the loop model of lipoprotein secretion in *E. coli* the importance of this charged region of the signal sequence.[3]

2. Following the charged segment there is a hydrophobic segment which maintains a constant length consisting of 18 to 20 amino acids. A minimum hydrophobic axis length has been suggested which is needed for the insertion of precursor molecules into the membrane.[151,152] The signal sequence of penicillinase indicates that the length of the Gram-positive presecretory protein molecules may have a longer stretch of signal sequence than the Gram-negative cells.[128]
3. It is possible to suggest from theoretical predictions that signal sequence has a helix conformation with minimum random coiling.[153] This may be energetically favorable for membrane insertion of signal sequence.[154] A mutation introducing helix breaker amino acid into signal sequence may stop secretion.[55]
4. The only general characteristic for the cleavage site by signal peptidase is that the side chains of amino acids are less bulky.

No clear data relating mature protein sequence and secretion are available. It has been noted from the studies on the secretion of hybrid proteins that in spite of the presence of the signal sequence at the amino terminal end of the β-galactosidase moiety it cannot be translocated through the membrane. It is possible that in the signal sequence an order of the amino acids is necessary for developing appropriate characteristics for secretion. Some examples of these characteristics are sizes of the amino acid side chains, a drop of polarity from the amino to the carboxy end of the protein, and significant interruption of the polar amino acids in the sequence with a stretch of hydrophobic amino acids. The sequence of the entire protein molecule has been suggested to be important for post-translational secretion, proposed in the membrane trigger hypothesis.[4] Therefore, the carboxy terminal must play a significant role. However, the studies on a large number of chain termination mutants show that the mutant proteins are bound to the membrane.[155] Hence, the carboxy terminal end may not play any specific role in secretion.

VII. SECRETORY APPARATUS

It is presumptuous to define the function of a structure which may not exist. However, an attempt will be made to accumulate some hypothetical functions which may act in sequential as well as synergistic fashion to complete the process of translocation of protein molecules from the site of synthesis through the membrane to the extracellular environment. Hence, secretory apparatus may be defined as an organization of membrane-associated and intramembranous molecules which participate in the translocation process of secretory protein molecules. Development of this concept certainly depends upon diverse experimental data, some of which, are discussed below.

A. Physiological Regulation

As discussed earlier, secretion of proteins in Gram-positive organisms may change in differing growth conditions, e.g., changes in the divalent cations, varying H^+ ion concentrations, specific nutritional condition, etc. Hence, translocation of nascent protein molecules is not an all-or-none process; a variety of physiological effector molecules may modulate this secretory activity. It is possible to argue that the site action of effectors must be exerted on a specific group of molecules in the membrane.

Penicillinase can be incorporated in only a limited number of sites in the membrane.[45] If the molecules occupying the membrane are removed new molecules may be incorporated in those sites until saturation is reached. These penicillinase-specific binding sites may be membrane regions functionally organized for secretion.

B. Morphological Demonstration

If there are organized regions in the membrane specialized for secretion these will bind with molecules destined for translocation. However, binding is not expected to cause instantaneous translocation because a number of molecular events are involved; therefore, molecules to be secreted should have a membrane residence time. Theoretically, these molecules could be visualized by labeling with specific probes. These labeled regions of the membrane should correspond to plasma membrane secretion sites. Demonstration of these membrane sites must fulfill some basic conditions: (1) labeling of the secretory molecules should maintain in vivo conditions when translocation of molecules is continuing, (2) the probe must be specific for the secretory molecules, and (3) resolution of the method should be such that the distribution of the molecules can be differentiated with respect to the outside and inside surfaces of the plasma membrane. This approach may provide new information which cannot be obtained by cell fractionation and determination of the concentration of the secretory molecules in one subcellular fraction or the other. High-resolution analysis of the different stages of translocation through cell envelope could not be done following cell fractionation, because the data may not represent in vivo conditions due to redistribution of molecules during cell rupture and fractionation of subcellular components.[58]

Electron microscopy has been used as a tool to visualize the secretory molecules in bacterial cells and examine their distribution. The use of ultra-structural cytochemistry has provided information on the distribution of alkaline phosphatase in both Gram-positive and Gram-negative cells.[156-159] Immobilization of enzyme molecules by some cross-linking agent (e.g., glutaraldehyde) is effective in preventing relocation of the molecules during cell processing. Microprecipitation of the enzyme reaction product and its visualization at the site of enzyme reaction is used as the marker of the enzyme molecule. Use of this method demonstrated that in the cells of *B. subtilis* strain SB15, which is phenotypically alkaline-phosphatase-secretion negative, the enzyme is bound to the inside surface of the membrane (Figure 25). There are only a few such binding sites per individual cell, suggesting that there are only a limited number of specific binding sites for alkaline phosphatase in the inside surface of the plasma membrane. Similar observations have been made in *B. licheniformis* nonsecretor and secretor strains.[150,157,158] Furthermore in *B. licheniformis* 749/C cytochemical data correspond to the results of labeling with a membrane-impermeable reagent. Frequent disorganization of the plasma membrane at the enzyme binding site shows considerable release of enzyme during protoplast formation; hence, relocation of enzyme molecules is possible during cell fractionation and the periplasmic enzyme may arise artifactually.

The cytochemical method is based upon indirect labeling of enzyme location by the accumulation of catalytically generated reaction product. Hence, any treatment which stimulates or inhibits catalytic activity of the enzymes or cause diffusion of the reaction product from the site of reaction may significantly affect the labeling. Furthermore, catalytically inert molecules cannot be labeled by this procedure. Immunoelectron microscopy technique has been used to examine characteristics of the binding sites for two secretory proteins, alkaline phosphatase and penicillinase. Ferritin-labeled anti-penicillinase antibody has been used to localize penicillinase in *B. licheniformis* 749/C cells.[43] The ferritin-IgG conjugate is distributed in both cytoplasm and cell envelope; these are characteristically present in clusters indicating the presence of aggregates of enzyme molecules. The clusters spanning the entire cell envelope may indicate the presence of specific secretion sites. These secretion sites can be demonstrated distinctly in freeze-dried replicas of labeled *B. licheniformis* cells (Figure 26A and B).[43] There is asymmetry in the distribution of label on the two surfaces of the membrane. Binding of the labeled ferritin with the inside surface is greater than that with the outside surface. Furthermore, there is significant binding of the label with the cytoplasm. This

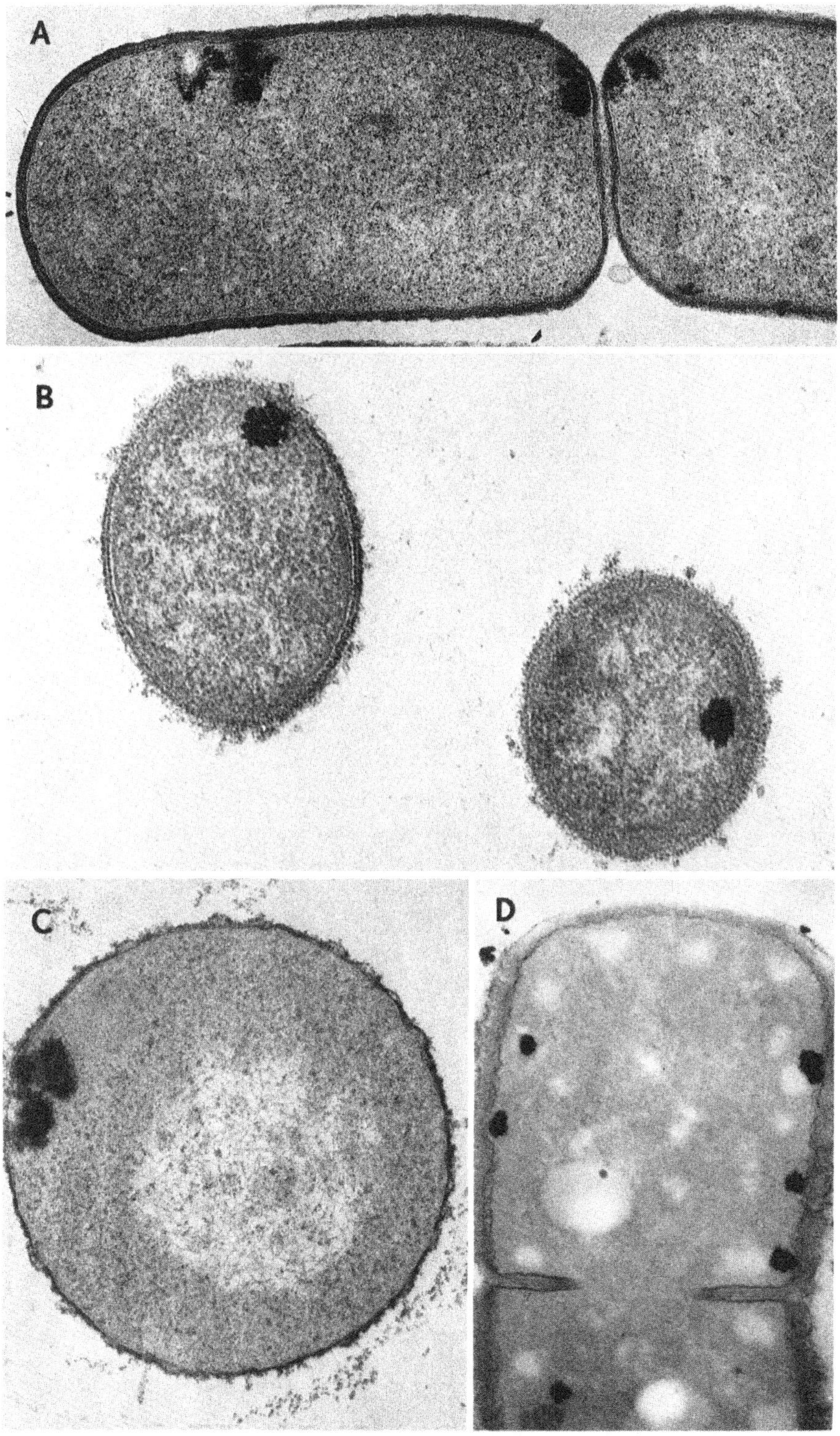

FIGURE 25. Cytochemical localization of alkaline phosphatase bound to the inside plasma membrane surface. (A) *Bacillus subtilis* SB15; (B) *B. licheniformis 749/C*; (C) protoplast of *B. licheniformis*; (D) the secretion negative mutant *B. licheniformis* SN105. (Based on the work of B. K. Ghosh and his associates.)

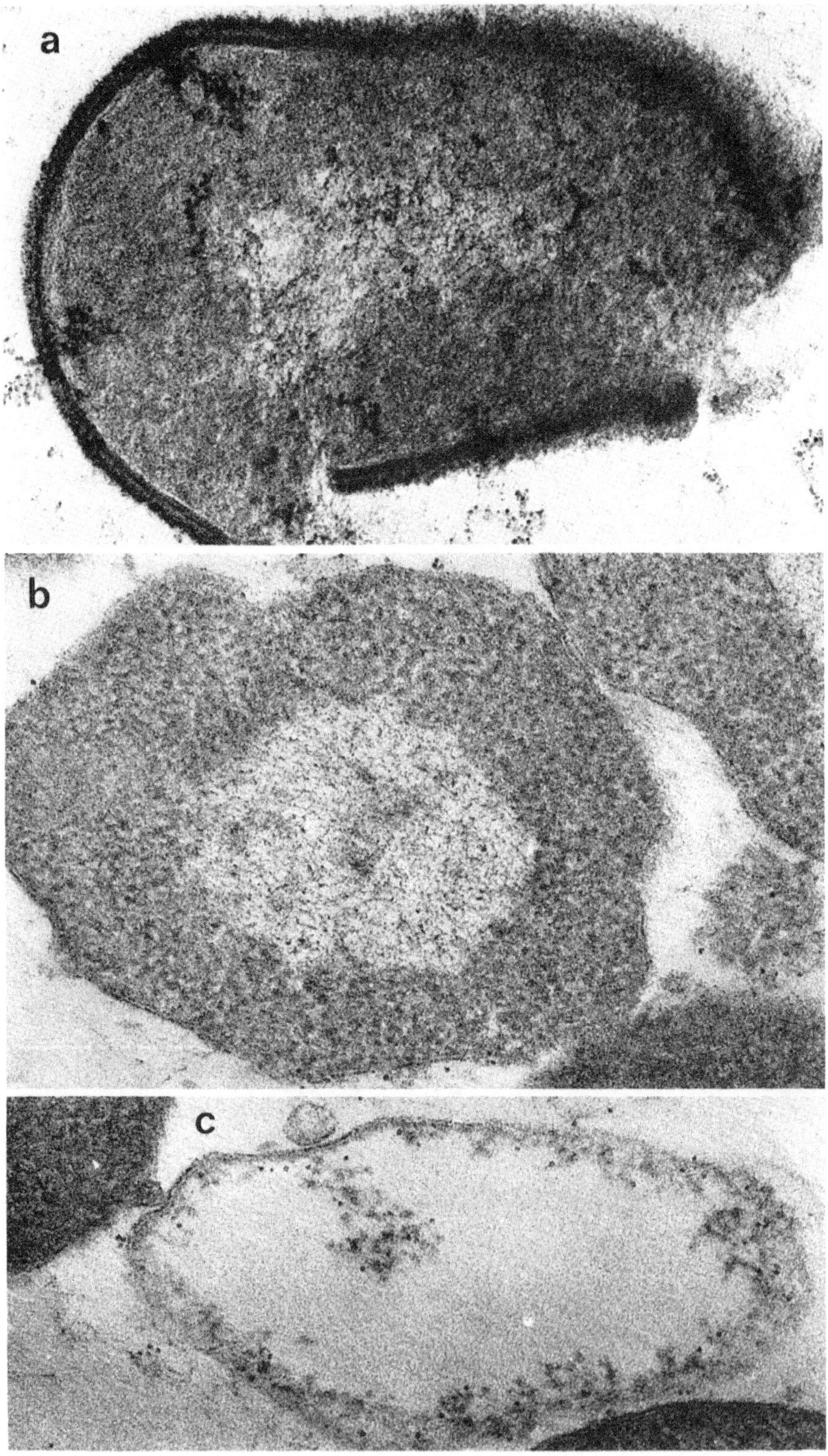

FIGURE 26A

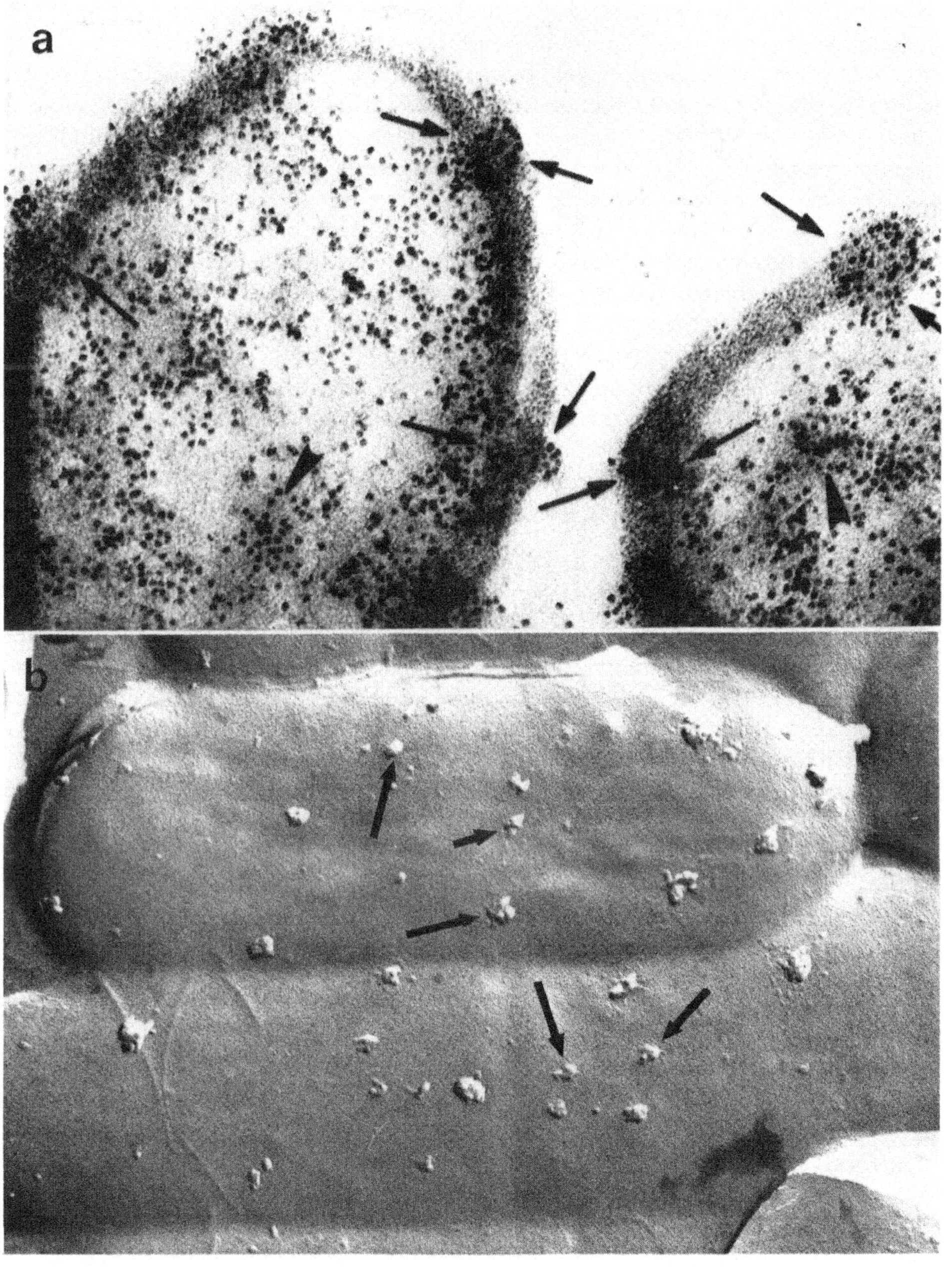

B

FIGURE 26. (A) Thin sections of ferritin-anti-penicillinase-IgG labeling of mildly fixed (a) homogenized whole cell, (b) protoplast, and (c) washed membrane show discrete aggregates of label in the cell envelope. (Based on the work of B. K. Ghosh and his associates.) (B) Ferritin-anti-penicillinase-IgG labeling of the (a) frozen thin sections show the extension of the clusters from the inside surface of the envelope to the outside surface (arrows) and also show the extensive labeling in the cytoplasm (arrowheads); (b) freeze-dried replica of whole cells of *B. licheniformis* 749/C after ferritin-IgG treatment show the label on the outside surface of the wall suggesting discrete secretion sites in the wall (arrows). (Based on the work of B. K. Ghosh and his associates.)

presence of penicillinase in the cytosol has been confirmed by extracting cytosolic enzyme and analyzing the extract by immunoelectrophoresis.[43] However, the cytosolic material has very poor catalytic activity and high sensitivity to protease degradation.

The presence of inactive penicillinase in the cytoplasm can be explained if one assumes that binding sites for penicillinase in the plasma membrane are limited. Thence, as a result of the synthesis continuing at a faster rate than the translocation through the membrane, the nascent penicillinase accumulates in the cytoplasm. It is possible that nascent molecules fold differently in the cytoplasm when these fail to bind to the plasma membrane binding sites and thus become highly sensitive to intracellular protease, assuring their eventual degradation.

The location of alkaline phosphatase has been demonstrated in *B. licheniformis 749/C* using anti-alkaline phosphatase antibody colloidal gold complex as label.[160] The label is primarily distributed as clusters distinctly bound to the inside surface of the membrane. However, there are a few wall-bound labels and a significant number of labels are present in the cytoplasm. It is notable that the label can be seen penetrating the membrane bilayer which suggests the existence of transmembrane secretion sites. Finally, with double labeling (small and large gold colloids) penicillinase and alkaline phosphatase labels bind to different locations in a discrete fashion with the *B. licheniformis* plasma membrane.[161] These data (Figure 27A and B) provide strong evidence that different secretory proteins may have specific binding sites in the plasma membrane.

C. Genetic Regulation

The concept of secretory apparatus has two broad aspects: (1) the unique characteristics of the secretory molecule itself, and (2) the unique organization of specific membrane molecules. Interaction of both of these components leads to the actual translocation through the membrane permeability barrier. One may use a lock and key analogy to describe the secretory apparatus. Extensive genetic analysis and biochemical characterization have demonstrated that secretory molecules have presecretory forms which allow these molecules to recognize and to bind to the secretion specific region of the membrane. Both the presecretory molecule and the membrane secretory sites are genetically regulated. It appears that the same membrane secretory sites may be shared by a range of secretory molecules (defined as subsets).[162] Furthermore, a genetic lesion affecting this membrane site would not only affect the secretion of molecules which belong to the same subset but, in addition, other cellular functions or properties closely tied to the secretion of these molecules are affected. This pleotropism of the secretory apparatus mutation has been reported.[121,162]

In *Bacillus,* mutations affecting exoenzyme synthesis and secretion are pleotropic either for more than one exoenzyme or for a variety of seemingly unrelated phenomena, e.g., motility, control of sporulation, etc.[121] Synthesis of α-amylase and protease is controlled by regulatory genes.[121,163] The regulatory gene *amyR* is closely linked to *amyE* the structural gene.[164] Chemically mutagenized *B. subtilis* yields strains which concomitantly increase the production of secreted α-amylase and protease. This mutation is a single-gene mutation and is not linked to *aro116* which is linked to the α-amylase gene.[165] If this mutation is transferred into another strain, the production of both α-amylase and protease is increased. Furthermore, if this mutation is present with mutations in the regulatory genes (*amyR* and *phoH*), there is a synergistic enhancement of both the enzymes. This mutation, defined as the "pap" gene mutation, may cause an alteration of membrane secretion sites shared by amylase and protease.[165] The *pap* gene regulates the synthesis of these two proteins through a coordinate mechanism. In addition the pap mutation causes a change in the membrane protein composition. Further studies on this pleotropic mutation show that three different classes of mutants, *sacU*^{h}, *amyB,* and *pap,* have been reported in the literature.[166] However, these three mutations appear to be identical because of common properties: (1) overproduction of α-amylase, lavansucrase, and protease; (2) they are devoid of flagella, and (3) they are poorly or not at all transformable. It has been suggested that in these mutants a regulatory gene has been affected. The product of this gene controls both the synthesis of cell envelope components and secretion. In fact, malfunction of a membrane secretory apparatus may explain all the above three changes. A comparable observation has been made in *B. licheniformis* 749/C.[150] It appears that either the *phoR* mutation or a mutation in the phoR region of the chromosome causes an alkaline phosphatase secretion block. The properties of the mutant enzyme (i.e., molecular weight, immunological identity, and substrate profile) are similar to those of the wild-type enzyme; furthermore, the enzyme is bound to the inside surface of the plasma membrane. The enzyme cannot be extracted with high magnesium salt and two new protein bands seem to have appeared in the *M* tris-hydroxymethyl-amino-methane extract of the membrane. Comparable mutation and selection technology have been applied to the *B. subtilis* SB15 strain which is phenotypically secretion negative. A strain has been selected which, unlike the parent organism, secretes 33% of the synthesized enzyme.[167] The enzyme is extractable with high magnesium salt and in addition, a prominent additional protein band is seen after gel electrophoresis of the magnesium salt extract of the mutant relative to the wild type. The data suggest that genetic loci may exist which regulate the production of proteins consitituting the components of a putative membrane secretory apparatus.

The question of secretory apparatus has been explored in *E. coli* through the application of gene fusion technology. The selection is based on the fact that if there is *malE-lacZ* gene fusion, substantial *lacZ* product is retained and the hybrid protein has β-galactosidase catalytic activity when located in the cytoplasm; but membrane-bound hybrid protein has poor catalytic activity.[162,168] Hence any *lac*⁻ strain should have predominantly membrane-bound hybrid protein. Although the signal sequence of the *malE* product causes the binding of nascent hybrid product to the plasma membrane, translocation does not occur presumably due to the nature of the β-galactosidase chain of the hybrid protein. This mutant strain of *E. coli* when grown in maltose accumulates precursors of several periplasmic or outer-membrane proteins. This effect could be detected early (15 min), at the onset of maltose growth, and the ratio of precursor to mature enzyme increases progressively for both the hybrid proteins as well as for the precursors of other proteins which could not be transported. However, a slow post-translational secretion occurs. Based on the assumption that all the membrane secretion sites are saturated by the hybrid protein at a very early stage of secretion it is calculated that there are 2×10^4 membrane secretion sites per cell.[168] Following the same technology, another mutant has been selected from *lac*⁻ *malE-lacZ* strains. This is a conditional lethal temperature-sensitive mutant defined as a *secA* mutant which accumulates precursors of λ receptor and many other periplasmic or outer-membrane proteins at the nonpermissive temperature. This temperature-sensitive gene maps at 2.5 min on the *E. coli* chromosomal map near a cluster of cell division and cell septation genes. It is assumed that the temperature-sensitive production of a plasma membrane component blocks the secretory apparatus which is shared by many other proteins.[162] There are a variety of other mutants, such as the *secB* mutant which maps at a different locus, 80.5 min near *cysE*.[169]

Suppression of receptor protein secretion by suppressor mutation provides further evidence for the existence of a secretory apparatus. The suppressor mutation by definition would affect a character other than the structure of the secretory protein molecule. The suppressor mutants are obtained from *lamB* gene secretion-blocked mutants which have only one amino acid alteration internal to the signal sequence. These mutations have been defined as protein localization *(prl)* mutants.[170] The *prl* mutants (A, B and C) possibly reduce the specificity of interaction between secretory apparatus and the signal sequence; thus, inhibition of secretion caused by a specific alteration of the signal sequence is reversed and the secretion is restored (Figure 28).[170] Furthermore, this effect appears to be specific only for the protein which has altered signal sequence.

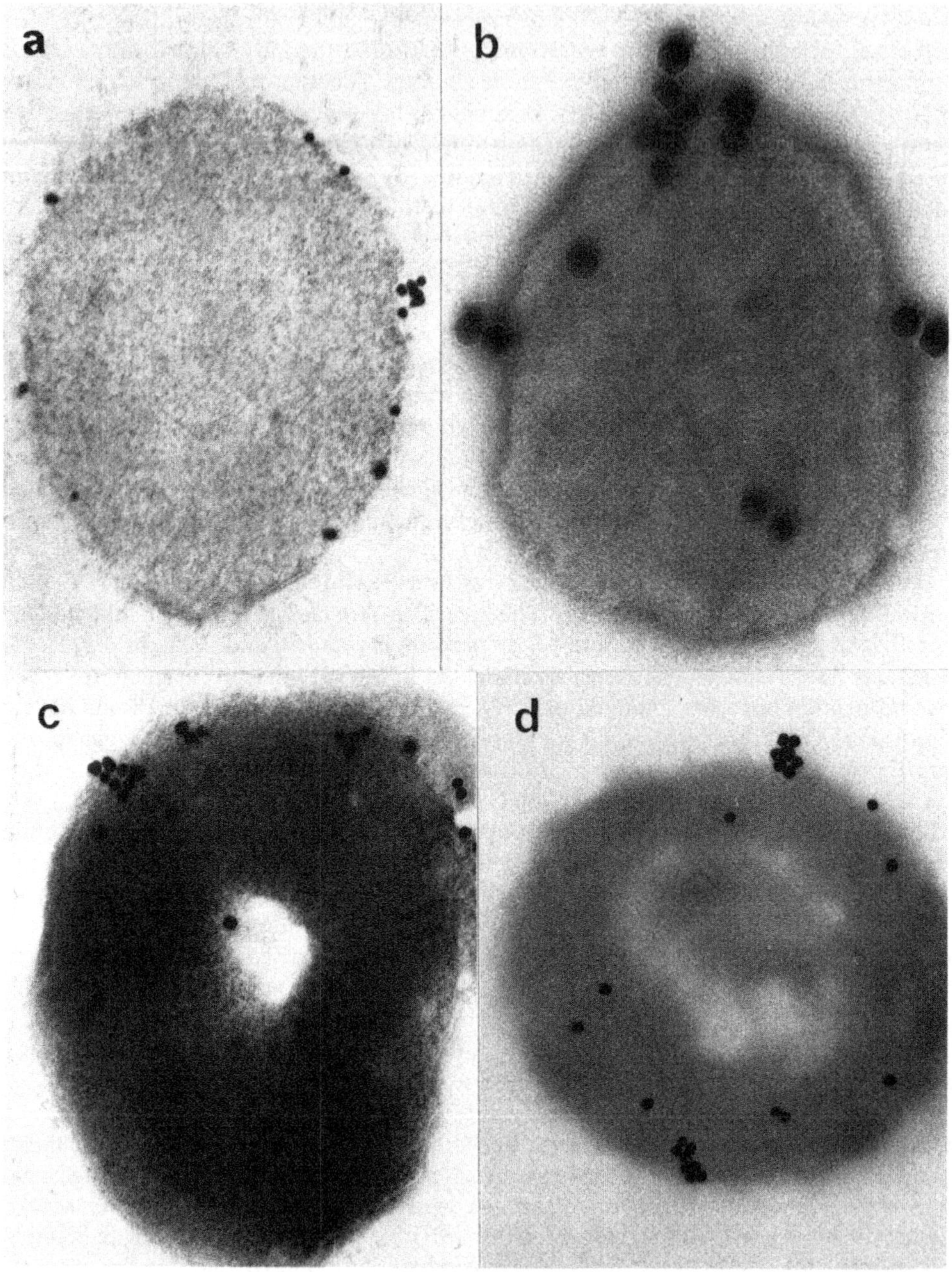

A

FIGURE 27. (A) Labeling of alkaline phosphatase sites of *Bacillus licheniformis* with colloidal gold-anti-alkaline phosphatase antibody label: (a) membrane labeling of the thin section of the protoplast, (b) labeling of negatively stained protoplast, (c) and (d), labeling of tannic acid-glutaraldehyde fixed frozen thin sections. Discrete clusters of membrane-bound labels suggest specific alkaline phosphatase binding sites in the membrane. (Based on the work of B. K. Ghosh and his associates.) (B) Colloidal gold-antibody double labeling of *B. licheniformis;* large gold, alkaline phosphatase; small gold, penicillinase. Note discrete cluster of labeling for alkaline phosphatase and single randomly distributed labels for penicillinase.

This mutation maps at the Pspc operon (Figure 29) which codes for ribosomal protein.[170-172] It has been suggested that the weak binding caused by a change in signal sequence is overcome by a mutationally altered ribosome which restores strong binding with the secretory apparatus.

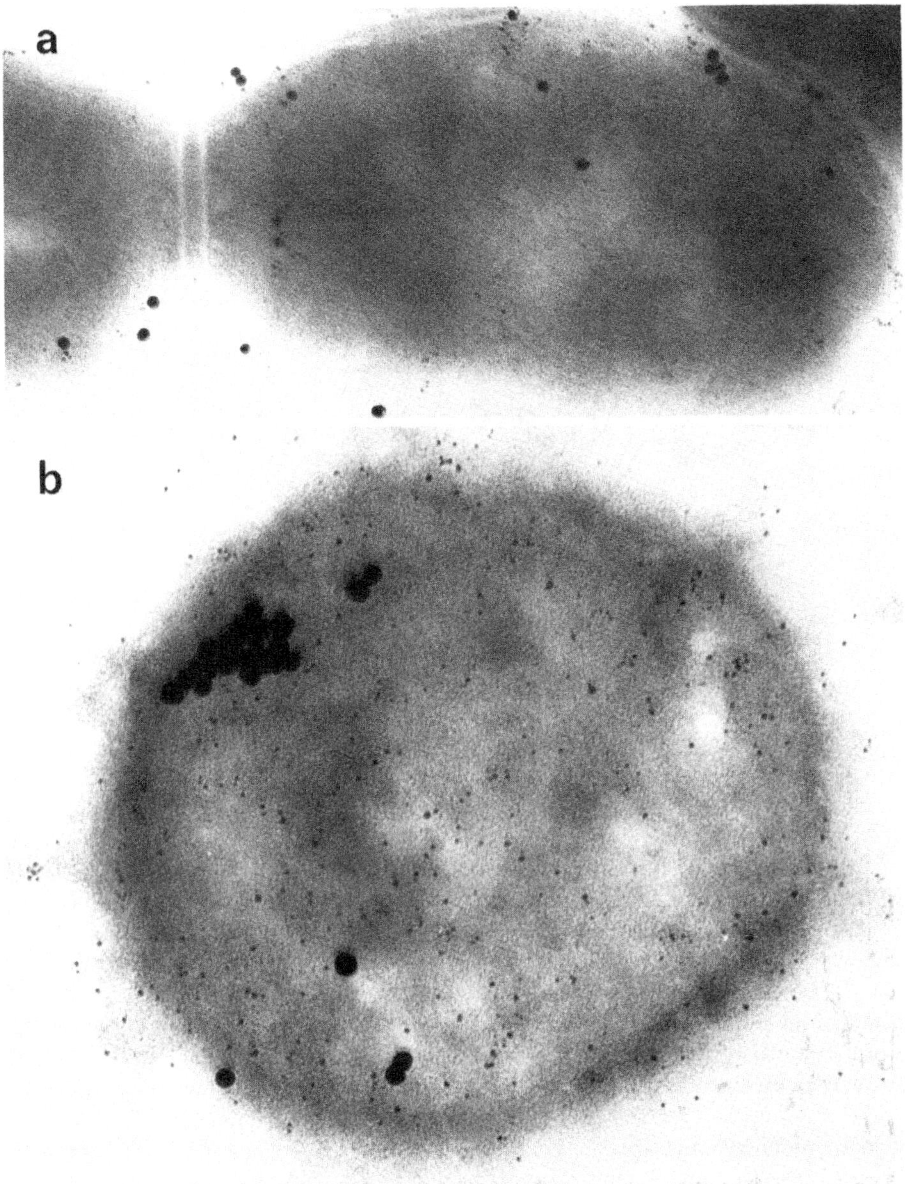

FIGURE 27B

The following is a discussion on the information generated from the above genetic data, which can be used to support the existence of a membrane secretory apparatus.

1. Mutations affecting single amino acids of secretory proteins may alter the conformation of protein molecules. Such changes may impair the interaction with the membrane secretory apparatus and consequently block the secretion.
2. Secretion defects are frequently relieved by suppressor mutation. Therefore, factors other than the secretory protein molecule itself must have played important roles in the secretory activity.
3. Many revertants of secretion-defective cells can be constructed. Frequently, the characters responsible for reversion are mapped at a chromosomal locus different

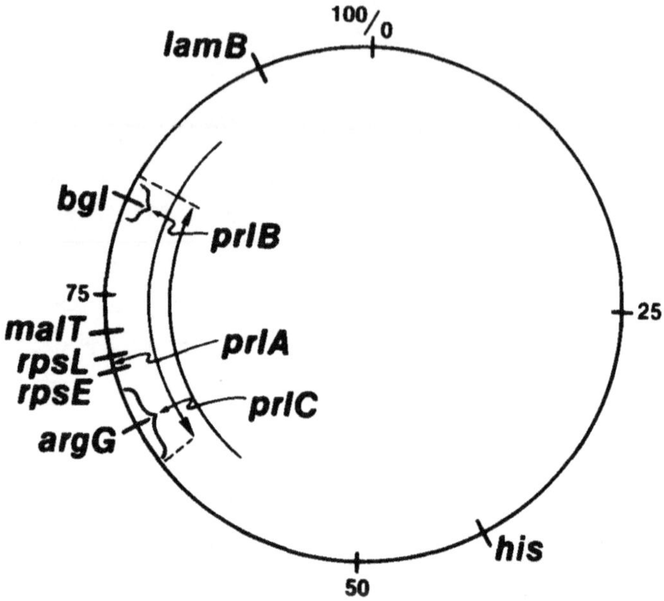

FIGURE 28. Position of relevant genetic markers on the *E. coli* linkage map. Used for mapping the three classes of suppressor mutations (*prlA*, *prlB*, and *prlC*). Lines with arrowheads indicate the origin and direction of transfer of the Hfr strains employed. The origin of transfer for these strains HfrG6 and KL228 are 66 min and 84 min, respectively. (From Emr, S. D., Hanley-Way, S., and Silhavy, T. J., *Cell*, 23, 79, 1981. Copyright M.I.T. With permission.)

from the locus of the structural gene. Therefore, secretion defects of these strains have not arisen from the defect in the structural gene, suggesting additional the role of genes other than the structural genes for secretion.

It should be noted, however, that these genetic data are at best suggestive. Unless detailed biochemical analysis and reconstitution experiments can be done, these data are valuable mostly to generate the experimental framework which could be exploited to show the existence and function of membrane secretory apparatus.

D. Biochemical Mechanism

The putative secretory apparatus has four major biochemical functions: (1) ribosomes containing extruded signal sequence bind with secretion specific regions of the plasma membrane; (2) intramembranous protein molecules, which form hydrophilic transmembrane channels when the signal sequence binds to the plasma membrane, may undergo reversible assembly; (3) energy transduction provides free energy required for the translocation of the secretory molecules across the putative plasma membrane channel; and (4) cotranslational processing modifies the secretory molecule and cleavage of the signal sequence releases the mature secretory protein from the putative secretory apparatus. Some of these functions await direct biochemical confirmation either by demonstrating specific biochemical reaction(s) or by isolating specific molecules having characteristics for the above-mentioned presumed functions. In eukaryotic cells evidence for the presence of the signal receptor particle (SRP) of the secretory apparatus and docking protein for SRP-specific membrane binding has been shown.[173-177] The SRP consists of both protein and a 6S RNA moiety; this is a cytosolic component which binds with a ribosome-extruding signal sequence; following binding

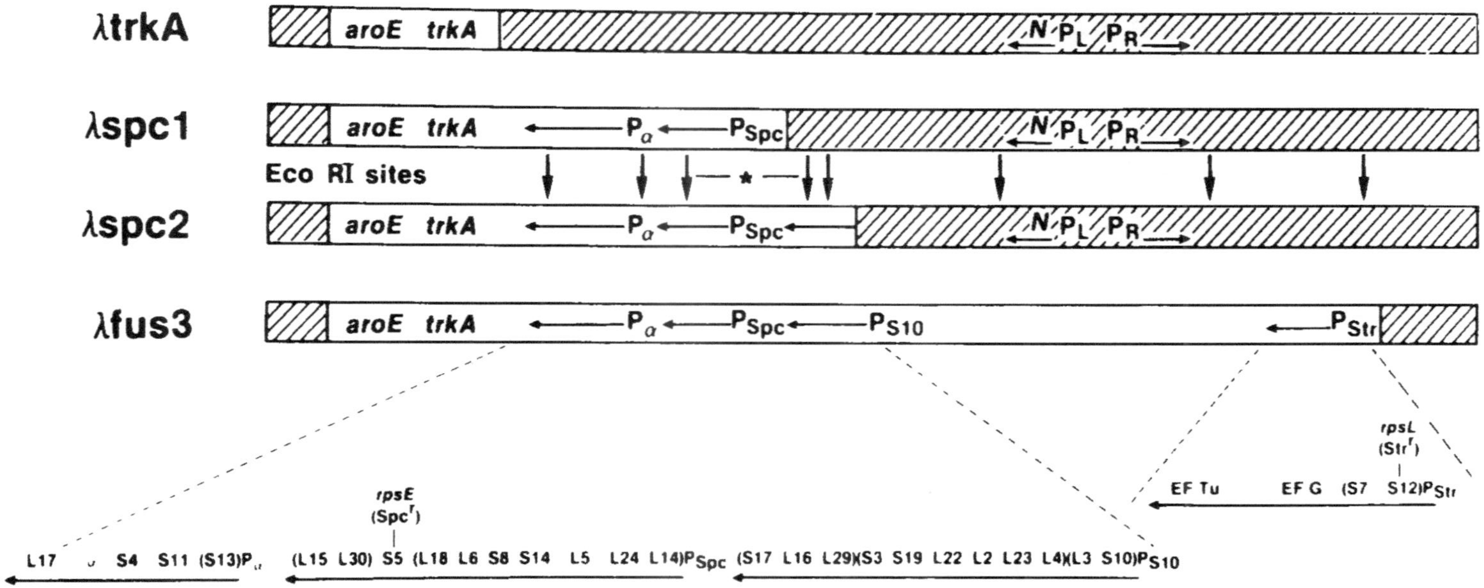

FIGURE 29. The position and organization of the ribosomal protein genes on the *λtrkA*, *λspc1*, *λspc2*, and *λfus3* defective transducing phages. $P_α$, P_{spc}, P_{s10} represent the approximate position of promoters. Arrows indicate the extent and direction of transcription. The protein products of genes contained in each of the different transcription units are indicated. Groups of gene products for which the relative order is unknown or less certain are given in parentheses. Mutations in the S5 gene *rpsE* can lead to Spc^r, and mutations in the S12 gene *rpsL* can lead to Str^r. An "S" or "L" before a number indicates that the protein is a part of either the small (30S) or the large (50S) ribosomal subunit. the position of Eco RI DNA restriction sites in the λspc2 DNA are shown. The DNA fragment containing the major portion of the P_{spc} operon is indicated by an asterisk. (From Emr, S. D., Hanley-Way, S., and Silhavy, T. J., *Cell*, 23, 79, 1981. Copyright M.I.T. With permission.)

the translation is arrested until the ribosome-SRP complex binds to a specific protein (docking protein) in the endoplasmic reticulum. Firm binding with ER is ensured through a specific ribosome-binding protein of the ER, ribophorin.[178,179] Translation resumes upon binding with ER; however, no definitive information is available on the activation of a transmembrane channel or energy transduction activity.

A heterologous transcription-translation-translocation system has been reconstituted using plasmid pBR322 DNA encoding β-lactamase and a wheat germ translation system with canine pancreas microsomal vesicles as the translocation system.[173] In this system, translated pre-β-lactamase can be processed to its mature form if incubated with dog pancreas microsome. This result implies that signal peptidase hydrolyzes a specific peptide bond irrespective of the signal peptide or secretory protein structure. The translation of β-lactamase is strongly arrested by the salt wash of dog pancreas microsome containing SRP; however, this inhibition is removed by treating with desalted microsomal membrane. Finally, in this reconstituted system the transcribed β-lactamase is translocated into microsomal vesicles. This study raises the intriguing possibility that the translocation-processing machinery is nonspecific and its function depends upon a few basic biochemical or biophysical characteristics, namely appropriate peptide bond, net charge, hydrophilic and hydrophobic property, etc. However, this study needs confirmation in another system besides β-lactamase.

Although hydrophobicity of the amino-terminal signal sequence of secretory proteins allows insertion of this initial nascent secretory protein chain into the plasma membrane, translocation of the rest of the polypeptide chain would require a supply of free energy. In bacterial cells the ribosome is not firmly attached to the membrane surface, so the free energy of protein systhesis obtained from GTP hydrolysis is expended to maintain ribosome-membrane association; hence, this energy may not be available for chain translocation.[180,181] It has been suggested that folding of the polypeptide chain to a minimum energy conformation may release enough free energy to power the translocation activity.[180]

This question of energy requirement has been further analyzed by studying the effect of uncouplers on the secretion of mature secretory proteins.[182,183] Intracellular accumulation of precursors of the omp protein, LamB, and a few periplasmic proteins in the presence of a variety of uncouplers (e.g., CCCP, dinitrophenol, dimethylsulfoxide, tosylarginyl methyl ester, and ethanol) demonstrates that these proteins are not being translocated. These uncouplers, which are known for their effect in dissipating proton motive force, inhibit lactose uptake by the same cells in a comparable condition. Furthermore, in a pair of isogenic strains, the strain with an *UncA* mutation lacks proton translocation-ATPase and accumulates precursors of the above-mentioned secretory proteins.[184] The *uncA* mutant cannot maintain proton motive force because it fails to hydrolyze ATP. Therefore, in an anaerobic condition, these mutants, unlike wild-type cells, accumulate secretory protein precursors inside the cell. Although this interpretation may be likely, other possibilities remain open because many of the uncouplers are also membrane-perturbing agents. One of the major perturbations is the inhibition of membrane fluidity (e.g., by ethanol treatments); the alteration of this basic property of the membrane would certainly affect secretion. The effects of uncouplers strongly indicate that secretion requires an energized membrane. A similar effect of CCCP has been shown on the transport of periplasmic arabinose-binding protein in *E. coli;* a potassium ionophore, valinomycin, also has the same effect.[183] This inhibition of protein transport can be reversed on removal of these inhibitory agents, suggesting the absence of any permanent damage to the membrane by any secondary action of these agents. This observation on the effect of membrane potential on secretion suggests that the vectorial translocation of secretory protein could be physiologically regulated by membrane potential.

The interrelationship of the energized membrane and secretion is reminiscent of gating phenomena and ion transport in biological membrane.[185] This is a complex electrophysiological activity and it is not possible to suggest how much of this information is applicable to the mechanisms of protein secretion. However, it must be pointed out that ion translocation coupled to potential change requires specific low-resistance (high-conductance) hydrophilic ion pores. Hence, discharge of membrane potential per se possibly cannot translocate the protein molecule; actual translocation processes must be coupled to a concomitant channel formation. This brings up the question of how much of an energy barrier there is for the translocation of a protein molecule through the membrane and whether or not this barrier can be overcome by reversible biochemical or biophysical changes. This question has been addressed by calculating ΔE of secretion and examining the relation of ΔE with membrane fluidity and presecretory protein processing.[186] In *E. coli,* alkaline phosphatase is almost totally secreted into the periplasm.[187] This secretion is decreased to 40% if the cells are converted to spheroplasts. This significant reduction of secretion is due to the loss of processing enzyme, which is most likely concentrated in the outer membrane.[81] The amounts of enzyme synthesized and secreted at different temperatures have been analyzed by an Arrhenius plot and the ΔE values for the synthesis and secretion have been calculated. Although this approach has been severely criticized the information from the relative measurement is meaningful.[188] The ΔE for alkaline phosphatase synthesis is 79 kJ at 22° C and 179 kJ below 22° C in both spheroplasts and whole cells.[189,190] The ΔE for secretion, on the other hand, is 125 kJ in the whole cells and 146 kJ in the spheroplasts; this difference arises from the ΔE of processing activity. The temperature dependence of secretion relates to the order/disorder transition of paraffin chains of the membrane lipid. In fatty acid auxotrophs changes in the membrane phase transition are induced by different unsaturated fatty acids; these changes cause alteration of the ΔE of secretion.[191] It has been shown that as long as 30% of the length of the paraffin chain of the fatty acids is maintained in an ordered state; however, the secretory activity may continue.[189] It has been suggested that in Gram-negative bacteria there are adhesion zones between outer and inner membrane.[192] These may function as hydrophobic bridges allowing transfer of the presecretory proteins into the outer membrane where they become accessible to processing enzymes and are secreted into the periplasm (Figure 30).[186] Since only a limited number of adhesion zones are sparsely distributed in the cell surface, after entry into the inner membrane pre-secretory material may move by lateral diffusion into these adhesion zones. Thus, membrane fluidity would play a significant role in the secretory activity.

The possibility that fatty acid synthesis may be related to exportable protein synthesis has been studied in *Bacillus subtilis* strains which carry a pleiotropic mutation affecting the regulation of biosynthesis of several exoenzymes, i.e., protease, α-amylase, and levansucrase.[144,193] Cerulanin, an inhibitor of fatty acid biosynthesis, inhibits or arrests the synthesis of levansucrase or α-amylase. However, this effect is not related to fatty acid biosynthesis because the concentration of this inhibitor needed to inhibit the exoenzyme synthesis is lower than that needed for inhibition of fatty acid synthesis. The effect of cerulanin possibly arises from the amphiphilic nature of the molecule allowing penetration into the membrane lipid bilayer; dodecanoic acid is another amphiphilic molecule which has the same effect as cerulanin. A large number of membrane-modifying agents including many amphiphilic agents cause a significant reduction of extracellular protein production.[194,195] How this effect is exerted at the molecular level is unknown. It is possible that the amphiphilic molecules, while penetrating a membrane lipid bilayer, may arrest the lateral diffusion of secretory protein molecules bound to a limited number of specific receptors. This lateral diffusion may be important for transbilayer translocation of secretory molecules. Thus, the mem-

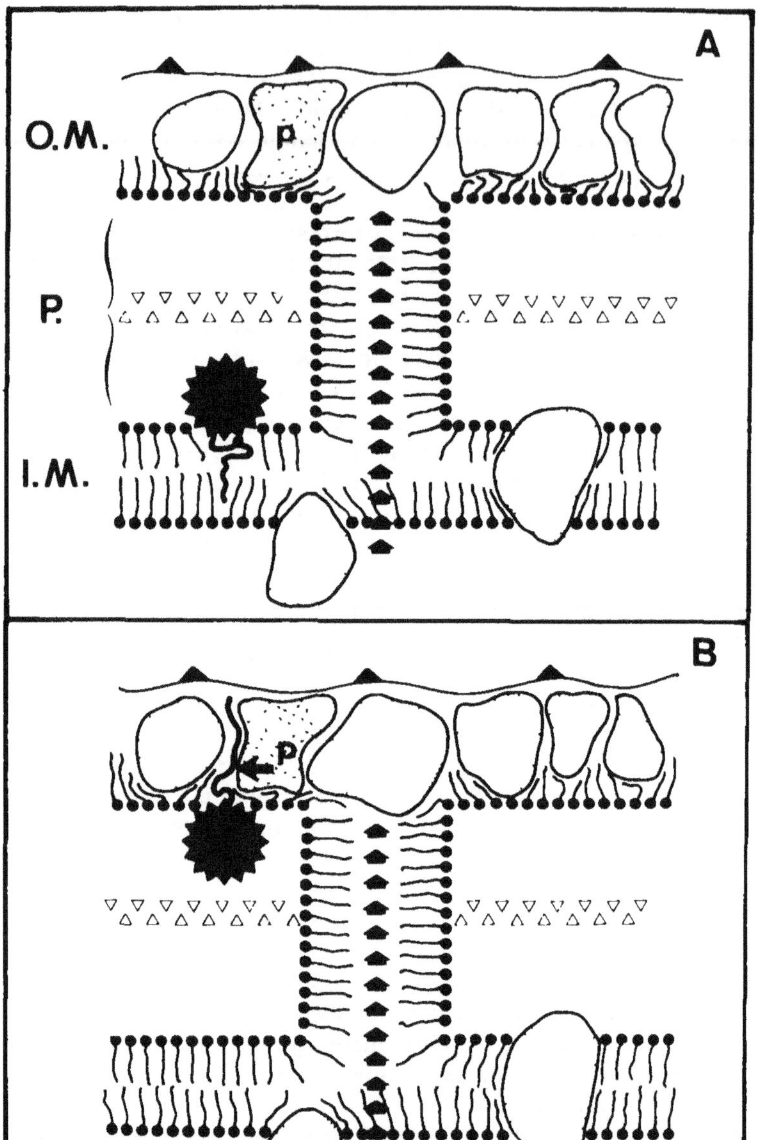

FIGURE 30. Hypothetical model for secretion of a periplasmic protein. Lipopolysaccharides are represented by (▲), arrows represent nascent lipopolysaccharide transferred through a hydrophobic pore to the outer membrane; (Δ), murein; (•), phospholipds. Spotted areas represent membrane proteins. That with a "p" is the protease which cleaves the hydrophobic "signal sequence" of the completed nascent chain of alkaline phosphatase (represented by a large star). (A) First associated with the inner (or cytoplasmic) membrane (I.M.); (B) then transferred by lateral diffusion to the outer membrane (O.M.) where the hydrophobic signal sequence becomes accessible to the protease, releasing it into the periplasm (P). (From Pages, J. M. et al., *Eur. J. Biochem.*, 86, 589, 1978. With permission.)

brane receptors saturated with secretory protein molecules may cause inhibition of extracellular enzyme synthesis preventing ribosome membrane association. Although the presence of specific membrane receptors for extracellular enzyme synthesis has been repeatedly suggested, these have not been demonstrated. The presence of a receptor in bacterial membrane is by no means a novel concept. Periplasmic histidine-binding protein (J) in *Salmonella typhimurium* interacts with a membrane protein P.[196] The same membrane protein also interacts with lysine-, arginine, and ornithin-binding proteins. By analogy one may assume that receptors for secretory proteins may exist in the membrane; it is possible that the receptors have limited specificity and thus interact with a subset of secretory proteins.

The possibility of channel formation could be suggested from cotranslational translocation of polypeptide chains through the plasma membrane. Labeling reagent which does not penetrate the plasma membrane has been used to label the polypeptide chain of secretory proteins growing through membranes of both Gram-positive and Gram-negative cells.[197,198] This labeling of the extended form of polypeptide can only occur if the protein molecule is cotranslationally translocated. The movement of the polypeptide chains, which consist of a variety of nonpolar as well as polar amino acids, will encounter an energy barrier unless the movement takes place through hydrophilic pores in the plasma membrane. Thus, one may assume that the plasma membrane contains binding proteins for signal sequence and secretory pore complexes. The results from labeling experiments have been interpreted as compelling evidence for cotranslational translocation of polypeptide chains. However, the data rest heavily on labeling during chain completion after puromycin treatment. Another interpretation of the data may be possible; labeling may occur if the protein molecule translocate post-translationally and after reaching the outside surface of the membrane its conformation is altered. The labeling may occur during this molecular reorganization although translocation of the molecule is post-translational. Such a possibility is strongly suggested by the following data which show that the translocation through the plasma membrane is likely to occur with the protein molecule as a domain rather than as an extended polypeptide chain.[199]

Maltose-binding protein and ribose-binding protein have been used as models for the study in *E. coli*. The experimental results are based upon the accessibility of signal peptide to the proteinase K. The author mentions that "The reader must first become familiar with the nature of the complex two-dimensional peptide patterns which are generated after proteinase K digestion."[199] The analysis of the peptide map is, in fact, difficult to comprehend. Hence, the validity of the argument based on this analysis awaits further confirmation in another system. However, the results appear to be convincing in showing that the signal peptide is available to proteinase K late in the translation. Maltose binding protein must be 80% translated and ribose binding protein must be fully translated before it becomes available to proteinase K digestion.

Processing of pre-secretory protein is an important activity which frequently requires multiple enzyme action; one of these is signal peptidase, which is universally present in secretory cells. This enzyme removes the signal peptide from the precursor form of the secretory protein and generates mature protein. Components of the secretory apparatus may include this processing enzyme. Signal peptide or leader peptide is a short polypeptide segment present at the amino terminal of the transient precursor of the secretory protein. This peptide segment is largely composed of hydrophobic amino acids, and is apparently removed by the hydrolysis of a specific peptide bond by a signal or leader peptidase. In eukaryotic cells this peptidase activity is present in the microsomal membrane fraction and can be released by nonionic detergent (Triton X-100®) treatment. In *E. coli*, signal peptidase for alkaline phosphatase co-purifies with the outer membrane;[81] M13 coat protein processing activity, on the other hand, is equally distributed in the outer and inner membrane.[200] Conversion of M13 pre-coat protein to coat protein has been used as an assay for determining the location of processing activity. The inner and outer membrane have been isolated by density gradient centrifugation; the membrane fractions have been identified using NADH oxidase activity as a marker for inner membrane, and typical polypeptide composition as a marker for the outer membrane.[201] The chromatographic behavior as well as catalytic properties are identical for the activities on both fractions. A DNA fragment from *E. coli* has been cloned in a high-copy-number plasmid.[202] A strain containing this plasmid overproduces the leader peptidase and the enzyme is found in equal concentrations in both membrane fragments. The purified enzyme from an overproducer strain is a monomer of 37 kDa.[66] This matches the calculated molecular weight of 323 amino acids deduced from the nucleotide sequence. The enzyme is likely to be a transmembrane protein asymmetrically exposed on both the membrane surfaces; the major portion of the protein is exposed on the outer surface of the plasma membrane. The enzyme appears to have a broad specificity that acts on signal peptides of precursors originating from both bacterial and animal cells.[200] The broad specificity of processing is suggested from the processing and integration of plasmid PBR322 encoded outer-membrane protein MRB in *E. coli* host.[203]

There are significant temporal differences in the processing of different extracellular proteins; some are processed cotranslationally and some post-translationally.[201] These temporal differences in the processing are not related to targeting the proteins to different subcellular compartments, because different temporal modes of processing may occur in the same protein under different conditions. However, both co- and post-translational modes of processing can occur after the proteins acquire a critical size. This may be related to the activation of dormant processing activity located in the secretory apparatus.[204]

Although the role of a secretory apparatus appears to be convincing,[55] a purely biophysical approach suggests that incorporation into and release from the plasma membrane may occur by a spontaneous self-assembly process.[154] It has been proposed that polypeptide segments of a protein molecule may be inserted into the plasma membrane as an antiparallel hairpin of α and 3_{10} helices. About 15 to 30 amino acid residues may span the 3.5-nm hydrophobic core of the plasma membrane. The major energetic factors favoring the partitioning of the helical hairpin into the membrane bilayer are hydrophobic forces which act to reduce surface area in contact with water. It is postulated that in the helical configuration, if two helices are in contact, the surface area in contact with the water would be a minimum. Therefore, a hydrophobic driving force will be generated and the helices will be partitioned into the membrane. In addition, a potential gradient between the inside and outside surfaces of the plasma membrane will further favor this insertion process. The helices will continue to form inside the cytoplasm and be inserted into the plasma membrane. Continuous presence of a membrane surface is required because the failure of the helix to be inserted into the membrane interior will result in lengthening of the polypeptide chain without any conversion to helix. Thus, the rate of transport must be greater or equal to the rate of protein synthesis. The growing peptide chain continuously grows and extrudes out of the membrane interior while the segment remaining inside the plasma membrane exists as a helix. The helix of the leader peptide remains in the membrane to orient the process and to prevent the peptide from coming out of the membrane; the amino terminal always faces the cytoplasmic surface of the membrane. The free energy arising from burying the hydrophobic helical surfaces exceeds the free energy cost of burying charged and hydrogen-bonding groups. A continuous drop in the polar character of the polypeptide chain from amino to carboxy terminal favors the export; additionally, energy derived from folding of the polypeptide chain on the exterior side of the membrane also supplies the free energy of export. Modifications of the molecule on the outside surfaces of the membrane by glycosylation, phosphorylation, or disulfide formation render this translocation process irreversible. Actual passage of the molecule through the membrane occurs by Brownian motion. It should be noted that a charged amino acid sequence with a continuous drop of polarity favors export, whereas a hydrophobic stretch arrests the translocation. This is a new approach; the data have been obtained from the studies on the insertion of bacteriorhodopsin. However, the general application of this concept requires quantitative analysis of the boundary conditions from a large number of secretory proteins.

This biophysical approach has been further pursued on a quantitative basis. On the basis of energy of interaction it has been suggested that binding of the signal sequence with the membrane should be followed by a firm and specific binding of the ribosome (e.g., mediated through ribosome-binding protein) releasing a significant amount of free energy which may force the hydrophilic nascent peptide into the membrane.[205] Finally, it has been shown through computer modeling that a variety of predictions can be made.[206] These approaches are novel and should be considered when interpreting genetic, biochemical, and physiological data. However, predictions from computer analysis will be limited if the data base does not relate to experimentally supported biological phenomena.

VIII. RELATION TO PATHOGENICITY AND TOXIN PRODUCTION

The following section has been developed to show how the secretory processes of bacterial cells and their pathogenicity are interrelated. It has been shown that pathogenicity depends on the expression of a complex cell surface phenomena. Starting from the adhesion to the host cell surface and continuing to the production and secretion of toxins, secretory activity plays a key role.[207] However, literature directly aimed toward secretion study in this respect is scant. Hence, some background material on pathogenicity has been integrated with studies on secretion. This will allow developing a focus in this field and generate future investigation.

Pathogenicity is the capacity of microorganisms to produce disease in man, animal, or plant hosts. The pathogenic character occurs rarely among microorganisms. The development of the pathogenic character depends upon the interactions between the microbial and host cells. The interaction may occur indirectly between substances secreted by the microbial and host cells, e.g., toxin/antitoxin. Direct interaction may involve secreted material and the cell surface or both cell surfaces.[208] A pathogen must (1) enter the host, (2) multiply in the host tissues, (3) resist host defenses, and (4) damage the host. The microbial products responsible for these processes are the determinants of pathogenicity or virulence, and many of them are surface components.[209] Secretory activity would be directly related to the pathogenicity of microorganisms because the products responsible for the pathogenic character should either be secreted outside the cell or be positioned on the surface of the cell. The process may be more complex than the usual secretory phenomenon, because in this case the process is tightly coupled to the stimulus. Microorganisms grown in vitro may be incomplete as regards all the determinants of virulence, since the genetic basis for virulence may be completely expressed only during growth in vivo.[210] The conditions of growth in vivo may influence the production of surface components important in pathogenicity. Many examples can be cited to substantiate this view. Subtle variations in the peptidoglycan structure caused by phosphate, ammonia, or magnesium limitation increase or decrease the suceptibility to lysozyme liberated from host cells.[211,212] The fast growth rate of *Aerobactor aerogenes* leads to the production of more toxic cell-wall components than slow growth rate.[213] Slow growing *Pseudomonas aeruginosa* is more resistant to the killing action of polymorphonuclear phagocytes and the cationic protein isolated from them. This resistance arises from the loss of some cell-wall antigens.[214] Pathogens obtain iron in iron-deficient conditions by synthesizing an iron chelator, siderochrome.[209,215,216] Host cells inhibit pathogenic cell growth by producing the iron-binding compounds transferrin or lactoferrin. Pathogens are unique among microorganisms in being able to overcome the host defense mechanisms by producing compounds or inducing processes that inactivate host defense mechanisms, resist their action, or even prevent their appearance. The bacterial products having this biological effect have been termed aggressins, auxiliary pathogenic factors, and impedins.[209] Some of the aggressins are extracellular, but many are surface components. They can act by (1) conferring resistance to humoral antimicrobial agencies in blood, milk, and other body fluids;[209,211,217] (2) stopping mobilization of phagocytes by inflammation;[211] (3) hindering contact with phagocytes;[211] (4) preventing ingestion by phagocytes;[209,217] (5) interfering with intracellular killing by phagocytes, leading to short-term survival in polymorphonuclear (PMN) or long-term survival in mononuclear cells;[218] and (6) inhibiting the immune response or reducing its effectiveness.[211,219] Pathogenic bacteria receive a variety of signals through the cell surface and respond to these signals appropriately; a balance between these two determine the pathogeneity of any microorganism.

Microbes damage tissues by two main processes: directly by the production of toxins,[220] and indirectly by evoking immunopathological reactions.[209] The most important surface toxins are the endotoxins of Gram-negative pathogens which play a major role in producing the host damage. First, the extent to which these toxins are released in vivo determines the degree to which they are responsible for host damage.[221] Second, endotoxins can be released from the cells during active growth as well as by cell lysis.[222] Third, they are not confined to pathogenic bacterial species only.[222] Fourth, the potency of the toxin may differ according to the source. Finally, toxicity derives mainly from damage to the host cell membranes and from triggering complement by an alternate pathway.[222,223]

In addition to endotoxins confined to the intact bacterial cells, some organisms (e.g., *Vibrio cholera*) produce exotoxins which diffuse freely from the cells and are found in supernatants of liquid cultures as well as within the cell. The toxin can be differentiated by their general properties and by differences in toxic activity. These several toxic activities fall into three groups on the basis of heat stability and dialyzability, and to a lesser extent on their location in the bacterial cell. Some are contained in the cell wall, others are intracellular, and some diffuse from the cell into the liquid culture media.[224] The type 1 toxins are contained within the bacterial cell and in contrast to the endotoxins of enteric bacilli, they occur both in the intracellular substance as well as in the cell wall. They are heat stable (to 100°C) and nondialyzable through cellophane. They are prepared in a high state of purity, as lipopolysaccharides and polysaccharides. The cholera toxins grouped together as type 2 are distinguished from type 1 by their heat lability (56°C, 15 min), their occurrence in the intracellular substance and in cell-free supernatants, their absence in the cell wall, and their nondialyzability. The type 2 toxins have two subgroups: (1) the choleragenic toxins which produce cholera-like disease and (2) the toxins whose activity is demonstrable in other systems. The toxins falling into type 3 are characterized by their heat stability and dialyzability.[224]

Cholera toxin (CT) and the *E. coli* heat-labile toxin (LT) are functionally, structurally, and immunologically similar enterotoxins.[225] Both toxins cause diarrhea by stimulating adenyl cyclase and elevating cylic AMP levels in gut epithelial cells.[226,227] Cyclic AMP catalyzes the NAD-dependent ADP ribosylation of membrane proteins.[228-230] Each toxin is composed of two dissimilar subunits A and B which share common antigenic determinants.[231,232] The subunit A has an enzymatic activity and is the adenylate-cyclase-activating component of the enterotoxin.[233] The subunit B recognizes membrane components and binds the holotoxin to the target cell, juxtaposing subunit A with its substrates. Binding studies and competition experiments indicate that the membrane receptors for cholera toxin subunit B and *E. coli* heat-labile subunit B are similar but not identical.[234] Monosialosylganglioside (GMI) is the receptor for the cholera toxin, and it probably constitutes part of the receptor for LT.

The heat-labile enterotoxin (LT) of *E. coli* has recently been purified to homogeneity and when activated by protease treatment has a specific activity equivalent to that of pure choleragen.[235] The two enterotoxins have remarkable physiochemical similarities, as well as unique features.[236] The LT has an isoelectric point of pH 8.0, as compared with choleragen and choleragenoid, which have isoelectric points of pH 6.75 and 7.75, respectively. Sedimentation equilibrium measurement has established an approximate molecular weight for LT of 91 kDa. LT has remarkably high affinity for agarose-containing gels, including ACA34 (LKB-Produktor AB). Relative to LT, the affinity of choleragen for agarose is much less. Galactose and lactose are highly efficient in removing ^{125}I-labeled LT from agarose-containing columns. LT dissociates into A and B subunits in the presence of 5 *M* guanidine. Although the subunits of LT and CT are immunologically distinct, there are shared antigenic determinants. During gel filtration of LT at pH 6.5 and room temperature, a spontaneously occurring toxoid of LT has been isolated and designated "coligenoid," analogous to choleragenoid, which is formed during purification of cholera toxin.[237] Coligenoid shows a high degree of antigenic homology with the holotoxin and the B subunit of the toxin; the amino acid sequence of the B subunit of LT has remarkable homology to the primary structure of the corresponding B subunit of cholera toxin. Cystine is located in the same position of both molecules and among the first 20 amino acids of the two chains. Only 5 amino acids are different; each of these differences may be attributable to a single base substitution.

A majority of LT synthesized by *E. coli* is cell associated and located in the outer membrane; only a small portion of LT (10%) is released into the culture supernatant.[238,239] The LT subunit B (LT-B) produced by a cloned LT-B gene *(tox B)* was released in amounts equal to the complete parent LT release. In contrast, no release of LT subunit A (LT-A) or its smaller derivatives was observed in strains containing cloned *tox A* genes. Cells can synthesize and secrete LT even if the LT-A and LT-B genes (*tox A* and *tox B*) are separately placed on two compatible cloning vehicles.[240] This suggests that clustering of these genes is not essential for the synthesis of LT, and perhaps subunits A and B are independently released across the inner membrane. LT-B is necessary, however, for the release of LT-A across the membrane. The gene which encodes the LT of *E. coli* has been isolated from a large natural plasmid. Using recombinant DNA technology, it was linked to the cloning vehicle, plasmid pBR313.[241,242] This plasmid when introduced into a minicell-producing strain of *E. coli* synthesizes the A and B subunits in *E. coli* strains carrying mutations in *minA* and *minB.* These cells divide abnormally, resulting in the production of minicells which do not contain chromosomal DNA and the plasmid DNA is segregated into the minicells.[243] Two polypeptide species related to the B subunit of the LT toxin are expressed in the minicells. The larger polypeptide (mol wt 13 kDa) is likely to be a precursor of the B subunit because it could be chased into the mature form. The other polypeptide (mol wt 11 kDa) was immunoprecipitated by the anitsera to CT. The B subunits of LT and CT have common antigenic sites, hence this species was probably the mature B subunit of LT. This conversion is inhibited by compounds which dissipate proton motive force, suggesting that the processing enzyme is inhibited. An *E. coli* strain of porcine origin (API) produces high levels of LT in the plasma and outer membrane.[244] The membranes isolated from a serologically related nonenterotoxigenic strain of *E. coli* of porcine origin (AP2) have no enterotoxigenic activity, but the cells contain the same O and K antigens. Although it has been reported that the periplasmic fraction isolated from the enterotoxigenic strains contain considerable enterotoxin activity, this activity is in fact associated with the outer-membrane fragments present in the periplasmic fraction of the cell lysate. The distribution of subcellular enterotoxin activity, is about 55, 15, and 30% in the outer membrane, plasma membrane, and the cell cytoplasm, respectively. The specific activity is 20 units per milligram of protein in the cytoplasm and 90 and 150 units per milligram of protein in the plasma and outer membranes, respectively.[244] Enterotoxin activity in the supernatant of enterotoxigenic strains is usually associated with large polysaccharide protein complexes.[245] Similar complexes are released by a variety of pathogenic and nonpathogenic Gram-negative organisms and in the case of normally growing *E. coli* K-12, they consist of outer-membrane fragments which are specially released from the areas where newly synthesized proteins are incorporated into the outer membrane.[246]

The polysaccharide-protein complexes with enterotoxin activity may have physiological significance. Specifically, Gram-negative bacteria may use the outer membrane as an offensive weapon by inserting protein toxins into it, and under appropriate conditions outer-membrane fragments containing high specific activities of toxins might be released before they have been thoroughly fixed in the membrane.[247] Such a process could explain why enterotoxin usually appears in the culture supernatants as large complexes.[245] The heat-labile (ST) toxin of *E. coli* has not been as extensively studied. It is a small molecule of approximately 5 kDa and has been shown to stimulate guanylate cyclase in vitro.[248,249] At least two species of ST have been identified, based on solubility in methanol and activity in particular animal test systems.[250] ST I is methanol soluble and can be assayed in the stomach of suckling mice or neonatal pigs, whereas ST II is methanol insoluble and reactive only in ligated intestines of pigs. The gene for ST I has been subcloned into pBR322 as restriction fragments ranging in size from 0.4 to 3.4 kb. Studies with these fragments indicate the ST I gene is encoded within a transposon (Tn 1681) flanked by inverted repeats of insertion sequence 1 (ISI).[251] The gene for the ST I toxin has been located within the transposon Tn 1681 by means of insertion mutagenesis and by subcloning various restriction endonuclease fragments. From this information, the complete nucleotide sequence of the central portion of Tn 1681 (i.e., part flanked by ISI) has been determined, and also identified as the coding sequence of the toxin; this gene has all the properties necessary for autonomous expression in bacteria.[252] From the nucleotide sequence, an amino acid sequence for ST I has been deduced. The amino-terminal portion of the sequence is extremely hydrophobic and bears a striking resemblance to the signal sequence of the fd phage minor coat protein.[252]

The above discussion suggests a considerable similarity in the mode of action of ST I and ST II on host cells, subunit structure, biochemical, and immunological properties. The structural genes of LT and CT show considerable homology (but they are not identical) and the genes are organized into small operons. However, only a small amount of LT is secreted as separate A and B subunits, but CT is synthesized as a large precursor molecule and secreted almost entirely from *V. cholerae*. The structural gene of CT is chromosomal and LT is on the plasmid.

The LT gene from *E. coli* on plasmid pCG86 has been transferred to a *V. cholerae* strain containing CT. Likewise, CT from *V. cholerae* has been transferred to *E. coli*. The synthesis and secretion of CT and LT have been examined in these reconstructed strains and compared with those in isogenic strains of *E. coli* and *V. cholerae*.[253] A specific competitive radioimmunoassay has been used for the study. The CT and LT from the *V. cholerae* reconstructed strain are secreted but in the *E. coli* reconstructed strain they remain cell bound. Synthesis of CT is regulated by two regulatory genes which control hyper- or hypotoxinogesis for CT, and these regulatory genes affect the LT production of reconstructed strains. Although it has not been fully explored, the presence of the plasmid PCG86 has only a small regulatory action on CT production.

Table 7
LARGE-SCALE PRODUCTION OF PROTEINS AND THEIR USE

Product	Potential use
Enzymes (e.g., amylase, protease, cellulase)	Brewing, baking, paper, textile and food industry, and leather manufacturing
Polypeptide hormones (e.g., growth hormone, insulin)	Biomedical
Vaccines and anti-infective agents (e.g., interferon)	Biomedical

Mutation of *V. cholerae* may block secretion of CT; if this mutant is reconstructed with PCG86 there is a 60% arrest of both CT and LT secretion, suggesting a common pathway for the secretion of both CT and LT. The following significant results of these studies suggest the existence of a secretory apparatus in *V. cholerae*: (1) the *V. cholerae* CT system is a rare Gram-negative system which produces a true extracellular protein; (2) the mechanism of this secretion exists only in *V. cholerae;* and (3) the structure of CT and LT alone cannot cause secretion.

IX. CONCLUDING REMARKS ON INDUSTRIAL APPLICATIONS

The above discussion on the secretory phenomenon clearly show that the process is complex. Multiple regulatory activities control the release of protein molecules from the bacterial cell body into the extracellular medium. This information is highly significant for the fermentation technology of proteins. The potential for the production of proteins and small polypeptides by fermentation is increasing at a fast rate because of the rapid developments in recombinant DNA technology. The production of proteins and polypeptides falls under three general categories (Table 7).

A quote from Debabov is appropriate at this stage: "We can now be sure that in the near future industrial microbiology will be even more involved in the solution of global problems such as those concerned with protection of environment, agriculture, and the production of foods for humanity and will contribute to solutions of the energy problem and increase the availability of the mineral resources."[120]

The production of enzymes by large-scale microbial fermentation is already in use. The worldwide sales of enzyme preparations have reached $180 million in 1978. Japan and the U.S. produce yearly 13,000 tons and 4500 tons of enzymes, respectively.[254] Besides the enzymes shown in Table 7, a variety of other enzymes are also produced by fermentation, e.g., β-galactosidase, glucose isomerase, renin substitute, etc.[121] Their potential use from the microbiological source will gradually increase. The enzymes are natural products of a microorganism; hence, industrial success depends upon the genetic enhancement of the natural potential through the use of standard mutation and selection technology. However, the advent of recombinant DNA technology has started a new era for the fermentation industry. Microorganisms can now be genetically programmed for the production of desired material. One may consider converting bacterial cells into "mini factories" for producing material which at present is obtained from preformed natural sources. By synthetic gene transfer into *E. coli*, large-scale productions of somatomedin, insulin, and interferon have become possible.[255,256] These successes show the bright future of fermentation technology using recombinant DNA.

Proteins or small peptides, while produced by the bacterial cells, may be retained within the cell body (cell bound) or may be secreted into the fermentation fluid (extracellular). If the product is cell bound, an extensive purification process is necessary for the recovery of the product. Furthermore, when the product is used for biomedical purposes, it must be highly purified from the large number of intracellular proteins, which may be toxic. This purification will be simple if the product is extracellular. In addition, accumulation of a product extracellularly may have an additional value in maintaining high productivity during fermentation. Accumulation of a specific product within a cell may eventually inhibit its synthesis. Whereas maintaining a long vegetative growth and constant physical and chemical conditions in the fermentation medium the potential for continued synthesis of a product by bacterial cells appears to be long if not limited. Therefore, secretion will be a major factor in the advance of fermentation technology. From the discussion on secretion in this chapter several points emerge which may be relevant to the future developments of fermentation technology:

1. Gram-negative organisms (e.g., *E. coli*) are not naturally suited for the production of large amounts of extracellular proteins. The envelope organization of these organisms supports the accumulation of secreted material in the periplasm or within the outer membrane. However, rich knowledge on the genetics of *E. coli* renders this organism accessible to extensive genetic programming.
2. Construction of a strain of *E. coli* allowing extracellular protein production appears to be possible. Regulatory mechanisms for the release of proteins into extracellular medium may be studied in the *V. cholerae* toxin secretion system. If this genetic regulation is introduced in *E. coli*, a strain of *E. coli* may be constructed with a high potential for the extracellular protein production.
3. Gram-positive cells naturally produce extracellular proteins and the secretory activity may be stimulated by physiological effectors (e.g., ions, nucleotides, etc.).
4. There is a rate-limiting step for the production of extracellular proteins in Gram-positive cells. A knowledge of the regulatory mechanisms for this regulation is essential for the construction of hypersecretor strains which may be used as the host of foreign genes.

The knowledge generated to date suggests that in order to construct a hypothetical "superbug" for the production of an extracellular protein the following three general points should be considered:

1. The structure of a protein molecule which confers secretability on the molecule, e.g., the presence of an appropriate signal sequence at the amino terminal end of the protein.
2. The status of the secretory apparatus inherently present in the membrane of a host organism.
3. Regulation of the activity of the membrane secretory apparatus for the continued secretion of a specific protein molecule through the membrane.

The future is promising for the construction of a "superbug" for protein production by fermentation technology.

ACKNOWLEDGMENTS

We express our sincere thanks to our colleagues, without whose help this chapter would still remain a proposal rather than a reality. They have given their patience and time to analyze published data and generate ideas. For this we thank Ranajit Kumar, Richard Nelson, Tinglu Guan, Marian Glenn, and Claudia Campbell. Preparation of this manuscript in an understandable form has been as important as the original writing. Gloria Trechock, Lori Morlock, and Ranjan Ghosh gave their untiring help to prepare this manuscript. The work in our laboratory included in this chapter was funded by Grants NSF PCM81-10613 and NSF PCM81-08665 and DOE DE-AS05-80ER10702.

REFERENCES

1. Davis, B. D. and Tai, P. C., The mechanism of protein secretion across membranes, *Nature (London)*, 283, 433, 1980.
2. Kreil, G., Transfer of proteins across membranes, *Annu. Rev. Biochem.*, 50, 317, 1981.
3. Inouye, M. and Halegoua, S., Secretion and membrane localization of proteins in *E. coli*, *CRC Crit. Rev. Biochem.*, 7, 339, 1980.
4. Wickner, W., Assembly of proteins into membranes, *Science*, 210, 861, 1980.
5. Blobel, G., Intracellular protein topogenesis, *Proc. Natl. Acad. Sci. U.S.A.*, 77, 1496, 1980.
6. Scheele, G., Biosynthesis, segregation, and secretion of exportable proteins by the exocrine pancreas, *Am. J. Physiol.*, 238, G467, 1980.
7. Michaelis, S. and Beckwith, J., Mechanism of incorporation of cell envelope proteins in *E. coli*, *Annu. Rev. Microbiol.*, 36, 435, 1982.
8. Hopkins, C. R., The secretory process in outline, *Symp. Soc. Exp. Biol.*, 33, 1, 1979.
9. Lampen, J. O., Movement of extracellular enzymes across cell membranes, *Symp. Soc. Exp. Biol.*, 28, 351, 1974.
10. Palade, G. E., Intracellular aspects of the process of protein synthesis, *Science*, 189, 347, 1975.
11. Costerton, J. W. and Cheng, K. J., The role of the bacterial cell envelope in antibiotic resistance, *J. Antimicrob. Chemother.*, 1, 363, 1975.
12. Bayer, M. E. and Leive, L., Effect of ethylenediaminetetraacetate upon the surface of *E. coli*, *J. Bacteriol.*, 130, 1364, 1977.
13. Ghuysen, J. M. and Schockman, G. D., Biosynthesis of peptidoglycan, in *Bacterial Membrane Walls*, Leive, L., Ed., Marcel Dekker, New York, 1983, 37.
14. Beveridge, T. J., Ultrastructure, chemistry, and function of the bacterial wall, *Int. Rev. Cytol.*, 72, 229, 1981.
15. Ghosh, B. K., The mesosome — a clue to the evolution of plasma membrane, *Sub Cell. Biochem.*, 3, 311, 1974.
16. Neu. H. C. and Heppel, L. A., The release of *E. coli* by osmotic shock and during the formation of spheroplasts, *J. Biol. Chem.*, 240, 3685, 1966.
17. Repaske, R., Lysis of gram-negative organisms and the role of versene, *Biochem. Biophys. Acta*, 30, 225, 1958.
18. Rosen, B. P. and Heppel, L. A., Present status of binding proteins that are released from gram-negative bacteria by osmotic shock, in *Bacterial Membranes and Walls*, Leive, L., Ed., Marcel Dekker, New York, 1973, 209.
19. Braun, V. and Hantke, K., Bacterial cell surface receptors, in *Organization of the Prokaryotic Cell Membrane*, Vol. 2, Ghosh, B. K., Ed., CRC Press, Boca Raton, Fla., 1981, 1.
20. Green, P., DiRienzo, J. M., Yamagata, H., and Inouye, M., The biosynthesis of bacterial outer membrane proteins, in *Organization of the Prokaryotic Cell Membrane*, Vol. 3, Ghosh, B. K., Ed., CRC Press, Boca Raton, Fla., 1984.
21. Ichihara, S., Hussain, M., and Mizushima, S., Mechanism of export of outer membrane lipoproteins through the cytoplasmic membrane in *Escherichia coli*, *J. Biol. Chem.*, 257, 495, 1982.
22. Costerton, J. W., Ingram, J. M., and Cheng, K. J., Structure and function of the cell envelope of gram-negative bacteria, *Bacteriol. Rev.*, 38, 87, 1974.
23. Salton, M. R., Owen, P., Bacterial membrane structure, *Annu. Rev. Microbiol.*, 30, 451, 1976.
24. Edidin, M., Two dimensional diffusion in membranes, *Symp. Soc. Exp. Biol.*, 28, 1, 1974.
25. Singer, S. J., The molecular organization of biological membranes, in *Structure and Function of Biological Membrane*, Rothfield, L. I., Ed., Academic Press, New York, 1971, 145.
26. Bretscher, M. S., A major protein which spans the erythrocyte membrane, *J. Mol. Biol.*, 58, 351, 1971.
27. Tanner, M. G. A., Isolation of integral membrane proteins and criteria for identifying carrier proteins, in *Current Topics in Membrane and Transport*, Bronner, F. and Kleinzeller, A., Eds., Academic Press, New York, 1979, 1.
28. Duthie, E. S., The production of penicillinase by organisms of the *Subtilis* group, *Br. J. Exp. Pathol.*, 25, 96, 1944.
29. Pollock, M. R., Exoenzymes, in *The Bacteria*, Vol. 4, Gunsalus, I. C. and Stanier, R. Y., Eds., Academic Press, New York, 1962, 121.
30. Citri, N. and Pollock, M. R., The biochemistry and function of β-lactamase (penicillinase), in *Advances in Enzymology*, Vol. 28, Nord, F. F., Ed., John Wiley & Sons, New York, 1966, 237.
31. Fawcett, D. W. *The Cell*, W. B. Saunders, Philadelphia, 1981, 372.
32. Pollock, M. R., The measurement of the liberation of penicillinase from *Bacillus subtilis*, *J. Gen. Microbiol.*, 26, 239, 1961.

33. Pollock, M. R., The mechanism of liberation of penicillinase from *Bacillus subtilis, J. Gen. Microbiol.*, 26, 267, 1961.
34. Kushner, D. J. and Pollock, M. R., Location of cell bound penicillinase in *Bacillus subtilis, J. Gen. Microbiol.*, 26, 255, 1961.
35. Lampen, J. O., Secretion of enzymes by microorganisms, *Symp. Soc. Gen. Microbiol.*, 15, 115, 1965.
36. Lampen, J. O., Movement of extracellular enzymes across cell membranes, *Symp. Soc. Exp. Biol.*, 28, 351, 1974.
37. Ghosh, B. K., Sargent, M. G., and Lampen, J. O., Morphological phenomena associated with penicillinase induction and secretion in *Bacillus licheniformis, J. Bacteriol.*, 96, 1314, 1968.
38. Sargent, M. G., Ghosh, B. K., and Lampen, J. O., Characteristics of penicillinase release by washed cells of *Bacillus licheniformis, J. Bacteriol.*, 96, 1231, 1968.
39. Sargent, M. G., Ghosh, B. K., and Lampen, J. O., Characteristics of penicillinase secretion by growing cells and protoplasts of *Bacillus licheniformis, J. Bacteriol.*, 97, 820, 1969.
40. Collins, J. F., The distribution and formation of penicillinase in a bacterial population of *Bacillus licheniformis, J. Gen. Microbiol.*, 34, 363, 1964.
41. Lampen, J. O., Release of penicillinase by *Bacillus licheniformis, J. Gen. Microbiol.*, 48, 261, 1967.
42. Aiyappa, P. S., Traficante, L. J., and Lampen, J. O., Penicillinase releasing protease of *Bacillus licheniformis:* purification and general properties, *J. Bacteriol.*, 129, 191, 1977.
43. Ghosh, A. and Ghosh, B. K., Immunoelectron microscopic localization of penicillinase in *Bacillus licheniformis, J. Bacteriol.*, 137, 1374, 1979.
44. Lampen, J. O., Nielson, J. B. K., Izui, K., and Caufield, M. P., *Bacillus licheniformis* β-lactamases: multiple forms and their roles, *Phil. Trans. R. Soc. London,* B289, 345, 1980.
45. Lampen, J. O., Bettinger, G. E., and Sharkey, L. J., The membrane bound forms of penicillinase in *Bacillus licheniformis* and their significance for the secretion process, in *Biomembranes,* Vol. 2, Manson, L. A., Ed., Plenum Press, New York, 1971, 211.
46. Nallin, M. and Ghosh, B. K., Alkaline phosphatase release from the *Bacillus licheniformis* 749/C cells in low phosphate casamino acid medium, *Abstr. Annu. Meet. Am. Soc. Microbiol.*, K32, 141, 1976.
47. Nallin, M. and Ghosh, B. K., Alkaline phosphatase release from *Bacillus licheniformis* 749/C, *Abstr. Annu. Meet. Am. Soc. Microbiol.*, K108, p. 144, 1978.
48. Ghosh, A., Nallin, M., and Ghosh, B. K., Physiological regulation of alkaline phosphatase secretion in *Bacillus licheniformis* 749/C, Basic Mechanism of Cellular Secretion Conference, Sept. 17 to 21, 1979, NIH, 3B, 1979.
49. Ghosh, A., Nallin, M., and Ghosh, B. K., Alkaline phosphatase secretion from *Bacillus licheniformis* 749/C, *J. Bacteriol.*, unpublished data.
50. Douglas, W. W., Stimulation-secretion coupling: the concept and clues from chromaffin and other cells, *Br. J. Pharmacol.*, 34, 451, 1968.
51. Burnstock, G., Past and current evidence for the purinergic nerve hypothesis, in *Physiological and Regulatory Functions of Adenosine and Adenine Nucleotides,* Ed., Baier, H. P. and Drummond, G. I., Eds., Raven Press, New York, 1979, 3.
52. Bernlohr, R. W., Maddox, M. K., and Goldberg, N. D., Cyclic guanosine 3′:5′-monophosphate in *E. coli* and *B. licheniformis, J. Biol. Chem.*, 249, 4329, 1974.
53. Sarner, N. Z., Bissell, M. J., Girolamo, M. D., and Gorini, L., Mechanism of excretion of a bacterial proteinase: demonstration of two proteolytic enzymes produced by a *sarcina* strain (Coccus P), *J. Bacteriol.*, 105, 1090, 1971.
54. Bissell, M. J., Tosi, R., and Gorini, L., Mechanism of excretion of a bacterial proteinase: factors controlling accumulation of the extracellular proteinase of a *sarcina* strain (Coccus P), *J. Bacteriol.*, 105, 1099, 1971.
55. Emr, S. D., Hall, M. M., and Silhavy, T. J., A mechanism of protein localization: the signal hypothesis and bacteria, *J. Cell Biol.*, 86, 701, 1980.
56. Casadaban, M. J., Transposition and fusion of the lac genes to selected promoters in *E. coli* using bacteriophage lambda and mu, *J. Mol. Biol.*, 104, 541, 1976.
57. Silhavy, T. J., Bassford, P. J., Jr., and Beckwith, J. R., A genetic approach to the study of protein localilzation in *E. coli,* in *Bacterial Outer Membranes. Biogenesis and Function,* Inouye, M., Ed., John Wiley & Sons, New York, 1979.
58. Bassford, P. J., Jr., Emr, S. D., Silhavy, T. J., Beckwith, J., Bedouelle, H., Clement, J-M., Hedgpeth, J., and Hofnung, M., The genetics of protein secretion in *E. coli,* in *Basic Mechanisms of Cellular Secretion: Methods In Cell Biology,* Vol. 23, Hand, A. and Oliver, C., Eds., Academic Press, New York, 1981, 28.
59. Silhavy, T. J., Casadaban, M. J., Shuman, H. A., and Beckwith, J. R., Conversion of galactosidase to a membrane-bound state by gene fusion, *Proc. Natl. Acad. Sci. U.S.A.*, 73, 3423, 1976.

60. Ito, K. and Beckwith, J. R., Role of the mature protein sequence of maltose-binding in its secretion across the *E. coli* cytoplasmic membrane, *Cell*, 25, 143, 1981.
61. Ito, K., Purification of the precursor form of maltose-binding protein, a periplasmic protein of *Escherichia coli*, *J. Biol. Chem.*, 257, 9895, 1982.
62. Bassford, P. J., Silhavy, T. J., and Beckwith, J. R., Use of gene fusion to study secretion of maltose-binding protein into *Escherichia coli* periplasm, *J. Bacteriol.*, 139, 19, 1979.
63. Ito, K., Bassford, P. J., and Beckwith, J., Protein localization in *E. coli:* is there a common step in the secretion of periplasmic and outer membrane proteins?, *Cell*, 24, 707, 1981.
64. Randall, L. L., Josefsson, L. G., and Hardy, S. J. S., Processing of exported proteins in *Escherichia coli*, *Biochem. Soc. Trans.*, 8, 413, 1980.
65. Bedouelle, H., Bassford, P. J., Fowler, A. V., Zabin, I., Beckwith, J., and Hofnung, M., Mutations which alter the function of the signal sequence of the maltose binding protein of *Escherichia coli*, *Nature (London)*, 285, 78, 1980.
66. Wolfe, P. B. and Wickner, W., Isolation and properties of leader peptidase from *E. coli*, *J. Cell Biochem.*, Suppl. 7B 353, Abst. 1429, 1983.
67. Oxender, D. L., in *Biomembranes*, Vol. 5, Manson, L. A., Ed., Plenum Press, New York, 1974, 25.
68. Oxender, D. L., Anderson, J. J., Daniels, C. J., Landick, R., Gunsalus, R. P., Zurawski,, G., Selker, E., and Yanofsky, C., Structural and functional analysis of cloned DNA containing genes responsible for bronchial-chain amino acid transport in *Escherichia coli*, *Proc. Natl. Acad. Sci. U.S.A.*, 77, 1412, 1980.
69. Bolivar, F., Rodriguez, R. L., Greene, P. J., Betlach, M., Heyneker, J. L., and Boyer, J. W., Construction and characterization of new cloning vehicles. II. A multipurpose cloning system, *Gene*, 2, 95, 1977.
70. Oxender, D. L., Anderson, J. J., Daniels, C. J., Landick, R., Gunsalus, R. P., Zurawski, G., and Yanofsky, C., Amino-terminal sequence and processing of the precursor of the leucine-specific binding protein, and evidence for conformational differences between the precursor and the mature form, *Proc. Natl. Acad. Sci. U.S.A.*, 77, 2005, 1980.
71. Anderson, J. J. and Oxender, D. L., *Escherichia coli* transport mutants lacking binding protein and other components of the branched-chain amino acid transport systems, *J. Bacteriol.*, 130, 384, 1977.
72. Brockman, R. W. and Heppel, L. A., On the localization of alkaline phosphatase and cyclic phosphodiestarase in *E. coli*, *Biochemistry*, 7, 2254, 1968.
73. Heppel, L. A., Selective release of enzymes from bacteria, *Science*, 156, 1451, 1967.
74. Cheng, K. J., Ingram, J. M., and Costerton, J. W., Alkaline phosphatase localization and spheroplast formation of *Pseudomonas aeruginosa*, *Can. J. Microbiol.*, 16, 1319, 1970.
75. Cheng, K. J., Day, D. F., Costerton, J. W., and Ingram, J. M., Alkaline phosphatase subunits in the culture filtrate of *Pseudomonas aeruginosa*, *Can. J. Biochem.*, 50, 268, 1972.
76. Bhatti, A. R. and Ingram, J. M., The binding and sectection of alkaline phosphatase by *Pseudomonas aeruginosa*, *FEMS Microbiology Letters*, 13, 353, 1982.
77. Roy, N. K., Ghosh, R. K., and Das, J., Monomeric alkaline phosphatase of *Vibrio cholerae*, *J. Bacteriol.*, 150, 1033, 1982.
78. Nesmeyanova, M. A., Motlokh, O. B., Kolot, M. N., and Kulaev, I. S., Mulltiple forms of alkaline phosphatase from *Escherichia coli* cells with repressed and derepressed biosynthesis of the enzyme, *J. Bacteriol.*, 146, 453, 1981.
79. Argast, M. and Boos, W., Co-regulation in *Escherichia coli* of a novel transport system for sn-glycerol-3-phosphate and outer membrane protein Ic (e,E) with alkaline phosphatase and phosphate-binding protein, *J. Bacteriol.*, 143, 142, 1980.
80. Inouye, H., Pratt, C., Beckwith, J., and Torriani, A., Alkaline phosphatase synthesis in a cell-free system using DNA and RNA templates, *J. Mol. Biol.*, 110, 75, 1977.
81. Inouye, K. and Beckwith, J., Synthesis and processing of an *Escherichia coli* alkaline phosphatase precursor *in vitro*, *Proc. Natl. Acad. Sci. U.S.A.*, 74, 1440, 1977.
82. Varenne, S., Piovant, M., Pages, J. M., and Lazdunski, C., Evidence for synthesis of alkaline phosphatase on membrane-bound polysomes in *Escherichia coli*, *Eur. J. Biochem.*, 86, 603, 1978.
83. Michaelis, S., Guarente, L., and Beckwith, J., *In vitro* construction and characterization of phoA-lacZ gene fusions in *Escherichia coli*, *J. Bacteriol.*, 154, 356, 1983.
84. Michaelis, S., Inouye, H., Oliver, D., and Beckwith, J., Mutations that alter the signal sequence of alkaline phosphatase in *Escherichia coli*, *J. Bacteriol.*, 154, 366, 1983.
85. Wilson, V. G. Hogg, R. W., The terminal sequence of a precursor form of the arabinose binding protein, *J. Biol. Chem.*, 14, 6745, 1980.
86. Randall, L. L., Hardy, S. J. S., and Josefsson, L., Precursors of three exported proteins in *Escherichia coli*, *Proc. Natl. Acad. Sci. U.S.A.*, 75, 1209, 1978.
87. Date, T., Zwizinski, C., Ludmerer, S., and Wichner, W., Mechanisms of membrane assembly: effects of energy poisons on the conversion of soluble M13 coliphage procoat to membrane-bound coat protein, *Proc. Natl. Acad. Sci. U.S.A.*, 77, 827, 1980.

88. Date, T., Goodman, J. M., and Wickner, W. T., Procoat, the precursor of M13 coat protein, requires an electrochemical potential for membrane insertion, *Proc. Natl. Acad. Sci. U.S.A.*, 77, 4669, 1980.
89. Zimmermann, R., Watts, C., and Wickner, W., The biosynthesis of membrane-bound M13 coat protein, *J. Biol. Chem.*, 257, 6529, 1982.
90. Asbeck, V. F., Beyreuther, K., Kohler, H., vonWettstein, G., and Braunitzer, G., *Hoppe-Seyler's Z. Physiol. Chem.*, 350, 1047, 1969.
91. Nakashima, Y. and Konigsberg, W., Reinvestigation of a region of the fd bacterio-phage coat protein sequence, *J. Mol. Biol.*, 88, 598, 1974.
92. Wickner, W., Assymetric orientation of a phage coat protein in cytoplasmic membrane of *E. coli*, *Proc. Natl. Acad. Sci. U.S.A.*, 72, 4749, 1975.
93. Smilowitz, H., Carson, J., and Robbins, P. W., Association of newly synthesized major fl coat protein with infected host cell inner membrane, *J. Supramol. Struct.*, 1, 8, 1972.
94. Mandel, G. and Wickner, W., Translational and post-translational cleavage of M13 procoat protein: extracts of both the cytoplasmic and outer membranes of *Escherichia coli* contain leader peptidase activity, *Proc. Natl. Acad. Sci., U.S.A.*, 76, 236, 1979.
95. Ito, K., Mandel, G., and Wickner, W., Soluble precursor of an integral membrane protein: synthesis of procoat protein in *Escherichia coli* infected with bacteriophage M13, *Proc. Natl. Acad. Sci, U.S.A.*, 76, 1199, 1979.
96. Ohkawa, I. and Webster, R., The orientation of the major coat protein of bacteriophage fl in the cytoplasmic membrane of *Escherichia coli*, *J. Biol. Chem.*, 256, 9951, 1981.
97. Chang, C. N., Blobel, G., and Model, P., Detection of prokaryotic signal peptidase in an *Escherichia coli* membrane fraction: endoproteolytic cleavage of nascent fl pre-coat protein, *Proc. Natl. Acad. Sci. U.S.A.*, 75, 361, 1978.
98. Chang, C. N., Model, P., and Blobel, G., Membrane biogenesis: cotranslational integration of the bacteriophage fl coat protein into an *Escherichia coli* membrane fraction, *Proc. Natl. Acad. Sci. U.S.A.*, 76, 1251, 1979.
99. Wickner, W., Asymmetric orientation of phage M13 coat protein in *E. coli* cytoplasmic membranes and in syntheitic lipid vesicles, *Proc. Natl. Acad. Sci. U.S.A.*, 73, 1159, 1976.
100. Snell, D. T. and Offord, R. E., The amino acid sequence of the B-Protein of bacteriophage ZJ-2, *Biochem. J.*, 127, 167, 1972.
101. Randall-Hazelbauer, L. and Schwartz, M., Isolation of the bacteriophage lambda receptor from *Escherichia coli*, *J. Bacteriol.*, 116, 1436, 1973.
102. Szmelcman, S. and Hofnung, M., Maltose transport in *Escherichia coli* K12: involvement of the bacteriophage Lambda receptor, *J. Bacteriol.*, 124, 112, 1975.
103. Hazelbauer, G. L., Role of the receptor for bacteriophage Lambda in the functioning of the maltose chemoreceptor of *Escherichia coli*, *J. Bacteriol.*, 124, 119, 1975.
104. Silhavy, T. J., Shuman, H. A., Beckwith, J., and Schwartz, M., Use of gene fusions to study outer membrane protein localization in *Escherichia coli*, *Proc. Natl. Acad. Sci. U.S.A.*, 74, 5411, 1977.
105. Emr, S. D., Schwartz, M., and Silhavy, T. J., Mutations altering the cellular localization of the phase λ receptor, an *Escherichia coli* outer membrane protein, *Proc. Natl. Acad. Sci. U.S.A.*, 75, 5802, 1978.
106. Marchal, C., Perrin, D., Hedgpeth, J., and Hofnung, M., Synthesis and maturation of λ receptor in *Escherichia coli* K-12: *in vivo* and *in vitro* expression of gene lamB under lac promoter control, *Proc. Natl. Acad. Sci. U.S.A.*, 77, 1491, 1980.
107. Hedgpeth, J., Clement, J., Marchal, C., Perrin, D., and Hofnung, M., DNA sequence encoding the NH_2-terminal peptide involved in transport of λ receptor, an *Escherichia coli* secretory protein, *Proc. Natl. Acad. Sci. U.S.A.*, 77, 2621, 1980.
108. Inouye, M., in *Bacterial Outer Membrane. Biogenesis and Function*, John Wiley & Sons, New York, 1979.
109. Nikaido, H. and Nakae, T., The outer membrane of gram-negative bacteria, in *Advances in Microbial Physiology, Vol. 20*, Rose, A. H. and Morris, J. G., Eds., Academic Press, London, 1979, 163.
110. Halegoua, S. and Inouye, M., Translocation and assembly of outer membrane proteins of *Escherichia coli*, *J. Mol. Biol.*, 39, 1979.
111. Inouye, S., Wang, S., Sekizawa, J., Halegoua, S., and Inouye, M., Amino acid sequence for the peptide extension on the prolipoprotein of the *Escherichia coli* outer membrane, *Proc. Natl. Acad. Sci. U.S.A.*, 74, 1004, 1977.
112. Lin, J. J. C., Kanazawa, H., Ozols, J., and Wu, H. C., An *Escherichia coli* mutant with an amino acid alteration within the signal sequence of outer membrane prolipoprotein, *Proc. Natl. Acad. Sci. U.S.A.*, 75, 4891, 1978.
113. Crowlesmith, I., Gamon, K., and Henning, U., Precursor proteins are intermediates *in vivo* in the synthesis of two major outer membrane proteins, the OmpA and OmpF proteins of *Escherichia coli* K-12, *Eur. J. Biochem.*, 113, 375, 1981.

114. Randall, L. L. and Hardy, S. J. S., Synthesis of exported proteins by membrane bound polysomes from *Escherichia coli, Eur. J. Biochem.*, 75, 43, 1977.
115. Randall, L. L., Hardy, S. J. S., and Josefson, L. G., Precursors of three exported proteins in *Escherichia coli, Proc. Natl. Acad. Sci. U.S.A.*, 75, 1209, 1978.
116. Movva, N. R., Nakamura, K., and Inouye, M., Amino acid sequence of the signal peptide of ompA protein, a major outer membrane protein of *Escherichia coli, J. Biol. Chem.*, 255, 27, 1980.
117. De Leij, L., Kingma, J., and Witholt, B., Insertion of newly synthesized proteins into the outer membrane of *Escherichia coli, Biochem. Biophys. Acta.*, 512, 365, 1978.
118. Lin. J. J. C., and Wu, H. C., Assembly of membrane proteins in *Escherichia coli, J. Biol. Chem.*, 255, 802, 1980.
119. Lin. J. J. C., Giam, C. Z., and Wu. H. C., Assembly of the outer membrane lipoprotein in *Escherichia coli, J. Biol. Chem.*, 255, 807, 1980.
120. Debabov, V. G., The industrial use of Bacilli, in *The Molecular Biology of the Bacilli*, Vol. 1., Dubnau, D. A., Ed., Academic Press, New York, 1982, 331.
121. Priest, F. G., Extracellular enzyme synthesis in the genus Bacillus, *Bacteriol. Rev.*, 41, 711, 1977.
122. Nielson, J. B. K. and Lampen. J. O., Membrane-bound penicilinase in gram-positive bacteria, *J. Biol. Chem.*, 257, 4490, 1982.
123. Ambler, R. P., The structure of β-lactamases, *Phil. Trans. R. Soc. London Ser. B*, 289, 321, 1980.
124. Jaurin, B. and Grundstrom, T., AmpC alphalosparinase of *Escherichia coli* K-12 has a different evolutionary origin from that of β-lactamases of the penicillinase type, *Proc. Natl. Acad. Sci. U.S.A.*, 78, 4897, 1981.
125. Jaurin, B., Grundstrom, T., Edlund, T., and Normark, S., The *Escherichia coli* β-lactamase attenuator mediates growth rate dependent regulation, *Nature (London)*, 290, 221, 1981.
126. Sarvas, M., Hirth, K. P., Fuchs, E., and Simons, K., A precursor form of the penicillinase from *Bacillus licheniformis, FEBS Lett.*, 95, 76, 1978.
127. Fuchs, E., The interdependence of magnesium with spermidine and phosphoenol pyruvate in an enzyme synthesizing system *in vitro, Eur. J. Biochem.*, 63, 15, 1976.
128. Chang, C. N., Nielson, J. B. K., Izui, K., Blobel, G., and Lampen, J. O., Identification of the signal peptidase cleavage site in *Bacillus licheniformis* prepenicillinase, *J. Biol. Chem.*, 257, 4340, 1982.
129. Nielsen, J. B. K., Caulfield, M. P., and Lampen, J. O., Lipoprotein nature of *Bacillus licheniformis* membrane penicillinase, *Proc. Natl. Acad. Sci. U.S.A.*, 78, 3511, 1981.
130. Lai, J. S., Sarvas, M., Brammar, W. J., Neugebauer, K., and Wu, H. C., *Bacillus licheniformis* penicillinase synthesized in *Escherichia coli* contain covalently linked fatty acid and glyceride, *Proc. Natl. Acad. Sci. U.S.A.*, 78, 3506, 1981.
131. Schmidt, M. F. G., Bracha, M., and Schlesinger, M. J., Evidence for covalent attachment of fatty acids to Sindbis virus glycoproteins, *Proc. Natl. Acad. Sci. U.S.A.*, 76, 1687, 1979.
132. Schlesinger, M. J., Magee, A. I., and Schmidt, M. F. G., Fatty acid acylation of proteins in cultured cells, *J. Biol. Chem.*, 255, 10021, 1980.
133. Smith, W. P., Tai, P. C., and Davis, B. D., *Bacillus licheniformis* penicillinase: cleavages and attachment of lipid during cotranslational secretion, *Proc. Natl. Acad. Sci. U.S.A.*, 78, 3501, 1981.
134. Nagata, Y., Yamaguchi, K., and Maruo, B., Genetic and biochemical studies on cell-bound α-amylase in *Bacillus subtilis* Marburg, *J. Bacteriol.*, 119, 425, 1974.
135. Rio, L. F. R. and Arroyo-Begovich, A., Evidence for the presence of α-amylase in the cell membrane of *Bacillus amyloliquefaciens, Biochem. Biophys. Res. Commun.*, 65, 161, 1975.
136. Welker, N. E. and Campbell, L. L., Unrelatedness of *Bacillus amyloliquefaciens* and *Bacillus subtilis, J. Bacteriol.*, 94, 1124, 1967.
137. Yamaguchi, K., Nagata, Y., and Maruo, B., Isolation of mutants defective in α-amylase from *Bacillus subtilis* and their genetic analysis, *J. Bacteriol.*, 119, 416, 1974.
138. Yuki, S., The chromosomal location of the structural gene for amylase in *Bacillus subtilis, Jpn. J. Genet.*, 50, 155, 1975.
139. Sekiguchi, J., Takada, N., and Okada, H., Genes affecting the productivity of α-amylase in *Bacillus subtilis* Marburg, *J. Bacteriol.*, 121, 688, 1975.
140. Yamaguchi, K., Nagata, Y., and Maruo, B., Genetic control of the rate of α-amylase synthesis in *Bacillus subtilis, J. Bacteriol.*, 119, 410, 1974.
141. Yoneda, Y. and Maruo, B., Mutation of *Bacillus subtilis* causing hyperproduction of α-amylase and protease and its synergistic effect, *J. Bacteriol.*, 124, 48, 1975.
142. Yoneda, Y., Yamane, K., and Maruo, B., Membrane mutation related to the production of extracellular α-amylase and protease in *Bacillus subtilis, Biochem, Biophys. Res. Commun.*, 50, 765, 1973.
143. Yoneda, Y. and Maruo, B., Mutation of *Bacillus subtilis* causing hyperproduction of μ-amylase and protease, and its synergistic effect, *J. Bacteriol.*, 124, 48, 1975.
144. Mantsala, P. and Zalkin, H., Membrane-bound and soluble extracellular α-amylase from *Bacillus subtilis, J. Biol. Chem.*, 254, 8540, 1979.

145. Palva, I., Pettersson, R. F., Kalkkinen, N., Lehtovaara, P., Sarvas, M., Soderlund, H., Takkinen, K., and Kaarianinen, L., Nucleotide sequence of the promoter and NH_2-terminal signal peptide region of the α-amylase gene from *Bacillus amyloliquefaciens, Gene,* 15, 43, 1981.
146. Gryczan, T., Contente, S., and Dubnau, D., Characterization of *Staphylococcus aureus* plasmids introduced by transformation into *Bacillus subtilis, J. Bacteriol.,* 134, 318, 1978.
147. Ghosh, R., Ghosh, A., and Ghosh, B. K., Properties of the membrane-bound alkaline phosphatase from glucose and lactate-grown cells of *Bacillus subtilis* SB15, *J. Biol. Chem.,* 252, 6813, 1977.
148. Ghosh, A., Vallespir, S., Manegus, M., Nelson, R., Kumar, R., and Ghosh B. K., Membrane bound and extracellular alkaline phosphatases of *Bacillus licheniformis* 749/C, *Eur. J. Biochem.* (communicated), 1984.
149. Nelson, R., Membrane Phenomena of Alkaline Phosphatase Secretion in *Bacillus licheniformis* 749/C, Ph.D. thesis, Rutgers, The State University, New Brunswick,, N.J., 1983.
150. Kumar, R., Ghosh, A., and Ghosh, B. K., Alkaline phosphatase secretion-negative mutant of *Bacillus licheniformis* 749/C, *J. Bacteriol.,* 154, 946, 1983.
151. Bedouelle, H. and Hofnung, M., in *Membrane Transport and Neuroreceptors,* Liss, New York, 1981, 399.
152. Bedouelle, H. and Hofnung, M., in *Intermolecular Forces,* Pullman, B., Ed., Reidel, 1981, 361.
153. Chou, P. Y. and Fasman, G. D., *Avd. Enzymol.,* 47, 45, 1978.
154. Engelman, D. M. and Steitz, T. A., The spontaneous insertion of proteins into and across membranes: the helical hairpin hypothesis, *Cell,* 23, 411, 1981.
155. Koshland, D. and Botstein, D., *Cell,* 20, 749, 1980.
156. Ghosh, B. K., Wouters, J. T. M., and Lampen, J. O., Distribution of the sites of alkaline phosphatase(s) activity in vegetative cells of *Bacillus subtilis, J. Bacteriol.,* 108, 928, 1971.
157. McNicholas, J. M. and Hulett, F. M., Electron microscopic histochemical localization of alkaline phosphatase(s) in *Bacillus licheniformis, J. Bacteriol.,* 129, 501, 1977.
158. Ghosh, A., Vallespir, S., and Ghosh, B. K., Specificity of subcellular distribution of alkaline phosphatase in *Bacillus licheniformis* 749/C, *Can. J. Microbiol.,* 30, 113, 1984.
159. Cheng, K. J. and Costerton, J. W., Localization of alkaline phosphatase in three gram-negative rumen bacteria, *J. Bacteriol.,* 116, 424, 1973.
160. Ghosh, B. K., Guan, T., and Ghosh, A., Immunoelectron microscopic localization of a secretory protein in bacteria, *J. Cell Biochem.,* 7B, 353 (1428), 1983.
161. Guan, T., Ghosh, A., and Ghosh, B. K., Collidal gold double labeling of two secretory proteins in bacterial cell, *Proc. 41st Annu. Mtg. Electron Microscopic Society of America,* p. 614, 1983.
162. Oliver, D. B. and Beckwith, J., *Escherichia coli* mutant pleiotropically defective in the export of secreted proteins, *Cell,* 25, 765, 1981.
163. Yuki, S. and Ueda, Y., Fine mapping analysis of the amylase gene in *Bacillus subtilis* by transformation, *Jpn. J. Genet.,* 43, 121, 1968.
164. Yamaguchi, K., Nagata, Y., and Maruo, B., Isolation of mutants defective in amylase from *Bacillus subtilis*: genetic analyses, *J. Bacteriol.,* 119, 414, 1974.
165. Yoneda, Y., Yamane, K., and Maruo, B., Membrane mutation related to the production of extracellular α-amylase and protease in *Bacillus subtilis, Biochem. Biophys. Res. Commun.,* 50, 765, 1973.
166. Steinmetz, M., Kunst, F., and Dedonder, R., Mapping of mutations affecting synthesis of exocellular enzymes in *B. subtilis, Mol. Gen. Genet.,* 148, 281, 1976.
167. Kumar, R., Ghosh, A., and Ghosh, B. K., Construction of alkaline phosphatase secretor mutants of *Bacillus subtilis* SB15, *Am. Soc. Microbiol.,* 83, 123, (Abstr. H105), 1983.
168. Ito, K., Bassford, P. J., Jr., and Beckwith, J., Protein localization on *E. coli*: is there a common step in the secretion of periplasmic and outer-membrane proteins?, *Cell,* 24, 707, 1981.
167. Kumamoto, C. A. and Beckwith, J., Mutations in new gene, *secB,* cause defective protein localization in *Escherichia coli, J. Bacteriol.,* 154, 253, 1983.
170. Emr, S. D., Hanley-Way, S., and Silhavy, T. J., Suppressor mutations that restore export of a protein with a defective signal sequence, *Cell,* 23, 79, 1981.
171. Jaskunas, S. R., Lindahl, L., and Nomura, M., Isolation of polar insertion mutants and the director of transcription of ribosomal protein genes in *Escherichia coli, Nature (London),* 256, 183, 1975.
172. Lindahl, L., Post, L., Zengel, J., Gilbert, S. F., Strycharz, W. A., and Nomura, M., Mapping of ribosomal protein genes by *in vitro* protein synthesis using DNA fragments of λ fus3 transducing phage DNA templates, *J. Cell Biol.,* 67, 835, 1977.
173. Muller, M., Ibrahimi, I., Chang, C. N., Walter, P., and Blobel, G., A bacterial secretory protein requires signal recognition particle for translocation across mammalian endoplasmic reticulum, *J. Biol. Chem.,* 257, 11860, 1982.
174. Walter, P. and Blobel, G., Purification of a membrane-associated protein complex required for protein translocation across the endoplasmic reticulum, *Proc. Natl. Acad. Sci. U.S.A.,* 77, 7112, 1980.
175. Walter, P., Ibrahimi, I., and Blobel, G., Translocation of proteins across the endoplasmic reticulum. I. Signal recognition protein (SRP) binds to *in vitro* assembled polysomes synthesizing secretory protein, *J. Cell Biol.,* 91, 545, 1981.

176. Walter, P. and Blobel, G., Translocation of proteins across the endoplasmic reticulum. III. Signal recognition protein (SRP) causes signal sequence-dependent and site-specific arrest of chain elongation that is released by microsomal membranes, *J. Cell Biol.*, 91, 557, 1981.
177. Walter, P. and Blobel, G., Translocation of proteins across the endoplasmic reticulum. II. Signal recognition protein (SRP) mediates the selective binding to microsomal membranes of *in vitro* assembled polysomes synthesizing secretory protein, *J. Cell Biol.*, 91, 551, 1981.
178. Kreibich, G., and Ulrich, B. L., and Sebatini, D. D., I. Identification of ribophorins I and II, membrane proteins characteristic of rough microsomes, *J. Cell Biol.*, 77, 464, 1978.
179. Kreibich, G., Ulrich, B. L., Pereyram, D., Sabatini, D., and Freinstein, C. M., II, Cross-linking of bound ribosomes to specific membrane proteins exposed at the binding sites, *J. Cell Biol.*, 77, 488, 1978.
180. Tai, P. C., Smith, W. P., and Davis, B. D., Direct demonstration of secretion of nascent polypeptide chains across membrane in bacteria, in *Limited Proteolysis in Microorganisms,* Cohen, G. N. and Holzer, H., Eds., Fogany Conf. NIH, 171-175, U.S. Govt. Printing Office, Washington.
181. Smith, W. P., Tai, P. C., and Davis, B. D., Nascent peptide as sole attachment of polysomes to membranes in bacteria, *Proc. Natl. Acad. Sci. U.S.A.*, 75, 814, 1978.
182. Enequist, H. G., Hirst, T. R., Harayama, S., Hardy, S. J. S., and Randall, L. L., Energy is required for maturation of exported proteins in *E. coli, Eur. J. Biochem.*, 116, 227, 1981.
183. Daniels, C. J., Bole, D. G., Quay, S. C., and Oxender, D. L., Role for membrane potential in the secretion of protein into the periplasm of *Escherichia coli, Proc. Natl. Acad. Sci. U.S.A.*, 78, 5396, 1981.
184. Butlin, J. D., Cox, G. B., and Gibson, F., Oxidative phosphorylation in *E. coli* K12, *Biochem. J.*, 124, 75, 1971.
185. Hille, B., Gating in sodium channels of nerve, *Annu. Rev. Physiol.*, 38, 139, 1976.
186. Pages, J. M., Piovant, M., Varenne, S., and Lazdunski, C., Mechanistic aspects of the transfer of nascent periplasmic proteins across the cytoplasmic membrane in *Escherichia coli, Eur. J. Biochem.*, 86, 589, 1978.
187. Torriani, A., Alkaline phosphatase subunits and their dimerization *in vivo, J. Bacteriol.*, 96, 1200, 1968.
188. Thilo, L., Trauble, H., and Overath, P., Mechanistic interpretation of the influence of lipid phase transitions on transport functions, *Biochemistry,* 16, 1283, 1977.
189. Overath, P., Schairer, H. U., and Stoffel, W., Correlation of *in vivo* and *in vitro* phase transitions of membrane lipids in *Escherichia coli, Proc. Natl. Acad. Sci. U.S.A.*, 67, 606, 1970.
190. Glew, R. H. and Heath, E. C., Studies on the extracellular alkaline phosphatase of *Micrococcus sodonensis*. II. Factors affecting secretion, *J. Biol. Chem.*, 246, 1566, 1971.
191. Kimura, K. and Izui, K., Importance of membrane fluidity in the induction of alkaline phosphatase, a periplasmic enzyme in *Escherichia coli, Biochem. Biophys. Res. Commun.*, 70, 900, 1976.
192. Bayer, M. E., Role of adhesion zones in bacterial cell surface function and biogenesis, in *Membrane Biogenesis,* Tzagoloff, A., Ed., Plenum Press, New York, 1975, 393.
193. Petit-Glatron, M. and Chambert, R., Levansucrase of *Bacillus subtilis:* convulsive evidence that its production and export are unrelated to fatty-acid synthesis but modulated by membrane-modifying agents, *Eur. J. Biochem.*, 119, 603, 1981.
194. Fishman, Y., Rottem, S., and Citri, N., Evidence linking penicillinase formation and secretion to lipid metabolism in *Bacillus licheniformis, J. Bacteriol.*, 134, 434, 1978.
195. Fishman, Y., Rottem, S., and Citri, N., Preferential suppression of normal exoenzyme formation by membrane modifying agents, *J. Bacteriol.*, 141, 1435, 1980.
196. Higgins, C. F. and Ames, G. F-L., Two periplasmic transport proteins which interact with a common membrane receptor show extensive homology: complete nucleotide sequences, *Proc. Natl. Acad. Sci. U.S.A.*, 78, 6038, 1981.
197. Smith, W. P., Tai, P. C., Thompson, R. C., and Davis, B. D., Extracellular labeling of nascent polypeptides traversing the membrane of *Escherichia coli, Proc. Natl. Acad. Sci. U.S.A.*, 74, 2830, 1977.
198. Smith, W. P., Tai, P. C., and Davis, B. D., Extracellular labeling of growing secreted polypeptide chains in *Bacillus subtilis* with diazoiodosulfanilic acid, *Biochemistry,* 18, 198, 1979.
199. Randall, L. L., Translocation of domains of nascent periplasmic proteins across the cytoplasmic membrane is independent of elongation, *Cell,* 33, 231, 1983.
200. Zwizinski, C., Date, T., and Wickner, W., Leader peptidase is found in both the inner and outer membranes of *Escherichia coli, J. Biol. Chem.*, 256, 3593, 1981.
201. Wickner, W. and Killick, T., Membrane-associated assembly of M13 phage in extracts of virus-infected *Escherichia coli, Proc. Natl. Acad. Sci. U.S.A.*, 74, 505, 1977.
202. Clarke, L. and Carbon, J., A colony bank containing synthetic col EL hybrid plasmids representative of the entire *E. coli* genome, *Cell,* 9, 91, 1976.

203. Ferrazza, D. and Levy, S. B., Biosynthesis of a plasmid-encoded outer membrane surface exclusion protein involves processing from a precursor polypeptide, *J. Biol. Chem.*, 255, 8955, 1980.
204. Josefsson, L. G. and Randall, L. L., Different exported proteins in *E. coli* show differences in the temporal mode of processing in vivo, *Cell*, 25, 151, 1981.
205. Heijne, G. V. and Blomberg, C., Transmembrane translocation of proteins, *Eur. J. Biochem.*, 97, 175, 1979.
206. Heijne, G. V., Transmembrane translocation of proteins: a detailed physico-chemical analysis, *Eur. J. Biochem.*, 103, 431, 1980.
207. Costerton, J. W., Irvin, R. T., and Cheng, K. J., The bacterial glycocalyx in nature and disease, *Annu. Rev. Microbiol.*, 35, 299, 1981.
208. Smith, H., Microbial surfaces in relation to pathogenicity, *Bacteriol. Rev.*, 41, 475, 1977.
209. Smith, H. and Pearce, J. H., Ed., Microbial pathogenicity in man and animals, *Symp. Soc. Gen. Microbiol.*, 22, 1, 1972.
210. Smith. H., Microbial behaviour in natural and artificial environments, *Symp. Soc. Gen. Microbiol.*, 11, 1, 1964.
211. Smith, H., Survival of vegetative bacteria in animals, *Symp. Soc. Gen. Microbiol.*, 26, 229, 1976.
212. Ellwood, D. C. and Tempest, D. W., Effects of environment on bacterial wall content and composition, *Avd. Microbiol. Physiol.*, 7, 83, 1972.
213. Ellwood, D. C., Growth environment and bacterial toxicity, *J. Med. Microbiol.*, 7, 391, 1974.
214. Finch, J. E. and Brown, M. R. W., The effect of growth environment on killing of chemostat grown *Pseudomonas aeruginosa* by phagocytes and cationic proteins, *Proc. Soc. Gen. Microbiol.*, 3, 82, 1976.
215. Kochan, I., The role of iron in bacterial infections, with special consideration of host tubercle bacillus interaction, *Curr. Top. Microbiol. Immunol.*, 60, 1, 1973.
216. Neilands, J. B., Microbial iron transport compounds (siderochromes), in *Inorganic Biochemistry, Vol. 1*, Eichhorn, G. L., Ed., Elsevier, Amsterdam, 1973, 167.
217. Schlesinger, D., Ed., *Microbiology*, American Society for Microbiology, Washington, D.C., 1975.
218. Parkinson, C. F. and Carter, P. B., Phagocytosis of *Mycoplasma sulivarium* by human polymorpholeukocytes and monocytes, *Infect. Inmunol.*, 11, 405, 1975.
219. Schwab, J. H., Suppression of the immune response by microorganisms, *Bacteriol. Rev.*, 39, 121, 1975.
220. Ajl, S. J., Kadis, S., and Montie, T. C., *Microbial Toxins*, Academic Press, New York, 1970.
221. Wardlaw, A. C., Parton, R., and Hooker, M. J., Loss of protective antigen, histamine-sensitising factor, and envelope polypeptides in cultural variants of *Bordetella pertussis*, *J. Med. Microbiol.*, 9, 89, 1976.
222. Wardlaw, A. C., Endotoxins in relation to disease, *Proc. Soc. Gen. Microbiol.*, 1, 50, 1974.
223. Urbaschek, B., Urbaschek, R., and Neter, E., *Gram-Negative Bacterial Infections and Mode of Endotoxin Actions. Pathophysiological, Immunological, and Clinical Aspects*, Springer-Verlag, Vienna, 1975.
224. Burrows, W., Cholera toxins, *Ann. Rev. Microbiol.*, 22, 245, 1968.
225. Dallas, W. S. and Falkow, S., Amino acid sequence homology between cholera toxin and *Escherichia coli* heat-labile toxin, *Nature (London)*, 288, 499, 1980.
226. Evans, D. J., Chin, L. C., Curlin, G. T., and Evans, D. G., Stimulation of adenyl cyclase by *Escherichia coli* enterotoxin, *Nature (London) New Biol.*, 236, 137, 1972.
227. Chen, L., Rhode, J. E., and Sharp, G. W. G., Properties of adenyl cyclase from human jejunal mucosa during naturally acquired cholera and convalescence, *J. Clin. Invest.*, 51, 731, 1972.
228. Gill, D. M., Evans, D. J., Jr., and Evans, D. G., Mechanism of activation of adenylate cyclase *in vitro* by polymyxin-released, heat labile enterotoxin of *Escherichia coli*, *J. Infect. Dis. Suppl.*, 133, S103, 1976.
229. Cassel, D. and Pfeuffer, T., Mechanism of cholera toxin action: covalent modification of the guanyl nucleotide binding protein of the adenylate cyclase system, *Proc. Natl. Acad. Sci. U.S.A.*, 75, 2669, 1978.
230. Moss, J. and Richardson, S. H. J., Activation of adenylate cyclase by heat-labile *Escherichia coli* enterotoxin, *J. Clin. Invest.*, 62, 281, 1979.
231. Lonnroth, I. and Holmgren, J., Subunit structure of cholera toxin, *J. Gen. Microbiol.*, 76, 417, 1973.
232. Clements, J. D. and Finkelstein, R. A., Immunological cross reactivity between a heat-labile enterotoxin(s) of *Escherichia coli* and subunits of *Vibrio cholerae* enterotoxin, *Infect. Innumol.*, 21, 1036, 1978.
233. Dallas, W. S., Gill, D. M., and Falkow, S., Cistrons encoding *Escherichia coli* heat-labile toxin, *J. Bacteriol.*, 139, 850, 1979.
234. Holmgren, J., Comparison of the tissue receptors for *Vibrio cholerae* and *Escherichia coli* enterotoxins by means of gangliosides and natural cholera toxoid, *J. Infect. Immunol.*, 8, 851, 1973.

235. Clements, J. D. and Finkelstein, R. A., Isolation and characterization of homogeneous heat-labile enterotoxins with high specific activity from *Escherichia coli*, *Infect. Immunol.*, 24, 760, 1979.
236. Clements, J. D., Yancey, R. J., and Finkelstein, R. A., Properties of homogeneous heat-labile enterotoxin from *Escherichia coli*, *Infect. Immunol.*, 29, 91, 1980.
237. Finkelstein, R. A., Peterson, J. W., and Lo-Spalluto, J. J., Conversion of cholera exo-enterotoxin (choleragen) to natural toxoid (choleragenoid), *J. Immunol.*, 1106, 868, 1971.
238. Yamamoto, T. and Yokota, T., Release of heat-labile enterotoxin subunits by *Escherichia coli*, *J. Bacteriol.*, 150, 1482, 1982.
239. Wensink, J., Genkema, H., Jansen, W. H., Guinee, P A. M., and Witholt, B., Isolation of the membranes of a enterotoxigenic strain of *Escherichia coli* and distribution of enterotoxin activity in different subcellular fraction, *Biochim. Biophys. Acta.*, 514, 128, 1978.
240. Yamamoto, T. and Yokota, T., *Escherichia coli* heat-labile enterotoxin genes are flanked by repeated deoxyribonucleic acid sequences, *J. Bacteriol.*, 145, 850, 1981.
241. Dallas, W. S. and Falkow, S., The molecular nature of heat-labile enterotoxin (LT) of *Escherichia coli*, *Nature (London)*, 277, 406, 1979.
242. So, M., Dallas, W. S., and Falkow, S., Characterization of an *Escherichia coli* plasmid encoding for synthesis of heat-labile toxin: molecular cloning of the toxin, *Infect. Immunol.*, 21, 405, 1978.
243. Frazer, A. C. and Curtiss, R., III, Production, properties, and utility of bacterial minicells, *Curr. Top. Microbiol. Immunol.*, 69, 1, 1978.
244. Wensink, J., Gankema, H., Jansen, W. H., Guinee, P. A. M., and Witholt, B., Isolation of the membranes of an enterotoxigenic strain of *Escherichia coli* and distribution of enterotoxin activity in different subcellular fractions, *Biochim. Biophys. Acta*, 514, 128, 1978.
245. Lariviere, S., Gyles, C. L., and Barnum, D. A., Preliminary characterization of the heat-labile enterotoxin of *Escherichia coli* F11 (P155), *J. Infect. Dis.*, 128, 312, 1973.
246. Hoekstra, D., van der Laan, J. W., de Leij, L., and Witholt, B., Release of outer membrane fragments from normally growing *E. coli*, *Biochim. Biophys. Acta.*, 455, 889, 1976.
247. Mug-Opstelten, D. and Witholt, B., Preferential release of new outer membrane fragments by exponetially growing *Escherichia coil*, *Biochim. Biophys. Acta.*, 508, 287, 1978.
248. Jacks, T. M. and Wu, B. J., Biochemical properties of *Escherichia coli* low-molecular-weight, heat stable enterotoxin, *Infect. Immunol.*, 9, 342, 1974.
249. Field, M., Graf, L. H., Laird, W. J., and Smith, P. L., Heat-stable enterotoxin of *Escherichia coli*: *in vitro* effects on guanylate cyclase activity, cyclic GMP concentration, and ion transport in small intestine, *Proc. Natl. Acad. Sci. U.S.A.*, 75, 2800, 1978.
250. Burgess, M. N., Bywater, R. J., Cowley, C. M., Millan, N. A., and Newsome, P. M., Biological evaluation of a methanol-soluble, heat-stable *Escherichia coli* enterotoxin in infant mice, pigs, rabbits, and calves, *Infec. Immunol.*, 21, 526, 1978.
251. So, M., Heffron, F., and McCarthy, B. J., The *Escherichia coli* encoding heat stable toxin is a bacterial transposon flanked by inverted repeats of ISI, *Nature (London)*, 277, 453, 1979.
252. So, M. and McCarthy, B. J., Nucleotide sequence of the bacterial transposon Tn1681 encoding a heat-stable (ST) toxin and its identification in enterotoxigenic *Escherichia coli* strains, *Proc. Natl. Acad. Sci. U.S.A.*, 77, 4011, 1980.
253. Neill, R. J., Ivins, B. E., and Holmes, R. K., Synthesis and secretion of the plasmid-coded heat-labile enterotoxin of *E. coli* in *Vibrio cholerae*, *Science*, 221, 289, 1933.
254. *Chem. Week*, 123, 111, 1978.
255. Phaff, H. J., Industrial microorganisms, *Sci. Am.*, 245, 76, 1981.
256. Hoopwood, A., The genetic programming of industrial microorganisms, *Sci. Am.*, 245, 90, 1981.

INDEX

A

Acetate-grown cells, 15
Acetone-precipitated membranes, 29
Acetone-treated membrane-PACT activity, 30—32
Acetone-treated membranes, 28, 30
N-Acetyl muramyl-alanine-amidase, 144, 217
Acholeplasma laidlawii, 117, 134
Acidic phospholipids, 130
Acinetobacter
  lwoffi, 15
  sp., see also *Acinetobacter* sp. H01-N, 2—3
    biochemical composition, 23—26
    cellular neutral lipids, 15—16
    cellular phospholipids, 15
    cytoplasmic membranes, 37—39
    extracellular lipids, 16—17
    extracellular membrane vesicles, 36—37
    hexadecane-grown, membranes derived from, 18
    hydrocarbon inclusions, 4—7, 18, 23, 36
    hydrocarbon metabolism, 10
    intracellular pools of hexadecane, 16
    intracytoplasmic membrane, 9—10, 15, 18, 36
    lipids of hexadecane-grown, 15—17
    membrane biogenesis, 40—41
    membranes of, 15—26
    nutrient broth-grown, membranes derived from, 18
    outer membrane, 18, 39—40
    phospholipid A_1 activities, 39—40
  sp. H01-N, see also *Acinetobacter* sp.
    cellular lipid composition, 16
    cellular phospholipid composition, 15
    cytochrome composition, 26
    diagrammatic model for alkane uptake and packaging, 40—41
    extracellular lipid accumulation, 17
    extracellular neutral lipid composition, 16
    hexadecane-grown, 20—22
    isolation of membranes from, 17—23
    mesosome-like membranes, 9
    phospholipid biosynthesis, 26—31
    phospholipid catabolism, 31—35
    phospholipid composition, 24
    phospholipid metabolism, 26—35
    physiological attributes, 3
    ultrastructural modifications, 3—4
Actinomyces sp., 140
Active transport, 147—148
Acylation, 110
1-Acylglycerol 3-phosphate acyltransferase, 111—112
Acyl-phosphatidylglycerol, 15, 24
Acyltransferase, 110
Adenine nucleotides, 196—197
Adenosinetriphosphatase, see ATPase
Adenylate kinase, 110
Adenyl cyclase, 240
Adhesion zones, 89, 110, 235
ADP, 196
ADP ribosylation, 240
Aerobacter
  aerogenes, 191, 239
  cloacae, 191
Aerobacter-Klebsiella group, 140
Agarase, 215
Aggressins, 240
Alanine, 139
D-Alanine-activating enzyme, 144
D-Alanine carboxypeptidase, 146
D-Alanine dehydrogenase, 130—131
D-Alanine membrane acceptor, 144
Alkaline phosphatase, 48
  comparison of amino acid composition in different organisms, 139
  functional organization, 136—137
  secretion
    apparatus, 224
    Gram-negative cells, 200—204
    Gram-positive cells, 216
    protein sequence, 220—221
  signal sequences, 75
Alkalophilic protease, 216
Alkane
  energy-dependent uptake, 13
  intracellular sequestering, 3
  membranous packaging, 3
  metabolism, 2—3
  molecule, 3
  physical partitioning, 13
  soluble, 37
  uptake, 13, 40—41
Alkane-grown yeasts, 13—15
Alkane-utilizing microorganisms, ultrastructural features of, 3
Allohydroxy-D-proline dehydrogenase, 130—131
Allosteric effectors, 108
D-Amino acid dehydrogenase, 131
31-Amino acid signal sequences, 219—220
Amino acids, comparisons of composition in alkaline phosphatase of different organisms, 139
Aminopeptidase, 216
Amino-terminal signal sequence, 202, 205—206, 209, 213—214
AMP, inhibitory effect of, 196
AmpC, 217
AmpC β-lactamase, 82, 221
  signal sequences, 75
Amphiphilic agents, 235
amyB, 228
amyE, 219, 228
α-Amylase, 215, 219—220, 228, 235
  signal sequences, 76
β-Amylase, 215
amyR, 219, 228

Antigenic determinants, 107, 241
Apocytochrome b, 133
Apolipoprotein, see Lipoprotein
Arabinase, 215
Arabinase-binding protein (ABP), 204—206, 221, 234
  signal sequences, 76
Arginine, 139
aro116, 228
Arthrobacter sp., 3
  cytoplasmic inclusions, 8—9
  hydrocarbon inclusions, 4, 8—9
Arthrobacter strain AK19, 9
asnS, 64
Aspergillus
  oryzae, 140
  sp., 3
Assembly, 58, 70, 84—85, 87—90
Assimilatory nitrate reductase, 125
Asymmetric distribution
  phospholipids, 41
  proteins, 41
AT content, 53, 59, 66, 68
ATP, 110, 144, 196
ATPase, 19, 115—125, 150
  channel, 124—125
  definition, 115
  demonstration in different bacteria, 117
  distribution in various bacteria, 115, 117
  F_1-F_0 complex, 124
  gate, 124—125
  multi-subunit enzyme, 117
  organization in membrane, 124—125
  physiological role, 115
  properties, 115—120
  proton gradients, 115
  proton translocation, 122—124
  pump, 124—125
  purification, 115—120
  subunit structure, 115—120
  TF_1-F_0 complex, 124
  uncoupler mutant, 120—121
ATPase phosphohydrolase, see also ATPase, 115
ATPase 3.6.1.3., see also ATPase, 115
AUT(ATG) codons, 53—60, 63, 66, 69
Auxilliary pathogenic factors, 240
Azotobacter vinelandii, 117
Bacillus
  acidocaldarius, 117
  amyloliquifaciens, 143
  anthracis, 140, 191
  cereus, 217
  cereus 5/B, 142, 191
  cereus 569, 142, 191
  cereus 569/H, 142
  cereus 569/11, 191
  licheniformis, 112
    alkaline phosphatase sites labeled with colloidal gold, 228, 230
    amino acid composition of alkaline phosphatase, 139
    DNA fragment, 217
    penicillin binding proteins, 152—153
    penicillinase, 188—189, 217—218
  licheniformis 94, 153
  licheniformis 749, 136, 191
  licheniformis 749/C
    alkaline phosphatase, 138, 220
    cytochemical localization of alkaline phosphatase, 224—225
    enzyme bound to inside surface of membrane, 137
    mutations, 229
    penicillinase distribution, 191
    plasma membrane, 184
    protein secretions, 188—190
    ribosome distribution, 180—181
  licheniformis 749/C(c), 142
  licheniformis 749/C(Sn), 142
  licheniformis 6346, 187
  licheniformis 6346/C, 142, 191
  licheniformis MC14, 137
  megaterium, 117—118, 152
  sp., 215—217
  stearothermophilus, 117—118
  subtilis, 112, 117
    D-alanine carboxypeptidase, 146
    alkaline phosphatase, 220
    amino acid composition, 139
    α-amylase, 219
    chromosome, 155
    DD-carboxypeptidase, 134
    fatty acid synthesis, 235
    penicillin binding proteins, 152
    plasma membrane bound, 137
    SDH mutants, 133
    serine protease, 135
    single-gene mutation, 228
  subtilis-mesentericus group, 140
  subtilis 138, 168
  subtilis SB15, 224—225, 229
  subtilis T98, 191
Bacteria, 2, 13
Bacterial cells, plasma membrane, 106
Bacteriolytic enzymes, 217
Bacteriophages, 156
Bacteriorhodopsin, 122—123
Band 1, 18—19
  negative stain, 20
  neutral lipids, 25
  phospholipids, 24
Bayer's junctions, 89
BiBi ping-pong kinetic mechanism, 112
Bicyclomycin, 86
Bilayer concept, 106—108
Bilayer structure, 184, 186, 190
Binding protein, 108, 146—156
  defined, 146
  DNA-, 154—156
  penicillin-, 152—154
  role in transport of branched chain amino acids, 151

secretion, 177
solute transport, 146—152
Binding sites, 192
penicillinase, 192
succinic dehydrogenase, 133
Biochemical characteristics
Acinetobacter sp H01-N, 19
derived membrane fractions, 29
Biochemical composition
Acinetobacter membranes, 23—26
hydrocarbon inclusions, 5
Biochemical mechanism of secretory apparatus, 232—239
Biogenesis, 58, 82
Biophysical approach, 238—239
Biosynthesis, 70, 81, 86, 91, 143—146
bacterial outer membrane proteins, 45—104
gene expression, 52—72
genes involved in, 46—52
post-transcriptional events and final assembly, 72—91
de novo fatty acid, 3
lipid, 110—115
phospholipids, 26—31, 110—113
secretion, 176—177
BLP, see Braun lipoprotein
Branched chain amino acids, role of binding protein in transport of, 151
Braun lipoprotein (BLP), 184, 212—213, 218
Brevibacterium sp., 3, 8
Brownian motion, 239
btuB, 47, 51

C

^{14}C antibiotic binding, 152
C-2 units, 14
C-4 units, 14
C55-isoprenoid alcohol phosphokinase, 146
Calcium ions, 193, 195
cAMP, 60, 64, 67—68, 196, 240
Candida
albicans IFO 0583, 14
guilliermondii IFO 0566, 14
intermedia NRRL Y-6328-1, 14
lipolytica
hydrocarbon inclusions, 4—5, 7, 11
intracytoplasmic inclusions, 13
lipolytica NRRL Y-6795, 14
107, 15
parapsilosis, 15
rugosa, 14
sp , 3, 14
tropicalis
alkane uptake, 13
hydrocarbon inclusions, 4, 7, 12—13
intracytoplasmic inclusions, 13
phospholipid content, 15
surface-localized polysaccharide, 14
tropicalis ATCC 32113, 13, 14
tropicalis IFO 0589, 14
tropicalis pk 233, 14
utilis, 8—9
Capsular layer, 179
Capsular polysaccharide, 72, 89
Capsular proteins, 186
Carbohydrases, 215—216
Carbohydrate, 29
DD-Carboxypeptidase, 134, 144
Cardiolipin, 15, 23—24, 28, 35, 109, 112, 130
hydrolysis by cardiolipin phospholipase, 35
turnover, 33
Cardiolipin phospholipase, 35
Cardiolipin phospholipase A_1, 34—35
Cardiolipin synthetase, 111—112
Carrier molecules, 147
Carrier protein, 111
CCCP, 82, 206—207, 234
CDP-diglyceride hydrolase, 114
Cell envelope, 13—14, 89
alterations to, 13
Cell fractionation, 137, 200, 224
Cell-free translation system, 54, 73
Cell lysis, 57—58
Cellular catalase, 14
Cellular lipid composition, 3
Acinetobacter sp H01-N, 16
Cellular neutral lipids, *Acinetobacter*, 15—16
Cellular phospholipid, 3
Acinetobacter sp. H01-N, 15
hexadecane grown *Acinetobacter*, 15
Cellular triglyceride, see Triglyceride
Cellulase, 215
Cell wall, 136, 143—146
Gram-negative bacteria, 183, 186
Gram-positive bacteria, 182
turnover, 144
Cell-wall hydrolase, 113
Cephalosporins, 217
Cerulanin, 235
Chain termination mutants, 223
Channel formation, 235, 237
Chaotropic agent, 108, 128
Charged segment, 220, 223
Chitinase, 215
Chitosanase, 215
Chloramphenicol inhibition of protein synthesis, 190, 192
Chlorophenylhydrazone, see CCCP
Chlorophyll, cytochrome reoxidization, 123
Cholate, 128
Cholate-deoxycholate, 150
Choleragen, 241
Cholera toxins, 240—243
Choline dehydrogenase, 130—131
Chromatophore membrane, 123
Chromosomal protein, 217
Chromosome, 154
Chromosome membrane attachment, 154
cir, 47, 51
Cisternal space, 179—180
citF gene, 133

Cladosporium sp., 3
Cleavage
secretory protein mutations affecting, 84
signal peptide, see Signal peptides
signal sequence, 232, 236
Clostridia, 117
Clostridium
pasturianum, 119
perfringens, 112, 153—154
CMP phosphatidyltransferase, 111—112
Coat proteins, 206—209, 221
Codon usage, 56, 61, 66, 68
Coenzyme Q10, 108
Coligenoid, 241
Coliphase M_{13} coat protein, 135
Colloidal gold
complex, 228, 230
labeling, 141
Computer modeling, 239
Conjugation process, 156
Corynebacterium sp., 3, 4
Cotranslational processing, 232
Cotranslational secretion, 77
Cotranslational translocation, 237
cpr, 70
cpxA, 47, 52
cpxB, 47, 52
Critical molecular weight, 82
Cross-wall formation, 113
CTP, 111
Cunninghamella sp., 3
Cyclic AMP, see cAMP
Cyclodextrin, 215
Cytidine 5′-diphosphate (CDP) diglyceride synthetase, 111
Cytochrome
Acinetobacter membranes, 25—26
composition, 14
reoxidation of chlorophyll, 123
Cytochrome a, 26
Cytochrome b, 15, 26
Cytochrome B_1 apoprotein, 126
Cytochrome d, 26
Cytochrome o, 26
Cytochrome P-450, 13—14
Cytoplasmic hydrocarbon inclusions, see also Hydrocarbon inclusions, 7, 11—12
Cytoplasmic membrane, 19, 27, 89
Acinetobacter, 37—39
alkane uptake, 13
asymmetric distribution of phospholipids and proteins, 41
carbohydrate, 29
DNA, 29
hexosamine, 29
KDO, 29
negative stain, 20—21
neutral lipids, 25
PACT, 29
phospholipids, 24, 29
RNA, 29
translocation across, 74—81

D

DD-Carboxypeptidase, 134, 144
Deacylation-reacylation cycle, 39
Debaryomyces quilliermondii, 140
Decarboxylase, 112
Dehydrogenase, 127—134
D-alanine, 130—131
allohydroxy-D-proline, 130—131
D-amino acid, 131
choline, 130—131
formate, 131
glycerol 3-phosphate, 111—112, 127—128
D-lactate, 108, 130
L-lactate, 128—130
malate, 131
miscellaneous, 134
NADH, 131—132
sarcosine, 130—131
succinic, 132—133
De novo fatty acid biosynthesis, 3
Density, 19
Deoxyribonuclease ribonuclease, 216
Derived membrane fractions, biochemical characteristics of, 29
Desulfovibrio gigas, 134
Detergents, 108, 127—128, 149
Dextranase, 215
Diacyl glyceryl group, 141
Diacyllysocardiolipin, 35
Dicarboxylic acids, 3
Digestion of protein secretions, 177
Diglyceride, 5, 16, 25
3-4-Dihydroproline, 149
Dimerization, 200—201
Dimethylsulfoxide, 234
Dinitrophenol, 206, 234
Dioxygenases, 3
Discontinuous sucrose gradient centrifugation, 18
Dissolved oxygen concentration, 25
Divalent cations, 193, 196
DNA, 29
DNA-binding protein, 154—156
DNA-membrane complex, 155—156
DNA-polymerases I, II, and III, 156
DNA-protein-phospholipid subcomplexes, 156
DNA receptors, 155
DNA sequence, 211, 217
Docking protein, 80, 232, 234
Dodecanoic acid, 235
Domain, 237

E

ΔE of processing, 235
ΔE of secretion, 235
Edman degradation, 86, 141, 199—201, 206, 208

EF-TU, 204
ϕ80d*phoA* DNA, 200
Electrogenic proton translocation, 122
Endo-*N*-acetylglucosaminidase, 217
Endo-*N*-acetylmuramidase, 217
Endopeptidase, 134
Endoplasmic reticulum (ER), 179—181, 234
Endotoxins, 240
Energized membrane, 147, 235
Energy barrier, 237
Energy coupling, 26
Energy for translocation, 80—81
Energy transduction, 232, 234
Enterotoxins, 240—242
  signal sequences, 76
Envelope organization relevant to secretion, 179—187
EnvZ, *envZ*, 47, 49—50, 67—68, 70
  AUG(ATG) codons, 53—60, 69
  codon usage, 61
  mRNA secondary structures, 69
  outer membrane protein synthesis, 48
Enzymes, 107
  function, 108—110
  lipid biosynthesis, 110—115
  production, 243
  wall biosynthesis, 143—146
Erwinia amylovora, 57
Escherichia coli, 108—156, 242
  alkaline phosphatase synthesis, 200—201
  amino acid compositions of alkaline phosphatase, 139
  D-amino acid dehydrogenase, 131
  ATPase, 115—125
  binding protein, 146—156
  cell-free system, 217
  cell wall model, 213
  chromosome, 155
  cloning of DNA fragment, 238
  DD-carboxypeptidase, 134
  dehydrogenase activities, 128—129
  distribution of ATPase, 117
  distribution of enzymes of phospholipid biosynthesis, 111
  envZ mutations, 68
  enzymes of lipid biosynthesis, 110—115
  extraction of nitrate reductase, 127
  fatty acid auxotrophs, 109
  fl coat protein, 207, 209
  fumarate reductase extracted from, 132
  gene fusion technique, 197—198
  genetic markers, 232
  heat-labile enterotoxin, 241
  heat-labile toxin, 240, 242
  industrial applications, 243—244
  lambda phage receptor, 209
  lipoprotein, 40, 136
    overproduction, 57, 64
  loop model of lipoprotein secretion, 223
  major outer-membrane proteins, 52
  maltose-binding protein, 237
  matrix proteins, 90
  membrane-bound dehydrogenase, 127—134
  membrane-bound enzymes for wall biosynthesis, 143—146
  membrane-bound esterases, 142
  membrane-bound hydrolase, 134—143
  membrane-bound polyribosomes, 77
  nitrate reductase, 125—127
  ompA protein, 49
  OmpT protein, 50
  penicillinase distribution, 140, 191
  phospholipase A_1 activities of outer membranes, 34
  phospholipid degrading enzyme, 114
  pre-coat proteins, 206
  pre-lipoprotein sequence, 218
  proline transport carrier extracted, 149
  properties of nitrate reductase, 128
  properties of penicillinase, 142
  proteases, 135
  protein, 2, 48—49
  PTD 101, 136
  purification of ATPase, 118—119
  respiratory loops, 122
  ribose-binding protein, 237
  role of binding protein in transport of branched chain amino acids, 151
  signal peptidases, 82—83
  temperature-sensitive gene, 229
  transport of periplasmic arabinose-binding protein, 234
ESR spectroscopy, 114
Esterases, 141—143, 216
Ethanol, 234
Eukaryotic cells, 106
Eukaryotic endoplasmic reticulum, 179
Eukaryotic microorganisms, 2
Evolution of porin genes, 64—66, 70—71
Evolutionary conservation, 133
Exo-*N*-acetylglucosaminidase, 217
Exo-*N*-acetylmuramidase, 217
Exogenous phospholipid, 28—32, 37
Exotoxins, 240
expA, 47, 52
Export apparatus, see Secretory machinery
Extracellular enzyme synthesis, 236
Extracellular enzymes, *Bacillus*, 215—217
Extracellular lipid accumulation, *Acinetobacter* sp. H01-N, 17
Extracellular lipid composition, 3
Extracellular membrane vesicles, 19, 41
  Acinetobacter, 36—37
  phospholipids, 24
Extracellular neutral lipid composition
  Acinetobacter sp., 16—17
  Acinetobacter sp. H01-N, 16
Extracellular self-assembly of protein secretions, 177

F

F_o ionophore, 124
F_1-F_o complex, 124—125
Facilitated diffusion, 147
Fatty acid auxtotrophs, 109
Fatty acid heterogeneity, 109
Fatty acids, 3
  fluidity, 109
  β-oxidation, 14
Fatty acid synthesis, 235
Fatty alcohols, 3, 5, 15—16, 25
Fatty aldehydes, 3
Fat vacuoles, 7
fecA, 47, 51
fecB, 47, 51
Feedback mechanism, 63
fepA, 47, 50—51
Fermentation technology, 243—244
Ferritin conjugated anti-ATPase, 124
Ferritin-labeled anti-penicillinase antibody, 224, 227
fl coat protein, 207—209
fl pre-coat protein, 135
Flavobacterium sp., 8
Flip-flop mechanism, 107, 192
Fluidity, see also Membrane fluidity, 107, 109—110
Formate dehydrogenase, 131
Fractionation, 89
Free fatty acid, 5, 16, 25
Free fatty alcohol, 15
Freeze-dried replicas, 224, 227
Freeze-etch, *Acinetobacter*, 6, 7
Freeze fracture
  Gram-negative organism, 185
  Gram-positive bacterial plasma membrane, 184
Fumarate reductase, 132
Functional organization of bacterial membrane proteins, see also specific topics, 105—174
  adenosinetriphosphatase (ATPase), 115—125
  binding protein, 146—156
  enzyme function, 108—110
  enzymes of lipid biosynthesis, 110—115
  general considerations, 106—110
  intramembranous movement of molecules, 110
  lipid protein association, 106—108
  membrane bilayer, 106—108
  membrane-bound dehydrogenase, 127—134
  membrane-bound enzymes for wall biosynthesis, 143—146
  membrane-bound hydrolase, 134—143
  membrane microenvironment, 108—110
  nitrate reductase, 125—128
Fungi, 2
Fusarium sp., 3
Fused *phoA-lacZ*, 202
Fusion, 197, 199

G

Galactanase, 215
Galactose, 241
β-Galactosidase, 243
  gene expression and, 52, 60—64, 66, 70
  post-transcriptional events and, 79, 88
  prolipoprotein fusion protein, 79
  secretion
    apparatus, 229
    Gram-negative cells, 197, 199, 202, 204, 209—210
Gating, 235
Gene dosage effect, 63
Gene expression
  lipoprotein, 52—58
  major outer membrane proteins, 52—72
  matrix proteins, 64—70
  ompA, 58—64
Gene fusion technique, 70, 197—198, 202, 209
Genes, see also specific types, outer membrane proteins, synthesis of , 46—52
Gene structures, matrix proteins, 64—66
Genetic regulation of secretory apparatus, 228—232
Gene *ucf*, 121
Globomycin, 50, 83, 86, 89, 213, 218
β-1,2-Glucanase, 215
β-1,6-Glucanase, 215
Glucanotransferase, 215
Glucose isomerase, 243
α-Glucosidase, 143
Glucosyltransferase, 142
Glutamic acid, 139
Glutamine transport, 150
α-Glycerol-CMP phosphatidyl transferase, 27
Glycerol 3-phosphate acyltransferase, 111—112
Glycerol 3-phosphate dehydrogenase, 111—112, 127—128
Glycine, 139
Glyoxylate cycle enzymes, 14
Gram-negative bacteria, see Gram-negative cells
Gram-negative cells, 117, 136, 179, 186
  adhesion zones, 235
  alkaline phosphatase, 200—204
  arabinose-binding protein, 204—206
  cell wall structure, 183, 186
  fl coat protein, 207—209
  industrial applications, 244
  lambda phage receptor, 209—212
  leucine-specific binding protein, 199—201
  lipoprotein, 212—214
  M13 coat protein, 206—208
  maltose, binding protein, 197—199
  molecular mechanism, see also other subtopics hereunder, 197—214
  outer-membrane protein, 209—214
  penicillinase distribution, 191
  periplasm, 184
  periplasmic protein, 197—206
  plasma membrane protein, 206—209
  signal sequence, 223
Gram-positive bacteria, see Gram-positive cells
Gram-positive cells, 117, 136, 179—180, 186
  alkaline phosphatase, 220

α-amylase, 219—220
cell wall structure, 182
Gram-negative cells distinguished, 214
industrial applications, 244
molecular mechanisms, 214—220
penicillinase, 214, 217—219
signal sequence, 223
Group translocation, 147
Growth temperature, 69

H

H^+ ion concentration, 193, 195
Hafnia sp., 140
Hairpin structures, 55, 66
Half-cystine, 139
Halobacterium, 138
halobium, 123
Halophilic protease, 216
Heat-labile toxins, 222, 240, 242—243
signal sequences, 76
Heat-stable toxin, 222
α-Helical content, 130
Helical hairpin, 238
Helix breaker, 223
Helix conformation, 223
Herellea sp., 140
Heterologous transcription-translation-translocation system, 234
Hexadecane, 5—6, 8, 10—11, 19
intracellular pools, 16
Hexadecane-grown *Acinetobacter*, see *Acinetobacter* sp.; *Acinetobacter* sp HO1-N
Hexadecane-grown cells
band 1, 19, 24—25
biochemical properties of membranes, 19
cellular lipid composition, 16
cytoplasmic membrane, 19, 24—25
extracellular membrane vesicle, 19, 24
extracellular neutral lipid composition, 16
hydrocarbon inclusion, 19, 24
membranes derived from, 18
neutral lipids, 25
outer membrane, 19, 24—25
phospholipid biosynthetic enzymes, 27
phospholipid composition, 15
phospholipids in, 24
physical properties of membranes, 19
Hexadecane metabolism, hexadecylpalmitate as end product of, 17
Hexadecanol, 25
Hexadecylpalmitate, 17
Hexosamine, 29
hisJ gene, 150
hisP gene, 150
Histidine, 139
Histidine-binding protein, 221, 237
signal sequences, 76
Histidine transport, 150
Host defense mechanism, 240
Hybrid products, 197—199, 202, 209
Hybrid proteins, 88, 209
Hydrocarbon inclusions, 3—12, 18—19, 23, 41
Acinetobacter, 36
appearance of, 13
biochemical composition, 5
intracytoplasmic membranes in direct contact with, 9
phospholipids, 24
Hydrocarbon metabolism, 41—42
Hydrocarbon microbiology, 2
Hydrocarbon oxidation, 2, 36
Hydrocarbons, 2—3
Hydrocarbon-utilizing microorganisms, see also specific topics, 1—44
membranes of *Acinetobacter*, 15—26
phospholipid metabolism in *Acinetobacter* sp. HO1-N, 26—35
ultrastructure of, 3—13
ultrastructure of alkane-grown yeasts, 12—15
Hydrogenase, 131
Hydrolases, 134—143
alkaline phosphatase, 136—139
cell-wall, 113
penicillinase, 138, 140—142
peptidase, 134—135
protease, 135—136
Hydrolysis, cardiolipin by cardiolipin phospholipase, 35
Hydrophobic barrier, 107
Hydrophobic domain, 130
Hydrophobic driving force, 238
Hydrophobic environment, 74
Hydrophobicity, 78
Hydrophobic sequence, 223
Hydroxy fatty acids, 3
Hypersecretor strains, 244
Hypertoxinogesis, 242
Hypothetic model, origin of internal membrane systems, 40
Hypotoxinogesis, 242

I

^{125}I-diazobenzine, 137
^{125}I-iodosulfanilic acid, 219
ilv gene cluster, 121
Immunocytochemistry, 137
Immunoelectron microscopy, 224
Immunopathological reactions, 240
Impedins, 240
Industrial applications, 243—244
Inhibitors of protein synthesis, 192—193
Inner-membrane proteins phage fd, 221
Inorganic phosphate, 192—195
Insulin, 243
Integral proteins, see also specific types, 107, 109, 152
Interferon, 243
Intracellular pooling of hexadecane, 16

Intracellular pooling of hydrocarbons, 3
Intracellular sequestering of alkane, 3
Intracytoplasmic inclusions, 7—8, 13
Intracytoplasmic membrane complex, 40
Intracytoplasmic membranes, 3—4, 9—10, 13, 27
  Acinetobacter sp., 15, 18, 36
  asymmetric distribution of phospholipids and proteins, 41
  hydrocarbon inclusions in close proximity to, 41
Intramembranous movement of molecules, 110
Intramembranous protein molecules, 232
In vitro biosynthesis of peptidoglycan, 144—145
Ionic groups, 179
Ion transport, 235
IPTG, 210
Isoamylase, 215
Isolation of membranes, *Acinetobacter* sp. H01-N, 17—23
Isoleucine, 139
Isoleucine-valine-binding protein, 76
Isopropyl-β-thiogalactoside (IPTG), 210

K

K_m values, 109
  phospholipase A_1 activities, 34
KCN, 206
KDO, 29
Ketones, 3
Klebsiella
  aerogenes, 128
  spp., 191

L

Labeling, 237
lac, 58, 202, 209, 229
lac carrier, 147
lac operon gene, 148
β-Lactam antibiotics, 154
β-Lactamase, 79, 187—188, 214, 216, 234
  signal sequences, 76
β-Lactam ring of penicillin, 138
D-Lactate, 147
D-Lactate dehydrogenase, 108, 130
D-Lactate oxidase, 108—109
L-Lactate dehydrogenase, 128—130
Lactobacillus
  arabinosus, 117
  casei, 144
  fermenti, 117
Lactoferrin, 240
Lactoperoxidase ^{125}I-labeling, 137
Lactose, 241
Lactose transport, 149
lac-Y protein, 147
LacZ, *lacZ*, 64, 88, 197
lacZ-lamB genes, 209—210
lamB, see Lambda receptors, *lamB*
Lambda phage receptor, 209—212
Lambda (λ) receptor, *lamB*, 50—51, 80, 88, 221, 229, 234
  AUG(ATG) codons, 53—60
  gene expression, 70—71
  Gram-negative cells, 209—211
  outer membrane protein synthesis, 46—48
Lambda (λ) receptor protein, 51, 88, 152
  signal sequences, 75, 79
Lateral diffusional flow, 110
Leader peptidase, see Signal peptidase
Leader peptide, 237—238
lep, 47, 83
Leucine, 139
Leucine-binding protein, signal sequences, 76
Leucine-isoleucine-valine-binding protein, 221
Leucine-specific binding protein (L-SBP), 199—210, 221
Levansucrase, 215, 235
Lichenase, 215
Ligase, 144
Linear translocation model, 77, 79—81
Linker protein, 156
Lipase, 217
Lipid biosynthesis, enzymes of, see also Enzymes of lipid biosynthesis, 110—115
Lipid phosphorus, 19
Lipid protein association, 106—108
Lipid/protein ratio, 113
Lipids, 23, 69
  fluid state, 109
  hexadecane-grown *Acinetobacter*, 15—17
Lipid vacuoles, 14
Lipopolysaccharide, LPS, 18—19, 23, 46, 64, 69—70, 90
  enzyme bound to, 137
Lipoprotein, *lpp*, 46—47, 49—50, 73, 78, 86, 89, 136, 184, 212—214, 221
  AUG(ATG) codons, 53—60
  codon usage, 61
  gene expression, 52—58
  new minor, 50, 83
  overproduction, 57—58, 64
  penicillinase, 217
  signal sequences, 75, 214
LIV-binding protein, 199
livK gene, 199
LIVT protein, 150
Local anesthetics, 69—70
Loop model, 77—80, 89
lpp, see Lipoprotein
LPS, see Lipopolysaccharide
LS protein, 150
Lysine, 139
Lysine-arginine-ornithine-binding protein, 221
  signal sequences, 76
Lysogeny, 70—71
Lysophosphatidylethanolamine, 24, 34
Lysophosphatidylglycerol, 34
Lysophospholipase, 114

M

M-complex, 155
M13 am8H1R6 phage, 207
M13 coat protein, 77, 81, 83, 206—208, 237
Macroevolution, penicillinase, 141
Magnesium ions, 193, 195
Magnesium salt, 138
Major coat protein, 221
  signal sequences, 76
Major outer membrane proteins, see also Outer membrane proteins, 46
  gene expression, 52—72
  regulatory genes, 46—51
  structural genes, 46—51
Malate dehydrogenase, 131
malB, 50
malB locus, 199, 210
malE, 80, 88, 199
malE-lacZ fused gene products, 209
malE-lacZ gene fusion, 229
malF protein, 152
malG protein, 152
malK, 71
malK product, 152
MalT, *malT*, 47, 50, 66, 71
Maltase, 215
Maltose-binding protein (MBP), 80—81, 197—199, 221, 237
  signal sequences, 76
Maltose regulon, 66
Maltose transport system, 197—198
Mannanase, 215
Marine pseudomonad, 117
Matrix protein regulation, 68—70
Matrix proteins, see also Outer membrane proteins, 46, 64—70, 90
  gene structures, 64—66
  ompB locus, 66—68
  other regulatory factors, 68—70
Mecillinum, 154
Membrane bilayer, 106—108
Membrane biochemistry, 3
Membrane biogenesis, 40—41
Membrane-bound cytoplasmic organelles, 14
Membrane-bound dehydrogenase, see Dehydrogenase
Membrane-bound endoproteases, 135
Membrane-bound enzymes for wall biosynthesis, 143—146
Membrane-bound esterases, 142
Membrane-bound hydrolase, see Hydrolases
Membrane-bound polyribosomes, 72—73, 77
Membrane-bound 31-kDa penicillinase, 136
Membrane-chromosome comples, 156
Membrane fluidity, see also Fluidity, 70, 73, 81, 83, 214, 234—235
Membrane-impermeable reagent, 224
Membrane-impermeant probes, 131
Membrane lipid bilayer, 235
Membrane-localized cytochromes, 25—26
Membrane microenvironment, 108—110
Membrane neutral lipids, 23, 25
Membrane perturbants, 82
Membrane phospholipids, 23—24
Membrane protein P, 237
Membrane proteins, 25
Membrane-receptor-mediated mechanism, 179
Membrane trigger hypothesis, 74, 79, 81
Membrane vesicles, 23
  known polarity, 147
  lipid, 23
  lipopolysaccharide, 23
  protein, 23
Membranous packaging of alkane, 3
Mesophilic bacteria, 117
Mesosome-like membranes, 9
Mesosome membrane, 188
Mesosomes, 155
Metal protease, 216
Methionine, 139
Methylotrophic bacteria, 9
Mg-sarkosyl crystals, 155
Michaelis-Menten kinetics, 39
Microbial degradation, hydrocarbons, 2
Microbial hydrocarbon metabolism, 2
Microbodies, 13—14
Micrococcus
  denitrificans, 128
  halodenitrificans, 128
  luteus, 144
  lysodeikticus, 117—118
    penicillinase distribution, 140
    release of ATPase from, 115
  sodonensis, 139
Microemulsion, 23
Microenvironment, 108—110
Microevolution, penicillinase, 141
Microsomal vesicles, 234
Microsomes, 13
minA, 241
minB, 241
Minimum inhibitory concentration (MIC), penicillin, 152
Minor coat protein, 76, 221
Minor outer membrane genes, 51
Mitochondria, transfer of C-2 units to, 14
Modification, 50, 83—87
Molecular oxygen, insertion into alkane molecule, 3
Monoglyceride, 5, 16, 25
Monosialosylganglioside (GMI), 241
Morganella morganii, 57
Morphological demonstration of secretory apparatus, 224—228
MRB, 238
mRNA 5′ leader region, 54, 60—62, 64, 66, 68
  ompA, 62
  ompC, 65
  ompF, 65
  phoE, 65

mRNA secondary structure, 54—55, 63, 69, 71
ompA, 62
mRNA stability, 52, 55, 62
Mucor sp., 3
Multi-subunit enzyme, 117
Mutant CT28, 155
Mutations, effects in *unc* genes, 121
Mycobacterium
album 7E4, 4
convolutum, 15
phlei, 117—118
penicillinase distribution, 140
phospholipase A_1 activities of outer membranes, 34
smegmatis, 140, 146
sp., 3, 140
tuberculosis strain BCG, 140
vaccae, 4, 15
Mycoderma valida, 140
Mycoplasma, 109
Mycoplasma laidlawii, 117, 134
Myxococcus xanthus, 137

N

NADH dehydrogenase, 131—132
NADH oxidase, 19, 109
NADH ubiquinone oxidoreductase, 131—132
NaN_3, 206
Negative stain
band 1, 20
cytoplasmic membrane, 20—21
outer membrane, 21, 22
Neisseria gonorrhoeae, 34, 152
Neutral lipids, *Acinetobacter* membranes, 23, 25
Nikkol, 208
Nitrate reductase, 125—128
nmpA, 47
NmpAB, see PhoE
nmpB, see *phoR*
nmpC, 47, 48
Nocardia
rubra, 4—5
sp., 3
cellular wax ester and glyceride, 15
cyclohexane-grown, 8—9
hydrocarbon inclusions, 4—5, 10
penicillinase distribution, 140
Nucleases, 216
Nuclei, 154
Nucleotidase, 143
3-Nucleotidase, 216
5-Nucleotidase, 216
Number of major membrane polypeptides, 19
Nutrient broth-grown cells
biochemical properties of membranes, 19
cellular lipid composition, 16
cytoplasmic membrane, 19, 24
extracellular neutral lipid composition, 16
membranes derived from, 18
neutral lipids, 25
outer membrane, 19, 24
phospholipid biosynthetic enzymes, 27
phospholipids in, 24
physical properties of membranes, 19
Nutrient broth medium, 4, 7—13

O

Oligonucleotide-guided site-specific mutagenesis, 78
omp, see Outer-membrane protein
ompA, *ompA*, 46—47, 49, 52, 68, 79—80, 82, 87, 214, 221
AUG(ATG) codons, 53—60
codon usage, 61
gene expression, 58—64
mRNA 5′ leader regions, 62
mRNA secondary structures, 62
overproduction, 70
signal sequences, 75
OmpB, *ompB*, see also EnvZ; OmpR, 47, 64
gene expression, 66—68
genes involved in synthesis, 46—48
ompB locus, 49
OmpC, *ompC*, 52, 67, 69—71
AUG(ATG) codons, 53—60
codon usage, 61
DNA sequence comparison, 64—66
evolution of porin genes, 65
genes involved in synthesis, 46—49
mRNA 5′ leader regions, 65
signal sequences, 75
OmpD, 49
OmpF, *ompF*, 52, 67, 69—71, 80, 82, 87, 214, 221
AUG(ATG) codons, 53—60
codon usage, 61
DNA sequence comparison, 64—66
evolution of porin genes, 65
genes involved in synthesis, 46—49
mRNA 5′ leader regions, 65
signal sequences, 75
OmpR, *ompR*, 47, 49, 67—68
AUG(ATG) codons, 53—60
codon usage, 61
outer membrane protein synthesis, 48
OmpT, *ompT*, 46—47, 50, 69
gene expression, 72
Order/disorder transition, 235
Osmoregulation, 66
Outer membrane, 19, 41, 136, 179—180, 184
Acinetobacter sp., 18, 39—40
isolation, 34
negative stain, 20, 22
neutral lipids, 25
phospholipases of *Acinetobacter*, 34—35
phospholipids, 24
purification, 34
Outer-membrane proteins, see also Major outer membrane proteins; Matrix proteins, 209—214, 221

assembly, 72—90
biosynthesis, see also Biosynthesis, 45—104
cleavage of signal peptide, 81—84
early stages of biosynthesis, 73—74
genes involved in synthesis, 46—52
lambda phage receptor, 209—212
lipoprotein, 212—214
modification, 84—87
signal sequences, 75
translocation across cytoplasmic membrane, 74—81
Overproduction of major outer-membrane proteins, 57—58, 64
Oxidases, 108—109, 134
Oxidation, sulfhydryl groups to disulfide, 200
β-Oxidation of fatty acids, 14
Oxidation mechanisms, 2, 3
Oxisomes, 9

P

Packaging, *Acinetobacter* sp. H01-N, diagrammatic model for, 40—41
PACT, see Phosphatidic acid cytidyl transferase
pap, 219, 228
Papulospora sp., 140
Paracolon group, 140
Partially lysed cell, 6
Passive diffusion, 147
Pasteurella sp., 140
Pathogenicity, 239—243
PBP1, 153
PBP 1B, 154
PBP3, 153—154
PBP 4, 154
PBP 5, 154
PBPs, 153
pBR313, 241
pBR322, 57, 238, 242
pCG86, 242—243
PEA, see Phenethylalcohol
Pectate lyase, 215
Penicillin, 217
Penicillin amidase, 216
Penicillinase, 138, 140—142, 214, 217—219
binding sites, 192
biochemical observation, 141
diffusion, 192
distribution, 140—141, 188—192
DNA sequence, 217
immunocytochemical observation, 141
lipophilic modification, 141
lipoprotein, 217
macroevolution, 141
microevolution, 141
properties, 142
secretion, 188—193
signal sequences, 76, 223
translocation, 192
Penicillinase-specific binding sites, 223
Penicillin-binding proteins, 144, 152—154
Penicillium sp., 3
Peptidase, 134—135
Peptide chain, 238
Peptides, 134
Peptide transport, 134
Peptidoglycan, 49—50, 179, 184, 239
in vitro biosynthesis, 144—145
Peptidoglycan-associated lipoprotein (PAL), 83, 212—213
Peptidoglycan cross-linking enzyme, 144
Peripherally located stacked lamellae (type II), 9
Peripheral proteins, see also specific types, 107—108
Periplasm, 136, 180, 184
Periplasmic membranes, signal sequences, 75—76
Periplasmic proteins, 88, 147, 150, 221, 234
alkaline phosphatase, 200—204
arabinose-binding protein, 204—206
leucine-specific binding protein, 199—201
maltose-binding protein, 197—199
Periplasmic space, 88, 179—180
Peroxisomes, 14
pH, optimum, phospholipase A_1 activities, 34
Phage fd, 221
Phenazine methosulfate, 147
Phenethylalcohol, PEA, 70, 73, 81—83, 87, 212—213, 217
Phenylalanine, 139
ϕ29 phage genes, 156
pho regulon, 48, 71—72
phoA, 71, 204
signal sequences, 204, 206
phoB, 48, 71—72
PhoE, *phoE*, 46—48
AUG(ATG) codons, 53—60
DNA sequence comparison, 64—66
evolution of porin genes, 65
gene expression, 71—72
mRNA 5′ leader regions, 65
outer membrane protein synthesis, 48
signal sequences, 75
phoH, 228
phoM, 47—48
phoR, 47—48, 72, 228
phoR mutation, 202
phoS, 48
Phosphatases, 216
Phosphatidic acid cytidyl transferase (PACT), 19, 26—27, 111
acetone-precipitated membranes, 29
cytoplasmic membrane, 29
enhancement, 38
enrichment, 27—29
stimulation of activity by exogenous phospholipid, 28—29
stimulation of phospholipid biosynthetic enzymes, 33
UAT membranes, 29
Phosphatidic acid phosphatase, 27, 114
Phosphatidylethanolamine, 15, 23—24, 28

turnover, 31
Phosphatidylethanolamine phospholipase A_1, 34
Phosphatidylglycerol, 15, 24, 28, 130
turnover, 32
Phosphatidylglycerol phospholipase A_1, 34
Phosphatidylglycerophosphate synthetase, 112
Phosphatidylserine, 15
Phosphatidylserine decarboxylase, 27, 111—112
Phosphatidylserine synthetase (PS synthetase), 30, 111
stimulation of, 30, 38
stimulation of phospholipid biosynthetic enzymes by, 33
Phosphodiesterase, 138
Phosphoenolpyruvate-glucose phosphotransferase, 142
Phospholipase, 19, 34—35
Phospholipase A, 114
Phospholipase A_1, 111—112
activities, 34, 39—40
Phospholipase C, 114, 217
Phospholipase D, 114
acetone-precipitated membranes, 29
Acinetobacter membranes, 23—24
asymmetric distribution, 41
cytoplasmic membrane, 29
translocation, 110
turnover, 113—115
UAT membranes, 29
Phospholipid bilayer, 106—107
Phospholipid biosynthesis, 26—31, 110—113
Phospholipid biosynthetic enzymes, 3, 27, 33
Phospholipid catabolism, 31—35
Phospholipid degradation, 110—113
Phospholipid fatty acids, 39
Phospholipid metabolism, *Acinetobacter* sp. H01-N, 26—35
Phospholipid to protein ratio, 23
Phosphomannase, 215
Phosphoprotein phosphatase, 143
Phosphotransferase, 141
phoT, 48
Photosynthetic apparatus, 122—123
Physical properties, membranes of *Acinetobacter* sp H01-N, 19
Physiological regulation of secretory apparatus, 223
Pinocytotic vesicles, 13
Plasma membrane, 14—15, 136, 179—180, 189, 232, 237
bacterial cells, 106
eukaryotic cells, 106
Plasma membrane protein, 206—209
Plasmic, 155, 220
pldA, 47, 51
Pleiotropic mutation, 228, 235
plsB gene, 111
Poly-β-hydroxybutyrate, 5
Polypeptide chain, 237—238
Polysaccharide-fatty acid complex, 14
Polysaccharide-protein complexes, 242
Polysaccharides, 186
Pore and binding activity, 150
Porins, see also Lambda (λ) receptor, OmpC; OmpF, PhoE, 47—49, 67—68
evolution of, see Evolution of porin genes
Potential change, 235
Pre-BLP, 213
Pre-coat protein 206—207
Precursor enzymes, 200, 213
ABP, 204, 206
L-SBP, 201
Precursor protein, 79
Precursors of outer-membrane proteins, see also Signal peptides, 74
Pre-lipoprotein, 136
Pre-MBP, 199
Pre-PAL, 213
Pre-penicillinase, 136, 217—219
Pre-secretory protein, 237
Pribnow box, 52, 71
Pribnow box sequence, 220
prl mutants, 229
prlA, 47, 51, 80, 232
prlB, 47, 51, 232
prlC, 47, 51, 232
Procaine, 70
Prokaryotic microorganisms, 2
Proline, 139
Proline transport carrier, 149
Prolipoprotein signal peptidase (PLP), see Signal peptidases
Promoters, 52—60, 63—64, 66, 71
pro-OmpA, see OmpA
Protease, 90, 135—136, 192, 216, 228
Protection by protein secretions, 177
Protein, see also Binding protein; specific types, 2, 5, 23, 70
Acinetobacter membranes, 25
asymmetric distribution, 41
carrier, 111
classification, 107—108
integral, 107, 109
large-scale production, 243
linker, 156
peripheral, 107—108
periplasmic, 147, 150, 197—206
receptor, 133
termination, 155
transport, 107
uses, 243
Protein 1, 152—153
Protein 2, 48—49
Proteinase K, 237
Protein-protein interaction, 37
Protein secretion in bacteria, see Secretion
Protein sequence, role of, 220—223
Protein synthesis inhibitors, 192—193
Protoliposomes, 150
Proteus
mirabilis, 191
mirabilis L-form, 146
morgani, 191

rettgeri, 191
sp., 117, 140
vulgaris, 191
Proton circulation, 123
Proton gradients, 115
Proton translocation, 122—124
PS3, 124
Pseudomonas
aeruginosa
D-alanine dehydrogenase, 131
alkaline phosphatase release, 200
penicillin binding proteins, 153
peptide transport, 134
putida, 2, 134
pyocyanea, 191
sp., 2—3
D-amino acid dehydrogenase, 131
hydrocarbon inclusions, 4—5
penicillinase distribution, 140
Pspc operon, 230, 233
PTD 101, 136
Pullulanase, 216
purA, 155
Pyrolidone carboxylpeptidase, 135
Pyrophosphatase system, 122
Pyruvic oxidase, 108

Q

Q gene, 150

R

Radioactive protein bands, 152
Random coiling, 223
Rate of transport, 238
Reaction product cytochemistry, 137—138
Receptor protein, 133
Receptors, 107
Receptors for plasmid, 155
Recombinant DNA technology, 243
Redox chains, 122
Reductase
fumarate, 132
nitrate, see Nitrate reductase
Regulatory genes, 228
major outer membrane proteins, 46—51
Releasing factor, 190
Renin substitute, 243
Respiratory activity, 13
Respiratory nitrate reductase, 125
Restriction endonuclease, 199
Rhodopseudomonas
capsulata, 122—123
spheroidea, 117
phototransferase system, 142
proton translocation, 123
Rhodospirillum
rubrum, 117
photosynthesis, 122—123
succinic dehydrogenase extracted from, 132—133
spheroides, 132
Ribitol-glycerol polyphosphate polymer, 179
Ribonucleoprotein complex, 78
Ribophorins, 78
Ribose-binding protein, 237
Ribosome-binding protein, 239
Ribosome-binding site, see Shine-Dalgarno sequence
Ribosomes, 232
RNA, 29
6S RNA, 232
7S RNA, 78

S

S-complex, 155
sacU^{h}, 228
Salmonella
sp., 140
typhimurium, 110, 117
D-amino acid dehydrogenase, 131
bound form of lipoprotein, 87
cell wall models, 213
cloning of *ompB* region, 68
distribution of enzymes of phospholipid biosynthesis in, 112
histidine transport, 150
major outer-membrane proteins, 49
ompA protein, 49
periplasmic histidine-binding protein interaction with membrane protein P, 237
signal sequences, 221
Sarcosine dehydrogenase, 130—131
Scattered membranous vesicles (type I), 9
SDH mutants, 133
SDS-polyacrylamide gel electrophoresis, 25
SecA, *secA*, 47, 51, 80
secB, 47, 51
secB mutant, 229
Secretion, see also Translocation, 77, 175—253
adenine nucleotides, effect of, 196—197
apparatus of, see also other subtopics hereunder, 223—239
basic nature of process, 177—179
binding, 177
biochemical mechanism, 232—239
biosynthesis, 176—177
components, 176—177
definition, 176—177, 206
digestion, 177
divalent cations, effect of, 193, 196
envelope organization relevant to, 179—187
extracellular self-assembly, 177
fermentation technology, 243—244
general considerations, 176—190
genes presumed to be involved in, 51—52
genetic regulation, 228—232

H^+ ion concentration, 193, 195
industrial applications, 243—244
inhibitors of protein synthesis, 192—193
inorganic phosphate, action of, 192—195
β-lactamase, 187—188
molecular mechanism in Gram-negative bacteria, see also Gram-negative cells, 197—214
molecular mechanism in Gram-positive cells, see also Gram-positive cells, 214—220
morphological demonstration, 224—228
pathogenicity, relation to, 239—243
penicillinase, 188—193
physiological regulation, 187—198, 223
physiological role, 177
processing of nascent protein molecules, 176—177
protection, 177
protein sequence, role of, 220—223
signal sequences, 75—76, 177, 192
toxin production, relation to, 239—243
translocation, 176—183
transport of nutrients, 177
vectorial movement, 178
Secretion negative, 229
Secretion sites, 224
Secretory activities, see Secretion
Secretory apparatus, see also Secretion, 223—239
defined, 223
Secretory machinery, 80
Secretory protein mutations affecting cleavage, 84
secY, 51, 80
Self-assembly, 81, 238
Sequencing, 141
Serine, 139
Serine CMP, 111
Serine-metal protease, 216
Serine proteases, 135, 216
L-Serine-SMP phosphatidyl transferase, 27
Serratia
marcescens, 57
sp., 140
Shigella
dysenteriae, 63
sp., 140
Shine-Dalgarno sequences, 53—60, 63, 66, 220
Shock wash, 115
Siderochrome, 240
Signal hypothesis, 74
Signal peptidases, 82—84, 86, 136, 206—207, 219, 223, 234, 237
Signal peptides, 74, 79—80, 206, 237
cleavage of, 53—60, 72, 81—84, 89
functions, 77
hydrophobicity, 78
Signal receptor particle (SRP), 232, 234
Signal recognition particle (SRP), 78—80
Signal sequences, 75—76, 177, 192, 221—222
altered, 229
31-amino acid, 219—220
amino-terminal, 202, 205—206, 209, 213—214
binding proteins for, 237
binding with membrane, 239
cleavage, 232, 236
defective, 80
extruded, 232
Gram-negative bacteria, 75—76
Gram-positive bacteria, 76
hydrophobicity, 78, 236
interaction between secretory apparatus and, 229
internalized, 80
lambda receptor protein, 79
lipoprotein, 214
penicillinase, 223
phoA, 204, 206
prolipoprotein, 89
stimulus for proper folding of protein, 74
structural homologies, 74
weak binding caused by change in, 230
Sn-glycerol-3-phosphate, 110, 112
Sodium chlorate, 128
Sodium perchlorate, 128
Sodium thiocyanate, 128
Solubilization, 138
water-insoluble hexadecane, 23
Soluble alkane, 37
Solute transport, proteins related to, 146—152
Somatomedin, 243
spc operon, 80
Specific activity, phospholipase A_1 activities, 34
Specific membrane proteins, 156
Stacked membranous vesicles (type I), 9
Staphylocoagulase, 143
Staphylococcus
albus, 140
aureus, 117
C55-isoprenoid alcohol phosphokinase from, 146
penicillinase distribution, 140
phosphatase, 137
properties of penicillinase, 142
spp., 140
Stop transfer sequences, 88
Stop translation sequences, 79
Streptococcus
faecalis, 117—119
penicillin-binding proteins, 144
pneumoniae, 156
Streptomyces
rimosus, 135
sp., 140
Structural genes, 228, 232
major outer membrane proteins, 46—51
Succinate oxidase, 109
Succinic dehydrogenase, 132—133
Superbug construction for protein production, 244
Suppressor mutation, 80, 229, 231—232

T

Teichoic acid, 144, 179
Teichuronic acid, 144

TEM β-lactamase, 82, 221
signal sequences, 75
Temperature-sensitive gene, 229
Temporal mode, 81—82
Terminal oxidases, 26
Termination protein, 155
Tetradecane, 12—13
TF_1-F_o complex, 124
Thermomicrobium fosteri, 7—9
Thermophilic bacteria, 117
Thiaminase, 217
Thin section
Acinetobactor, 4—6
Arthrobacter, 8—9
Bacillus licheniformis 749/C, 188—190
Candida lipolytica, 11
Candida tropicalis, 12—13
Nocardia sp., 10
Threonine, 139
Titration
acetone-treated membrane-PACT activity, 30—32
UAT membrane-PACT activity, 29
tolC, 47, 48
Toluene, 73
tonA, 47, 51
Tosylarginyl methyl ester, 234
Toxin production, 239—243
Toxins, 221
Transbilayer, 107
Transbilayer movement, 110
Transbilayer translocation, 235
Transcription, 48, 52—53, 60, 64, 67, 70
Transcription termination, 55, 68, 70—71
Transferase, miscellaneous, 141—143
Transferrin, 240
Translation, 77, 79, 82, 234
lipoprotein and, 53—54, 56
matrix proteins and, 66, 69
OmpA protein and, 59
OmpT protein and, 72
Translation termination, 70
Translocation, see also Secretion, 51, 58, 70, 91, 110, 176—183, 224, 235
across cytoplasmic membrane, 74—81
energy for, 80—81
penicillinase, 192
Translocation models, 74—80
Transmembrane, 89
Transmembrane channels, 232, 234
Transpeptidation, 86
Transport of nutrients, 177
Transport proteins, 107
Transposon (Tn 1681), 242
Triacyl-lysocardiolipin, 15, 23—24, 34—35
Tricarboxycillin, 153
Trichoderma reesei, 180—181
Triglyceride, 5, 15—16, 25
Trypsin, 90
Trypsin-sensitive covalent bond, 141
tsx, 47, 51
Turnover
cardiolipin, 33
cell wall, 144
phosphatidylethanolamine, 31
phosphatidylglycerol, 32
phospholipid, 113—115
fatty acids, 39
ϕ29 phage genes, 156
Tyrosine, 139

U

UAT membrane-PACT activity, titration of, 29
UAT membranes, 27—31
ugp, 47
Ultrastructural cytochemistry, 136—137, 200, 224
Ultrastructure
alkane-grown yeasts, 13—15
alkane-utilizing microorganisms, 3
hydrocarbon-utilizing microorganisms, 3—13
Unc mutant, 120—121
UncA, 81
UncA mutation, 234
Uncoupler, 82, 87, 234
Undifferentiated membrane structure, 106
Urea-$(NH_4)_2SO_4$-treated membranes, see UAT membranes

V

Valine, 139
Valinomycin, 234
Vectorial movement, 178
Vibrio
cholerae, 117, 240, 242
industrial applications, 244
mutation, 243
parahaemolyticus, 117
Viral proteins, signal sequences, 76
Vitreoscilla sp., 117
V_{max} values, phospholipase A_1 activities, 34

W

Wall biosynthesis, enzymes for, 143—146
Wall membrane cooperative functions, 144
Wall polymer, 179
Wax esters, 3, 5, 15—16, 25

X

Xylanase, 216

Y

Yeast, 2, 13
Yeast microsomal fraction, 13

Z

Zinc binding, 200—201

For Product Safety Concerns and Information please contact our EU representative GPSR@taylorandfrancis.com Taylor & Francis Verlag GmbH, Kaufingerstraße 24, 80331 München, Germany

Batch number: 10731074

Printed by Printforce, the Netherlands